COSMIC ABUNDANCES

A SERIES OF BOOKS ON RECENT DEVELOPMENTS IN ASTRONOMY AND ASTROPHYSICS

A.S.P. CONFERENCE SERIES
PUBLICATIONS COMMITTEE

Dr. Sallie L. Baliunas, Chair
Dr. John P. Huchra
Dr. Roberta M. Humphreys
Dr. Catherine A. Pilachowski

© Copyright 1996 Astronomical Society of the Pacific
390 Ashton Avenue, San Francisco, California 94112

All rights reserved

Printed by BookCrafters, Inc.

First published 1996

Library of Congress Catalog Card Number: 96-84712
ISBN 1-886733-20-1

D. Harold McNamara, Managing Editor of Conference Series
Box 63 KMB
Brigham Young University
Provo, UT 84602-4463
801-378-2298

pasp@astro.byu.edu
Fax 801-378-2665

A SERIES OF BOOKS ON RECENT DEVELOPMENTS IN ASTRONOMY AND ASTROPHYSICS

Vol. 1-Progress and Opportunities in Southern Hemisphere Optical Astronomy: The CTIO 25th Anniversary Symposium
ed. V. M. Blanco and M. M. Phillips   ISBN 0-937707-18-X

Vol. 2-Proceedings of a Workshop on Optical Surveys for Quasars
ed. P. S. Osmer, A. C. Porter, R. F. Green, and C. B. Foltz   ISBN 0-937707-19-8

Vol. 3-Fiber Optics in Astronomy
ed. S. C. Barden   ISBN 0-937707-20-1

Vol. 4-The Extragalactic Distance Scale: Proceedings of the ASP 100th Anniversary Symposium
ed. S. van den Bergh and C. J. Pritchet   ISBN 0-937707-21-X

Vol. 5-The Minnesota Lectures on Clusters of Galaxies and Large-Scale Structure
ed. J. M. Dickey   ISBN 0-937707-22-8

Vol. 6-Synthesis Imaging in Radio Astronomy: A Collection of Lectures from the Third NRAO Synthesis Imaging Summer School
ed. R. A. Perley, F. R. Schwab, and A. H. Bridle   ISBN 0-937707-23-6

Vol. 7-Properties of Hot Luminous Stars: Boulder-Munich Workshop
ed. C. D. Garmany   ISBN 0-937707-24-4

Vol. 8-CCDs in Astronomy
ed. G. H. Jacoby   ISBN 0-937707-25-2

Vol. 9-Cool Stars, Stellar Systems, and the Sun. Sixth Cambridge Workshop
ed. G. Wallerstein   ISBN 0-937707-27-9

Vol. 10-The Evolution of the Universe of Galaxies. The Edwin Hubble Centennial Symposium
ed. R. G. Kron   ISBN 0-937707-28-7

Vol. 11-Confrontation Between Stellar Pulsation and Evolution
ed. C. Cacciari and G. Clementini   ISBN 0-937707-30-9

Vol. 12-The Evolution of the Interstellar Medium
ed. L. Blitz   ISBN 0-937707-31-7

Vol. 13-The Formation and Evolution of Star Clusters
ed. K. Janes   ISBN 0-937707-32-5

Vol. 14-Astrophysics with Infrared Arrays
ed. R. Elston   ISBN 0-937707-33-3

Vol. 15-Large-Scale Structures and Peculiar Motions in the Universe
ed. D. W. Latham and L. A. N. da Costa   ISBN 0-937707-34-1

Vol. 16-Atoms, Ions and Molecules: New Results in Spectral Line Astrophysics
ed. A. D. Haschick and P. T. P. Ho   ISBN 0-937707-35-X

Vol. 17-Light Pollution, Radio Interference, and Space Debris
ed. D. L. Crawford   ISBN 0-937707-36-8

Vol. 18-The Interpretation of Modern Synthesis Observations of Spiral Galaxies
ed. N. Duric and P. C. Crane   ISBN 0-937707-37-6

Vol. 19-Radio Interferometry: Theory, Techniques, and Application, IAU Colloquium 131
ed. T. J. Cornwell and R. A. Perley   ISBN 0-937707-38-4

Vol. 20-Frontiers of Stellar Evolution, celebrating the 50th Anniversary of McDonald Observatory
ed. D. L. Lambert   ISBN 0-937707-39-2

Vol. 21-The Space Distribution of Quasars
ed. D. Crampton   ISBN 0-937707-40-6

Vol. 22-Nonisotropic and Variable Outflows from Stars
ed. L. Drissen, C. Leitherer, and A. Nota   ISBN 0-937707-41-4

Vol. 23-Astronomical CCD Observing and Reduction Techniques
ed. S. B. Howell   ISBN 0-937707-42-4

Vol. 24-Cosmology and Large-Scale Structure in the Universe
ed. R. R. de Carvalho   ISBN 0-937707-43-0

Vol. 25-Astronomical Data Analysis Software and Systems I
ed. D. M. Worrall, C. Biemesderfer, and J. Barnes   ISBN 0-937707-44-9

Vol. 26-Cool Stars, Stellar Systems, and the Sun, Seventh Cambridge Workshop
ed. M. S. Giampapa and J. A. Bookbinder   ISBN 0-937707-45-7

Vol. 27-The Solar Cycle
ed. K. L. Harvey   ISBN 0-937707-46-5

Vol. 28-Automated Telescopes for Photometry and Imaging
ed. S. J. Adelman, R. J. Dukes, Jr., and C. J. Adelman   ISBN 0-937707-47-3

Vol. 29-Workshop on Cataclysmic Variable Stars
ed. N. Vogt   ISBN 0-937707-48-1

Vol. 30-Variable Stars and Galaxies, in honor of M. S. Feast on his retirement
ed. B. Warner   ISBN 0-937707-49-X

Vol. 31-Relationships Between Active Galactic Nuclei and Starburst Galaxies
ed. A. V. Filippenko   ISBN 0-937707-50-3

Vol. 32-Complementary Approaches to Double and Multiple Star Research, IAU Collouquium 135
ed. H. A. McAlister and W. I. Hartkopf   ISBN 0-937707-51-1

Vol. 33-Research Amateur Astronomy
ed. S. J. Edberg   ISBN 0-937707-52-X

Vol. 34-Robotic Telescopes in the 1990s
ed. A. V. Filippenko   ISBN 0-937707-53-8

Vol. 35-Massive Stars: Their Lives in the Interstellar Medium
ed. J. P. Cassinelli and E. B. Churchwell   ISBN 0-937707-54-6

Vol. 36-Planets and Pulsars
ed. J. A. Phillips, S. E. Thorsett, and S. R. Kulkarni   ISBN 0-937707-55-4

Vol. 37-Fiber Optics in Astronomy II
ed. P. M. Gray   ISBN 0-937707-56-2

Vol. 38-New Frontiers in Binary Star Research
ed. K. C. Leung and I. S. Nha   ISBN 0-937707-57-0

Vol. 39-The Minnesota Lectures on the Structure and Dynamics of the Milky Way
ed. Roberta M. Humphreys   ISBN 0-937707-58-9

Vol. 40-Inside the Stars, IAU Colloquium 137
ed. Werner W. Weiss and Annie Baglin   ISBN 0-937707-59-7

Vol. 41-Astronomical Infrared Spectroscopy: Future Observational Directions
ed. Sun Kwok   ISBN 0-937707-60-0

Vol. 42-GONG 1992: Seismic Investigation of the Sun and Stars
ed. Timothy M. Brown   ISBN 0-937707-61-9

Vol. 43-Sky Surveys: Protostars to Protogalaxies
ed. B. T. Soifer   ISBN 0-937707-62-7

Vol. 44-Peculiar Versus Normal Phenomena in A-Type and Related Stars
ed. M. M. Dworetsky, F. Castelli, and R. Faraggiana   ISBN 0-937707-63-5

Vol. 45-Luminous High-Latitude Stars
ed. D. D. Sasselov   ISBN 0-937707-64-3

Vol. 46-The Magnetic and Velocity Fields of Solar Active Regions, IAU Colloquium 141
ed. H. Zirin, G. Ai, and H. Wang   ISBN 0-937707-65-1

Vol. 47-Third Decinnial US-USSR Conference on SETI
ed. G. Seth Shostak   ISBN 0-937707-66-X

Vol. 48-The Globular Cluster-Galaxy Connection
ed. Graeme H. Smith and Jean P. Brodie   ISBN 0-937707-67-8

Vol. 49-Galaxy Evolution: The Milky Way Perspective
ed. Steven R. Majewski   ISBN 0-937707-68-6

Vol. 50-Structure and Dynamics of Globular Clusters
ed. S. G. Djorgovski and G. Meylan   ISBN 0-937707-69-4

Vol. 51-Observational Cosmology
ed. G. Chincarini, A. Iovino, T. Maccacaro, and D. Maccagni ISBN 0-937707-70-8

Vol. 52-Astronomical Data Analysis Software and Systems II
ed. R. J. Hanisch, J. V. Brissenden, and Jeannette Barnes   ISBN 0-937707-71-6

Vol. 53-Blue Stragglers
ed. Rex A. Saffer   ISBN 0-937707-72-4

Vol. 54-The First Stromlo Symposium: The Physics of Active Galaxies
ed. Geoffrey V. Bicknell, Michael A. Dopita, and Peter J. Quinn   ISBN 0-937707-73-2

Vol. 55-Optical Astronomy from the Earth and Moon
ed. Diane M. Pyper and Ronald J. Angione   ISBN 0-937707-74-0

Vol. 56-Interacting Binary Stars
ed. Allen W. Shafter   ISBN 0-937707-75-9

Vol. 57-Stellar and Circumstellar Astrophysics
ed. George Wallerstein and Alberto Noriega-Crespo   ISBN 0-937707-76-7

Vol. 58-The First Symposium on the Infrared Cirrus and Diffuse Interstellar Clouds
ed. Roc M. Cutri and William B. Latter   ISBN 0-937707-77-5

Vol. 59-Astronomy with Millimeter and Submillimeter Wave Interferometry
ed. M. Ishiguro and Wm. J. Welch   ISBN 0-937707-78-3

Vol. 60-The MK Process at 50 Years: A Powerful Tool for Astrophysical Insight
ed. C. J. Corbally, R. O. Gray, and R. F. Garrison   ISBN 0-937707-79-1

Vol. 61-Astronomical Data Analysis Software and Systems III
ed. Dennis R. Crabtree, R. J. Hanisch, and Jeannette Barnes ISBN 0-937707-80-5

Vol. 62-The Nature and Evolutionary Status of Herbig Ae / Be Stars
ed. P. S. Thé, M. R. Pérez, and E. P. J. van den Heuvel   ISBN 0-937707-81-3

Vol. 63-Seventy-Five Years of Hirayama Asteroid Families: The role of Collisions in the Solar System History
ed. R. Binzel, Y. Kozai, and T. Hirayama   ISBN 0-937707-82-1

Vol. 64-Cool Stars, Stellar Systems, and the Sun, Eighth Cambridge Workshop
ed. Jean-Pierre Caillault   ISBN 0-937707-83-X

Vol. 65-Clouds, Cores, and Low Mass Stars
ed. Dan P. Clemens and Richard Barvainis   ISBN 0-937707-84-8

Vol. 66- Physics of the Gaseous and Stellar Disks of the Galaxy
ed. Ivan R. King   ISBN 0-937707-85-6

Vol. 67-Unveiling Large-Scale Structures Behind the Milky Way
ed. C. Balkowski and R. C. Kraan-Korteweg   ISBN 0-937707-86-4

Vol. 68-Solar Active Region Evolution: Comparing Models with Observations
ed. K. S. Balasubramaniam and George W. Simon   ISBN 0-937707-87-2

Vol. 69-Reverberation Mapping of the Broad-Line Region in Active Galactic Nuclei
ed. P. M. Gondhalekar, K. Horne, and B. M. Peterson   ISBN 0-937707-88-0

Vol. 70-Groups of Galaxies
ed. Otto G. Richter and Kirk Borne   ISBN 0-937707-89-9

Vol. 71-Tridimensional Optical Spectroscopic Methods in Astrophysics
ed. G. Comte and M. Marcelin   ISBN 0-937707-90-2

Vol. 72-Millisecond Pulsars—A Decade of Surprise, ed. A. A. Fruchter
M. Tavani, and D. C. Backer   ISBN 0-937707-91-0

Vol. 73-Airborne Astronomy Symposium on the Galactic Ecosystem: From Gas to Stars to Dust
ed. M. R. Haas, J. A. Davidson, and E. F. Erickson   ISBN 0-937707-92-9

Vol. 74-Progress in the Search for Extraterrestrial Life,
ed. G. Seth Shostak   ISBN 0-937707-93-7

Vol. 75-Multi-Feed Systems for Radio Telescopes
ed. D. T. Emerson and J. M. Payne   ISBN 0-937707-94-5

Vol. 76-GONG '94: Helio- and Astero-Seismology from the Earth and Space
ed. Roger K. Ulrich, Edward J. Rhodes, Jr., and Werner Däppen   ISBN 0-937707-95-3

Vol. 77-Astronomical Data Analysis Software and Systems IV
ed. R. A. Shaw, H. E. Payne, and J. J. E. Hayes   ISBN 0-937707-96-1

Vol. 78-Astrophysical Applications of Powerful New Databases
ed. S. J. Adelman and W. L. Wiese   ISBN 0-937707-97-X

Vol. 79-Robotic Telescopes: Current Capabilities, Present Developments, and Future Prospects for Automated Astronomy
ed. Gregory W. Henry and Joel A. Eaton   ISBN 0-937707-98-8

Vol. 80-The Physics of the Interstellar Medium and Intergalactic Medium
ed. A. Ferrara, C. F. McKee, C. Heiles, and P. R. Shapiro   ISBN 0-937707-99-6

Vol. 81-Laboratory and Astronomical High Resolution Spectra
ed. A. J. Sauval, R. Blomme, and N. Grevesse   ISBN 1-886733-01-5

Vol. 82-Very Long Baseline Interferometry and the VLBA,
ed. J. A. Zensus, P. K. Diamond, and P. J. Napier   ISBN 1-886733-02-3

Vol. 83-Astrophysical Applications of Stellar Pulsation, IAU Colloquium 155
ed. R. S. Stobie and P. A. Whitelock   ISBN 1-886733-03-1

Vol. 84-The Future Utilisation of Schmidt Telescopes, IAU Colloquium 148
ed. Jessica Chapman, Russell Cannon, Sandra Harrison, and Bambang Hidayat   ISBN 1-886733-05-8

Vol. 85-Cape Workshop on Magnetic Cataclysmic Variables
ed. D. A. H. Buckley and B. Warner   ISBN 1-886733-06-6

Vol. 86-Fresh Views of Elliptical Galaxies
ed. Alberto Buzzoni, Alvio Renzini, and Alfonso Serrano   ISBN 1-886733-07-4

Vol. 87-New Observing Modes for the Next Century
ed. Todd Boroson, John Davies, and Ian Robson   ISBN 1-886733-08-2

Vol. 88-Clusters, Lensing, and the Future of the Universe
ed. Virginia Trimble and Andreas Reisenegger   ISBN 1-886733-09-0

Vol. 89-Astronomy Education: Current Developments, Future Coordination
ed. John R. Percy   ISBN 1-886733-10-4

Vol. 90-The Origins, Evolution, and Destinies of Binary Stars in Clusters
ed. E. F. Milone and J. -C. Mermilliod   ISBN 1-886733-11-2

Vol. 91-Barred Galaxies, IAU Colloquium 157
ed. R. Buta, D. A. Crocker, and B. G. Elmegreen   ISBN 1-886733-12-0

Vol. 92-Formation of the Galactic Halo--Inside and Out
ed. H. L. Morrison and A. Sarajedini   ISBN 1-886733-13-9

Vol. 93-Radio Emission from the Stars and the Sun
ed. A. R. Taylor and J. M. Paredes   ISBN 1-886733-14-7

Vol. 94-Mapping, Measuring, and Modelling the Universe
ed. Peter Coles, Vicent Martinez, and Maria-Jesus Pons-Borderia   ISBN 1-886733-15-5

Vol. 95-Solar Drivers of Interplanetary and Terrestrial Disturbances
ed. K.S. Balasubramaniam, S. L. Keil, and R. N. Smartt   ISBN 1-886733-16-3

Vol. 96-Hydrogen-Deficient Stars
ed. C. S. Jeffery and U. Heber   ISBN 1-886733-17-1

Vol. 97-Polarimetry of the Interstellar Medium
ed. W. G. Roberge and D. C. B. Whittet   ISBN 1-886733-18-X

Vol. 98- From Stars to Galaxies: The Impact of Stellar Physics on Galaxy Evolution
ed. Claus Leitherer, Uta Fritze-von Alvensleben, and John Huchra ISBN 1-886733-19-8

Vol.99- Cosmic Abundances
ed. Stephen S. Holt and George Sonneborn   ISBN 1-886733-20- 1

Inquiries concerning these volumes should be directed to the:
Astronomical Society of the Pacific
CONFERENCE SERIES
390 Ashton Avenue
San Francisco, CA 94112-1722
415-337-1100
e-mail asp@stars.sfsu.edu

ASTRONOMICAL SOCIETY OF THE PACIFIC
CONFERENCE SERIES

Volume 99

COSMIC ABUNDANCES
Proceedings of the Sixth Annual October
Astrophysics Conference in College Park, Maryland
9-11 October 1995

Edited by
Stephen S. Holt and George Sonneborn

TABLE OF CONTENTS

PREFACE 1
    Holt, S. S.

INTRODUCTION: Cosmic Abundances: Past, Present, and Future 3
    V. Trimble

PRIMORDIAL NUCLEOSYNTHESIS

Primordial Nucleosynthesis 36
    D. N. Schramm

Testing Big Bang Nucleosynthesis 48
    G. Steigman

Assessing Big-Bang Nucleosynthesis 59
    C. J. Copi, D. N. Schramm, and M. S. Turner

Big-Bang Nucleosynthesis and a New Approach to
Galactic Chemical Evolution 63
    C. J. Copi, D. N. Schramm, and M. S. Turner

Cosmic Deuterium and Baryon Density 67
    C. J. Hogan

Future Cosmic Microwave Background Constraints to the Baryon Density 74
    M. Kamionkowski, G. Jungman, A. Kosowsky, and D. N. Spergel

Interstellar Abundances of the Light Elements 78
    R. Ferlet and M. Lemoine

QSO Absorption Lines from Primordially Produced Elements 90
    E. B. Jenkins

Primordial D/H from Q0014+813 100
    M. Rugers and C. Hogan

The Chemical Enrichment History of Damped Lyman-alpha Galaxies   105
  L. Lu, W. L. W. Sargent, and T. A. Barlow

Understanding the Deuterium Abundance: Measurements
with the FUSE Satellite   109
  S. Friedman, W. Moos, W. Oegerle, and D. York

The Galactic Center Abundances of Deuterium, Lithium, and Boron   114
  D. Lubowich

THE SOLAR SYSTEM

Standard Abundances   117
  N. Grevesse, A. Noels, and A. J. Sauval

Solar Coronal Abundance Anomalies 127   127
  J.-P. Meyer

Isotopic Abundances in Stars as Inferred from the Study of
Presolar Grains in Meteorites   147
  E. Zinner

Radioisotope Production in the Early Solar Nebula by
Local High Energy Plasma Winds   162
  M. S. Spergel

STARS

Lithium in Stars   165
  J. A. Thorburn

Be Abundances in the Alpha Centauri System   175
  F. Primas, D. K. Duncan, R. C. Peterson, and J. A. Thorburn

The Evolution of Boron in the Galaxy   179
  *D. K. Duncan, F. Primas, K. A. Coble, L. M. Rebull, A. M.
  Boesgaard, C. P. Deliyannis, L. M. Hobbs, J. R. King, and S. Ryan*

Boron Abundance of BD -13 3442   184
  *L. M. Rebull, D. K. Duncan, A. M. Boesgaard, C. P. Deliyannis,
  L. M. Hobbs, J. R. King, and S. Ryan*

Carbon and Oxygen Nucleosynthesis in the Galaxy:
Problems and Prospects   188
  S. C. Balachandran

Carbon Stars and Elemental Abundances 196
    W. K. Rose

Spectroscopic Constraints on the Helium Abundance
in Globular Cluster Stars 199
    W. B. Landsman, A. P. S. Crotts, I. Hubeny, T. Lanz,
    R. W. O'Connell, J. Whitney, and T. P. Stecher

Selected Elemental Abundances in Five Oxygen-Poor Stars
in Omega Centauri 203
    D. Zucker, G. Wallerstein, and J. A. Brown

Abundance Anomalies in Globular Cluster Red Giant Stars. I.
Synthesis of the Elements Na and Al 207
    R. M. Cavallo, R. A. Bell, and A. W. Sweigart

Chemical Composition of Supergiants in the Magellanic Clouds 211
    V. Hill

X-Ray Measurements of Coronal Abundances 215
    S. A. Drake

ACSA Measurements of Coronal Elemental Abundances
in an Active K0 Dwarf Star 227
    K. P. Singh, S. A. Drake, and N. E. White

NOVAE AND SUPERNOVAE

The r-, s-, and p-Processes 231
    B. S. Meyer, J. S. Brown, and N. Luo

Nucleosynthesis and the Nova Outburst 242
    S. Starrfield, J. W. Truran, M. Wiescher, and W. M. Sparks

Nucleosynthesis in Supernovae 253
    S. E. Woosley

Observational Evidence for Nucleosynthesis by Supernovae 263
    R. P. Kirshner

Inferring Abundances from the Spectra of Supernovae 273
    R. McCray

X-Ray Observations of Supernova Remnants 284
    R. Petre

The Cassiopeia A Supernova Remnant: Dynamics
and Chemical Abundances   294
    K. J. Borkowski, J. M. Blondin, A. E. Szymkowiak,
    and C. L. Sarazin

GALAXIES

Chemical Evolution of Galaxies   298
    F. X. Timmes

Abundances and Globular Cluster Ages   307
    B. E. J. Pagel

Abundances in the Galactic Halo Gas   315
    B. D. Savage and K. R. Sembach

The Composition of Interstellar Dust   327
    J. S. Mathis

Abundances in Gaseous Nebulae   337
    H. L. Dinerstein

Temperature Fluctuations in the Planetary Nebula NCG 6543   350
    R. L. Kingsburgh, J. A. Lopez, and M. Peimbert

Abundances and Stellar Populations in Giant H II Regions   354
    W. H. Waller, J. W. Parker, and E. M. Malumuth

Abundance Measurements in the Outer Galaxy   358
    A. L. Rudolph, J. P. Simpson, M. R. Haas, E. F. Erickson,
    and M. Fich

Measureing ISM Molecular Abundances in the Direction of Cassiopeia A
by Comparing X-Ray and Radio Absorption Studies   362
    J. W. Keohane

Energetic Particles and LiBeB   366
    E. Vangioni-Flam and M. Cassé

Abundance Determinations from Gamma Ray Spectroscopy   377
    R. Ramaty

Anomalous Cosmic Rays: A Sample of Interstellar Matter   381
    R. A. Mewaldt, R. A. Leske, and J. R. Cummings

Galactic Cosmic Ray Source Elemental Compsition   385
    M. A. DuVernois

The O/H Abundance Distribution in the Large Spiral Galaxy NGC 4258   389
   Y. Dutil

Abundances in Elliptical Galaxy Hot Interstellar Media   393
   M. Loewenstein

Galaxies in Clusters: Implications for Abundances   397
   J. Silk

Metal Enhancements in the X-Ray Gas Around Central Cluster Galaxies   405
   A. Reisenegger

Abundances in the Intra-Cluster Medium   409
   K. A. Arnaud

RAPPORTEUR: From Solar Flares to the Big Bang   419
   R. Ramaty

CONFERENCE PROGRAMME   425

CONFERENCE PARTICIPANTS   427

PHYSICAL/ASTROPHYSICAL CONSTANTS   433

AUTHOR INDEX   436

SUBJECT INDEX   438

Preface

This is the sixth in a series of annual October Astrophysics Conferences in Maryland. These conferences are organized by astrophysicists at the Goddard Space Flight Center and the University of Maryland. The topic for each conference is selected by a permanent committee of senior scientific staff with the help of an International Advisory Committee, the current membership of which is:

Marek Abramowicz, Göteborg	*Sir Martin Rees*, Cambridge (UK)
Roger Blandford, Pasadena	*Vera Rubin*, Washington
Claude Canizares, Cambridge (US)	*Joseph Silk*, Berkeley
Arnon Dar, Haifa	*David Spergel*, Princeton
Alan Dressler, Pasadena	*Rashid Sunyaev*, Moscow
Guenther Hasinger, Potsdam	*Alex Szalay*, Budapest
Steve Holt, Greenbelt	*Yasuo Tanaka*, Tokyo
Dick McCray, Boulder	*Scott Tremaine*, Toronto
Jim Peebles, Princeton	*Simon White*, Garching

The subject chosen for this conference was "Cosmic Abundances," with its program developed by the Scientific Organizing Committee:

Chuck Bennett	*Rob Petre*
Arnon Dar,	*Reuven Ramaty*
Steve Holt	*Ed Salpeter*
Dick McCray	*David Schramm*
Jean-Paul Meyer	*George Sonneborn*
Jonathan Ormes	*Virginia Trimble*
Bernard Pagel	*Stan Woosley*

"Cosmic Abundances" is a subject that touches upon elemental and isotopic origins ranging from the early universe to the youngest stars. Light element abundances are generally supportive of big-bang primordial nucleosynthesis, and heavy element abundances are sensitive probes of the stellar and post-stellar histories of matter in various settings.

We began the conference with an invited review by *Virginia Trimble* on the twentieth anniversary of her classic paper in the Reviews of Modern Physics with a title similar to that of the conference. The conference then proceeded through the next two days with a series of non-parallelled sessions, each devoted to a specific topic and led by a distinguished session chair. A typical featured two or three invited talks and an extensive discussion period, and may have also included one or two short contributions "promoted" from the poster papers by the session chair.

All the chairs should be commended for keeping the activities lively and the speakers to their allotted times:

Chuck Bennett   *John Mather*   *Vera Rubin*
Doug Duncan   *Jonathan Ormes*   *George Sonneborn*
Neil Gehrels   *Art Poland*   *David Spergel*

Special thanks are due *John Mather* for substituting for *Arnon Dat* when the latter missed his plane, and to *Doug Duncan* for delivering the invited paper of his colleague *Julie Thorburn* when she was suddenly taken ill.

At previous meetings in this series, the conference banquet has featured a speaker who has generally been one of the early heroes of the discipline. This year we invited a speaker who seems to be one of the early heroes of every discipline, and the conference attendees were delighted by the reminiscences of *Ed Salpeter*.

Similarly, this series has been blessed with thoughtful and fair (if sometimes strongly opinionated) rapporteurs. This year's rapporteur met all of these conditions; thanks to *Reuven Ramaty* for a job well done.

Finally, I note with some pride that these published Proceedings will include the written versions of the talks of all (without exception) of the invited speakers. I am gratified that each of the Proceedings from these annual October meetings have been quite successful insofar as they have been close-to-complete current summaries of their subject disciplines, which were available to the community well before the next annual October meeting.

Steve Holt
January 1996

Cosmic Abundances: Past, Present, and Future

Virginia Trimble

Astronomy Department, University of Maryland, College Park, MD 20742

and

Physics Department, University of California, Irvine, CA 92717

Abstract. To achieve a full grasp of cosmic abundances – that is, what the universe is made of and how it got to be that way – we need considerable knowledge in each of six areas. These are (a) a complete inventory of elements and isotopes, (b) the nuclear properties of each, (c) the observed abundances of stable, decaying, and extinct nuclides as a function of time and place, (d) reaction chains and networks, (e) sites for each, and (f) galactic chemical evolution. Each of these topics is traced through part of its history to our current understanding and on to some possibilities for the future.

1. Introduction

The organizers had originally planned two introductory talks for the workshop, one providing an historical perspective (to be given by me) and one addressing the current situation and future prospects (which it was hoped *A.G.W. Cameron would give). When Al decided not to participate, the two talks were glued together, thereby saving 15 minutes or so extra for poster viewing and coffee, but also mixing historical and current ideas in the sections that follow. My credentials for covering the topics of origins and abundances of the elements consist largely of having reviewed the subject before (Trimble 1975, 1991). The first of these publications is now probably old enough to count as part of the history of the subject.

The sections that follow address what I see as the main subject areas we need to understand. Section 2 covers the problem of achieving a complete inventory of elements and isotopes (generically, nuclides), both stable ones and unstable ones along reaction paths to stable nuclei, and the internal structure of atoms. A long standing problem in this area was figuring out whether any nuclide with 5 or 8 particles was stable. Section 3 is concerned with the properties of the elements and isotopes, their masses, spin and parity values, cross sections and lifetimes against decays, captures, reactions, spallation, and so forth. An ancient problem in this area is the correct low-energy cross section for $^{12}C(\alpha,\gamma)^{16}O$, which determines the ratio of carbon to oxygen that comes out of helium burning and, therefore, the subsequent course of nucleosynthesis. Section 4 summarizes some of what we know about the abundances of the nuclides in the solar system and in other places and times and how the anomalies (meaning differences from the solar system average) are correlated with each other and with stellar population types. A long-standing problem was forcing solar and meteoritic abundance to agree for elements, especially iron, that ought not to be enhanced or missing either place. Section 5 includes two closely related topics. First is the identification of the reaction chains and networks that are primarily responsible for the production of each of the nuclides. *Hans Bethe, *Edwin Salpeter, and *Fred Hoyle will appear in the following pages for important contributions in this area. Second is the problem of finding suitable sites for each of the reactions. This requires understanding stellar structure and evolution (including mass loss), the physics of the early universe, cosmic ray spallation, and probably other astrophysical entities. Just where most of the r-process (rapid capture of neutrons by iron peak seeds) occurs is an old question in this territory.

Finally, in Section 6, we look at efforts to calculate galactic chemical evolution. The goal here is to guess or deduce the initial conditions in a typical galaxy, transform gas into stars at some variable rate and with some variable mix of stellar masses (the initial mass function), deal with inflow and outflow of gas from the region being evolved, all as a function of gas composition and other parameters, and to make the end product come out looking like the Milky Way or some other galaxy. This chemical evolution must also be coupled with dynamical evolution of galaxies and with whatever pre-galactic nucleosynthesis occurred. Because we have no real theory of many of the processes, especially star formation, all such models of chemical evolution suffer from the Curse of the Adjustable Parameter.

2. The Inventory of Nuclides and Atomic Structure

The science textbooks of an earlier generation invariably began by telling you what the ancient Greeks had thought about the subject. One might, therefore, imagine a periodic table compiled by Aristotle (-384 to -322, people lived backwards in those days) as looking something like Figure 1.

In fairness, though, one should also mention the atomists, Leucippus, Democritus (fl. -430), and Epicurus (author of the Elements of Dining?), who believed that complex entities were the result of many very small, identical atoms interacting. Their views were put into poetic form by Lucretius (-96 to -55), an extract from whose De Rerum Natura still hangs in the seminar room of what was once Fred Hoyle's Institute of Theoretical Astronomy. The ancients recognized copper, carbon, gold, iron, lead, mercury, silver, sulphur, and tin as interesting, distinguishable substances, though not as elements in the modern sense.

During the middle ages, arbitrary transmutability of substances was regarded as reasonable and possible. The search for the philosopher's stone or an Al-iksir (elixir, from the Arabic) that would facilitate the processes was pursued by Roger Bacon, Albertus Magnus, and Paracelsus, among those whose names have come down to us. Additional substances that came to be regarded as discrete and interesting included arsenic (Magnus, about 1250), zinc, (in India, about 1250), antimony (before 1600), phosphorus (1669), cobalt and platinum

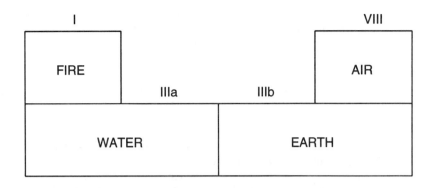

Figure 1.   Early Periodic Table

(1735), nickel (1751), bismuth as distinct from tin and lead (1753), and magnesium (1755).

Then came phlogiston and an era of confusion ushered in by Becker and Stahl in Germany in the 17th and 18th centuries. The idea was that metals were compound substances, when heated, released phlogiston, leaving behind a calx or ash. The phlogiston and ash were then the pure substances or elements, as first defined by Robert Boyle (1627-1691) to mean something that could not be decomposed into anything simpler. The year he said this, 1669, was the era of Charles II in England, Louis XIV in France, and between Shakespeare and Bach in culture. Pierre Gassendi (1592-1655) revived the idea of atomism in the same time frame.

"The rise of modern chemistry" begins in 1774 with Joseph Priestly's recognition of "respirable air" as a discrete substance. Antoine Lavoisier (1743-1794) gave it the name oxygen, and Cavendish demonstrated in 1783 that water was composed of "inflammable air" and oxygen. Lavoisier is the real hero here. His *Elements of Chemistry* (published in French in 1789 and in English in 1790) established the notion of "elements" as Boyle had defined them, with the inventory therefore subject to change. The title of his book was presumably a live pun at the time, at least in English.

The next vital step was taken by Dmitri Ivanovich Mendeleev (1834-1907; he died the year that *Hans Bethe and *Dorrit Hoffleit were born, so that we still can just barely make contact with his epoch). He was not quite the first to attempt to put the elements in some sensible order or pattern, but his 1869 grouping by chemical properties, perpendicular to increasing atomic weight, was essentially our modern periodic table. He looked at the gaps in 1871 and predicted the existence of ekaboron (scandium, found in 1875), ekasilicon (germanium, found in 1876), and eka-aluminum (gallium, found in 1875).

Fifteen naturally-occurring elements were added to the table in the next fifty years: Ho (1878 for a salt; the pure metal not until 1911), Sm and Tm (1879), Gd (1880), Nd and Pr (distinguished in 1885), Dy (1886, but the pure metal only in

1950), Ar (1894), Kr and Xe (1898), these three particularly important because they added a whole new column to the periodic table, suggesting briefly that it might expand forever, Ac (1899), Eu (1901), Er (1905), Lu (distinguished from Yb in 1907), Hf (separated from Zr in 1922 and the first to be identified from its X-ray spectrum), and Re (distinguished from Pt, the last, in 1925). A plot of the number of known elements as a function of time looks like slightly ragged stairs (Masterton & Slowinski 1973), with sharp rises when new theoretical or experimental tools (like Humphrey Davy's photovoltaic cell) entered the arena, and plateaus in between.

False alarms of identification of elements 43 (Masurium) and 61 (Illinium) preceeded their actual creation as artificial, radioactive elements by Emilio Segre and his colleagues in 1937 and 1941-45, respectively. And so, after WWII, onward to Seaborgium, if the powers that be allow it to keep that name while Glenn is still alive. I have some personal interest in the issue, since *Seaborg and my father, *Lyne Starling Trimble, were 2/3's of the graduating class in chemistry from UCLA in 1935. Seaborg is now thought of primarily as a physicist. My father remained a chemist, but always said that the physicists had stolen an enormous mount of territory, and that the whole topic of the internal structure of atoms and their constituent parts should have been part of chemistry, leaving the physicists with classical mechanics, electromagnetism, and so forth.

Curiously, E.O. Lawrence (1935) raised the question of chemistry vs. physics in the same time frame. His last sentence in the proceedings of a 1935 workshop (sponsored by Sigma Xi of Michigan) was, "Shall we call it nuclear physics, or shall we call it nuclear chemistry?"

Our modern notion of atoms goes back only to John Dalton (1761-1844), who, in 1803 (the year of the Louisiana Purchase), suggested, that one could define an element as a set of identical atoms, and compounds would then consist of the sum of a few atoms and have definite atomic weights. Joseph Proust (1754-1826) provided the closely-related idea that elements would combine in fixed proportions by weight to make compounds. The idea of fixed (often equal) volumes for gases combining belongs, of course, to Amadeo Avogadro, who was born the year of the American revolution and put forward his best-known idea in 1811.

A suggestion that an oxygen atom might be rather like sixteen hydrogens glued closely together came from William Prout (1816) and strikes me as exceedingly important and deserving of having made his name better known. We reach the threshold of atoms ceasing to be "not dividable" in 1887 when Svant Arrhenius showed that a Faraday of electricity could deposit at most one gram-atomic-weight of a substance, and so must contain an Avogadro's number of unit charges. G.J. Stoney (1826-1911) had already provided the name "electron" for this unit charge in the year Garfield was assassinated (1881).

These unit charges became real particles in 1897, when J.J. Thomson (1856-1940) showed that cathode rays could be bent by both electric and magnetic fields. His image of atomic structure has been dubbed the plum pudding model, with electrons, like raisins, studded through a diffuse blob of positive charge. Ernest Rutherford (1871-1937) entered the picture in 1909 with the demonstration that alpha rays were the same as helium nuclei. He placed the positive charge of atoms in a dense central knot through his experiments on scattering of positively charged particles by thin gold foils in 1911. Two years later, Philipp Lenard (1862-1947) wanted to put electrons and pluses in compact pairs with empty spaces in between, a picture that even then must have been somewhat at variance with the implications of Rutherford scattering.

Rutherford and Frederick Soddy (1877-1956) recognized the existence of isotopes and coined the name in 1913. For Soddy, at least, chemical identity was an essential part of the definition. And we bid farewell to Rutherford in 1920, when he proposed the name proton, as part of a model of the nucleus in which oxygen, for instance, would have 16 positive charges and 8 negative charge units (electrons) in its nucleus, with 8 more negative charges at a distance.

Meanwhile, in 1913, Henry Moseley (born the same year as my maternal grandmother, 1887, and idiotically sent to die at Gallipoli) had provided an absolutely vital idea. He said that atomic number was simply the number of discrete positive charge units in a nucleus. This required that the periodic table must be finite, not at all clear when the whole new column of noble gases had just been added and the rare earths were continuing to proliferate.

The years 1931-33 were enormously fruitful for atomic and molecular physics. *Harold Urey (1893-1981) separated deuterium from normal hydrogen in 1931. It is arguably the most important isotope from a nucleosynthetic point of view (because it is the essential bridge from protons to nuclei with neutrons), though its chemical distinctiveness caused Soddy to deny that it was an isotope to his dying day. The same year, John Cockcroft (b. 1897) and Ernest Walton (1903-1995) produced the first laboratory nuclear reaction triggered by artificially accelerated particles. They bombarded ^{7}Li with protons and found themselves with a bunch of alpha particles (helium nuclei). The first Van de Graaff accelerator, the first cyclotron (built by Lawrence), and the first linear accelerator also belong to 1931.

James Chadwick (b. 1891) was, meanwhile, carrying on a slightly older form of nuclear alchemy, using alpha particles from naturally radioactive substances as his projectiles. When, in 1932, he turned his beam on ^{9}Be, something came out that was both non-ionizing (uncharged) and capable of knocking protons out of paraffin (massive). Called the neutron, the new particle provided the key to a correct understanding of nuclei as compact assemblages of protons and neutrons (Werner Heisenberg 1933).

The portion of the cosmic abundances program described in this section is complete. We know about all possible stable nuclides and their (Z, A) values and the unstable ones near enough to them to get involved in nucleosynthesis. One of the last outstanding issues in this field was the possible existence of an island of relative stability for elements around $Z = 112$, the next magic number or closed shell for protons after Pb($Z = 82$). The evidence was meteoritic xenon that seemed, on the basis of the preponderance of the heaviest isotopes, to have come from fission of such a superheavy nuclide. It was a false alarm (Anders and Zinner 1991). There was an even briefer false alarm in connection with giant halos in mica, which the discoverer attributed to decay of superheavy radioactive nuclides in situ. I won a bottle of red wine from Al Cameron by betting instantly against the discovery on first hearing (based mostly on the location of the discoverer at a fundamentalist college; and indeed he was primarily interested in establishing that mesozoic rocks were very young).

Most of the historical material here has been taken from CRC (1949, 1987), McKie (1951), Asimov (1966), and a long-out-of-print history of chemistry by J.H. Moore, the 1922 edition of which lived on the bookshelves at home for many years.

3. Nuclear Properties and Cross Sections

Fredrick Aston (1920, 1927) pioneered the use of the mass spectrograph to measure the first atomic/nuclear masses accurate enough to reveal, for instances, the four hydrogens add up to more than one helium, and that many elements in their common forms do not have integral atomic weight (later largely explained as the effect of mixtures of isotopes). Very shortly before, Rutherford had triggered the first man-made nuclear reaction by firing alpha particles from a natural source of radioactivity at ^{14}N. Hydrogen was liberated and oxygen remained in a reaction, ^{14}N$(\alpha, p)^{17}$O confirmed by P.M.S. Blackett in 1925.

For most of us, this whole topic is associated inextricably with the names of *William A. Fowler and his associates Charles and *Tommie Lauritsen at Kellogg Lab (California Institute of Technology). Fowler (1984, 1992) has told the story himself, and it is far above our poor power to add or subtract, except that he does not seem to have included in either place the description of his very first cross section measurement. Attempting to learn something about energy dependence by repeating the same experiment "with copper shield and without copper shield", he was told by the elder Lauritsen that, if you aren't using a shield, it doesn't matter what it is made of.

The bible of reaction rates of astrophysical importance has been through almost as many editions as there are politically correct modern translations of the King James testaments (Fowler et al. 1967, 1975; Harris et al. 1983; Caughlan & Fowler 1988). Since the last of these, several groups have been attempting to maintain up-to-date listings of the most important rates in electronic form. These include F.K. Thielemann and M. Wiescher in Basle, S.E. Woosley and R. Hoffman at University of California, Santa Cruz, and C. Rolfs and M. Arnoud as part of a collaboration funded by the European Union (Thielemann 1995).

Through the various editions, rates of some of the critical reaction, including ^{15}N$(p, \gamma)^{16}$O have changed by factors of two or three, and others, like the triple alpha and $p + p$ by 20-30%. We still, of course, await an actual measurement of this last cross section at stellar energies, the rates in use being calculated from the lifetime of the neutron.

Three topics of historical interest in this area are the stability of $A = 5$ and 8, the existence of an essential excited state in ^{12}C that makes helium burning go smoothly, and the correct low-energy cross section for ^{12}C$(\alpha, \gamma)^{16}$O. None of the stories can be told in an entirely linear, straightforward way.

The very short lifetime of ^{8}Be was clear in a series of pre-war German experiments (Kirchner et al. 1937; Fink 1939), with the analysis corrected by *John A. Wheeler (1941) to reveal the first excited state (of some importance in helium burning). But a seemingly much more accurate experiment (Allison et al. 1939) found ^{8}Be to be bound relative to two alpha particles by about 0.3 MeV. Everyone was then very busy with war work for the next six years, and the unboundedness of ^{8}Be had to be re-established both theoretically (Fermi &

Terkevich 1949) and experimentally (Hemmendinger 1948, and work at Kellogg by *Tollestrup, Lauritsen, and Fowler) before everyone was fully persuaded.

The excited state of ^{12}C that is essential for helium burning to work at stellar temperatures has a similarly spotted history. It is *There* in the ^{14}N$(d,\alpha)^{12}$C results of Holloway & Moore (1939) at 7.62 MeV above ground (as the second excited state and with the right spin and parity to give a large cross section for ^{8}Be + ^{4}He), and equally *Not There* in seemingly better data for the same reaction by Malm & Buechner (1951). It had correspondingly moved into and out of standard tables before Hoyle came to Kellogg in 1953. Hoyle's theoretical conclusion that the state must nevertheless exist or we wouldn't be here to argue about it was, therefore, drawn not exactly in the absence of data, but in the face of the data. That he persuaded *Ward Whaling to put together a group (of which Fowler was not a member) to repeat the experiment yet again and find the critical level (Dunbar et al. 1953) is in some ways, therefore, the more impressive. Many of the references in this paragraph and the previous one have been traced from privately supplied manuscripts by Brown (1984), for which I am most grateful.

^{12}C$(\alpha, \gamma)^{16}$O is sneaky enough, even without human frailty, to cast doubts on Einstein's good opinion of God. It goes through a resonance that is 7.15 MeV above the ground state of ^{16}O – but slightly *below* the threshold for ^{12}C and ^{4}He approaching each other at zero relative velocity. Properties of the level and the resulting reaction rate must, therefore, be determined by examining related nuclides and reactions. B^{2}FH suspected that one critical component was not known better than to a factor 100. This had narrowed to only(?) a factor 40 by the time of the conference reported in Trimble (1975). If the rate is low, helium burning makes lots of carbon. If the rate is high, most of the carbon burns through to oxygen. Buchmann et al. (1993) and Zhao et al. (1993) recently matched nature in sneakiness by looking at the alpha decay of ^{16}N. They found a rate at the upper end of the previously allowed range for one of the two branches leading to ^{16}O consistent with the implications of the large amount of oxygen $(1-3M_\odot)$ expelled by supernova 1987A.

4. Abundances of the Elements and Isotopes

It is customary to look first at the solar system (on the grounds that "normal" means "a lot like me") and then at the differences to be found in astronomical objects of other ages and at other locations.

4.1. Abundances Here and Now

A pedant might insist (in fact a pedant at the workshop did insist) that this should say "here and 4.5 Gyr ago", but you know what I mean. *Kuchowitz (1967) has provided a very complete annotated bibliography of the history of nucleosynthesis and can be consulted for many details missing here.

The first recorded cut at "how much of what" was an examination of the earth and meteorites by Kleiber (1885). He recognized the existence of the iron peak and remarked that light elements (meaning oxygen and silicon) were generally commoner than heavy ones (meaning silver and gold). Soon after, Clarke (1889) concluded that there were no periodicities in abundance to be

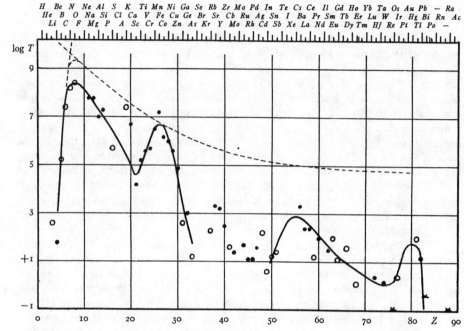

Figure 2. Abundances of the elements as they were known to Atkinson (1931), largely from Russell's (1929) analysis of the sun, but with odd Z artificially enhanced by $\log N = 0.6$, thereby obscuring an important nucleosynthetic fact (the odd-even effect). CNO and Fe peaks are present, and some structure associated with the $N = 82$ and 126 closed neutron shells (though the neutron was not yet known).

seen when looking along the rows and columns of the periodic table. Oddo (1914) and Harkins (1917) recognized that you must look for correlations with nuclear properties rather than with chemical ones. They concluded that both even Z and even A were commoner than nearby odd ones.

The standard work on solar abundances was (and for a few elements still is) Russell (1929). His numbers were the ones available to *R. d'E. Atkinson (1931) when he first attempted to account for the solar system composition with nuclear reactions distantly related to the CNO cycle. A sharp-eyed observer of his plot (Fig. 2) can see the iron peak, peaks at neutron number 82 and 126 (but not 50), and the illusory elements Ma and Il. Hydrogen and helium do not appear at all.

The best available numbers for elemental abundances evolved gradually from Goldschmidt (1937) through *Brown (1949), Suess & *Urey (1956), and Cameron (1968, 1973) to *Anders & *Grevasse (1989). Small modifications continue to appear, but, on a log scale, the most recent plots are indistinguishable from Fig. 3 (taken from Trimble 1975). In fact, apart from beryllium, which has wandered from $\log N = 1$ to $\log N = -1$ (on the scale where $\log N(H) = 12$), none of the elements Goldschmidt was brave enough to evaluate has changed by more than a factor 3-4, at least in the meteorite data.

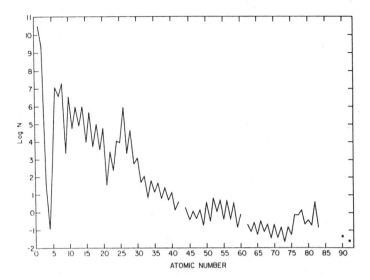

Figure 3. Abundances of the elements from Trimble (1975), normalized to $\log N(\text{Si}) = 6$. Of the elements tabulated by Goldschmidt (1937), none except Be changed by more than a factor of 3-4. Changes in the last 20 years have been even smaller.

A long-standing problem was how to merge the meteoritic scale (based on silicon $= 10^6$) and the solar one (based on hydrogen $= 10^{12}$). Iron was particularly troublesome, with the solar photospheric abundance derived by Russell and his successors more than an order of magnitude below the meteoritic (and solar coronal) value. The meteoriticists were right, the photospheric determinations having been bedeviled by inaccurate transition probabilities.

Conspicuous features in any modern compilation of elemental abundances include the enormous preponderance of H and He (*Payne 1925; Russell 1929), the scarcity of Li, Be, and B (recognized by Russell and Goldschmidt), the peaks at CNO and Fe, the odd-even effect (which Atkinson, 1931, obscured by arbitrarily enhancing odd Z abundances by $\log z = 0.6$), and moderate enhancements around $Z = 55$ and $76 - 82$.

Greater insight comes from examining abundances of individual nuclides. Some decisions are required. Do you add up all the nuclides at a given value of A, for instance ^{180}Hf, ^{180}Ta, and ^{180}W (the latter two of which are quite rare) and of ^{186}Re and ^{186}Os (with comparable abundances)? Or, if not, how do you separate them? Brown (1949) tried showing only odd-A nuclides (for which duplication is less common), because he thought the plot looked smoother. This turned out to be a poor choice; much informative structure has been obscured, and great peaks appear at, e.g., $A = 25$ and 57, without telling us much about Mg or Fe. *Suess and *Urey (1956) tried all sorts of combinations of adding, separating, and being guided by theoretical considerations.

The Suess and Urey (1956) compilation reveals a number of interesting features. First, the dominant magic numbers (closed nucleon shells) are determined by different physics for inner and outer shells. Thus, of the two sequences, {2, 8, 20, 40, 112} and {2, 6, 14, 28, 50, 82, 126}, we see high abundances associated with $Z = 2, 8$, and 20 (He, O, Ca) and with 28 (Ni), 50 (Sn, the element with the largest number of stable isotopes), 82 (Pb), and for neutron shells also $N = 126$. Second, they pointed out that, among light elements, the isotopes with the lowest stable n/p ratios were commonest, while high n/p ratios predominate among heavier elements. Third, they recognized that some nuclides might not be synthesized as themselves, the natural pathway to ^{56}Fe from ^{28}Si, for instance, going through the unstable ^{56}Ni. This last point is implicit in Hoyle's (1946) discussion of nuclear statistical equilibrium and subsequent beta decays.

The first thoroughly modern-looking plot of abundances of the nuclides is that of Suess and Urey data "as told to" Burbidge et al. (1957; B^{2}FH throughout the folklore), reproduced as Figure 4. Guided by confidence (well placed) that they had identified the correct production mechanisms for virtually all stable nuclides, they smoothed over observed values to reveal the physically important features of $N(A)$, including the excesses of nuclides that act like the sums of alpha particles, the double peaks at the closed neutron shells $N = 50, 82$, and 126, and the extreme sparsity of nuclides with the lowest stable n/p ratios among the heavier elements.

Extinct or fossil radioactivities are isotopic (and sometimes chemical) anomalies in meteorite grains whose chemical context says that some unstable nuclide must have been incorporated in a solid before it had time to decay. A classic example is an excess of ^{26}Mg in phases where Mg is rare but Al is common, meaning that the stuff was still ^{26}Al when it condensed. Depending on whether solidification occurred near the nucleosynthetic event (supernova or whatever) or in the early solar system, we can learn either about particular synthesis sites and reactions or about the history of solar system formation, including the possibility of a supernova or other trigger for the event.

Another interesting example is ^{22}Ne (called Ne-E by meteoriticists) which was once ^{22}Na. Since the half life is all of 2.6 yr, this must be a relic of the synthesis site in the form of pre-solar grains that survived, unvaporized, to be incorporated in the host meteorites. ^{26}Al has a half life of 0.72 Myr, and one can, therefore, imagine its being either a pre-solar grain component or still alive at the formation of the meteorites. Expert opinion favors the latter because of the potential of ^{26}Al for heating the meteorite parent bodies to permit their chemical fractionation early in the history of the solar system. Because decaying ^{26}Al leaves the ^{26}Mg in an excited state that de-excites radiatively, you can now also see ^{26}Al live at a gamma-ray observatory near you (Diehl et al. 1995).

Grains that are clearly pre-solar (with elemental as well as isotopic anomalies) have come from both carbon rich material (e.g. graphite grains) and oxygen-rich material (e.g. Al$_2$O$_3$) representing either different supernovae or different zones. Among the fossil radioactive nuclides that must have been alive at solar system formation are ^{129}I and ^{244}Pu, which decay and fission respectively to xenon with half lives near 10^8 yr. They are sporadically regarded as evidence for supernova-triggering of the formation of the solar system. For more about this fascinating but rather specialized topic, see *Wasserburg (1987) and Trimble (1994). Still longer-lived radionuclides, especially ^{232}Th, ^{235}U, ^{238}U and the isotopes of rhenium and osmium are the chronometers that tell us both the length of time since the solar system (meteorites, moon rocks, earth rocks) solidified and something about the time since the galaxy began making heavy elements. The topic is called nucleocosmochronology and has been excellently reviewed by Cowan et al. (1991a).

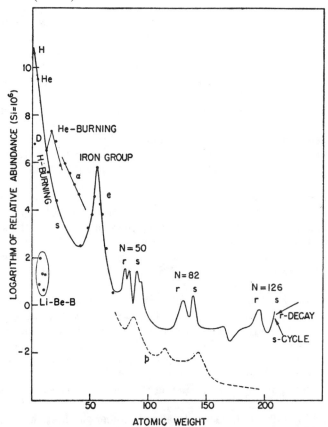

Figure 4. The abundances of the stable nuclides from Suess & Urey (1956) "as told to" Burbidge, Burbidge, Fowler, & Hoyle (1957). Making use of their conclusions about which nuclear processes have produced which nuclides, they were able to produce a memorable plot that still looks right. Normalization is again to $\log N(\text{Si}) = 6$. Purely solar or stellar data are normally shown with $\log N(\text{H}) = 12$, which moves everything up by about 1.5 dex.

4.2. Abundances There and Then

Let's look first at the Milky Way. In a general sort of way, most stars have about the same chemical composition, dominated by hydrogen and helium (Payne 1925 and others over the years). Not very many people solve a fundamental scientific problem as part of a doctoral dissertation (though you will meet another one in the galactic evolution section). *Cecilia Payne did, applying the then-new Saha equation to the spectra of coolish giant stars. It took the astronomical community a few years to absorb the dominance of hydrogen and helium, but the idea of uniformity quickly became so deeply imbedded that even her own efforts to modify it (below) were largely doomed to failure!

The most significant deviations from this uniformity are spatial and temporal gradients, with the bulge more metal rich than the disk (which also has a radial gradient, and perhaps a vertical one) and with the oldest, halo stars distinctly deficient in heavy element's. The overall pattern of $Z(R, z, t)$ is, however, hazed around with very large scatter that is not very well understood.

Some detailed differences in abundance patterns can cast light on the events leading up to the present. For instance, the ratio $^{13}C/^{12}C$ is about twice the solar value (1/90) in the present interstellar medium, at least in the inner galaxy. This suggests continuing operation of the CNO cycle. The ratio O/Fe is high (though both elements are deficient) in globular clusters, indicating a change in the dominant site of heavy element production in the galaxy from type II supernovae alone over the first 10^{6-7} yr to a significant input from SNe Ia after 10^{8-9} yr.

Some anomalies correlate in interesting ways. For instance, the trend of [O/Fe] decreasing as [Fe/H] rises persists to higher overall metallicity among bulge stars (and probably elliptical galaxies) than in the disk. This implies more rapid enrichment in the bulge. And the regressions of [Li, Be, B/H] against [C, N, O, or Fe/H] provide primordial values of the light nuclides to be used in calculations of big bang nucleosynthesis. None of these correlations is established with absolutely enormous firmness.

How did we come to even this limited level of knowledge? One always begins the history of any astronomical topic by asking what *Jan Oort thought about it. Oort (1926) established the existence of high velocity stars. He declared that they (a) had the same correlation of absolute magnitude, M_v, with spectral type as low velocity stars, (b) had no spectral peculiarities (admittedly there were no hydrogen lines in his data base), and (c) included no binaries. All three are wrong. He included globular clusters among the high velocity stars, which is, of course, correct.

Adams & *Joy (1922) had actually already found the first three examples of what they called "intermediate white dwarfs," that is stars falling on the HR diagram between the main sequence and the white dwarfs (a deviant correlation of M_v with spectral type!). They placed the stars too early because they used metal lines as their temperature,indicator. Adams et al. (1935) expanded the sample to six stars, which they described as displaying spectra with narrow, sharp H lines, faint metal lines (yes), and resemblances to the spectra of Sirius B and o Eri B (no!). From the proper motions and parallaxes they tabulate, I calculated an average transverse velocity of 234 km/sec for the six stars. They did not, perhaps because the average is so very "high velocity" that they distrusted it (this is a guess).

The first hints of chemical peculiarity are contemporaneous. *Lindblad (1922) noted that the giants of M13 had remarkable weak CN lines. This was confirmed by *Popper (1947) for the K giants of M13 and M3. *Morgan, *Keenan, & Kellman (1943) mentioned the weakness of CN features of high velocity stars in their classic atlas of stellar spectral types.

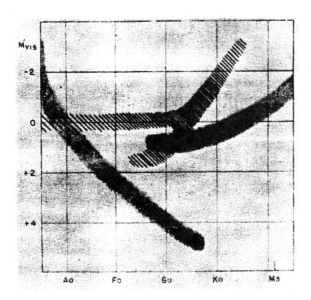

Figure 5. Baade's (1944) symbolic representation of HR diagrams of the two stellar populations. Because the globular cluster data did not reach down to the main sequence, the essential similarity of stellar evolution in the two contexts could not be seen.

The much-cited paper by Baade (1944) that defined stellar populations I and II on the basis of their color magnitude diagrams (Fig. 5) included a "prediction" that high velocity stars should display weak CN. The first large sample of high velocity, weak lined stars was collected by *Roman (1950, 1955), a participant in the present workshop. That line weakness implied genuine deficiency of calcium and iron was bravely enunciated by Chamberlain & *Aller (1951) somewhat before the world was prepared for it (though they moderated their deficiency factor from 100 to only 10 in deference to prevailing winds). They called their stars "subdwarfs or intermediate white dwarfs," and the former name has, of course, prevailed. Morgan (1956) added that the weak-lined globular clusters must also be metal poor.

The third part of the equation, "high velocity = metal poor = old" required evolutionary tracks for globular cluster stars. *Hoyle & *Schwarzschild (1955) and *Haselgrove & *Hoyle (1956) provided the first of these. They assumed initial stellar abundances $X = 0.93, Y = 0.07, Z = 0.007$, and $Z(CN) = 0.0025$, and, looking at an HR diagram for M3, derived an age of 6.5×10^9 yr. Thus, when B^{2}FH wrote in 1957, they were aware that globular cluster stars and high velocity stars were metal poor and that the former at least were old.

*Helfer, *Wallerstein, & *Greenstein (1959) dropped Z down to 1% of solar for M13 and M92, and *Wallerstein (1962) established the first correlation of anomalies, recognizing that the high velocity stars, though deficient in all metals, had Mg, Sc, Ca, and Ti less deficient than iron. These are the classic "alpha nuclei," the ones you can think of as being made up of integral numbers of helium nuclei. They are important products of Type II supernovae.

The preceeding anomalies and correlations reflect the initial compositions of the stars displaying them (that is, the progress of chemical evolution before the stars formed). Some much weirder stars reveal nucleosynthesis in progress, since their surfaces show the effects of nuclear reactions, mixing, and mass loss in the stars themselves. Examples include the R and N stars with reversal of the normal C/O ratio (Curtiss 1926), the hydrogen-free Wolf-Rayet stars (*Aller 1943) and R Corona Borealis stars (*Berman 1935), the helium stars (*Greenstein 1940; *Popper 1942), some of which are stripped binaries (*Bidelman 1950), and the S-type stars, enriched in zirconium, barium, and other products of slow neutron capture (*Merrill 1947; Keenan & Aller 1951). Some of these (apparently the non-binary ones) flout technitium, requiring nucleosynthesis to have occurred within the last few million years (Merrill 1952). The report of promethium in HR 465 (*Aller & *Cowley 1970) has never been confirmed, but also never repudiated (Cowley 1995). The longest-lived Pm isotope has a half life of 17.7 years. And then there is FG Sge, whose surface composition as well as color and luminosity have wandered all over the map in the past century (*Herbig & *Boyarchuk 1968; *Wallerstein 1990).

Among globular cluster giants, there are correlated anomalies in oxygen, sodium, and aluminum that clearly require reactions that proceed considerably beyond the CNO cycle (*Kraft et al. 1995). Curiously, the most metal poor globular clusters cut off at [Fe/H] = −2.5, while field stars extend down to -4.0 (Sarajedini & Milone 1995; McWilliam et al. 1995; von Winckel et al. 1995).

Abundance anomalies resulting from in situ nucleosynthesis are also to be found in the ejecta of supernovae (*Whipple and *Payne-Gaposchkin 1941), novae (Aller & *Payne-Gaposchkin 1942), and planetary nebulae, though the mainstream astronomical community accepted these differences with painful slowness (see footnote 45 of Aller 1943 and Aller 1994). Of those who objected to the nova results in preprint form, the most vociferous opponents were *D.H. Menzel, O. Struve, and P. Swings (Aller 1995).

There are also stars of strange surface composition which do not belong in this chapter because the origins are not primarily nuclear. Thus if you want to know about Ap stars and their ilk, you will have to go elsewhere.

Stepping outside the Milky Way, we find that most spirals are rather like it, with relatively metal rich bulges, metal poor (old) halos, and disks with composition gradients easier to see in gas than in stars (Aller 1942 on M33). The nitrogen gradient is sometimes steeper than the oxygen or carbon ones, indicating that it is a secondary nuclide, whose production (in the CNO cycle) requires that there already be some C or O present.

The giant elliptical galaxies display strong lines and weak gradients, with their globular clusters normally bluer than field stars at the same radius (opposite to the Milky Way pattern). Disentangling the effects of age and metallicity remains a problem in this and any other context where individual stars are not resolved.

Magellanic spiral and irregular galaxies are systematically metal poor. For the clouds themselves, the average values of [Fe/H] are −0.5 and −0.8, and the current gas abundances are [Fe/H]= −0.3 and −0.6 (with the SMC being more deficient). The older, more metal poor stars have [O/Fe] > 0 and, probably, others of the patterns seen in galactic metal-poor stars. I Zw 18, with [O/H]= −1.6, remains the least polluted galaxy that has enough gas to permit measurements of helium to be correlated with O, N, Fe (or whatever you think most appropriate) as an indicator of the primordial helium abundance. The goal, naturally, is to decide whether the value is one that big bang nucleosynthesis can produce and, if so, at what baryon density.

The dwarf spheroidals come as pristine as [Fe/H]= −1.9 for Draco and Leo II (but, in the absence of ionized gas, cannot be used to measure Y_p). Many have had several epochs of star formation, but none very recently. There is a clear correlation of mean metallicity with the mass or luminosity of galaxies, but also evidence for some second parameter, perhaps related to the local density of galaxies or other environmental conditions.

The best way to probe how current conditions came about might seem to be examination of stars and gas at large redshift. This has not proven quite as informative as you might expect. You see emission from very distant entities only when they are very bright and presumably not typical. Thus one does not quite know what to make of the apparent metal overabundances in QSO emission line regions or the approximate normality implied by emission line ratios in IRAS, radio, and emission line galaxies at $z \sim 3$. The gas producing QSO absorption lines is probably more typical of $z = 1 - 4$ material of some kind. Unfortunately, we are not quite clear about just what kind. That is, are we seeing mostly outskirts of galaxies; clouds that are in clusters but not part of galaxies; pieces of proto-galaxies; intergalactic clouds, or what? The majority of line systems in which heavy elements are seen have abundances of C, Mg, Zn and other tracers 1-10% of solar. You might expect a trend in metallicity vs. redshift, $Z(z)$, or even if you could locate clouds relative to the planes of their host galaxies a trend in $Z(z[Z])$. No strong one is seen. And, of course, QSO absorption lines tell us that the primordial value of D/H is .....

5. The Reaction Chains, Cycles, and Sites of Nucleosynthesis

The core of a correct understanding of cosmic abundances is figuring out which reaction(s) produce(s) each known stable and unstable nuclide and where each occurs. Historically we can approach the problem in four stages: the prehistoric, the golden age of *Cameron (1957) and *Burbidge et al. (1957), progress from then to the present, and prospects for the future.

5.1. The Eocene (Dawn of the Recent)

Two opposing ideas had already appeared by the end of the last century. Clarke (1889) as part of his attempt to describe the pattern of abundances had suggested that light elements, making up a primitive substance or "protyl", might be assembled into heavier ones, while Vernon (1890) had proposed a sort of primordial atom made of heavy things that would decompose into the lighter elements. In the years just after the first world war, J. Perrin, H.N. Russell, and A.S. Eddington recognized some sort of connection between accounting for the abundances of the elements and accounting for the sources of stellar energy.

The next decade saw the establishment of a number of basic ideas. The mass of a helium atom is less than the mass of four hydrogen atoms (Aston 1927). Bringing hydrogen and helium into equilibrium would require very high temperature (Tolman 1922), but assembling the full range of elements requires both a range of conditions and non-equilibrium (*Urey & Bradley, 1931; Pokrowski 1931; Farkas & Hartock 1931). Walke (1934) and *Gamow (1935) both drew attention-to the importance of (non-equilibrium) neutron captures with intervening beta decays in creating the heaviest elements. Though we normally associate *Gamow with the idea of cosmological nucleosynthesis, this first of his papers on the subject in fact addresses only events in stars. He was, briefly, committed to the idea that the main source of stellar energy was the contraction of normal material first to degenerate electron densities and then to nearly pure neutrons, with nuclear reactions more or less incidental to the process. Landau briefly suffered from the same delusion; both recovered promptly and, it would seem, completely.

*Atkinson & Houtermans (1929) and Atkinson (1931) were the first to consider stellar nuclear reactions with barrier penetration (which, of course, greatly reduces the temperatures needed). They had in mind a catalytic, recycling process in which an atom of moderate weight would sequentially capture 4 protons and 2 electrons and spin off a helium nucleus (not far from what we now call the CNO cycle), but, with no knowledge of neutrons, they could not quite see where the first catalyst nuclei were to come from. The last effort to get from hydrogen to zinc in a single equilibrium marathon (with gradually increasing density) came from Sterne (1933).

We enter the modern era for hydrogen fusion with Von Weizsächer (1937, 1938) who, knowing about both protons and neutrons, emphasized that everything simply had to start with $p + p$, *Bethe & *Critchfield (1938) wrote down the details correctly, as *Bethe (1939) did the next year for the basic CN cycle. Neither set of reactions shows neutrinos explicitly.

Details of nuclear reactions in stars on beyond helium belong to the post-WWII years. In an ideal world, helium burning would have come first, then carbon burning, and so forth. In fact, *Hoyle (1946) was first off the mark with what we know now to be a much later stage, nuclear statistical equilibrium of the elements of the iron peak in highly evolved stars. Helium burning came next, in work by Opik (1951). *Ernst Opik, a greatly undersung pioneer of many astrophysical ideas, ended up with far too high a temperature for his triple-alpha reaction because of an inadequate appreciation of barrier penetration and incomplete knowledge of the properties of ^{8}Be and ^{12}C. That, plus publishing in Irish journals, was sufficient to keep helium burning from being primarily associated with his name. A more correct, and more appreciated, calculation soon came from *Salpeter (1952), who went on to demonstrate the possibility of additional alpha captures leading to ^{16}O and beyond (Salpeter 1953). *Hoyle (1954) looked at what would happen when two ^{12}C nuclei or two ^{16}O's came together at high enough energy. The dominant products are not, as you might expect, ^{24}Mg and ^{32}S, because of the need to conserve energy, momentum, and angular momentum simultaneously. Rather you get products like ^{20}Ne + ^{4}He and ^{23}Na + p from the former and ^{28}Si + ^{4}He from the latter.

Gamow's (1935) brief discussion of neutron capture reactions assumed they would occur at a stage when stellar cores already consisted mostly of neutrons. *Greenstein (1954) and *Cameron (1954, 1955), by demonstrating that $^{13}C(\alpha,n)^{16}O$ could provide free neutrons during helium burning, moved neutron capture nucleosynthesis forward into the period of hydrostatic burning in stars.

What we now call cosmological nucleosynthesis began in a sea of pure neutrons, or ylem (Gamow 1946; *Alpher, *Bethe, & *Gamow 1948; Alpher & Hermann 1950, 1953; *Alpher, *Follin, & *Hermann 1953). The correct initial condition, a proton-electron-neutron soup expanding away from a high temperature equilibrium was the inspiration of *Hayashi (1950).

Other sites that we now expect to make some contribution were also explored early. These include pycnonuclear reactions on white dwarfs in novae, (*Schatzman 1947), cosmic ray spallation (Gurevich 1954), and reactions on active stellar surfaces (*Biermann 1956).

5.2. The Golden Age

Burbidge et al. (1957) and Cameron (1957) not only contributed an enormous number of new ideas and calculations to our knowledge of nucleosynthesis but, of equal importance, superbly synthesized what had come before. Table 1 shows the processes and products proposed by each, with B²FH in the left column and Cameron in the middle. "Fe peak" means Ti to Zn or thereabouts. The most stable heavy isobars, made by the s process, are also, apart from bypassed (p-process) nuclides, the lowest mass isotopes of each element. The excluded or bypassed nuclides do not dominate the make up of any elements, and, as a result, we have no evidence for their existence or abundances outside the solar system. By "photonuclear" Cameron meant mostly (γ,n) reactions.

Fig. 6 shows how the products of the three heavy-element reactions can be separated into r-, s-, and p-process products, including some made by s + r. A few might also be made by s + p, but the p component will always be swamped.

5.3. From B²FH to the Present

Whether 1957 is infinitely long ago depends entirely on how old you are. For me it is not. I was in the 9th grade at Joseph Le Conte Junior High School, studying algebra under *Mitsunori Kawagoye (still a good friend) and Christmas caroling with the Mixed Glee and Troubadours directly by *Mae Nightingale (long dead; but I can still sight read, clean piano keys, and tie a four-in-hand). Father (whose tie was always Windsor-knotted) was working for Papermate Pen, a brief, three-year respite in two decades of constant job changes that permitted a pool in 1956, a two-tone brand-new Chevy in 1957 (our first new car, ever), and a vacation in 1958 (four days in Yosemite). In many ways, it seems very close.

A good deal has, however, happened in the area of nucleosynthesis, without much disturbing the basic pattern shown in Table 1. Perhaps most important, the early universe has been admitted as a full partner with stars, responsible for making "all the elements up to helium" plus a bit of ^7Li. Second, the alpha process, in which ^{12}C, ^{16}O, ^{20}Ne, ^{24}Mg, etc. sequentially captured helium nuclei, fractured itself against a ^{20}Ne barrier. It has no low lying states with the right angular momentum and parity for $^{16}O(\alpha,\gamma)^{20}Ne$ to have a reasonable cross section. The alpha process thus broke into discrete stages of C, Ne, O, and Si burning (Table 2), which, however, are now beginning to blur again, because the full assortment of possible nuclides and reactions are included in the network

Table 1. Processes and products as proposed by B²FH and Cameron

	Processes	Products
H burning		^{4}He, ^{13}C, N, ^{16,17}O, F, ^{21,22}Ne, Na
He burning	hydrogen and helium thermonuclear reactions in orderly evolution of stellar interiors	^{12}C, ^{16}O, ^{20}Ne, ^{24}Mg He, C, N, O, Ne
α process	heavy-ion thermonuclear reactions in orderly evolution of stellar interiors + neutron captures on slow time scale + hydrogen and helium thermonuclear reactions in supernova explosions	^{24}Mg, ^{28}Si, ^{32}S, ^{36}Ar, ^{40,44}Ca, ^{48}Ti Ne to Ca
e-process	statistic equilibrium in pre-supernovae and supernovae	Fe peak
r-process	neutron capture on fast time scale in Type I supernovae	unshielded isobars $A \geq 62$ including actinides
s-process	neutron capture on slow time scale in orderly evolution of stellar interiors	most stable isobars $A \geq 62$
p-process	proton capture and photonuclear reactions in Type II supernovae + photonuclear reactions on slow time scale in orderly evolution of stellar interiors	excluded/bypassed isobars $A \geq 62$
x-process	possibly made by nuclear reactions in stellar atmospheres	D, Li, Be, B

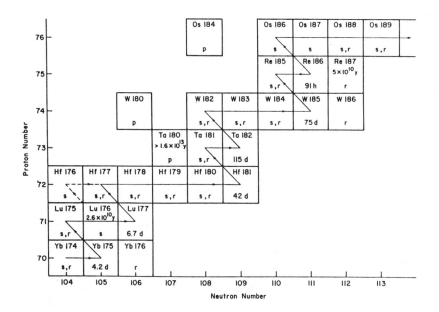

Figure 6. The region $A = 174 - 189$ showing the progress of the s-process through the stable (or very long lived) nuclides. Nuclides with n/p larger than those on the s-process path can all be reached by beta decays of unstable, neutron rich nuclides formed by the r- (rapid capture) process. The orphans of lower n/p ratio to the left of the s-process path are all very rare and must be derived from s- or r-process progenitors through addition of protons (hence the name p-process), removal of neutrons, or both.

at every stage. The lines in the table can be thought of as "the seven ages of a 20 $M_\odot$ star", by analogy with the seven ages of man. The eighth is also rather similar for both: "My, but you're looking well!" and "Gracious, you've become a pulsar!"

Hot hydrogen burning has acquired an identity of its own, extending upward in A from the CNO cycle to a Ne-Na cycle and probably a Mg-Al one. It contributes to isotopes that are not sums of alpha particles. For nova explosions on ONeMg white dwarfs, these reactions will be the dominant ones. The ongoing puzzle of just where most of the ^{26}Al comes from is part of this picture if the main reaction is ^{25}Mg$(p,\gamma)^{26}$Al. Sites proposed in the past include asymptotic giant branch stars, Wolf-Rayets, and novae. The distribution in the sky of the gamma ray line from the decay to ^{26}Mg (Diehl et al. 1995) suggests that production in supernovae is the most important. In this case, ^{26}Al may be made by the neutrino process discussed by S.E. Woosley elsewhere in this volume. Other potential ν-process products include ^{7}Li, ^{11}B, ^{19}F, ^{138}La, ^{180}Ta, ^{10}B, ^{22}Na, and the odd isotopes of Cl, K, Sc, Ti, V, Mn, Co, and Cu.

Table 2. The Seven Ages of a 20 $M_\odot$ Star (from Arnett et al. 1987)

Fuel	Central Density g/cm^3	Central Temperature K	Photon Luminosity erg/sec	Neutrino Luminosity erg/sec	Duration years
H	5.6	4×10^7	3×10^{38}	small	10^7
He	940	2×10^8	5×10^{38}	small	10^6
C	3×10^5	8×10^8	4×10^{38}	7×10^{39}	300
Ne	4×10^6	1.7×10^9	4×10^{38}	1×10^{43}	0.38
O	6×10^6	2.1×10^9	4×10^{38}	7×10^{43}	0.50
Si	5×10^7	4×10^9	4×10^{38}	3×10^{45}	2 days
core collapse	10^{9-15}	4×10^{10}	10^{42-44}	$\sim 10^{52}$	10 sec

Explosive processes in general look more important (or anyhow more calculable than in 1957. The first discussions of reactions in supernova explosions (as opposed to in pre-SN massive stars) came in 1960 (*Hayakawa, *Hayashi & Nishida 1960; Hoyle & Fowler 1960). The next round of calculations treated explosive nucleosynthesis as a sort of fine tuning of abundances established by hydrostatic processes (Schramm & Arnett 1973). Current supernova models typically have all reactions available at all stages and let the star decide what it wants to do.

Some refinement of understanding of the s-, r-, and p-processes has occurred. For the s-process, there is a second possible neutron source during the double shell burning phase of intermediate mass stars, when convection zones move up and down, bringing the products of hydrogen and helium burning into contact: $^{14}N(\alpha,\gamma)^{18}O(\alpha,\gamma)^{22}Ne(\alpha,n)^{25}Mg$. But the ^{13}C source is probably still more important (Lambert et al. 1995). Many more accurate cross sections have been measured, many by groups working with H. Beer and F. Kappeler (e.g. Kappeler et al. 1994). A process called n, intermediate between s and r (in the sense that the time scales for beta decay and neutron capture are about the same) has been considered sporadically.

The nuclear data needed for r-process calculations have also improved greatly. Cowan et al. (1991b), for instance, tabulate cross sections and such clear up to $^{137}113$ (which I am inclined to call Cameronium, though this violates IUPAC rules). The "best buy" site remains core-collapse supernovae (Woosley et al. 1994), with the core, carbon burning, and helium burning zones all having been considered. Helium flash, novae, and mergers of neutron star binaries have also been proposed. In the carbon detonation site proposed by Panov et al. (1995), there is somehow no need for intervening beta decays.

Supernovae of Type II remain also the most promising site for the p-process (Rayet et al. 1994). Lambert (1992) has provided an excellent review of what we know about these rarest of nuclides.

The so-called x-process, responsible for deuterium, lithium, beryllium, and boron, is clearly not a single entity. At minimum, it includes the early universe (responsible for all the ^{2}H and 10% or so of the ^{7}Li), cosmic ray spallation (for which the clearest evidence is that the cosmic rays are themselves greatly enriched in Li, Be, and B, as well as in rare odd isotopes), and probably one or two other contributors, like the neutrino process in supernovae, red giants (at least the lithium-rich ones), flares, and novae.

Finally, it has become clear that the several types of supernovae make very different contributions to nucleosynthesis. It is less clear just how many types (plus their rates, products, etc.) need to be considered. The current commonest assumption (to which I have no fundamental objections) is that there are two basic physical processes that make observable supernovae – nuclear explosions in degenerate material and core collapse in evolved massive stars.

Type Ia supernovae occur only about once per century in the Milky Way, show no hydrogen features in their spectra (the definition) and are blamed on explosive burning of carbon and oxygen in degenerate dwarfs. It takes 0.5 – 1.0 $M_\odot$ of C and O burning through to iron peak elements to match light curves and spectral evolution. If you have to start with at least a Chandrasekhar mass (M_{Ch}) of degenerate stuff to make it happen, this leaves 0.5 – 1.0 $M_\odot$ to be expelled as partly burned Ne, Si, S, Ar, etc. and unburned C and O. Observations do not, at any rate, contradict this. Type Ia supernovae therefore provide an additional source of iron and some other elements that kicks in after a stellar population is at least 10^{8-9} yr old (the time taken to get to the relevant degenerate objects). The actual progenitors are essentially unknown. Stripped CO cores of moderately massive stars, cataclysmic binaries with massive white dwarfs (so that a little bit of mass transfer drives them over the edge; some recurrent novae may fit), and mergers of white dwarf pairs all have their advocates. The third is probably most popular. It requires a white dwarf binary with total mass at least M_{Ch} and orbital period less than about 12 hours (so that angular momentum loss in gravitational radiation will bring the stars together in less than the age of the universe). The number of such pairs in our catalogues after several years of intense searching remains precisely zero. The short period white dwarf binaries are all low mass; the few massive ones all have long periods.

Core collapse supernovae are somewhat commoner (at least here and now) and are generally regarded as including type II (with hydrogen lines) and types Ib and Ic (no hydrogen, helium optional, blamed on collapse of stripped cores, binary or single). Production of oxygen is the best nucleosynthetic signature for type II SNe. The progenitors are well known – all the stars you observe with main sequence masses of $8-12$ $M_\odot$ or more. Until very recently the main problem was getting the things to explode. Thus most nucleosynthetic calculations simply hit the base of the stellar envelope with a numerical hammer or piston. Two- and three-dimensional calculations of neutrino-driven turbulence, convection, and instabilities appear to have revealed the solution (Herant et al. 1994; Burrows et al. 1995).

Mass loss before the explosion and similar mass loss by smaller stars during their asymptotic giant branch and pre-planetary-nebula phases can add carbon, nitrogen, and s-process material to the galactic inventory in a peaceful fashion. Not all models of chemical evolution yet incorporate this source.

5.4. Topics to be Reviewed in 2015

Trimble (1975) ended with a set of residual puzzles, worries, etc. Time has actually taken partial care of some of them. We do now, for instance, have empirical evidence for newly-produced iron in supernovae and their remnants. One can similarly hope that some of the items listed above as unclear, disputed, or just plain messy will be sorted out in the next decade or two. First among these is figuring out the correct relative contributions of multiple sites to synthesis of elements like lithium and zinc that can be made in many different ways. Another issue to be resolved is just how many physically distinct kinds of supernovae exist and what each contributes to nucleosynthesis.

Another unanswered question is the source composition of cosmic rays, which is associated with the dispute about whether initial acceleration takes place in flare stars, supernova-driven shocks, or someplace else. This might or might not be part of mainstream cosmic chemical evolution, depending on what the answer is. Finally, cosmologists will wait with anxiety to hear (or pronounce) an answer to whether the standard hot big bang (presumably with small baryon density) can "predict" the observed prestellar abundances of hydrogen, helium, and lithium. This issue has two parts, both discussed extensively elsewhere in this volume: deciding what the observed abundances are, and doing the calculations correctly.

6. Galactic Chemical (and Chemo-Dynamical) Evolution

The final task is to put all the reactions and sites together over the history of the galaxy, add up their products, and see whether the sum as a function of time and place agrees with the data we have on abundances vs. time and place, subject to the constraint that the evolved galaxy must have luminosity, color, and residual gas fraction matching the real galaxy you are trying to model. This is far and away the most immature part of our discipline, with even the basic ideas going back 40 years or less, and the first major step toward a synthesis dating from *Beatrice M. Tinsley's 1967 PhD dissertation (Tinsley 1968: remember, I promised you would meet someone else who had solved a fundamental problem in a thesis). With the wisdom of hindsight, it is obvious that the most important aspects of her work were the demonstrations that galactic evolution is (a) calculable and (b) important. I have approached the subject plonkingly, year by year.

6.1. The First 25 Years

1955. *Salpeter (1955) determined that star formation acts so as to produce numbers of stars as a function of mass proportional to $M^{-2.35}$ over the range 0.3 to 30 $M_\odot$. He called this the original mass function. We now call it the initial mass function, or IMF. Modern forms are often Gaussians, peaked at $0.3 - 0.5 M_\odot$ whose declining right edges look quite a lot like a Salpeter function.

1957. *Von Hoerner (1957) attempted to predict the number of stars, integrated over all ages, that would have a particular metallicity, $N(Z)$, on the assumption that the ratio of mass of metals coming out of stars to mass of gas going into stars was a constant. We would now call this the constant yield approximation. He handled the effect of decreasing mass in interstellar gas incorrectly and predicted a flat $N(Z)$, thereby missing the discovery of the G dwarf problem (of which more shortly).

1958. *Van den Bergh (1958) corrected the Von Hoerner calculation to allow for the increase of metallicity of the ISM as its mass decreases. *G.R. Burbidge (1958) pointed out that only 10% of the known helium could be made in stars over the age of the universe or galaxies would look much brighter than they do. This is a very important argument in favor of the universe having gone through a hot, dense early state (big bang) that clearly predates the discovery of the 3K background radiation.

1959. *Schmidt (1959) examined numbers of stars being formed under various conditions and concluded that normal star formation rates scale as (gas density)^2. Thus the rate in the young Milky Way was five times the present one, and enough gas remains for 10^{10} more years of star formation. He spoke of the "initial luminosity function."

1961. *Sandage (1961) estimated how much the luminosity of a giant elliptical galaxy would have changed since $z = 0.46$ (the largest then known) due to stellar evolution, and concluded that, if $N(M)$ were flat, it would be only 0.38 *mag* in bolometric magnitude (enough to make a true $q = 0.2$ universe look like $q = 1.0$ in a Hubble diagram) and much less if $N(M)$ rises to small masses (since there are more red giants coming along with time to make up for each one being somewhat fainter). The effect is considerably larger in any one observed wavelength band because you are looking at photons originally emitted at shorter wavelengths in large redshift galaxies.

This was also the year of IAU Symposium 15 (*McVittie 1962) on problems in extragalactic research. Among the highlights there, Allen Sandage announced a best value for H_0 of 98 ± 9 km/sec/Mpc (at least the error bars haven't changed), Fred Hoyle said that the luminosity evolution of an elliptical galaxy depends primarily on its IMF (true), and *Ivan King said that the present age also matters (also true). *Robert Christy asked whether some seemingly young globular clusters in the outer halo might be material recently added to the Milky Way (could be!), and *H.C. (Chip) Arp showed the time evolution of galactic metallicity as then understood (Fig. 7). Some of the individual points have probably moved around in the intervening 35 years, but the basic picture hasn't changed much.

1962 was something of a banner year. *Sandage (1962) provided a picture of the rate of increase of metallicity in the Milky Way that fed directly into the classic *Eggen, *Lynden-Bell, & *Sandage (1962) model of galaxy formation as a monolithic and fairly rapid collapse, leaving metal poor stars behind in the halo, not sharing the rotation of the disk. An unpublished manuscript by *Schmidt (1962) suggested that the low metallicity in dwarf elliptical galaxies might be a result of outflow carrying away the metals made by their first generations of stars. Finally, *van den Bergh (1962) concluded that the rate of metal creation in the galaxy has decreased faster than the rate of star formation. This is the first explicit description of what becomes called the G dwarf problem. He also noted that, on the basis of data from *Wallerstein (1962), enrichment in the alpha nuclei went faster than that in iron.

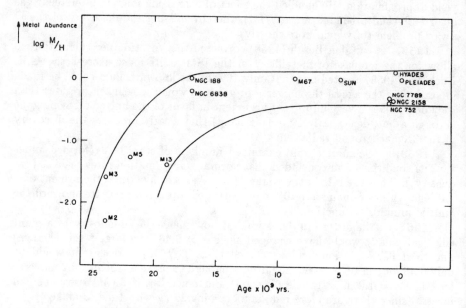

Figure 7. Arp's (1962) version of the gradual increase of metallicity in the Milky Way, based on globular and open clusters (from IAU Symposium 15). The current version of the plot would not look very different because the real scatter is large and only partly explained by spatial gradients.

1963. This was the year that Schmidt concluded, from counts of G dwarfs with different ultraviolet excesses, that "relatively more bright stars formed in the past." This both enunciates the G dwarf problem and suggests a variable IMF as a solution. For better or for worse, 1963 was also the Year of the Quasar (3C 273). Many astronomers, including Schmidt, were thereby diverted from work on the Milky Way and other normal galaxies, ushering in a few years of "dark ages" in the field.

1964. The 13th Solvay conference was devoted to structure and evolution of galaxies (Solvay 1964), but no new ideas can be discerned in the proceedings, for which J. Robert Oppenheimer provided the concluding remarks. A portion of that year's IAU General Assembly was also concerned with structure of galaxies. In a much under-appreciated calculation, McVittie (1964) demonstrated that the large mass to light ratios already then known for spiral and elliptical galaxies would require that a power law (Salpeter) IMF continue down to 0.01 $M_\odot$, or that there be some other sort of invisible stuff. He suggested objects inside their Schwarzschild radii.

1966. *Lequeux (1969) produced another volume entitled *Structure and Evolution of Galaxies* and containing no new ideas. The three year delay in publication makes the printed version seem even more backward-looking. At IAU Symposium 26 (Hubenet 1967), on abundance determinations in stellar spectra, *J.L. Greenstein delivered an uncharacteristic sort of fin-de-siecle message expressing a dislike of supernovae and neutron stars as part of galactic chemical history and doubting the possibility of ever learning anything useful about the primordial helium abundance.

1967. The year 1967 saw the completion of Tinsley's thesis, but the real renaissance of galactic astronomy (in our narrow sense) was still a few years away.

1970. *Sandage, *Freeman, & Stokes (1970) proposed that continuous infall (in contrast to a single, rapid epoch of collapse) might be important in the growth of the Milky Way.

1971. Prompt Initial Enhancement, or PIE, became the first solution to the G dwarf problem with a cute nickname (*Truran & *Cameron 1971). The problem, shown in Fig. 8, is simply that the local population of stars old enough to trace the entire history of disk star formation (G dwarfs and later types) includes far fewer low-metallicity stars than would be there if the disk had evolved as a homogeneous, closed system with constant initial mass function (hence constant yield). The S's represent the data available to Schmidt (1963) and van den Bergh (1962), when they first called attention to the discrepancy. PIE is the idea that star formation in our part of the galaxy began with a sudden burst of massive objects which enriched the gas that, only later, made stars of lower mass that survive to the present time. A pre-galactic generation of supermassive objects would also work as the "initial enhancer."

1972. *Larson (1972), *Fowler (1972), and *Quirk & *Tinsley (1972) proposed an alternative solution based on infall (Sandage et al. 1970). The idea is that local gas metallicity has been kept at essentially its current value for billions of years because fresh hydrogen and helium gas continuously rains down into the disk, mixing with the material enriched by previous generations of stars. Fowler described this at a 1974 conference as: "Here these stars are burning their hearts out trying to make heavies, and those bastards keep diluting it." Outflow, if it removes metal-rich gas preferentially, will have much the same effect.

1973. Metal-enhanced star formation (MESF, *Talbot & *Arnett 1973) was another G dwarf solution, in which the first few supernovae enriched gas in their immediate surroundings, and future star formation (at least of low mass stars) occurred only in the enriched gas. This is not unreasonable because higher metallicity means more rapid cooling of gas, facilitating star formation. In the same year, *Leonard Searle and *Bernard Pagel (in not very accessible venues) showed that you can get an analytical solution for number of stars as a function of metallicity for the homogeneous, closed, constant yield system, provided that you make one additional approximation, called instantaneous recycling. This means you pretend that the new metals come out as soon as the massive stars form, (not far wrong at least for type II supernovae). The answer is

$$\frac{N(Z)}{N(Z_0)} = \frac{1 - \mu_0{}^{Z/Z_0}}{1 - \mu_0}$$

where Z_0 is the metallicity in the gas now, Z is any lower value, $N(Z)$ is the number of stars with metallicity Z or less, and μ_0 is the present residual gas fraction. The derivation is reproduced in Trimble (1975) and breaks down for large Z (more than about 10%). The continuous curve in Fig. 8 is this expression. That the data points for stars with much less than solar metallicity fall well below the curve is fairly obvious.

1974. Early attempts at coupling dynamical and chemical evolution together came from Talbot & Arnett (1974 for disk galaxies) and *Larson & *Tinsley (1974 for spheroids). *Ostriker & *Thuan (1975) considered the transformation of spheroids into disks both dynamically and chemically. This was also the year of the three-week NATO workshop "Origin and Abundances of the Chemical Elements" that was my introduction to the subject (Trimble 1975). The transformation of my lecture notes from the workshop into a review article happened primarily because a participant who was not a native speaker of English asked for a copy of my notes (which meant typing them up), and I would like to thank *Anna Zytkow, very belatedly, for her unintended but pivotal role in the process.

For the first 15 years or so, discussions of galactic chemical evolution focussed largely on passive processes. That is, you started with a given amount of gas, turned it into stars, and let the stars die away, allowing, at most, moderate amounts of gas flow into or out of your system. The modern study of active evolution, in which galaxies or parts of galaxies interact, merge, trigger each other's star bursts, and generally fail to mind their own business began with a classic numerical simulation of galaxy mergers by *Toomre & *Toomre (1972), which produced a very persuasive simulacrum of the pair NGC 4038/4039, complete with the insect-like antennae. (Radio astronomers have antennas; insects have antennae.) The paper is also, I think, one of the best-written ever in the astronomical literature.

*Press & *Schecter (1974) proposed a complete "bottom up" scenario for galaxy formation, in which nearly all large galaxies had been assembled from much smaller units, perhaps like the remaining dwarf irregulars. This became nearly everybody's favorite model during the recent heyday of cold dark matter, which produces its first bound structures on small scales.

1977. Another important conference took place at Yale (Tinsley & Larson 1977) on galaxies and stellar populations. Pre-galactic stars were widely blamed for setting a metallicity floor in the disk and for leaving behind brown dwarfs in the halo to produce the large mass to light ratios that were by then widely accepted. The latter, at least, has now been ruled out by the sparsity of gravitational microlensing events that can be attributed to compact halo objects in the brown dwarf mass range.

1978. *Searle & *Zinn (1978) suggested that even the halo of the Milky Way had been gradually assembled out of many entities, in contrast to the monolithic model of Eggen et al. (1962).

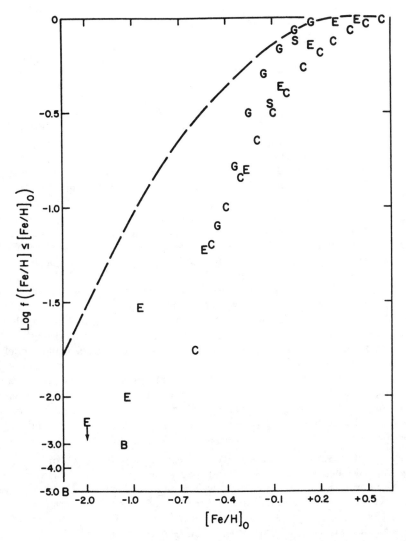

Figure 8. A visual presentation of the "G dwarf problem" from Trimble (1975). The solid curve is the prediction of fraction of all stars with metallicity equal to or less than Z as a function of Z in the simplest possible model, for a single, closed system with constant IMF or yield and instantaneous recycling. The letters show measured values of the fraction at different metallicities. S's are the data available to Schmidt and van den Bergh in the early 1960's. Of the other letters, B = Bond, G = Gliese & Pagel, C = Clegg and Bell, E = Eggen (references in Trimble 1975).

6.2. Current and Future Issues

First the G dwarf problem. It remains true that the simplest model is not a good fit to local disk stars. Current models of galactic chemical evolution tend to incorporate at least several of the traditional solutions (variable IMF, gas flows, perhaps some initial enhancement provided by thick disk stars...). But the simple model does fit a number of other systems, including the halo globular clusters, disk globular clusters, K giants in Baade's window (Pagel 1987), other bulge stars (Ibata & Gilmore 1995) and local halo stars with metallicities below [Fe/H]= -2.5 (Beers 1992), provided that you choose a suitable value of the yield for each population. Other issues that one might expect to see sorted out (not necessarily very soon) include the existence and nature of population III (that is, where did the very first metals come from?) and, closely related, what is the source of the metals in X-ray cluster gas and in the clouds producing QSO absorption lines. A number of people know the answers to these, but they don't all quite know the same answer.

Next, just how important are mergers in galactic evolution? Are all giant ellipticals the result of multiple mergers of disk systems? Is most star formation, including globular cluster formation, triggered in such events, or are they a fairly minor perturbation on isolated evolution. And, when you add up both active and passive evolution, do big galaxies turn out to be brighter or fainter at redshifts of $0.5 - 1.0$? This information is essential if the deceleration parameter, q_0, is ever to be determined by looking at a Hubble diagram for distant galaxies.

A very old puzzle is the ratio of helium to heavy element production by a generation of stars. Regressions on the data for metal-poor, gas-rich galaxies seem to suggest quite high values of $\Delta Y/\Delta Z = 3$ or even 6, while model stars produce ratios of 1 or 2.

Finally, of course, one would like to triumph over the curse of the variable parameter and learn to calculate or predict star formation rates, the initial mass function, the numbers and properties of binaries, and so forth, as a function of the amount of residual gas in a galaxy, its distribution, turbulence, and whatever else matters.

Acknowledgments. Sharp-eyed readers may have been wondering about the asterisks peppering the preceding pages. They are attached to the names of the people that I know or knew well enough to have heard them lecture at least once or to have had at least one real conversation with on nucleosynthesis or related topics. Some are obvious. Ed Salpeter and Nancy Roman, after all, were at this workshop. Some others, like George Gamow, R. d'Eath Atkinson, and Cecilia Payne-Gaposchkin are less so, but to have known these people is some considerable compensation for being no longer precisely young oneself! I am grateful to all of them for their contributions to the understanding of cosmic chemical evolution and my appreciation of the history of the subject. Special thank yous go to Drs. Lawrence H. Aller, Louis Brown, Charles R. Cowley, Vera Rubin, Sumner Starrfield, F.-K. Thielemann, and Stan Woosley for asking provocative questions and/or providing answers that could not readily have been obtained elsewhere.

References

Adams, W., & Joy, A. 1922, ApJ, 56, 262
Adams, W.S., et al. 1935, ApJ, 81, 187
Aller, L.H. 1942, ApJ, 95, 52
——————. 1943, ApJ, 97, 135
——————. 1994, ApJ, 432, 427
——————. 1995, Private communication
Aller, L.H., & Payne-Gaposchkin, C.H. 1942, Unpublished
Aller, M.F., & Cowley, C.R. 1970, ApJ, 162, L145
Allison, S.K., et al. 1939, Phys. Rev., 55, 107L and 624
Alpher, R.A., & Hermann, R.L. 1950, Rev. Mod. Phys., 22, 153
——————. 1953, Ann. Rev. Nucl. Sci., 2, 1
Alpher, R.A., Bethe, H.A., & Gamow, G. 1948, Phys. Rev., 73, 863
Alpher, R.A., Follin, J.W., & Hermann, R.L. 1953, Phys. Rev., 92, 1347
Anders, E., & Grevasse, N. 1989, Geochim. Cosmochim. Acta, 53, 197
Anders, E., & Zinner, E. 1991, Meteoritics, 28, 490
Asimov, I. 1966, The History of Physics (New York: Walker)
Aston, F.W. 1927, Proc. Roy. Soc., 115, 510
Atkinson, R. d'E. 1931, ApJ, 73, 250
Atkinson, R. d'E., & Houtermans, F.G. 1929, Z. Phys., 54, 656
Baade, W. 1944, ApJ, 100, 137
Beers, T.C. 1992, in M.G. Edmunds & R.J. Terlevich, eds., Elements and the Cosmos (Cambridge: Cambridge Univ. Press) p. 122
Berman, L. 1935, ApJ, 81, 369
Bethe, H.A. 1939, Phys. Rev., 55, 434
Bethe, H.A., & Critchfield, C.L. 1938, Phys. Rev., 54 31 248
Bidelman, W. 1950, ApJ, 111, 313
Biermann, L. 1956, Z. f. Ap., 41, 46
Brown, H. 1949, Rev. Mod. Phys., 21, 625
Brown, L. 1984, Unpublished manuscript
Buchmann, L. et al. 1993, Phys.Rev.Lett, 70, 726
Burbidge, G.R. 1958, PASP, 70, 83
Burbidge, E.M., Burbidge, G.R., Fowler, W.A., & Hoyle, F. 1957, Rev. Mod. Phys., 29, 547
Burrows, A., Hayes, J., & Fryxell, B.A. 1995, ApJ, 450, 830
Cameron, A.G.W. 1954, Phys. Rev., 93, 932
——————. 1955, ApJ, 121, 144
——————. 1957, Chalk River Report CRL-41
——————. 1968, in Origin and Distribution of the Elements, ed. L.H. Ahrens (Oxford: Pergamon) p. 125
——————. 1973, Space Sci.Rev., 15, 121

Caughlan, G., & Fowler, W.A. 1988, Atomic & Nuclear Data Tables, 40, 283
Chamberlain, J.W., & Aller, L.H. 1951, ApJ, 114, 52
Cowan, J.J., Thielemann, F.-K., & Truran, J.W. 1991a, ARA&A, 29, 447
───────────. 1991b, Phys. Reports, 208, 267
Cowley, C.R. 1995, Private communication
CRC 1949, Handbook of Chemistry and Physics, 31st edition (Cleveland: Chemical Rubber Co.)
CRC 1987, Handbook of Chemistry and Physics, 68th edition (Cleveland: Chemical Rubber Co.)
Curtiss, R.H. 1926, quoted in Astronomy by Russell, Dugan, & Stewart, (Boston: Ginn & Co.) p. 865
Diehl, R., et al. 1995, A&A, 298, L25 and 445
Dunbar, D.N.F., et al. 1953, Phys. Rev., 92, 649
Eggen, O.J., Lynden-Bell, D., & A. Sandage 1962, ApJ, 136, 748
Farkas, L., & Harteck, P. 1931, Naturwiss, 19, 705
Fermi, E., & Terkevich, A. 1949, cited by G. Gamow 1949, Rev. Mod. Phys., 21, 367
Fink, K. 1939, Ann. Physik, 34, 717
Fowler, W.A. 1972, in Cosmology, Fusion, and Other Matters, F. Reines ed., (U. Colorado Press)
───────────. 1984, Rev. Mod. Phys., 56, 149; Science, 226, 922
───────────. 1992, ARA&A, 30, 1
Fowler, W.A., Caughlan, G., & Zimmerman, B.1967, ARA&A, 5, 528
───────────. 1975, ARA&A, 13, 69
Gamow, G., 1935, Ohio J. Sci., 35, 406
───────────. 1946, Phys. Rev., 70, 372
───────────. 1949, Rev. Mod. Phys., 21, 367
Goldschmidt, V.M. 1937, Skr. Norske Vid. Acad. I. Mat. Nat. Klasse, No. 4
Greenstein, J.L. 1940, ApJ, 91, 435
───────────. 1954, in Modern Physics for the Engineer, ed. L.N. Ridenour (McGraw Hill, NY) p. 235
Gurevich, L.E. 1954, Trudy 3. Soveshch. pr. Vopr. Kosmogenii (Moscow) p. 191-202
Harkins, W.D. 1917, J. Am. Chem. Soc., 39, 856
Harris, M.J., et al. 1983, ARA&A, 21, 16
Hayakawa, S., Hayashi, C., & Nishida, M. 1960, Supl. Prog. Theor. Phys., 16,169
Hayashi, C. 1950, Prog. Theor. Phys., 5, 224
Haselgrove, B.C., & Hoyle, F. 1956, MNRAS, 116, 527
Helfer, H.L., Wallerstein, G., & Greenstein, J.L. 1959, ApJ, 129, 700
Hemmendinger, A,. 1948, Phys. Rev., 73, 706
Herant, M., et al. 1994, ApJ, 435, 339

Herbig, G.H., & Boyarchuk, A.A. 1968, ApJ, 153, 397
Holloway, M.G., & Moore, B.L. 1940, Phys. Rev., 58, 847
Hoyle, F. 1946, MNRAS, 106, 232
─────────── . 1954, ApJS, 1, 121
Hoyle, F., & Fowler, W.A. 1960, ApJ, 132, 565
Hoyle, F., & Schwarzschild, M. 1955, ApJS, 2, 1
Hubenet, H., ed. 1966, IAU Symp. 26, Structure and Evolution of Galaxies
Ibata, R., & Gilmore, G. 1995, MNRAS, 275, 605
Kappeler, F., et al. 1994, ApJ, 467, 396
Keenan, P., & Aller, L.H. 1951, ApJ, 113, 72
Kirchner, F., et al. 1937, Ann. Physik, 30, 521
Kleiber, I.A. 1885, J. Ross. Fis-Khim. Obschch., 17, 147
Kraft, R.P., et al. 1995, AJ, 109, 2586
Kuchowicz, B. 1967, Nuclear Astrophysics (Gordon & Breach)
Lambert, D.L 1992, Astr. & Ap. Rev., 3, 201
Lambert, D.L., et al. 1995, ApJ, 450, 312
Larson, R.B. 1972, Nat. Phys. Sci., 236, 7
Larson, R.B., & Tinsley, B.M. 1974, ApJ, 192, 293
Lequeux, J., ed. 1969, Structure and Evolution of Galaxies (Gordon & Breach)
Lindblad, B. 1922, ApJ, 55, 85
Malm, R., & Buechner, W.W. 1951, Phys. Rev., 81, 519
Masterton, W.L., & Slowinski, E.J. 1973, Chemical Physics (W.B. Saunders) Fig. 1.11
McKie, D. 1951, in A Short History of Science (Garden City: Doubleday) p. 69
McVittie, G.C., ed. 1962, IAU Symp. 15, Problems of Extragalactic Research (MacMillan)
─────────── . 1964, Trans. IAU 12B, Evolution Of Galaxies (Dordrecht: Reidel) p. 280
McWilliams, A., et al. 1995, AJ, 109, 2757
Merrill, P.W. 1947, ApJ, 104, 360
─────────── . 1952, Science, 115, 484
Morgan, W.W. 1956, PASP, 68, 509
Morgan, W.W., Keenan, P.C., & Kellman, E. 1943, An Atlas of Stellar Spectra (Chicago: U. Chicago Press)
Oddo, G. 1914, Z. f. Anorg. Allgem. Chem., 87, 253
Oort, J.H. 1926, Publ. Kapteyn Astron .Lab. Gron. 40, 59
Opik, E. 1951, Proc. R. Irish Acad., A54, 49
Ostriker, J.P., & Thuan, T.X. 1975, ApJ, 202, 353
Pagel, B.E.J. 1987, in The Galaxy, G. Gilmore & R.F- Carswell eds. (Dordrecht: Reidel) p. 341
Panov, I.V., et al. 1995, Astron. Lett., 21, 185
Payne, C.H. 1925, Stellar Atmospheres (Cambridge: W. Heffer & Sons)

Pokrowski, G.I. 1931, Phys. Z., 32, 374
Popper, D.M. 1942, PASP, 54, 160
———————. 1947, ApJ, 105, 204
Press, W.H., & Schechter, P. 1974, ApJ, 187, 425
Quirk, W.J., & Tinsley, B.M. 1974, ApJ, 192, 293
Rayet, M., et al. 1994, A&A, 298, 517
Roman, N.G. 1950, ApJ, 112, 554
———————. 1955, ApJS, 2, 198
Russell, H.N. 1929, ApJ, 70, 11
Salpeter, E.E. 1953, ApJ, 115, 326
———————. 1953, Ann. Rev. Nucl. Sci., 2, 41
———————. 1955, ApJ, 121, 161
Sandage, A.R. 1961, ApJ, 134, 916
———————. 1962, ApJ, 135, 333
Sandage, A., Freeman, K.C., & Stokes, N.K. 1970. ApJ, 160, 831
Sarajedini, A., & Milone, A.A.E. 1995, AJ, 109, 209
Schmidt, M. 1959, ApJ, 129, 243
———————. 1962, quoted by D. Lynden-Bell in Elements in the Cosmos, ed. M.G. Edmunds & R.J. Terlevich (Cambridge: Cambridge Univ. Press) p. 270
———————. 1963, ApJ, 137, 758
Schramm, D.N., & W.D. Arnett, eds. 1973, Explosive Nucleosynthesis (Austin: U. Texas)
Searle, L., & Zinn, R. 1978, ApJ, 225, 357
Solvay 13th Conf. 1965, Structure and Evolution of Galaxies (Interscience)
Sterne, T.E. 1933, MNRAS, 73, 736, 763 & 771
Suess, H.E., & Urey, H.C. 1956, Rev. Mod. Phys., 28,53
Talbot, R.J., & Arnett, W.D. 1973, ApJ, 170, 409
———————. 1974, ApJ, 190, 605
Tinsley, B.M. 1968, ApJ, 151, 547
Tinsley, B.M. & Larson, R.B. eds. 1977, The Evolution of Galaxies and Stellar Populations (New Haven: Yale Univ. Obs.)
Thielemann, F.-K. 1995, private communication
Tolmani R.C. 1922, J. Am. Chem. Soc., 44, 1902
Toomre, A., & Toomre, J. 1973, ApJ, 178, 623
Trimble, V. 1975, Rev. Mod. Phys., 47, 877
———————. 1991, Astr. & Ap. Rev., 3, 1
———————. 1994, SLAC BeamLine 24, No. 2, 28
Truran, J.W., & Cameron, A.G.W. 1971, Ap&SS, 14, 179
Urey, H.C., & Bradley, C.A. 1931, Phys. Rev., 38, 718
van den Bergh, S. 1958, AJ, 63, 492
———————. 1962, AJ, 67, 486

Vernon, H.M. 1890, Chem. News (London), 61, 51
von Hoerner, S. 1957, ApJ, 126, 592
von Winckel, H. et al. 1995, A&A, 2937, L25
Walke, H.J. 1934, Phil. Mag., 18, 795
Wallerstein, G. 1962, ApJS, 6, 407
—————————— . 1990, ApJS, 74, 758
Wasserburg, G.J. 1987, Earth & Plan. Sci., 86, 129
Wheeler, J.A. 1941, Phys. Rev., 59, 27
Whipple, F., & Payne-Gaposchkin, C. 1941, Proc. Amer. Phil. Soc., 84, 1
Woosley, S.E. et al. 1994, ApJ, 433, 229
Zhao, Z. et al. 1993. Phys.Rev.Lett, 70, 2066

Primordial Nucleosynthesis

David N. Schramm

University of Chicago, 5640 S. Ellis Avenue, Chicago, IL 60637

NASA/Fermilab Astrophysics Center, Fermilab, Box 500, Batavia, IL 60510, USA

Abstract. This review focuses on one of the most active topics in physical cosmology: the light element abundances and their impact on the issue of dark matter. The agreement between the Big Bang Nucleosynthesis (BBN) predictions and the observed abundances is discussed. It is noted that the basic conclusions of BBN on baryon density are remarkably robust. However, detailed questions of ^{3}He evolution and potential ^{4}He systematics remain unresolved, but in no way do these issues lead to any doubt about the basic success of BBN itself. The recent extragalactic deuterium observations as well as the other light element abundances are discussed. The BBN constraints on the cosmological baryon density are reviewed and demonstrate that the bulk of the baryons are dark and also that the bulk of the matter in the universe is non-baryonic. Arguments from recent MACHO/EROS observations of halo dark matter seem to imply that our Galaxy's halo is < 30% baryonic; hence, we seem to need non-baryonic dark matter even on that scale. Comparison of baryonic density arguments with recent x-ray data on clusters of galaxies is also made. Discussion of the interface of density and age arguments is also presented.

The study of the light element abundances has undergone a recent burst of activity on many fronts. New results on each of the cosmologically significant abundances have sparked renewed interest and new studies. The bottom line remains: primordial nucleosynthesis has joined the Hubble expansion and the microwave background radiation as one of the three pillars of Big Bang cosmology. Of the three, Big Bang Nucleosynthesis probes the universe to far earlier times ($\sim$ 1 sec) than the other two and led to the interplay of cosmology with nuclear and particle physics. Furthermore, since the Hubble expansion is also part of alternative cosmologies such as the steady state, it is BBN and the microwave background that really drive us to the conclusion that the early universe was hot and dense.

Recent heroic observations of ^{6}Li, Be and B, as well as ^{2}D, ^{3}He and new ^{4}He determinations, have all gone in the direction of strengthening the basic picture of cosmological nucleosynthesis. It will be shown that theoretical calculations of cosmic ray production of ^{6}Li, Be and B have fit the observations remarkably well, thus preventing these measurements from disturbing the standard scenario (Olive et al. 1990). The recent reports of D/H in quasar absorption systems at redshift $Z \sim 3$ are particularly interesting and will be discussed (Songaila et al.

1994, Carswell et al. 1994, Tytler & Fann 1995, Rogers & Hogan 1995) since BBN requires that fragile deuterium be found in primitive material. Although there is still significant scatter, it does appear that primordial D/H is beginning to be seen (Rogers & Hogan 1995). Furthermore, recent theoretical calculations have confirmed that quark-hadron inspired inhomogenous Big Bang Nucleosyntheis does not significantly alter the basic conclusions of standard BBN. We will also briefly discuss the possible impact on BBN of the recent ROSAT and ASCA x-ray satellite cluster results and the recent halo microlensing results. This summary will attempt to put it all together within an historical framework. The bottom line that emerges is how dramatically robust BBN is.

Let us now briefly review the history, with special emphasis on the remarkable agreement of the observed light element abundances with the calculations. This agreement works only if the baryon density is well below the cosmological critical value. This summary draws on the reviews of Walker et al. (1991); Schramm (1995); Copi, Schramm and Turner (1994); and Schramm, Copi and Shi (1995).

It should be noted that there is a symbiotic connection between BBN and the 3K background dating back to Gamow and his associates, Alpher and Herman. The initial BBN calculations of Gamow's group (Alpher et al. 1948) assumed pure neutrons as an initial condition and thus were not particularly accurate, but their inaccuracies had little effect on the group's predictions for a background radiation.

Once Hayashi (1950) recognized the role of neutron-proton equilibration, the framework for BBN calculations themselves has not varied significantly. The work of Alpher, Follin and Herman (1953) and Taylor and Hoyle (1964), preceeding the discovery of the 3K background, and of Peebles (1966) and Wagoner, Fowler and Hoyle (1967), immediately following the discovery, and the more recent work of our group of collaborators (Walker et al. 1991, Copi et al. 1994, Olive et al. 1990, Schramm & Wagoner 1977, Olive et al. 1981, Boesgaard & Steigman 1985, Yang et al. 1984, Kawano et al. 1988) all do essentially the same basic calculation, the results of which are shown in Figure 1. As far as the calculation itself goes, solving the reaction network is relatively simple by the standards of explosive nucleosynthesis calculations in supernovae, with the changes over the last 25 years being mainly in terms of more recent nuclear reaction rates as input, not as any great calculational insight, although the current Kawano code (Kawano et al. 1988) is somewhat streamlined relative to the earlier Wagoner code (Wagoner et al. 1967). In fact, the earlier Wagoner code is, in some sense, a special adaptation of the larger nuclear network calculation developed by Truran (Truran 1965, Truran & Cameron 1966) for work on explosive nucleosyntheis in supernovae. With the exception of Li yields and non-yields of Be and B, to which we will return, the reaction rate changes over the past 25 years have not had any major affect [see Yang et al. (1984) and Krauss and his collaborators (Krauss & Romanelli 1990, Kernan & Krauss 1994) or Copi, Schramm and Turner (1994) for discussion of uncertainties]. The one key improved input is a better neutron lifetime determination (Mampe et al. 1989, 1993). There has been much improvement in the $t(\alpha, \gamma)$ ^{7}Li reaction rate (Mathews 1995), but as the width of the curves in Figure 1 shows, the ^{7}Li yields are still the poorest determined, both because of this reaction and even more because of the poorly measured ^{3}He (α, γ) ^{7}Be.

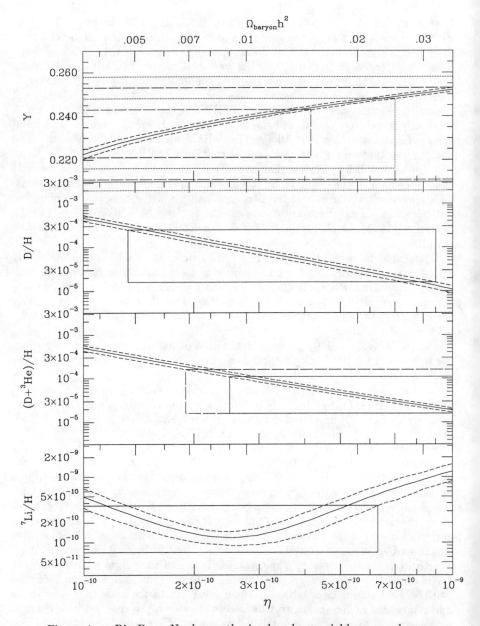

Figure 1. Big Bang Nucleosynthesis abundance yields versus baryon density (Ω_b) and $\eta \equiv \frac{n_b}{n_\gamma}$ for a homogeneous universe. ($h \equiv H_0/100$ km/sec/Mpc; thus, the concordant region of $\Omega_b h^2 \sim 0.015$ corresponds to $\Omega_b \sim 0.06$ for $H_0 = 50$ km/sec/Mpc.) Figure is modified from Copi, Schramm and Turner (1994). Note concordance region is slightly larger than Walker et al. (1991) due primarily to inclusion of possible systematic errors on Li/H and to allowing for additional ^{3}He evolution uncertainties. Constraint from ^{4}He shows sensitivity to various assumed systematic errors.

With the exception of the effects of elementary particle assumptions, to which we will also return, the real excitement for BBN over the last 25 years has not really been in redoing the basic calculation. Instead, the true action is focused on understanding the evolution of the light element abundances and using that information to make powerful conclusions. In the 1960's, the main focus was on ^{4}He which is very insensitive to the baryon density. The agreement between BBN predictions and observations helped support the basic Big Bang model but gave no significant information, at that time, with regard to density. In fact, in the mid-1960's, the other light isotopes (which are, in principle, capable of giving density information) were generally assumed to have been made during the T-Tauri phase of stellar evolution (Fowler et al. 1962), and so, were not then taken to have cosmological significance. It was during the 1970's that BBN fully developed as a tool for probing the universe. This possibility was in part stimulated by Ryter et al. (1970) who showed that the T-Tauri mechanism for light element synthesis failed. Furthermore, ^{2}D abundance determinations improved significantly with solar wind measurements Geiss & Reeves 1971, Black 1971) and the interstellar work from the Copernicus satellite (Rogerson & York 1973). (Recent HST observations reported by Linsky et al. (1993) have compressed the ^{2}D error bars considerably.) Reeves, Audouze, Fowler and Schramm (1973) argued for cosmological ^{2}D and were able to place a constraint on the baryon density excluding a universe closed with baryons. Subsequently, the ^{2}D arguments were cemented when Epstein, Lattimer and Schramm (1976) proved that no realistic astrophysical process other than the Big Bang could produce significant ^{2}D. This baryon density was compared with dynamical determinations of density by Gott, Gunn, Schramm and Tinsley (1974). See Figure 2 for an updated $H_0 - \Omega$ diagram.

By the late 1970's, a complimentary argument to ^{2}D had also developed using ^{3}He. In particular, it was argued (Rood et al. 1976) that, unlike ^{2}D, ^{3}He was made in stars; thus, its abundance would increase with time. Since ^{3}He like ^{2}D monotonically decreased with cosmological baryon density, this argument could be used to place a lower limit on the baryon density (Yang et al. 1979) using ^{3}He measurements from solar wind (Ryter et al. 1979, Geiss & Reeves 1970) or interstellar determinations (Wilson et al. 1983). Since the bulk of the ^{2}D was converted in stars to ^{3}He, the constraint was shown to be quit restrictive (Yang et al. 1984). Rood, Bania and Wilson (1992) showed that ^{3}He was indeed enriched in planetary nebulae; hence, the argument that ^{3}He increases with time was strengthened. However, there is nonetheless the worry that interstellar ^{3}He measurements (Balser et al. 1994) do vary with location more than one might expect for an isotope produced primarily in low mass stars. Also, the mix of high to low mass stars in the initial mass function might have been different in the past, so the degree of early ^{3}He destruction is quite model dependent. In early work, we did not allow for significant high mass star processing and destruction. However, in Copi et al. (1995) we did, and the lower bound on η from ^{3}He + D dropped from ~ 3 to $\sim 2 \times 10^{-10}$.

Independent of the ^{3}He details, it is interesting that the lower boundary from ^{3}He and the upper boundary from ^{2}D yield the requirement that ^{7}Li be near its minimum of ^{7}Li/H $\sim 10^{-10}$, which was verified by the Pop II Li measurements of Spite and Spite (1982, Rebolo et al. 1988, Hobbs & Pilachowski 1988), hence yielding the situation emphasized by Yang et al. (1984) that the light element

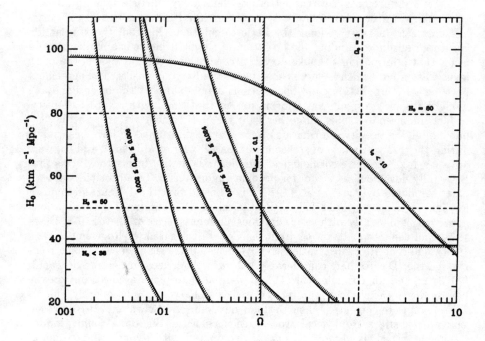

Figure 2. An updated version of $H_0 - \Omega$ diagram of Gott, Gunn, Schramm and Tinsley (1974) showing that Ω_b does not intersect $\Omega_{VISIBLE}$ for any value of H_0 and that $\Omega_{TOTAL} > 0.1$, so non-baryonic dark matter is also needed.

abundances are consistent over nine orders of magnitude with BBN, but only if the cosmological baryon density, Ω_b, is constrained to be around 6% of the critical value (for $H_0 \simeq 50$ km/sec/Mpc). The Li plateau argument was further strengthened with the observation of ^{6}Li in a Pop II star by Smith, Lambert and Nissen (1982). Since ^{6}Li is much more fragile than ^{7}Li, and yet it survived, no significant nuclear depletion of ^{7}Li is possible (Olive & Schramm 1992, Steigman et al. 1993). This observation of ^{6}Li has now been verified by Hobbs and Thorburn (1994), and a detection in a second Pop II star has been reported. Lithium depletion mechanisms are also severly constrained by the recent work of Spite et al. (1995), showing that the lithium plateau also is found in Pop II tidally locked binaries. Thus, meridonal mixing is not causing lithium depletion.

The other development of the 70's for BBN was the explicit calculation of Steigman, Schramm and Gunn (1977) showing that the number of neutrino generations, N_ν, had to be small to avoid overproduction of ^{4}He. [Earlier work (Taylor & Hoyle 1964, Schvartzman 1969, Peebles 1971) had commented about a dependence on the energy density of exotic particles but had not done an explicit calculation probing N_ν.] This will subsequently be referred to as the SSG limit. To put this in perspective, one should remember that the mid-1970's also saw the discovery of charm, bottom and tau, so that it almost seemed as if each new detector produced new particle discoveries, and yet, cosmology was arguing against this "conventional" wisdom. Over the years, the SSG limit on N_ν improved with ^{4}He abundance measurements, neutron lifetime measurements, and with limits on the lower bound to the baryon density, hovering at $N_\nu \lesssim 4$ for most of the 1980's and dropping to slightly lower than 4 just before LEP and SLC turned on (Olive et al. 1990, Walker et al. 1991, Schramm & Kawano 1989, Pagel 1990). This was verified by the LEP results (ALEPH 1993) where now the overall average is $N_\nu = 2.99 \pm 0.02$.

An exciting new observation has been reported by Songaila et al. (1994) and by Carswell et al. (1994) of a possible D/H measurement in a QSO absorption system at $Z = 3.3$. Such a detection would be an extremely important confirmation of the basic BBN argument that deuterium is primordial. Subsequent work by Rogers and Hogan (1995) has shown this feature to consist of two deuterium lines, each consistent with D/H $\sim 1 - 2 \times 10^{-4}$. It should be noted that such high values are hard to fit with relatively low ^{3}He values (few $\times 10^{-5}$) observed in the solar system and in the ISM today unless more ^{3}He destruction occurs than assumed by Hata et al. (1995). The high value is consistent with the less restrictive assumptions of Copi et al. (1995). A recent report by Tytler and Fann (1995) of another extragalactic D/H determination, this time at $Z = 3.6$ with a metalicity of only 1/500 of solar, appears to disagree with the Rogers and Hogan value, so the final word on the primordial D/H is certainly not in yet. If extragalactic D/H determination can eventually be confirmed in several directions, it would provide the firmest determination of the baryon density and may enable a collapse of the present range for ρ_b to an even narrower band.

The power of homogeneous BBN comes from the fact that essentially all of the physics input is well determined in the terrestrial laboratory. The appropriate temperature regimes, 0.1 to 1 MeV, are well explored in nuclear physics laboratories. Thus, what nuclei do under such conditions is not a matter of guesswork, but is precisely known. In fact, it is known for these temperatures far better than it is for the centers of stars like our sun. The center of the sun is only a little over 1 keV, thus, below the energy where nuclear reaction rates yield significant results in laboratory experiments, and only the long times and higher densities available in stars enable anything to take place.

The success and robustness of BBN have given renewed confidence to the limits on the baryon density constraints. Let us convert this density regime into units of the critical cosmological density for the allowed range of Hubble expansion rates. This is shown in Figure 2. In particular, $\Omega_b = 0.01h^{-2}$ to $0.02h^{-2}$ where h is the Hubble constant in units of 100km/sec/Mpc. Figure 2 also shows the lower bound on the age of the universe of 10 Gyr from both nucleochronology and from globular cluster dating (Shi et al. 1995) and a lower bound on H_0 of 38 from extreme type IA supernova models with pure 1.4 $M_\odot$ carbon white dwarfs being converted to ^{56}Fe. The constraint on Ω_b means that the universe *cannot be closed with baryonic matter*. [This point was made over twenty years ago (Reeves et al. 1970) and has proven to be remarkably strong.] If the universe is truly at its critical density, then nonbaryonic matter is required. This argument has led to one of the major areas of research at the particle-cosmology interface, namely, the search for non-baryonic dark matter. In fact, from the lower bound on Ω_{TOTAL} from cluster dynamics of $\Omega_{TOTAL} > 0.1$, it is clear that non-baryonic dark matter is required unless $H_0 < 50$. The need for non-baryonic matter is strengthened on even larger scales (Fisher 1992, Davis & Nusser 1995). Figure 2 also shows the range of $\Omega_{VISIBLE}$ and shows that there is no overlap between Ω_b and $\Omega_{VISIBLE}$. Hence, the bulk of the baryons are dark.

Another interesting conclusion (Gott et al. 1974) regarding the allowed range in baryon density is that it is in agreement with the density implied from the dynamics of single galaxies, *including their dark halos*. The recent MACHO (Alcock et al. 1993) and EROS (Aubourg et al. 1993) reports of halo microlensing may well indicate that at least some of the dark baryons are in the form of brown dwarfs in the halo. However, Gates, Gyuk and Turner (1995), and Alcock et al. (1993) show that the observed distribution of MACHOs (4 towards LMC, 45 towards Galactic Bulge) cannot be fit if more than 30% of the halo is MACHOs. Thus, the bulk of the halo must be in some other form such as cold dark matter.

For dynamical estimates of Ω one estimates the mass from $M \sim \frac{v^2 r}{G}$ where v is the relative velocity of the objects being studied, r is their separation distance, and G is Newton's constant. The proportionality constant out front depends on orientation, relative mass, etc. For large systems such as clusters, one uses averaged quantities. For single galaxies v would represent the rotational velocity and r the radius of the star or gas cloud. It is this technique which yields the cluster bound on Ω shown on Figure 2. It should be noted that the value of $\Omega_{CLUSTER} \sim 0.2$ is also obtained in those few cases where alignment produces giant gravitational-lens arcs. Recent work using weak gravitational lensing by Kayser (1995) also supports large Ω. As Davis and Nusser (1995) show, if the large scale velocity flows measured from the IRAS survey are due to gravity, then $\Omega_{IRAS} \gtrsim 0.3$. Similar arguments have been obtained using the Great Attactor study or the Potent technique. All imply $\Omega > \Omega_{BARYON}$, hence the need for non-baryonic dark matter. However, there is still considerable uncertainty of the exact value of Ω determined in this way as discussed by Szalay (1995). But all groups agree it is greater than 0.3. However, as Figure 2 illustrates, except for $H_0 < 50$, $\Omega_{CLUSTER}$ already required $\Omega_{TOTAL} > \Omega_{BARYON}$ and hence the need for non-baryonic dark matter.

An Ω of unity is, of course, preferred on theoretical grounds since that is the only long-lived natural value for Ω, and inflation (Guth 1981, Linde 1990) or something like it provided the early universe with the mechanism to achieve that value and thereby solve the flatness and smoothness problems. Note that our need for exotica is not dependent on the existence of dark galatic halos and that high values of H_0 increase the need for non-baryonic dark matter.

Non-baryonic matter can be divided (Bond & Szalay 1992) into two major categories for cosmological purposes: hot dark matter (HDM) and cold dark matter (CDM). Hot dark matter is matter that is relativistic until just before the epoch of galaxy formation, the best example being low mass neutrinos with $m_\nu \sim 20\text{eV}$. Cold dark matter is matter that is moving slowly at the epoch of galaxy formation. Because it is moving slowly, it can clump on very small scales, whereas HDM tends to have more difficulty in being confined on small scales. Examples of CDM could be massive supersymmetric particles with masses, M_x, greater than several GeV or the lightest super-symmetric particle which is presumed to be stable and might also have masses of several GeV. Following Michael Turner, all such weakly interacting massive particles are called "WIMPS." Axions, while very light, would also be moving very slowly (Turner et al. 1993) and thus would clump on small scales. Note that CDM would clump in halos, thus requiring the dark baryonic matter to be out between galaxies, whereas HDM would allow baryonic halos. The MACHO and EROS events now favor at least some CDM. Obviously, mixed models with some HDM and some CDM have even more flexibility and have thus become quite popular as data constraints increase.

Some baryonic dark matter must exist since we know that the lower bound from Big Bang Nucleosynthesis is greater than the upper limits on the amount of visible matter in the universe. However, unless the microlensing efficiencies turn out to be in error, dark baryons cannot all be in halos. If the baryonic dark matter is not in the halo, it could be in hot intergalactic gas, hot enough not to show absorption lines in the Gunn-Peterson test, but not so hot as to be seen in the x-rays. The exciting report by Jakobsen et al. (1994) of a Gunn-Peterson effect observed with HST for He-II at high Z showed that at least some hot IGM exists (and verifies that He seems primordial).

Another possible hiding place for the dark baryons would be failed galaxies, large clumps of baryons that condensed gravitationally but did not produce stars. Such clumps are predicted in galaxy formation scenarios that include large amounts of biasing where only some fraction of the clumps shine. Evidence for some hot gas is found in clusters of galaxies from the ROSAT and ASCA satellites. In particular, Mushotzky (1993) and White et al. (1993) have discussed how certain observed rich clusters have $M_{HOTGAS}/M_{TOTAL} \sim 1/5$, which, with $\Omega_{CLUSTER} \sim 0.2$ and $\Omega_b \sim 0.05$, would imply no conflict with BBN, but that clusters are not fair samples of the baryon to non-baryon ratio in the universe. Apparent variation in M_{HOT}/M_{TOTAL} for small groups relative to clusters seems to support the point of view that megaparsec scales are not always fair samples. If true, this is difficult to reconcile with cold dark matter models but can be fit with topological defects and HDM or with mixed models.

It seems clear that we need the bulk of the matter in the universe to be in some non-baryonic form. In particular, significant cold dark matter is needed to satisfy the halo arguments now that MACHO and EROS have shown that baryons don't work. There may also need to be an admixture of hot dark matter (low mass neutrinos).

This latter point may also be motivated by the need for neutrino masses from the solar neuitrino problem. The best fit solar ν mass hints at $m_{\nu_\mu} \sim 10^{-3}$ eV which in a single see-saw model may imply $m_{\nu_\tau} \sim 10$ eV, that is, great hot dark matter.

Particularly exciting are the experimental searches for non-baryonic dark matter. These searches include accelerator searches for supersymmetry (or other Weakly Interacting Massive Particles, WIMPS) and for neutrino oscillations. They also include direct underground searches for WIMPS and axions and even satellite searches for WIMP annihilation products.

Hopefully, with all this activity, we will find the dark matter before the turn of the century.

Acknowledgments. I would like to thank my collaborators, Craig Copi, David Dearborn, Brian Fields, Dave Thomas, Gary Steigman, Brad Meyer, Keith Olive, Angela Olinto, Bob Rosner, Michael Turner, George Fuller, Karsten Jedamzik, Rocky Kolb, Grant Mathews, Bob Rood, Jim Truran and Terry Walker for many useful discussions. I would further like to thank Poul Nissen, Jeff Linsky, Julie Thorburn, Doug Duncan, Lew Hobbs, Evan Skillman, Bernard Pagel and Don York for valuable discussion regarding the astronomical observations.

This work is supported by the NASA and the DoE(nuclear) at the University of Chicago, and by the DoE and by NASA grant NAG5-2788 at Fermilab.

References

Alcock, C. et al. 1993, Nature, 365, 621

ALEPH, L3, OPAL, DELPHI results, 1993, 1993 Lepton-Photon meeting at Ithaca, NY

Alpher, R.A., Bethe, H., & Gamow, G. 1948, Phys. Rev. 73, 803

Alpher, R.A., Follin, J.W., & Herman, R.C. 1953, Phys. Rev., 92, 1347

Aubourg, E. et al. 1993, Nature, 365, 623

Balser, D.S. et al. 1994, ApJ, submitted

Black, D. 1971, Nature, 234, 148

Boesgaard, A., & Steigman, G. 1985, Ann. Rev. of Astron. and Astrophys., 23, 319

Bond, R., & Szalay, A. 1992, in Proc. Texas Relativistic Astrophysical Symposium, Austin, Texas

Carswell, R.F. et al. 1994, MMRAS, in press

Copi, C., Schramm, D.N., & Turner, M. 1994, Science, 267, 192

Copi, C., Schramm, D.N., & Turner, M.S. 1995, Phys. Rev. Lett., 75, 3981; Copi, C., Schramm, D.N., & Turner, M.S. 1995, ApJ, in press (Dec. 20, 1995)

Davis, M., & Nusser, A. 1995, in Proc. Maryland Symposium on Dark Matter, this volume

Epstein, R., Lattimer, J., & Schramm, D.N. 1976, Nature 263, 198

Fisher, C. 1992, Ph.D. Thesis, University of California at Berkeley, and references therein

Fowler, W.A., Greenstein, J., & Hoyle, F. 1962, Geophys. J.R.A.S., 6, 6

Gates, E., Gyuk, G., &. Turner, M. 1995, Phys. Rev. Lett., 74, 3724; Bennett, D. et al. 1995, LLNL preprint

Geiss, J., & Reeves, H. 1971, A&A, 18, 126

Gott, J.R., III, Gunn, J., Schramm, D.N., & Tinsley, B.M. 1974, ApJ, 194, 543

Guth, A. 1981, Phys. Rev. D, 23, 347

Hata, N., Scherrer, R.J., Steigman, G., Thomas, D., Walker, T.P., Bludman, S., & Langacker, P. 1995, Phys. Rev. Lett., 75, 3977

Hayashi, C. 1950, Prog. Theor. Phys., 5, 224

Hobbs, L., & Pilachowski, C. 1988, ApJ, 326, L2

Hobbs, L., & Thorburn, J. 1994, ApJ, 428, L25

Jakobsen, P., Boksenberg, A., Deharveng, J.M., Greenfield, P., Jedrzewski, R., & Paresce, F. 1994, Nature, in press

Kawano, L., Schramm, D.N., & Steigman, G. 1988, ApJ, 327, 750

Kayser, N. 1995, in Proc. of the Texas Symposium on Relativistic Astrophysics, Munich, December 1994, in press

Kernan, P., & Krauss, L. 1994, Phys. Rev. Lett., 72, 3309

Krauss, L.M., & Romanelli, P. 1990, ApJ, 358, 47

Linde, A. 1990, Particle Physics and Inflationary Cosmology, New York: Harwood Academic Publishers

Linsky, J. et al. 1993, ApJ, 402, 694

Mampe, W. et al., 1993, JETP Lett., 57, 82

Mampe, W., Ageron, P., Bates, C., Pendlebury, J.M., & Steyerl, A. 1989, Phys. Rev. Lett. 63A, 593

Mathews, G. 1995, in Proc. Snowmass Workshop on Particle Astrophysics in the Next Millenium, in press

Mushotsky, R. 1993, in Relativistic Astrophysics and Particle Cosmology: Texas PASCOS 92, C. W. Akerlof and M. A. Srednicki, Annals of the N.Y. Academy of Sciences 688, 184

Olive, K., & Schramm, D.N. 1992, Nature 360, 434

Olive, K., Schramm, D.N., Steigman, G., Turner, M., & Yang, J. 1981, ApJ, 246, 557

Olive, K., Schramm, D.N., Steigman, G, & Walker, T. 1990, Phys. Lett.

Pagel, B., 1990, in Proc. of 1989 Rencontres de Moriond

Peebles, P.J.E., 1966, Phys. Rev. Lett., 16, 410

Peebles, P.J.E. 1971, Physical Cosmology, Princeton: Princeton University Press

Rebolo, R., Molaro, P., & Beckman, J. 1988, A&A, 192, 192

Reeves, H., Audouze, J., Fowler, W.A., & Schramm, D.N., 1973, ApJ, 179, 909

Reeves, H., Fowler, W.A., & Hoyle, F. 1970, Nature, 226, 727
Rogers, M., & Hogan, C. 1995 ApJ, in press B, 236, 454
Rogerson, J., & York, D. 1973, ApJ, 186, L95
Rood, R.T., Bania, T., &. Wilson, J. 1992, Nature, 355, 618
Rood, R.T., Steigman, G., & Tinsley, B.M. 1976, ApJ, 207, L57
Ryter, C., Reeves, H., Gradstajn, E., & Audouze, J. 1970, A&A, 8, 389
Schramm, D.N. 1995, in Proc. of Yamada Conf. XXXVII - Evolution of the Universe and Its Observational Quest, Tokyo, June 1993, Tokyo: Universal Academic Press, in press. See also Schramm, D.N. 1995, in The Light Element Abundances: Proc. of ESO/EIPC Workshop on the Light Element Abundances, Elba, May 1994, P. Crane, Heidelberg: Springer-Verlag, 51
Schramm, D.N., Copi, C., & Shi, X. 1995, in Proc. of Physics Summer School on Cosmology, Australian National University, January 1995, Canberra, Australia, B.A. Robson, Singapore: World Scientific, in press
Schramm, D.N. & Kawano, L. 1989, Nuc. Inst. and Methods A, 284, 84
Schramm, D.N., & Wagoner, R.V. 1977, Ann. Rev. of Nuc. Sci., 27, 37
Schvartzman, V.F. 1969, JETP Letters, 9,184
Shi, X., Schramm, D.N., Dearborn, D., & Truran, J.W. 1995, Comments on Astrophys., in press
Smith, V.V., Lambert, D.L., & Nissen, P.E. 1982, ApJ, 408, 262
Songaila, A., Cowie, L.L., Hogan, C., & Rogers, M. 1994, Nature, 368, 599
Spite, M., Pasquini, F., & Spite, F. 1995, A&A, in press
Spite, J., & Spite, M. 1982, A&A, 115, 357
Steigman, G., Fields, B., Olive, K., Schramm, D.N., &. Walker, T. 1993, ApJ, 415, L35
Steigman, G., Schramm, D.N., & Gunn, J. 1977, Phys. Lett. B, 66, 202
Szalay, A. 1995, in Proc. of the Australian National University Summer School on Cosmology, Canberra, Australia, January 1994, Singapore: World Scientific
Taylor, R., & Hoyle, F. 1964, Nature, 203, 1108
Truran, J.W. 1965, Doctoral Thesis, Yale University
Truran, J.W., Cameron, A.G.W., & Gilbert, A. 1966, Can. Jour. of Phys., 44, 563
Turner, M., Wilczek, F., &. Zee, A. 1993, Phys. Lett. B, 125, 35; 125, 519
Tytler, D., & Fann, J., 1995, private communication
Wagoner, R., Fowler, W.A., & Hoyle, F. 1967, ApJ, 148, 3
Walker, T., Steigman, G., Schramm, D.N., Olive, K., & Kang, H.-S. 1991, ApJ, 376, 51
White, S.D.M., Navarro, J.F., Evrard, A.E. &. Frenck, C.S. 1993, Nature, 366, 261
Wilson, R., Rood, R.T., & Bania, T. 1983, in Proc. of the ESO Workshop on Primordial Healing, P. Shaver and D. Knuth, Garching: European Southern Observatory

Yang, J., Schramm, D.N., Steigman, G.,& Rood, R.T. 1979, ApJ, 227, 697
Yang, J., Turner, M., Steigman, G., Schramm, D.N., & Olive, K. 1984, ApJ, 281, 493

Testing Big Bang Nucleosynthesis

Gary Steigman

Departments of Physics and Astronomy, The Ohio State University, 174 West 18th Avenue, Columbus, OH 43210, USA

Abstract. Big Bang Nucleosynthesis (BBN), along with the cosmic background radiation and the Hubble expansion, is one of the pillars of the standard, hot, big bang cosmology since the primordial synthesis of the light nuclides (D, ^{3}He, ^{4}He, ^{7}Li) must have occurred during the early evolution of a universe described by this model. The overall consistency between the predicted and observed abundances of the light nuclides, each of which spans a range of some nine orders of magnitude, provides impressive support for the standard models of cosmology and particle physics. Here, the results of recent, statistically consistent tests of BBN are described. This new confrontation between theory and data challenges the standard model. The crises confronting BBN are identified and several possible resolutions are outlined.

1. Introduction

The discovery of the Cosmic Background Radiation (CBR) by Penzias & Wilson (1965) transformed forever the study of Cosmology from an exercise in philosophy to the pursuit of science. The presence of the CBR in an expanding Universe favors the hot big bang cosmology. A Universe described by this model was very hot and very dense during early epochs in its evolution. As a consequence, it is a prediction of this "standard" cosmological model that, briefly, the early Universe was a primordial nuclear reactor in which the light nuclides D, ^{3}He, ^{4}He and ^{7}Li were synthesized in astrophysically interesting abundances (for details and references see, e.g., Boesgaard & Steigman 1985; Walker et al. 1991). Thus, along with the CBR and the "Hubble" expansion, Big Bang Nucleosynthesis (BBN) provides one of the three pillars supporting the standard model of cosmology. The standard hot big bang model is, in principle, falsifiable. In contrast to cosmology as theology, this empirical model is not a matter of faith but, rather, demands our eternal vigilance and critical scrutiny. The success of the model is gauged by the degree to which the BBN predictions are consistent with the primordial abundances of the light nuclides inferred from observational data. Over the years BBN has emerged unscathed from the confrontation between theory and observations, providing strong support for the standard, hot big bang cosmological model (e.g., Yang et al. 1984; Boesgaard & Steigman 1985; Walker et al. 1991). This success has, however, not spawned complacency, and the testing continues. In recent years, as the astronomical data has become more precise, hints of a possible crisis have emerged (Copi, Schramm & Turner 1995; Olive &

Steigman 1995; Hata et al. 1995). It is my goal here to describe the impressive success of BBN and to map out the paths leading to the current challenges to the standard model. To better appreciate these challenges and the opportunities they present, an historical analogy may be instructive.

1.1. Three Crises For The 19th Century Standard Model

The gravitational theory described by Newton (1686) was outstandingly successful in explaining the motion of the moon and planets. Perhaps one of the most thoroughly tested physical theories in history, Newtonian gravity had become the standard model of 19th Century Physics. Soon, however, some challenges to the standard model emerged. The nature of these challenges and their different resolutions provide some interesting lessons for the emerging crisis in BBN.

(i) Perturbations to the Orbit of Uranus

Deviations in the orbit of Uranus from the predictions of Newtonian gravity (the standard model) led Adams and LeVerrier to predict the existence and location of Neptune. The standard model was used to discover something new, verifying the accuracy of the data (the orbit of Uranus) and providing spectacular support for Newtonian gravity.

(ii) Perturbations to the Orbit of Neptune

Observations of the newly discovered Neptune suggested that its orbit, too, was being perturbed away from the standard model predictions. So began the long search which culminated in the discovery of Pluto. Pluto, however, is not responsible for measurable perturbations to the orbit of Neptune - it is too small. Rather, here we have a case of insufficiently accurate data. The discovery of Pluto was serendipitous; more accurate observations of Neptune's orbit are entirely consistent with the predictions of Newtonian gravity.

(iii) Precession of the Perihelia of Mercury

By the mid-19th century LeVerrier had noted a discrepancy between the predicted and observed precession of the perihelia of Mercury. LeVerrier and others proposed one or more planets (Vulcan) between Mercury and the Sun to resolve this crisis. None were found. Alternately, it was proposed by Newcomb and others that the perturbing mass might be in a ring of dust or asteroids. This, however, would have perturbed the orbits of Mercury and Venus in conflict with observational data. A more radical solution, modifying the inverse square law, was proposed by Newcomb (1895). This, however, is in conflict with the accurately observed lunar orbit.

As is so well known, the resolution of this crisis confronting the 19th century standard model was new physics! Einstein's General Theory of Relativity (1916) predicts a precession in beautiful agreement with that observed.

Three crises, three different resolutions: the standard model preserved and a new discovery (Neptune); the standard model preserved and insufficently accurate data (Neptune/Pluto); the standard model replaced (perihelia of Mercury).

2. Consistency of the Standard Model

2.1. Predictions

Now let us turn to the standard model of cosmology and the predictions of primordial nucleosynthesis. Employing measured weak interaction rates and nuclear reaction cross sections the primordial abundances of the light nuclides are predicted by BBN as a function of only one adjustable parameter, η, the universal ratio of nucleons (baryons) to photons ($\eta = N_B/N_\gamma; \eta_{10} = 10^{10}\eta$). The predicted abundances of ^{4}He (Y is the ^{4}He mass fraction) D and ^{7}Li ($y_2 = N_D/N_H, y_7 = N_{Li}/N_H$) are shown for $1 \leq \eta_{10} \leq 10$ in Figure 1 from Hata et al. (1995). For clarity of presentation the predicted abundance of ^{3}He, very similar to that of D, is not shown.

The predicted abundances depend on the universal expansion rate, t^{-1}, during the epoch of BBN ($\sim 3MeV \gtrsim T_{BBN} \gtrsim 30keV; 0.1 \lesssim t_{BBN} \lesssim 10^3 sec$). For the early Universe $t^{-1} \propto \rho_{TOT}^{1/2}$, where ρ_{TOT} is the total mass-energy density. For the "standard" model (SBBN), ρ_{TOT} is dominated by photons, electron-positron pairs and three flavors of light, left-handed neutrinos (ν_e, ν_μ, ν_τ).

$$\rho_{TOT}^{SBBN} = \rho_\gamma + \rho_e + 3\rho_\nu^0. \tag{1}$$

In (1), ρ_ν^0 is the contribution from one flavor of light ($m_\nu << T_{BBN}$) neutrinos. To account for a possibly massive τ-neutrino and/or for other, new particles beyond the standard model, it is convenient to modify eq. (1) by introducing N_ν, the "effective" number of equivalent light neutrinos (Steigman, Schramm & Gunn 1977).

$$\rho_{TOT}^{BBN} = \rho_\gamma + \rho_e + N_\nu \rho_\nu^0. \tag{2}$$

For SBBN, $N_\nu = 3$; for $N_\nu \neq 3$ the universal expansion rate at BBN is modified. For $N_\nu \geq 3$, the universe expands more rapidly leaving less time for the conversion of neutrons to protons. Since most neutrons are incorporated in ^{4}He, Y_{BBN} increases with N_ν (and, vice-versa). Therefore, it is convenient to use N_ν as a second parameter to explore deviations from SBBN and extensions of the standard model of particle physics (Steigman, Schramm & Gunn 1977). The results in Figure 1 are for SBBN ($N_\nu = 3$).

For $1 \leq \eta_{10} \leq 10$, the predicted abundances of the light nuclides span a range of some 9 orders of magnitude from $\sim 10^{-10} - 10^{-9}$ for Li/H, to $\sim 10^{-5} - 10^{-4}$ for D/H and ^{3}He/H, to ~ 0.1 for ^{4}He/H.

2.2. Observations

Primordial abundances are, of course, not observed. Rather, they are inferred from astronomical data. Some, such as D and ^{3}He, have been mainly observed "here and now" (in the solar system and the interstellar medium (ISM) of our own Galaxy). For these nuclides it is necessary to extrapolate from here and now to "there and then" to derive their universal primordial abundances. ^{4}He and ^{7}Li are observed (in addition to here and now) in regions where much less chemical processing has occurred (low metallicity, extragalactic HII regions for ^{4}He; very metal-poor halo stars for ^{7}Li). For these nuclides the extrapolations to primordial abundances are smaller.

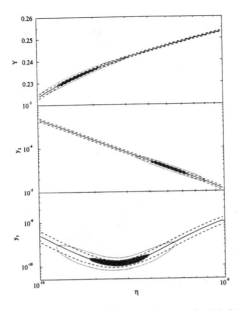

Figure 1. The SBBN predicted abundances (solid lines) of ^{4}He (Y is the ^{4}He mass fraction), D ($y_2 = $ D/H), and ^{7}Li ($y_7 = $ Li/H) as a function of the nucleon-to-photon ratio η. The dashed lines are the 1σ theoretical uncertainties from Monte Carlos. The shaded (dashed) contours are the regions constrained by the observation at the 68% (95%) CL.

In addition to observational uncertainties and those associated with the extrapolations to primordial abundances, systematic effects in deriving abundances from data may contribute to the overall uncertainties. The bad news is that such systematic uncertainties are difficult to constrain. The good news is that the sources of possible systematic errors are different for the different nuclides.

2.3. Testing SBBN

The relatively strong and monotonic y_2 vs. η relation visible in Figure 1 points to D (and, to a lesser extent, ^{3}He) as an ideal baryometer. If the primordial abundance of D were known, for example, to $\sim 40\%$, the universal density of baryons would be known to $\sim 25\%$. The large extrapolation from here and now to there and then has inhibited the implementation of this approach. Rather, to avoid this large extrapolation, a more conservative approach has been adopted. Since any D incorporated in a star is burned (to ^{3}He) and there are no significant astrophysical sources of post-BBN D, the abundance of D observed anywhere at anytime provides a lower bound to its primordial abundance (e.g., $y_{2P} \geq y_{2\odot}$, $y_{2P} \geq y_{2ISM}$). From Figure 1 it is clear that a lower bound to y_{2P} leads to an upper bound to η.

It is difficult to avoid the uncertainties of chemical evolution models in using observations of D to infer an upper bound to y_{2P}. However, Yang et al. (1984)

noted that since D is burned to ^{3}He and some ^{3}He survives stellar processing, the primordial abundances of D + ^{3}He are strongly correlated with the evolved abundances of D + ^{3}He. Burying the stellar and evolution model uncertainties in one parameter, g_3, the ^{3}He survival fraction, Yang et al. (1984; also, Walker et al. 1991) used solar system data to place an upper bound on primordial D (and/or on D + ^{3}He). An upper bound on y_{2P} provides a lower bound on η (see Fig. 1).

Due to the "valley" shape in the BBN prediction of Li vs. η (see Fig. 1), an upper bound to y_{7P} will provide both lower and upper bounds to η. The lithium abundance also offers a key test of the standard model since its primordial value must not lie below the minimum predicted ($y_{7BBN} \gtrsim 1 \times 10^{-10}$).

One test of the consistency of SBBN is to use D, ^{3}He and ^{7}Li to infer lower and upper bounds to η (η_{MIN}, η_{MAX}) and to check that $\eta_{MIN} < \eta_{MAX}$. If SBBN passes this test, the "^{4}He test" may be applied. The predicted ^{4}He mass fraction, Y_{BBN}, is a very weak function of η, increasing from $Y_{BBN} = 0.22$ at $\eta_{10} = 1$ to $Y_{BBN} = 0.25$ at $\eta_{10} = 10$. Thus, it is key to the success of SBBN ($N_\nu = 3$) that for $\eta_{MIN} < \eta < \eta_{MAX}$, Y_P (the inferred primordial abundance) is consistent with $Y_{BBN}(\eta)$ (the predicted abundance).

2.4. Consistency

Yang et al. (1984) were among the first to carry out a detailed analysis of the observational data and to implement the tests described above. From D, ^{3}He and ^{7}Li (with $g_3 \geq 1/4$, see Dearborn, Schramm & Steigman 1986) they found consistency: $3 \lesssim \eta_{10} \lesssim 7$, leading to a predicted range for ^{4}He: $0.24 \lesssim Y_{BBN} \lesssim 0.26$. Comparing with the rather sparse data available, they derived $0.23 \lesssim Y_P \lesssim 0.25$ and concluded that SBBN passed the ^{4}He test. They did note that SBBN is, in principle, falsifiable and pointed out that if future comparisons should increase η_{MIN} and/or decrease Y_P, consistency would require $N_\nu < 3$, modifying the standard model.

By 1991 uncertainties in the neutron lifetime (as well as its central value) had been reduced considerably permitting a very accurate prediction of Y_{BBN} vs. η (at the 2σ level, Y_{BBN} is known to $\lesssim \pm 0.001$; see Thomas et al. 1995). At the same time there was extensive new data on lithium (in halo stars) and helium-4 (in extragalactic HII regions). Applying the above tests, Walker et al. (1991) found $2.8 \leq \eta_{10} \leq 4.0$ and $0.236 \leq Y_{BBN} \leq 0.243$. From the HII region data, Walker et al. (1991) derived $Y_P = 0.23 \pm 0.01$ and concluded that SBBN passed the ^{4}He test. However, they did emphasize, "that if our lower bound on η were increased from $\eta_{10} = 2.8$ to $\eta_{10} = 4.0$, the window on N_ν would be closed (for $Y_P \lesssim 0.240$)."

2.5. Crisis?

Recent applications of the two consistency tests ($\eta_{MIN} < \eta_{MAX}$? $Y_P = Y_{BBN}$?) have provided hints of a possible crisis (Copi, Schramm & Turner 1995; Olive & Steigman 1995; Hata et al. 1995). The two "weak links" are the lower bound on η inferred from D and ^{3}He observations and the upper bound on Y_P derived from the extragalactic HII region data.

It has long been known that the D + ^{3}He analysis of Yang et al. (1984) and Walker et al. (1991) is likely overly conservative. In both analyses the synthesis

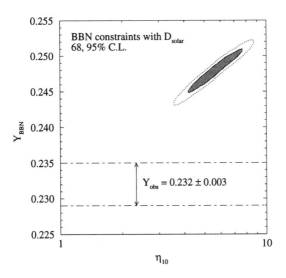

Figure 2. The predicted 68% and 95% CL contours for the ^{4}He primordial mass fraction with η constrained by D, ^{3}He and ^{7}Li. Also shown is the $\pm 1\sigma$ range for Y_P inferred from the data.

of new ^{3}He in low mass stars (Iben 1967; Rood 1972; Iben & Truran 1978) was neglected. But, Rood, Steigman & Tinsley (1976) had demonstrated that such production might dominate the primordial (D + ^{3}He) contribution. Even neglecting this contribution, Steigman & Tosi (1992) had followed the evolution of ^{3}He in a variety of chemical evolution models and found more ^{3}He survival ($g_3 \gtrsim 1/2$ rather than $g_3 \gtrsim 1/4$) leading to a higher lower bound to η. More recently, Steigman & Tosi (1995) revisited the "generic" evolution of D and ^{3}He and, using updated solar system data (Geiss 1993) inferred (for $g_3 \geq 1/4$) $\eta_{10} \geq 3.1$. In a more sophisticated implementation of the "generic" approach, Hata et al. (1996) found (for $g_3 \geq 1/4$) $\eta_{10} \geq 3.5$.

Although it is still true that $\eta_{MIN} < \eta_{MAX}$, the increasing lower bound to η increases the lower bound to Y_{BBN}. For $\eta_{10} \geq 3.1$ (3.5), $Y_{BBN} \geq 0.241(0.242)$. An accurate determination of Y_P from observations of ^{4}He in low metallicity extragalactic HII regions is required for the ^{4}He test. From their analysis of this data, Olive & Steigman (1995) derive $Y_P = 0.232 \pm 0.003$ where 0.003 is the 1σ statistical uncertainty. Thus, at 2σ, $Y_P^{MAX} < Y_{BBN}^{MIN}$, failing the ^{4}He test. It should be noted that η_{MIN} and Y_{BBN}^{MIN} have already been pushed to their "2σ" lower bounds so this discrepancy is at greater than the 95% confidence level. The crisis emerges!

Indeed, Olive & Steigman (1995) used all the data (D, ^{3}He, ^{4}He, ^{7}Li) to infer $N_\nu = 2.17 \pm 0.27$ which deviates from the standard model value ($N_\nu = 3$) by $\sim 3\sigma$. This crisis for SBBN is reflected in Figure 2 where D, ^{3}He ($g_3 \geq 1/4$) and ^{7}Li have been used to bound η, leading to predictions of Y_{BBN} at the 68%

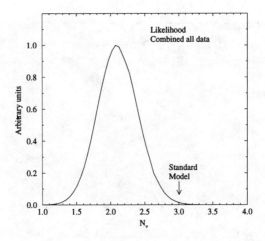

Figure 3. The likelihood function (arbitrary normalization) for the combined fit (D,^{3}He, ^{4}He, ^{7}Li) of the data and BBN as a function of N_ν^{BBN}. At each value of N_ν^{BBN} the likelihood is maximized for η.

and 95% CL. Unless the primordial abundance of ^{4}He has been systematically underestimated, the evidence signals a potential crisis for SBBN.

3. A Statistical Analysis of BBN

To explore more carefully the consistency of SBBN, my colleagues and I (Hata et al. 1995; Thomas et al. 1995) have undertaken the first comprehensive statistical analysis of the confrontation between theory and observation. We have reexamined the nuclear and weak interactions and their uncertainties and have performed a Monte Carlo analysis of the BBN predictions. Indeed, the curves in Figure 1 reflect the $\pm 1\sigma$ uncertainties in the predictions (for $N_\nu = 3$) of Y_{BBN}, y_{2P}, y_{7P} vs. η. From our Monte Carlos we derive $P(A)_{BBN}$, the probability distributions for the predicted BBN abundances (A). We have also reexamined the observational data, accounting for the statistical uncertainties as well as attempting to allow for various systematic uncertainties which may arise in using the data to infer the distribution, $P(A)_{OBS}$, of primordial abundances. These latter uncertainties are not necessarily modelled by gaussian distributions. In contrast to previous approaches which treated each element one at a time, we may use the information on all light nuclides to form a likelihood function (as a function of η and N_ν) from $P(A)_{BBN}$ and $P(A)_{OBS}$. The likelihood function, maximized with respect to η at each N_ν is shown in Figure 3 (from Hata et al. 1995). We derive $N_\nu = 2.1 \pm 0.3$, consistent with Olive & Steigman (1995). It is clear that SBBN ($N_\nu = 3$) provides a poor fit to the primordial abundances inferred from the data.

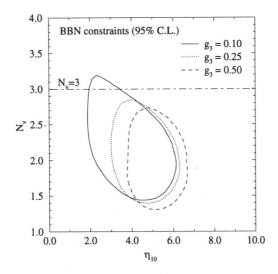

Figure 4. The 95% CL contours in the N_ν^{BBN} vs. η plane for several choices of the ^{3}He survival fraction g_3.

As with the 19th century standard model, SBBN is challenged. As with the challenges to the 19th century standard model, there are several options for the resolution of this crisis.

3.1. Is The Chemical Evolution Extrapolation Wrong?

One source of the challenge to SBBN is the relatively high lower bound to η imposed by the relatively low primordial abundances of D and ^{3}He inferred from solar system and interstellar observations. These stringent upper bounds to primordial D and ^{3}He are suggested by many specific chemical evolution models (Steigman & Tosi 1992) as well as the "generic" model for the evolution of D and ^{3}He (Steigman & Tosi 1995; Hata et al. 1996). In the latter case, the crisis worsens with increasing g_3 and/or if stellar production of ^{3}He is allowed for. The crisis could be ameliorated if g_3 is less than the lower bound ($g_3 \geq 1/4$) adopted in the above analyses. In Figure 4 (from Hata et al. 1995), 95% CL contours are shown in the N_ν vs. η plane for several choices of g_3. If the "effective" g_3 (averaged over stars of all masses and the evolution history of the ISM) is ~ 0.1, consistency of SBBN is reestablished.

3.2. Is The Primordial ^{4}He Abundance Larger?

An alternate source of the challenge to SBBN is the relatively low abundance of primordial ^{4}He inferred from the observations of extragalactic HII regions. By allowing only for statistical uncertainties perhaps we've underestimated the true uncertainty in Y_P. A larger value for Y_P could reestablish the consistency of ^{4}He

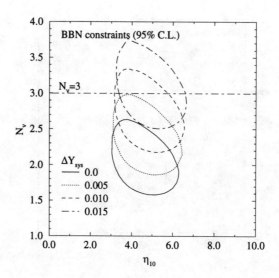

Figure 5. The 95% CL contours in the N_ν^{BBN} vs. η plane for several choices of the systematic error (ΔY_{sys}) in the ^{4}He abundance inferred from HII region data.

with D and ^{3}He. Many sources of possible systematic uncertainty in Y_P have been identified and some have been studied (Davidson & Kinman 1985; Pagel et al. 1992; Skillman & Kennicutt 1993; Skillman et al. 1994; Copi, Schramm & Turner 1995; Sasselov & Goldwirth 1995; Olive & Steigman 1995). In Figure 5 (from Hata et al. 1995) are shown 95% CL contours in the N_ν vs. η plane for several choices of ΔY_{sys}, where $Y_{BBN} = 0.232 \pm 0.003 + \Delta Y_{sys}$. If Y_P is shifted up by $\gtrsim 0.010$, SBBN may be consistent at the 95% CL. It should, however, be emphasized that ΔY_{sys} may be negative as well as positive; a negative ΔY_{sys} exacerbates the crisis for SBBN.

3.3. Is There New Physics?

By employing N_ν as a second parameter, we have allowed for a class of modifications of the standard model. If, in addition to three flavors of light, left-handed neutrinos ($N_\nu = 3$, SBBN) there are additional light neutrinos or other new particles, $N_\nu > 3$ (Steigman, Schramm & Gunn 1977) and the crisis worsens. However, although ν_e and ν_μ are known to be "light" ($m_\nu < T_{BBN}$), accelerator data on ν_τ (ALEPH Collaboration) permits $m_{\nu_\tau} \leq 24 MeV$. As Kawasaki et al. (1994) have shown, the presence of a massive, unstable tau neutrino with $5 - 10 \lesssim m_{\nu_\tau} \leq 24 MeV$ and $0.01 \lesssim \tau_{\nu_\tau} \lesssim 1$ sec. would correspond to an "effective" $N_\nu < 3$. Perhaps the crisis for SBBN is teaching us about extensions of the standard model of particle physics.

4. Summary and Conclusions

Primordial nucleosynthesis must have occurred during the early, hot, dense evolution of a Universe described by the hot big bang model. Therefore, BBN offers a test of standard cosmology as well as a probe of particle physics. As with the standard model of 19th century physics, over many years SBBN has provided support for the standard model of cosmology. Indeed, the success of SBBN in predicting the abundances of the light nuclides with only one adjustable parameter η ($N_\nu = 3$) restricted to a narrow range ($3 \lesssim \eta_{10} \lesssim 4$) while the abundances range over some 9 orders of magnitude, is impressive indeed.

However, as with the 19th century standard model, some clouds have now emerged on the horizon. Recent analyses (Copi, Schramm & Turner 1995; Olive & Steigman 1995; Hata et al. 1995) point to a crisis unless the data are in error, or the extrapolations of the data are in error, or there is new physics. Some analogies with the crises which confronted the 19th century standard model may be instructive.

(i) Perhaps our extrapolations of the observations of D and ^{3}He from here and now to there and then have been naive. Chemical evolution models in which more D is cycled through stars and destroyed (without a concommitant overproduction of ^{3}He) would permit a higher primordial abundance of D, allowing a lower η and Y_{BBN} consistent with Y_P. Thus, as with the discovery of Uranus, the crisis for SBBN may teach us something new about galactic evolution.

(ii) Perhaps our estimates of the primordial abundance of ^{4}He are in error because we have overlooked some large systematic error in the abundance determinations. If Y_P is larger than the value inferred form the observational data, η may be as large as inferred from D and ^{3}He (see Figure 2) and still Y_{BBN} and Y_P may be consistent. Then, as with the discovery of Pluto, our crisis may have been a false alarm from which, nonetheless, we learn something new.

(iii) The most exciting possibility, of course, would be that the data are accurate, the systematic errors small and the extrapolations true. Then, this crisis may point us to new physics beyond the standard models of particle physics or cosmology. The "window" on a massive, unstable τ-neutrino is accessible to current accelerators.

To summarize, then, SBBN ($N_\nu = 3$; $g_3 \geq 1/4$; $\Delta Y_{sys} \leq 0.005$) provides a poor fit to the primordial abundances of the light nuclides inferred from current observational data (Hata et al. 1995). This crisis is not a cause for alarm, but an opportunity to learn something new about astronomy, cosmology, or particle physics.

Acknowledgments. The work described here has been done in collaboration with S. Bludman, N. Hata, P. Langacker, K. Olive, R. Scherrer, D. Thomas, M. Tosi and T. Walker. I thank them for all that I've learned in our collaborations, and for permission to present our joint work here. The research of the author is supported at Ohio State by the DOE (DE-AC02-76-ER01545). I am pleased to thank Steve Holt and his colleagues and staff for all their support and patience.

References

Boesgaard, A. M. & Steigman, G. 1985, ARA&A, 23, 318
Copi, C., Schramm, D. N. & Turner, M. S. 1995, Science, 267, 192
Davidson, K. & Kinman, T. D. 1985, ApJS, 58, 321
Dearborn, D. S. P., Schramm, D. N. & Steigman, G. 1986, ApJ, 302, 35
Einstein, A. 1916, Ann. d. Phys., 49, 769
Geiss, J. 1993, in Origin and Evolution of the Elements (N. Prantzos, E. Vangioni-Flam & M. Casse, eds.; Cambridge Univ. Press) p. 89
Hata, N., Scherrer, R. J., Steigman, G., Thomas, D., Walker, T. P., Bludman, S. & Langacker, P. 1995, Phys.Rev.Lett, 75, 3977
Hata, N., Scherrer, R. J., Steigman, G., Thomas, D. & Walker, T. P. 1996, ApJ, in press, (Feb. 20, 1996)
Iben, I. 1967, ApJ, 147, 624
Iben, I. & Truran, J. W. 1978, ApJ, 220, 980
Newcomb, S., Encyc. Brit. (11th ed.), XVIII, 155
Newton, I. 1686, Phil. Nat. Princ. Math.
Olive, K. A. & Steigman, G. 1995, ApJS, 97, 49
Pagel, B. E. J., Simonson, E. A., Terlevich, R. J. & Edmunds, M. 1992, MNRAS, 255, 325
Penzias, A. A. & Wilson, R. W. 1965, ApJ, 142, 419
Rood, R. T. 1972, ApJ, 177, 681
Rood, R. T., Steigman, G. & Tinsley, B. M. 1976, ApJ, 207, L57
Sasselov, D. & Goldwirth, D. 1995, ApJ, 444, L5
Skillman, E. & Kennicutt, R. C. 1993, ApJ, 411, 655
Skillman, E., Terlevich, R. J., Kennicutt, R. C., Garnett, D. R. & Terlevich, E. 1994, ApJ, 431, 172
Steigman, G., Schramm, D. N. & Gunn, J. E. 1977, Phys. Lett. B66, 202
Steigman, G. & Tosi, M. 1993, ApJ, 401, 150
Steigman, G. & Tosi, M. 1995, ApJ, 453, 173
Walker, T. P., Steigman, G., Schramm, D. N., Olive, K. A. & Kang, H. S. 1991, ApJ, 376, 51
Yang, J., Turner, M. S., Steigman, G., Schramm, D. N. & Olive, K. A. 1984, ApJ, 381, 493

Cosmic Abundances
ASP Conference Series, Vol. 99, 1996
Stephen S. Holt and George Sonneborn (eds.)

Assessing Big-Bang Nucleosynthesis

Craig J. Copi

Department of Physics, The University of Chicago
NASA/Fermilab Astrophysics Center

David N. Schramm and Michael S. Turner

Department of Physics, The University of Chicago
NASA/Fermilab Astrophysics Center
Department of Astronomy & Astrophysics, The University of Chicago

Abstract. Systematic uncertainties in the light-element abundances and their evolution make a rigorous statistical assessment difficult. However, using Bayesian methods we show that the following statement is robust: the predicted and measured abundances are consistent with 95% credibility only if the baryon-to-photon ratio is between 2×10^{-10} and 6.5×10^{-10} and the number of light neutrino species is less than 3.9. Our analysis suggests that the ^{4}He abundance may have been systematically underestimated.

1. Introduction

The predictions of big-bang nucleosynthesis depend upon the baryon-to-photon ratio ($\equiv \eta$) as well as the number of light ($\lesssim 1\ MeV$) particle species, often quantified as the equivalent number of massless neutrino species ($\equiv N_\nu$). For a decade it has been argued that the abundances of all four light elements can be accounted for provided η is between 2.5×10^{-10} and 6×10^{-10} and $N_\nu < 3.1 - 4$ (Yang etal. 1984; Walker etal. 1991; Krauss & Kernan 1995; Copi, Schramm, & Turner 1995a).

However, these conclusions were not based upon a rigorous statistical analysis. Because the dominant uncertainties in the light-element abundances are systematic such an analysis is difficult and previous work focussed on concordance intervals. Given the importance of big-bang nucleosynthesis it is worthwhile to try to use more rigorous methods. Here we apply Bayesian likelihood and identify the the conclusions which are insensitive to the systematic errors (see Copi, Schramm, & Turner 1995b for more details).

The dominant uncertainties in comparing the predicted and measured light-element abundances are systematic: the primeval abundance of ^{4}He; the chemical evolution of D and ^{3}He; and whether or not ^{7}Li in the oldest stars has been reduced significantly by nuclear burning. Systematic error is difficult to treat as it is usually poorly understood and poorly quantified. This is especially true for astronomical *observations*, where the observer has little control over the object being observed.

There are at least three kinds of systematic error. (1) A definitive, but unknown, offset between what is measured and what is of interest. (2) A random source of error whose distribution is poorly known. (3) An important source of error that is unknown. The first kind of systematic error is best treated as an additional parameter in the likelihood function. The second kind of systematic error is best treated by use of a distribution, or by several candidate distributions. The third type of systematic error is a nightmare.

Several sources of systematic error for ^{4}He have been identified which can reduce or increase the measured abundance (Sasselov & Goldwirth 1995; Skillman etal. 1995). If the same effect dominates in each measurement use of an offset parameter in the ^{4}He abundance would be appropriate. On the other hand, if different effects dominate different measurements enlarging the statistical error would be appropriate. We allow for both: the statistical error σ_Y is permitted to be larger than 0.003, and an offset in the ^{4}He abundance, ΔY, is a parameter in the likelihood function ($Y_P = 0.232 + \Delta Y$).

Finally, there is the systematic uncertainty associated with the chemical evolution of D and ^{3}He. Based upon a recent study of the chemical evolution of D and ^{3}He (Copi, Schramm, & Turner 1995c) we consider three models that encompass the broadest range of possibilities: Model 0 is the plain, vanilla model; Model 1 is characterized by extreme ^{3}He destruction; and Model 2 is characterized by minimal ^{3}He destruction.

2. Discussion

Systematic errors of the first kind are treated as additional (nuisance) parameters in the likelihood function which can be determined by the experiment itself or can be eliminated by marginalization; we treat ΔY as such. We also allow σ_Y to vary to study how results depend upon the assumed uncertainty in the ^{4}He abundance. Because we are interested in setting a limit to N_ν, it too is taken to be a parameter. Values of N_ν greater than three describe extensions of the standard model with additional light degrees of freedom.

A likelihood function that is not compact must be treated with care, because no information about the parameters can be inferred independently of what was already known (the priors). For example, the likelihood function $\mathcal{L}(N_\nu)$, which is needed to set limits to N_ν, is obtained by integrating over ΔY and depends upon the limits of integration. To derive limits to N_ν we do the following: integrate from $-|\Delta Y|$ to $|\Delta Y|$; normalize $\mathcal{L}(N_\nu)$ to have unit likelihood from $N_\nu = 2$ or $N_\nu = 3$ to ∞; the limit is the value of N_ν beyond which 5% of the total likelihood accumulates. The dependence of the limit upon $|\Delta Y|$ is shown in table 1.

In a recent paper the likelihood function $\mathcal{L}(N_\nu)$ obtained by integrating from $\Delta Y = -0.005$ to 0.005 was used in an attempt to assess the viability of the standard theory (Hata etal. 1994). This likelihood is peaked at $N_\nu = 2.1$ with gaussian $\sigma_{N_\nu} = 0.3$. On this basis it was claimed that the standard theory of nucleosynthesis is ruled out with 98.6% confidence. Equal weight was implicitly given to all values of N_ν (flat priors). The prior for $N_\nu = 3$ (standard model of particle physics) is certainly greater than that for $N_\nu < 3$ (e.g., massive, short lived tau neutrino), and this, together with the dependence of $\mathcal{L}(N_\nu)$ upon the

Table 1. Limits to N_ν for Models 0, 1, 2 and $Li/H = (1.5 \pm 0.3) \times 10^{-10}$ for priors $N_\nu > 2$ (first number) and $N_\nu > 3$ (second number).

$|\Delta Y|$	Model 0	Model 1	Model 2
0	2.5/3.1	2.8/3.2	2.5/3.1
0.005	2.6/3.2	2.9/3.3	2.6/3.2
0.010	2.9/3.3	3.1/3.5	2.9/3.3
0.015	3.2/3.5	3.4/3.7	3.2/3.4
0.020	3.5/3.7	3.8/3.9	3.5/3.7

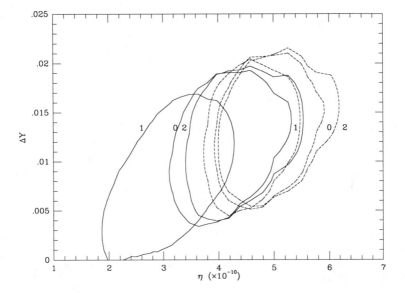

Figure 1. The likelihood function $\mathcal{L}(\Delta Y, \eta, \sigma_Y = 0.003)$ (solid curves = low ^{7}Li, broken curves = high ^{7}Li).

prior for ΔY (here $|\Delta Y| < 0.005$), casts strong doubt on the above assessment of the standard theory.

In figure 1 we show the 95% contours of the likelihood function $\mathcal{L}(\eta, \Delta Y)$ for Models 0, 1, and 2 and both values of the central ^{7}Li abundance. The 95% credibility contours in the $\Delta Y - \eta$ plane suggest that the primeval ^{4}He abundance has been systematically underestimated, by an amount $\Delta Y \approx +0.01$. (Though it should be noted that Model 1 and the lower ^{7}Li abundance are just consistent with $\Delta Y = 0$ at 95% credibility.) Put another way, D, ^{3}He, and ^{7}Li are concordant and ^{4}He is the outlayer. When the likelihood function is marginalized with respect to ΔY, the 95% credibility interval is $\eta \simeq (2-6.5) \times 10^{-10}$ (allowing again for the uncertainty both in astration of ^{7}Li and in the chemical evolution of D and ^{3}He).

3. Conclusions

The fact that systematic uncertainties dominate precludes crisp statistical statements. The lack of a viable alternative to the standard theory of nucleosynthesis complicates matters further as the most powerful statistical techniques assess relative viability. However, the rigorous techniques that we have applied point to several conclusions that are insensitive to assumptions made about systematic uncertainty. The predictions of the standard theory of primordial nucleosynthesis are only consistent with the extant observations with 95% credibility provided $\eta \simeq (2-6.5) \times 10^{-10}$. Our analysis suggests that the primordial ^{4}He abundance has been systematically underestimated ($\Delta Y \approx +0.01$) or that the random errors have been underestimated ($\sigma_Y \approx 0.01$). Only for Model 1 (extreme destruction of ^{3}He) are $\Delta Y = 0$ and $\sigma_Y = 0.003$ in the 95% credibility region (cf., figure 1). The limit to N_ν depends upon the systematic uncertainties in the ^{4}He abundance (cf., table 1); taking $|\Delta Y| \leq 0.02$, which is four times the estimated systematic error and also encompasses the 95% likelihood contour in the $\Delta Y - \eta$ plane, leads to the 95% credible limit $N_\nu < 3.9$.

4. Future Measurements

There are two measurements that should reduce the systematic uncertainties significantly, permitting a sharper test of big-bang nucleosynthesis. The first is a determination of the primeval D abundance by measuring D-Lyα absorption due to high-redshift hydrogen clouds. The second is a determination of the primeval ^{7}Li abundance by studying short period, tidally locked pop II halo binaries; astration is believed to involve rotation-driven mixing astration and is minimized in these stars because they rotate slowly (Pinsonneault etal. 1992).

Acknowledgments. We thank Donald Q. Lamb for many valuable discussions and comments. This work was supported by the DoE (at Chicago and Fermilab) and by the NASA (at Fermilab by grant NAG 5-2788 and at Chicago by a GSRP (CJC)).

References

Copi, C. J., Schramm, D. N., & Turner, M. S. 1995a, Science, 267, 192.
Copi, C. J., Schramm, D. N., & Turner, M. S. 1995b, PRL, to appear (November 13, 1995).
Copi, C. J., Schramm, D. N., & Turner, M. S. 1995c, these proceedings.
Hata, N. etal., hep-ph/9505319.
Krauss, L. M. & Kernan, P. J. 1995, Phys. Lett. B, 347, 347.
Pinsonneault, M. H., Deliyannis, C. P., & Demarque, P. 1992, ApJS, 78, 179.
Sasselov, D. & Goldwirth, D. S. 1995, ApJ, in press.
Skillman, E. D., Terlevich, R., & Garnett, D. R. 1995, ApJ, in press.
Walker, T. P., etal. 1991, ApJ, 376, 51.
Yang, J. etal. 1984, ApJ, 281, 493.

Big-Bang Nucleosynthesis and A New Approach to Galactic Chemical Evolution

Craig J. Copi

Department of Physics, The University of Chicago
NASA/Fermilab Astrophysics Center

David N. Schramm and Michael S. Turner

Department of Physics, The University of Chicago
NASA/Fermilab Astrophysics Center
Department of Astronomy & Astrophysics, The University of Chicago

Abstract. Big-bang production of deuterium is the best indicator of the baryon density; however, only the present abundance of D is known (and only locally) and its chemical evolution is intertwined with that of ^{3}He. Because galactic abundances are spatially heterogeneous, mean chemical-evolution models are not well suited for extrapolating the pre-solar D and ^{3}He abundances to their primeval values. We introduce a new approach which explicitly addresses heterogeneity by statistically tracing the history of the pre-solar material back to its primeval beginning. We show that the decade-old concordance interval $\eta \approx (2-8) \times 10^{-10}$ based on D and ^{3}He is well founded.

1. Introduction

Big-bang nucleosynthesis provides the best determination of the density of ordinary matter. Of the light elements D has the most potential as a "baryometer" because its production depends sensitively upon η. On the other hand, its interpretation is challenging because D is burned in virtually all astrophysical situations.

Because D is so readily destroyed, it is not possible to obtain a lower bound to η based upon D alone. D and ^{3}He together define a big-bang consistency interval, $\eta \simeq (2.5-9) \times 10^{-10}$. Here we provide a firmer basis for the concordance interval by introducing a new approach to the chemical evolution of D and ^{3}He which allows their primeval abundances to be determined statistically and explicitly addresses the heterogeneity of galactic abundances.

2. Stochastic Histories

Our new approach allows for heterogeneity in a most fundamental way: we follow the history of the material in the pre-solar nebula through stars back to its primeval beginning (see Copi, Schramm, & Turner 1995 for more details). We use a stochastic algorithm for generating histories; from each history the primeval D and ^{3}He abundances can be determined from pre-solar abundances. Taking an ensemble of histories and allowing for the uncertainty in the pre-solar abundances as discussed above, we construct "a fuzzy map" from local D and ^{3}He abundances to primeval D and ^{3}He abundances.

Histories are generated by a diagrammatic technique and set of rules. We suppose that the pre-solar material came from the primeval mix (fraction f_P) and from N other stars (fractions f_i, $i = 1, \cdots, N$). The fraction f_P is drawn from a linear distribution whose mean is $1 - \epsilon \sim 0.5$. The number of "first-tier stars" N is drawn from a flat distribution whose mean is $N_0 \sim 10$; if $N < 1$, there is no material from other stars and f_P is set equal to one. The fractions f_i are drawn from a flat distribution whose mean is $(1 - f_P)/N$.

A star is assumed to do the following: (i) burn all its D to ^{3}He; (ii) return a fraction g_3 of its ^{3}He to the ISM; and (iii) possibly add some ^{3}He and heavy elements to the material it returns to the ISM. The amount of ^{3}He returned to the ISM by a star is related to the D and ^{3}He from which it is made

$$\left(\frac{D}{H}\right)_{IN} = f_P \left(\frac{D}{H}\right)_P ; \qquad (1)$$

$$\left(\frac{^3He}{H}\right)_{IN} = f_P \left(\frac{^3He}{H}\right)_P + \sum_i f_i \left(\frac{^3He}{H}\right)_{OUT} ; \qquad (2)$$

$$\left(\frac{^3He}{H}\right)_{OUT} = g_3 \left[\left(\frac{D}{H}\right)_{IN} + \left(\frac{^3He}{H}\right)_{IN}\right] + h_3 . \qquad (3)$$

The quantities g_3 and h_3 are chosen from distributions that are adjusted to reflect the mix of stars and our knowledge about their processing of ^{3}He.

We use oxygen as a surrogate for the heavy elements. Massive stars produce oxygen quantified by the mass fraction $h_{16} \sim 0.10$ of the material they return to the ISM; low-mass stars preserve oxygen. We require that the oxygen mass fraction in the pre-solar material is between 0.5% and 2%. The oxygen constraint ensures that some—but not too much—of the material in the pre-solar nebula has been processed through massive stars.

Crucial to this approach are the parameters and distributions that specify the statistical properties of the histories. The parameter ϵ controls the fraction of material that has undergone stellar processing; conventional wisdom has it that about 50% of the pre-solar material has undergone stellar processing. The parameter N_0 controls the number of stars that contribute to the material from which a given star is made; we have tried values from 5 to 15. The distribution $f(g_3)$ determines the amount of ^{3}He that survives stellar processing; it in turn depends upon the stellar mass function and the rate of return of material from stars of a given mass to the ISM. We parameterize $f(g_3)$ by a minimum value, $g_3^{min} \sim 0.15$, and a power-law index m, $f(g_3) \propto g_3^m$ for $1 \geq g_3 \geq g_3^{min}$. A standard mass function and conventional stellar models correspond roughly to $m = 0$ (Truran 1995). The distribution $f(h_3; g_3)$ determines the amount of stellar ^{3}He production. It is parameterized by g_{3*}: only stars with $g_3 \geq g_{3*} \sim 0.8$ are assumed to produce ^{3}He, and the amount of ^{3}He production, h_3, which is chosen from a flat distribution with $0.5 \times 10^{-5} \leq h_3 \leq 2 \times 10^{-5}$. Lastly, the distribution $f(h_{16}; g_3)$ quantifies heavy-element production by massive stars. The distribution is characterized by $g_3^{max} \sim 0.3$, only stars with $g_3 \leq g_3^{max}$ are

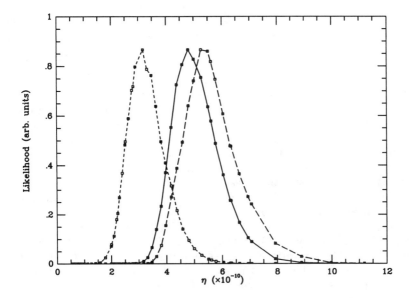

Figure 1. Likelihood functions for η based upon D and ^{3}He abundances for models 0 (solid), 1 (short dash) and 2 (long dash).

assumed to produce ^{16}O, and $h_{16} = 0.025 - 0.20$, the mass fraction of oxygen produced by massive stars which is returned to the ISM. The range for these two parameters is based upon models for the yields of type II supernovae (Timmes, Woosley, & Weaver 1995).

We have explored many models; here we present three models that serve to span the extreme range of plausible possibilities. Model 0 is chosen to be the plain, vanilla model for chemical evolution, model 1 has extreme stellar processing and ^{3}He destruction, and model 2 has less stellar processing by massive stars, more primeval material (e.g., due to infall), and more stellar ^{3}He survival/production.

3. Discussion

By Monte Carlo we constructed around 300,000 histories for each of our three chemical-evolution models. For each history, we draw pre-solar ^{3}He and D + ^{3}He abundances from a distribution with a gaussian statistical error and top-hat systematic uncertainty. About half of the histories are acceptable: satisfy the oxygen constraint (pre-solar mass fraction between 0.5% and 2%) and have positive primeval ^{3}He abundance. In addition, to ensure that the primeval D abundance is large enough to account for that in the ISM today we weight each point with the probability that the primordial D abundance is greater than $(1.6 \pm 0.1) \times 10^{-5}$.

In the many other models for chemical evolution we have explored, the likelihood function always drops precipitously at a value of η no smaller than

2×10^{-10}—at such low values of η a massive amount of D is produced and some ^{3}He necessarily survives. From all this we conclude that there is a robust concordance interval for D and ^{3}He, $\eta \approx (2-8) \times 10^{-10}$ (see figure 1).

This consistency interval encompasses those derived by others based upon a variety of chemical-evolution models (see e.g., Hata et al., 1994; Olive, 1995). Our results strongly suggest that the "generic," mean chemical evolution model of Hata et al. (1994), which is supposed to encompass the full range of possibilities for the chemical evolution of D and ^{3}He, is less generic than the authors claim: their 95% confidence interval corresponds to our Model 0.

A determination of the primeval D abundance by measuring D-Lyα absorption by high-redshift hydrogen clouds could both shed light on the chemical evolution of D and ^{3}He as well as accurately determine the baryon density. At the moment, there are conflicting measurements and upper limits, and the situation is unsettled. However, it seems likely that a definitive determination of the primeval D abundance will be made.

The D abundance measured in the nearby ISM along different lines of sight varies significantly (Linsky 1995), and the ^{3}He abundance measured in different HII regions in the Galaxy varies by almost an order of magnitude (Bania, Rood and Wilson 1994). For this reason mean chemical evolution cannot be trusted to accurately represent the history in a specific location.

Acknowledgments. We thank Jim Truran, Frank Timmes, Brian Fields, and Keith Olive for useful conversations about galactic chemical evolution and Roy Lewis for his help in understanding pre-solar D and ^{3}He abundances. This work was supported in part by the DOE (at Chicago and Fermilab) and NASA (at Fermilab through grant NAG 5-2788 and at Chicago through a GSRP Fellowship for CJC).

References

Bania, T., Rood, R., & Wilson, T. 1994, Proc. of the 1994 ESO Meeting on Light-element Abundances (Springer-Verlag).

Copi, C. J., Schramm, D. N., & Turner, M. S. 1995 ApJL, to appear (December 20, 1995).

Hata, N., Scherrer, R. J., Steigman, G., Thomas, D., & Walker, T. P. 1994, astro-ph/9412087, preprint.

Linsky, J. L. 1995, presented at the Aspen Center for Physics Workshop on Big-bang Nucleosynthesis.

Olive, K. A. 1995, The Light Element Abundances, ed. P. Crane (Springer-Verlag) 40.

Timmes, F. X., Woosley, S. E., & Weaver, T. A. 1995, ApJS, 98, 617.

Cosmic Abundances
ASP Conference Series, Vol. 99, 1996
Stephen S. Holt and George Sonneborn (eds.)

Cosmic Deuterium and Baryon Density

Craig J. Hogan

Astronomy and Physics Departments, PO Box 351580, University of Washington, Seattle, WA 98195

Abstract.
Quasar absorption lines now permit a direct probe of deuterium abundances in primordial material, with the best current estimate $(D/H) = 1.9 \pm 0.4 \times 10^{-4}$. If this is the universal primordial abundance $(D/H)_p$, Standard Big Bang Nucleosynthesis yields an estimate of the mean cosmic density of baryons, $\eta_{10} = 1.7 \pm 0.2$ or $\Omega_b h^2 = 6.2 \pm 0.8 \times 10^{-3}$, leading to SBBN predictions in excellent agreement with estimates of primordial abundances of helium-4 and lithium-7. Lower values of $(D/H)_p$ derived from Galactic chemical evolution models may instead be a sign of destruction of deuterium and helium-3 in stars. The inferred baryon density is compared with known baryons in stars and neutral gas; about two thirds of the baryons are in some still-unobserved form such as ionized gas or compact objects. Galaxy dynamical mass estimates reveal the need for primarily nonbaryonic dark matter in galaxy halos. Galaxy cluster dynamics imply that the total density of this dark matter, while twenty or more times the baryon density, is still well below the critical value, unless both baryons and galaxies are concentrated in galaxy clusters relative to the dark matter.

1. Introduction

The mere presence of deuterium in the universe confirms the conceptual framework of the Hot Big Bang, since unlike other primordial nuclei there is no other known source. Deuterium is the unique relic of the Big Bang: the total history of its cosmic production is over in only a few minutes, followed by billions of years of slow destruction in stars. Because it is relatively sensitive to the baryon/photon ratio η ($\equiv 10^{-10}\eta_{10}$), the primordial deuterium abundance is also the best way of measuring the amount of matter in the universe.

Until recently, deuterium was only measured within our highly chemically evolved Galaxy, which has destroyed a large but uncertain fraction of its initial deuterium. Measurements of deuterium abundance in QSO absorbers now allow a much more direct estimate of primordial abundance, in pristine material that has suffered little stellar processing. The Big Bang prediction is approximately fitted by

$$(D/H)_p \approx 4.6 \times 10^{-4} \eta_{10}^{-5/3} = 1.9 \times 10^{-4}(\eta_{10}/1.70)^{-5/3},$$

so an estimate of primordial D/H with an absolute accuracy of ±20% yields an estimate of primordial baryon/photon ratio accurate to ±12%, a very good precision by the standards of cosmic bookkeeping.[1] This accuracy is attainable from measurements in QSO absorption line systems. I survey here the strengths and limitations of the QSO technique, compare early QSO results with Galactic estimates of $(D/H)_p$, and summarize cosmological implications of the low value of baryon density estimated from the high $(D/H)_p$.

2. Deuterium in Quasar Absorbers

Certainly in the long run, and I argue here even at the present, the best way to measure primordial D/H is through absorption lines of distant quasars. The principles are discussed by Jenkins in this volume. Profiles of the Lyman series lines of HI and DI and the Lyman limit optical depth combine to give good absolute column densities for both species in many situations, which in principle yield an absolute abundance almost free of ionization, temperature or density corrections in atomic gas. Accurate absolute column density information comes from optically thin or damped absorption lines of each species, and from the Lyman limit; ionization and recombination processes are nearly identical for both, so little detailed physical or geometrical modeling is required. Metal lines from the same clouds provide an independent measure of the amount of chemical processing, and in some cases the metal abundances are so small that significant deuterium depletion can essentially be ruled out. Eventually, many clouds will be measured along different lines of sight, testing the universality of the abundance in different places all over our past light cone, confirming (or not) that the past worldlines of matter were similar over an enormous spacetime volume, testing both the large scale homogeneity of spacetime (the "Cosmological Principle") and the small scale homogeneity of matter.

The greatest weakness with the QSO technique is the ambiguity caused by the unknown distribution of material in velocity, or the problem of interlopers—clouds of hydrogen gas which happen by chance to lie at the redshift expected for deuterium associated with some other hydrogen cloud. Nature does not provide us with a system cleanly arranged in velocity, so any detection of deuterium must be regarded as suspect until this possibility is dealt with.

The first way to deal with interlopers is by seeking those relatively rare narrow-line systems where the turbulent component of linewidth is small and thermal width itself is also small, close to the minimum temperature consistent with photoionization heating. Because deuterium is heavier, the Doppler parameter $b \equiv \sqrt{\frac{2kT}{m}}$ is smaller; in a purely thermally broadened profile, $b = 13T_4^{1/2}$ for hydrogen, but only $9T_4^{1/2}$ for deuterium. Gas under equilibrium conditions in these Lyman Limit absorbers is not expected to be cooler than 10^4 K (usually, $T_4 = T/10^4 K \approx 1 \ to \ 3$, over a very wide range for the ionization parameter; see Donahue and Shull, 1991), so lines with $b \leq 10$ are unlikely to be hydrogen

---

[1]The accuracy of the fitting formula is about 7% in this range, which is almost as good as the estimated theoretical error (Krauss and Kernan 1995). Note that the conversion to a physical density also requires a knowledge of CBR temperature: $\Omega_b h^2 = 3.631 \times 10^{-3} \eta_{10} T_{2.726}^3$.

interlopers. The ratio of the fitted Doppler parameter for D candidates to that of their H counterparts is also an important clue. Turbulent broadening tends to give both lines the same profile (so different shapes are not always expected), but a situation where the deuterium line is narrower than the hydrogen by a factor between $1/\sqrt{2}$ and 1 is again unlikely to be an interloper. Rugers and Hogan (1995) argue that the candidate deuterium absorber in Q0014+813 in fact displays both of these signatures of real deuterium, and so is likely to be a real measurement of D/H. It yields an abundance estimate $D/H = 1.9 \pm 0.4 \times 10^{-4}$ (Songaila et al 1994, Carswell et al 1994, Rugers and Hogan 1996 and this volume).

The errors in this estimate are real measured errors, in the sense that they reflect the total uncertainty in the fitted column densities, including ambiguities in Doppler parameter and velocity. They do not include systematic "model errors", the most extreme example being a hydrogen interloper, and for this reason it is risky to rely on only one example, however clean. Also, one should bear in mind that although these are true "1σ" statistical uncertainties, the distribution of allowed values is highly nongaussian, so we do not have a good estimate of the probability of larger 2σ excursions from the fitted values. The ranges quoted throughout must be taken only as current best guesses, with the possibility of large departures not well constrained. This is main reason why more clean absorbers are urgently required.

Unfortunately clean conditions are not the rule. One can invoke statistical arguments based on larger samples of less trustworthy candidates— is there a statistical excess of lines at 82 km/sec to the blue of hydrogen (either relative to the red or relative to adjacent velocities)? Is there a statistical tendency of fits to improve with deuterium? There are several examples[2] where this is the case, but so far the samples are not large enough to make statistical tests meaningful, although the dozen or so absorbers in two spectra we have studied so far indicate that that within the large errors the data are consistent with a universal abundance of the order of 10^{-4}. The most trustworthy system to make a measurement remains the pair of absorbers in Q0014+813.

We now make a major leap of reasoning, and take the Q0014 measurement as an estimate of the primordial abundance. The justifications for this leap are (1) The Big Bang is the only known source of this deuterium, so any measurement is a reasonable lower limit on $(D/H)_p$; (2) The Q0014 cloud is extremely metal-poor and is unlikely to have cycled an appreciable faction of its material through stars (that is, D destruction by stellar cycling should be accompanied by noticeable enrichment), so it gives a reasonable upper limit on $(D/H)_p$; (3) Although it is only one site, it is still the best one we have at this writing— the most accurately measured and most pristine. For this value of $(D/H)_p$, Big Bang nucleosynthesis implies $\eta_{10} = 1.7 \pm 0.2$, which gives excellent concordance of Big Bang predictions with with the most straightforward interpretations of

---

[2] One possible counterexample with $D/H \approx 2 \times 10^{-5}$ may have been found by Tytler and Fan (1994) on the line of sight to Q1937-1009. Carswell (1995) quotes a lower limit in a system in Q0420-388 of $D/H \geq 2 \times 10^{-5}$, with a best guess of 2×10^{-4}. Songaila and Cowie have a good fit for 2×10^{-4} in Q0956+122. See the contribution by Rugers and Hogan in this volume for another example in GC0636+68, which yields $\log(D/H) = -3.95 \pm 0.54$.

helium-4 and lithium-7 abundance data (Copi et al 1995ab, Hata et al 1995, and Schramm and Steigman, this volume.) This gives us the confidence to trust in the Big Bang picture enough to use D/H as a probe of cosmic baryon density.

3. Deuterium in Galactic Chemical Evolution

The value $(D/H)_p = 2.0 \pm 0.4 \times 10^{-4}$ is surprisingly high. Smaller values have previously been quoted as *upper* limits on $(D/H)_p$, giving *lower* limits on $\Omega_b h^2$, based on abundances in our Galaxy, together with the assumption that the sum $(D + {}^3He)/H$ cannot decrease.

In the Galaxy today, the deuterium abundance is less than 2×10^{-5} (Linsky et al 1993, 1995, and Lemoine, this volume). This can be understood if the interstellar gas has almost all been processed through at least the outer envelopes of stars. A high primordial value $(D/H)_p \geq 10^{-4}$ requires that stellar processing in the Galaxy destroys not only deuterium but also its principal burning product, helium-3, in order to agree with the low values of this isotope found for the presolar material: analysis of the solar wind and meteorites reveals that the presolar nebula had about ${}^3He/H \approx 1.5 \pm 0.5 \times 10^{-5}$, $D/H \approx 2.7 \pm 2 \times 10^{-5}$, and $(D + {}^3He)/H \approx 4 \pm 2 \times 10^{-5}$ (Copi et al 1995ab, Hata et al 1995). It is probably necessary to have some helium-3 destructive mechanism also in order to explain the interstellar observations of helium-3, which are highly variable and sometimes very low, of the order of $0.6 - 6 \times 10^{-5}$ (Balser et al 1994; Wilson and Rood 1994).

While deuterium is destroyed by stars even before they enter the main sequence, and there is no doubt that helium-4 increases due to stellar processing, helium-3 is both created and destroyed by stars. It is fairly abundant in the interiors of main sequence stars in the temperature range of H burning, but is destroyed at higher temperatures. The helium-3 ejected back into the ISM when a star dies depends in detail on what the material of the star does after it leaves the main sequence and before it throws off its hydrogen envelope. Although the nuclear physics is well understood, details of the movement of material between different temperatures, and how it is ejected into the interstellar medium, are murky.

In galactic chemical evolution models, low mass stars (one or two solar masses) dominate stellar cycling and destroy the bulk of the pregalactic deuterium, so it is these stars that must destroy the helium-3 (Galli et al 1995, Olive et al 1996). There is a way this could happen: if the material in the stellar envelope on the giant branch is brought to high temperature (1.5×10^7K or so) before it is ejected, the helium-3 would be destroyed. There is evidence for just such a process ("Giant Branch Mixing", "Cool Bottom Processing" or "extra-mixing"; Hogan 1995, Wasserburg et al 1995, Charbonnel 1995)— from the change in C and N isotopic composition as stars leave the main sequence, from O isotope mixtures in meteorites, and from the continued decrease in lithium after cool giants leave the main sequence. There is thus at least a plausible way the helium-3 could be destroyed. Charbonnel (1995) has recently shown a detailed model which destroys helium-3 between 1 and 2 solar masses, but not above 2— thereby plausibly explaining the above facts. The three nearby planetary nebulae which have very high helium-3 (indeed, so high that they must be atypical; Wilson and Rood 1992) could simply have progenitor masses above 2 solar masses. Although we do not have a clear constraint on the integrated stellar population ejecta, it is clearly not safe to assume that the Galaxy cannot reduce its helium-3 abundance with time. This is why the QSO measurements, even in their present unsettled state, are a more reliable measure of primordial D/H than extrapolating from present Galactic values.

4. Cosmic Baryon Inventory

The inventory of things which must be made of baryons includes HI absorbers ($\Omega h \geq 0.003$, Wolfe 1993), galactic stars ($\Omega \approx 0.002$), and cluster gas ($\Omega h^{0.5} \approx 0.001$) (Persic and Salucci 1992), with errors on these quantities typically ±30%. Elliptical and spiral galaxies contribute about equally to the stellar mass, although spirals are more numerous and with their younger populations dominate the blue and visible light by about a factor of two (Schechter and Dressler 1987). In addition there is an ionized gas component, including for example the Lyα forest clouds, whose density could be larger than all of these or smaller than any of them. The HI gas density is only this large at high redshift (that is, it might convert into stars at low redshift), so the minimum requirement is to count things at only one redshift. Taking $h = 0.75$, it could be that all known baryonic things could be accounted for with as little as $\Omega_b \approx 0.004$. The deuterium estimate of $\Omega_b = 0.011$ thus provides amply sufficient baryons to make all the things that need to be made of baryons. There must still be some unaccounted baryons, probably in the form of ionized gas and/or compact objects (Carr 1994). The baryon to galaxy mass ratio is about 5, and the baryon to (galaxy+known gas) ratio is about 3. Thus within the errors, the uncounted baryons could comprise up to, but probably not more than about twice that already seen. This is quite different from the previous situation, where lower estimates of $(D/H)_p$ required more than 90% of the baryons to be dark or in an ionized IGM. It leads to a tidy model of galaxy formation, which accounts for most of the facts about galaxies and QSO absorbers from $z = 0$ to $z = 3$, where baryons reside for the most part in gas, stars and compact objects in the vicinity of galaxies today (Fukugita et al 1996). The abundance of MACHOs in the Galactic halo (Alcock et al. 1995) is consistent with such a low density of baryons.

5. Nonbaryonic Dark Matter in Galaxies

On the other hand the estimated baryon density is not enough to account for the known dark matter in the universe. This problem is well known, although the need for nonbaryonic dark matter is greater than before, and now extends even to galactic halos.

Using the observed integrated blue luminosity density of galaxies, $L_B = 1.93^{+0.8}_{-0.6} \times 10^8 h$ solar units per Mpc^3 (Efstathiou et al. 1988), we can write the physical baryon density estimate as a global baryon mass density to blue luminosity density ratio, $\rho_b/L_B \approx 9h^{-1}$ in solar units. For our Galaxy, the mass to blue light ratio is about 60 if the mass is 10^{12} solar masses (Binney and Tremaine 1987), a typical dynamical estimate (Zaritsky et al 1993, Peebles

1995). The mass to light ratio from galaxy rotation curves is often inferred in typical spirals— the same types that dominate the luminosity density— to be at least $30h^1$ (Rubin 1993). If this ratio is universal for galaxies,[3] the global density of galactic dark matter is $\Omega \geq 0.02$.

Galactic dark matter is thus probably not made mostly of baryons— a stronger statement about nonbaryonic dark matter than was possible with the earlier higher baryon density estimates.

6. Global Dark Matter Density

The need for nonbaryonic dark matter is much greater if rich galaxy clusters, with dynamical mass to galaxy light ratios of about $300h^1$, fairly represent the global M/L, implying $\Omega \approx 0.2$ (Binney and Tremaine 1987). It is possible however that galaxy formation is more or less efficient in clusters than in the field, so that their M/L is not the universal one. The appeal to greater efficiency is necessary especially in models with an overall dark matter density close to $\Omega = 1$.

Physical models of this "biasing" are constrained by the fact that the baryon to galaxy ratio in clusters, as measured directly, is close to its global value, as inferred from nucleosynthesis. White et al. (1993) estimate for example that the Coma cluster[4] has a ratio of baryons to dark matter within the virial radius of $(0.05 \pm 0.01)h^{-1.5}$ in the form of gas, and 0.009 ± 0.002 in the form of stars; the higher baryon to galaxy mass ratio of 8 (for $h = .75$, compared to the above global estimate of 5) indicates if anything that galaxy formation was less efficient there than in the field, unless the gas mass is overestimated. This point does not depend on estimates of cluster mass. This argues against classical biasing (i.e., protocluster galaxies forming earlier and more efficiently than protofield galaxies) as a way of reconciling cluster M/L with $\Omega = 1$: the only way to much higher cosmic density is to increase the ratio of baryons to dark matter in clusters relative to the cosmic mean, by about a factor of five.

White et al. dispense with the galaxies altogether, and use the baryon to dark matter ratio directly. They use simulations and physical models of cluster collapse to show that composition within the virial radius ought to be representative of the global baryon/dark matter mix for collisionless cold dark matter models. The above estimate for $\Omega_b h^2$ yields $\Omega = 0.12 h^{-0.5}$— again implying an open universe, or else a flat universe with an unclustered mass density, such as a cosmological constant.

Acknowledgments. This work was supported at the University of Washington by NSF grant AST 9320045 and NASA grant NAG-5-2793.

---

[3] In some gas-rich, star-poor dwarfs the rotation curve can be measured out to many scale lengths; there are cases (eg, DDO 154) where M/L is more than $50h^1$ (Rubin 1993). Since these are not typical galaxies (the luminosity per baryon is known to be less than usual), it is hazardous to use their M/L as a universal value (their bias is to overestimate Ω).

[4] Although they used Coma specifically, similar numbers are obtained in other clusters, using other techniques; for example, Squires et al.1995 derive from weak lensing mass estimators an upper limit of $M_{gas}/M \leq (0.04 \pm 0.02)h^{-3/2}$ and $M/L_B = (440 \pm 80)h$ for the inner 400kpc of A2218

References

Alcock, C., et al. 1995 *Phys. Rev. Lett.* **74**, 2867
Balser, B. S., Bania, T. M., Brockway, C. J., Rood, R. T., Wilson, T. L. 1994, *ApJ* 430, 667
Binney, J. and Tremaine, S. 1987, *Galactic Dynamics*, Princeton
Carr, B. J. 1994, *ARA&A*, **32**, 531
Carswell, R. F., Rauch, M., Weymann, R. J., Cooke, A. J., and Webb, J. K. 1994, *MNRAS* **268**, L1
Carswell, R. F. 1995, preprint.
Charbonnel, C. 1995, *ApJ* 453, L41
Copi, C. J., Schramm, D. N., and Turner, M. S. 1995a, *Science* **267**, 192.
Copi, C. J., Schramm, D. N., and Turner, M. S. 1995b, *Phys. Rev. Lett.*, in press
Efstathiou, G., Ellis, R. S., and Peterson, B. A. 1988, *MNRAS* 232, 431
Fukugita, M., Hogan, C. J. and Peebles, P. J. E., 1995 *Nature*, submitted.
Galli, D., Palla, F., Ferrini, F. and Penco, U., 1995, *ApJ*, **443**, 536.
Hata, N., Scherrer, R. J., Steigman, G., Thomas, D., Walker, T. P., Bludman, S., Langacker, P., 1995, *Phys. Rev. Lett.*, in press.
Hogan, C. J. 1995 *ApJ* **441**, L17.
Krauss, L. M. and Kernan, P. J. 1995, *Phys. Lett.*, B 347, 347.
Linsky, J. L., Brown, A., Gayley, K., Diplas, A., and Savage, B. D., et al 1993, *ApJ* 402, 694
Linsky, J. L., Diplas, A., Wood, B. E., Brown, A., Ayres, T. R. and Savage, B. D., 1995, *ApJ*, in press.
Olive, K. A., Rood, R. T., Schramm, D. N., Truran, J. W., Vangioni-Flam, E., 1996, *ApJ*, submitted.
Peebles, P. J. E. 1995, *ApJ* 449, 52
Persic, M. and Salucci, P. 1992, *MNRAS* 258,14P
Reeves, H., Audouze, J., Fowler, W.A., and Schramm, D.N. 1973, *ApJ*, **179**, 909.
Rubin, V. C., 1993, *Proc. Natl. Acad. Sci. USA* 90, 4814
Schechter, P. L., and Dressler, A. 1987, *AJ* 94, 563
Songaila, A., Cowie, L. L., Hogan, C. J., and Rugers, M. 1994, *Nature* **368**, 599.
Steigman, G., 1994, *MNRAS* **269**, L53.
Tytler, D., and Fan, X.M. 1994, Bull. of the AAS, Vol. 26 No. 4, 1424.
Wasserburg, G. J., Boothroyd, A. I., Sackmann, I.-J. 1995, *ApJ Lett* **447**, L37.
White, S. D. M., et al. 1993, *Nature* 366,429
Wilson, T. R., and Rood, R. T. 1994, *ARA&A*, **32**, 191.
Wolfe, A. M. 1993 *Ann. N.Y. Acad. Sci.* 688, 281
Zaritsky, D., Smith, R., Frenk, C. S., White, S. D. M. 1993, *ApJ* 405, 464

Future Cosmic Microwave Background Constraints to the Baryon Density

M. Kamionkowski[1], G. Jungman[2], A. Kosowsky[3,4], and D. N. Spergel[5,6]

[1] *Department of Physics, Columbia University, New York, NY 10027*

[2] *Department of Physics, Syracuse University, Syracuse, NY 13244*

[3] *Harvard-Smithsonian Center for Astrophysics, Cambridge, MA 02138*

[4] *Department of Physics, Harvard University, Cambridge, MA 02138*

[5] *Department of Astrophysical Sciences, Princeton University, Princeton, NJ 08544*

[6] *Department of Astronomy, University of Maryland, College Park, MD 20742*

Abstract. We discuss what can be learned about the baryon density from an all-sky map of the cosmic microwave background (CMB) with sub-degree angular resolution. With only minimal assumptions about the primordial spectrum of density perturbations and the values of other cosmological parameters, such a CMB map should be able to distinguish between a Universe with a baryon density near 0.1 and a baryon-dominated Universe. With additional reasonable assumptions, it is conceivable that such measurements will constrain the baryon density to an accuracy similar to that obtained from BBN calculations.

The current range for the baryon-to-photon ratio allowed by big-bang nucleosynthesis (BBN) is $0.0075 \lesssim \Omega_b h^2 \lesssim 0.024$ (Copi et al. 1995). This gives $\Omega_b \lesssim 0.1$ for the range of acceptable values of h, which implies that if $\Omega = 1$, as suggested by inflationary theory (or even if $\Omega \gtrsim 0.3$ as suggested by cluster dynamics), then the bulk of the mass in the Universe must be nonbaryonic. On the other hand, X-ray–cluster measurements might be suggesting that the observed baryon density is too high to be consistent with BBN (see, e.g., Felten & Steigman 1995 and references therein); this becomes especially intriguing given the recent measurement of a large primordial deuterium abundance in quasar absorption spectra (Hogan & Ruger 1995). The range in the BBN prediction can be traced primarily to uncertainties in the primordial elemental abundances. There is, of course, also some question as to whether the X-ray–cluster measurements actually probe the universal baryon density. For these reasons and more, it would clearly be desirable to have an independent measurement of $\Omega_b h^2$.

Here, we evaluate the precision with which the baryon-to-photon ratio, $\Omega_b h^2$, can be determined with high-resolution CMB maps (Bennett et al. 1995; Janssen et al. 1995; Bouchet et al. 1995). We work within the context of models with adiabatic primordial density perturbations, although similar arguments apply to isocurvature models as well, and we expect the power spectrum to distinguish clearly the two classes of models (Crittenden & Turok 1995). (More details may be found in Jungman et al. 1995a,b.)

A given cosmological theory makes a statistical prediction about the distribution of CMB temperature fluctuations, expressed by the angular power spectrum

$$C(\theta) \equiv \langle [\Delta T(\hat{m})/T_0][\Delta T(\hat{n})/T_0]\rangle_{\hat{m}\cdot\hat{n}=\cos\theta} \equiv \sum_\ell (2\ell+1)C_\ell P_\ell(\cos\theta)/(4\pi), \quad (1)$$

where $\Delta T(\hat{n})/T_0$ is the fractional temperature perturbation in the direction $\hat{n}$, P_ℓ are the Legendre polynomials, and the brackets represent an ensemble average over all angles and observer positions. Since we can observe from only a single location in the Universe, the observed multipole moments C_ℓ^{obs} will be distributed about the mean value C_ℓ with a "cosmic variance" $\sigma_\ell \simeq \sqrt{2/(2\ell+1)}C_\ell$.

We consider an experiment which maps a fraction f_{sky} of the sky with a gaussian beam with full width at half maximum θ_{fwhm} and a pixel noise $\sigma_{pix} = s/\sqrt{t_{pix}}$, where s is the detector sensitivity and t_{pix} is the time spent observing each $\theta_{fwhm} \times \theta_{fwhm}$ pixel. We adopt the inverse weight per solid angle, $w^{-1} \equiv (\sigma_{pix}\theta_{fwhm}/T_0)^2$, as a measure of noise that is pixel-size independent (Knox 1995). Current state-of-the-art detectors achieve sensitivities of $s = 200\,\mu K\sqrt{sec}$, corresponding to an inverse weight of $w^{-1} \simeq 2 \times 10^{-15}$ for a one-year experiment. Realistically, however, foregrounds and other systematic effects may increase the noise level; conservatively, w^{-1} will likely fall in the range $(0.9 - 4) \times 10^{-14}$. Treating the pixel noise as gaussian and ignoring any correlations between pixels, the C_ℓ^{obs} will be distributed about the C_ℓ with a standard error

$$\sigma_\ell = [(2\ell+1)f_{sky}/2]^{-1/2} \left[C_l + (wf_{sky})^{-1}e^{\ell^2\sigma_b^2}\right], \quad (2)$$

where $\sigma_b = 7.4 \times 10^{-3}(\theta_{fwhm}/1°)$.

Given a spectrum of primordial density perturbations, the C_ℓ are obtained by solving the coupled equations for the evolution of perturbations to the space-time metric and perturbations to the phase-space densities of all particle species in the Universe. We consider models with initial adiabatic density perturbations filled with photons, neutrinos, baryons, and collisionless dark matter; this includes all inflation-based models. The CMB power spectrum depends upon many parameters. Here, we include the following set: the total density Ω; the Hubble constant, $H_0 = 100\,h\,km\,sec^{-1}\,Mpc^{-1}$; the density of baryons in units of the critical density, $\Omega_b h^2$; the cosmological constant in units of the critical density, Λ; the power-law indices of the initial scalar- and tensor-perturbation spectra, n_S and n_T; the amplitudes of the scalar and tensor spectra, parameterized by Q, the total CMB quadrupole moment, and $r = Q_T^2/Q_S^2$, the ratio of the squares of the tensor and scalar contributions to the quadrupole moment; the optical depth to the surface of last scatter, τ; the deviation from scale invariance of the scalar perturbations, $\alpha \equiv dn/d\ln k$; and the effective number of light-neutrino species at decoupling, N_ν. Thus for any given set of cosmological parameters, $s = \{\Omega, \Omega_b h^2, h, n_S, \Lambda, r, n_T, \alpha, \tau, Q, N_\nu\}$, we can calculate the mean multipole moments $C_\ell(s)$.

We now wish to determine the precision with which CMB maps will be able to determine $\Omega_b h^2$ without making any assumptions about the values of

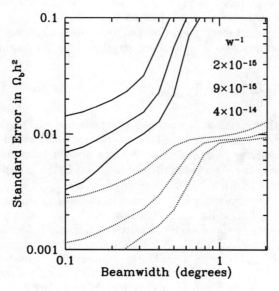

Figure 1. The standard error on $\Omega_b h^2$.

the other undetermined parameters. The answer will depend on the measurement errors σ_l, and on the underlying cosmological theory. If the actual parameters describing the Universe are s_0, then the probability distribution for observing a CMB power spectrum which is best fit by the parameters s is $P(s) \propto \exp\{-\frac{1}{2}(s-s_0) \cdot [\alpha] \cdot (s-s_0)\}$ where the curvature matrix $[\alpha]$ is given approximately by

$$\alpha_{ij} = \sum_\ell \frac{1}{\sigma_\ell^2} \left[\frac{\partial C_\ell(s_0)}{\partial s_i} \frac{\partial C_\ell(s_0)}{\partial s_j} \right] \qquad (3)$$

with σ_ℓ as given in Eq. (2). The covariance matrix $[\mathcal{C}] = [\alpha]^{-1}$ is an estimate of the standard errors that would be obtained from a maximum-likelihood fit to data: the error in measuring the parameter s_i (obtained by integrating over all the other parameters) is approximately $\mathcal{C}_{ii}^{1/2}$. If some of the parameters are known, then the covariance matrix for the others is determined by inverting the submatrix of the undetermined parameters.

Fig. 1 displays the standard error in $\Omega_b h^2$ as a function of the beam width θ_{fwhm} for different noise levels and for $f_{sky} = 1$. The underlying model assumed here for the purpose of illustration is "standard CDM," given by $s = \{1, 0.01, 0.5, 1, 0, 0, 0, 0, 0, Q_{COBE}, 3\}$, where $Q_{COBE} = 20\,\mu K$ is the COBE normalization (Górski et al. 1994). The solid curves show the $\mathcal{C}^{1/2}_{\Omega_b h^2, \Omega_b h^2}$ obtained by inversion of the full 11×11 curvature matrix $[\alpha]$ for $w^{-1} = 2 \times 10^{-15}, 9 \times 10^{-15}$, and 4×10^{-14}. These are the sensitivities that can be attained at the given noise levels with the assumption of uniform priors (that is, including no information about any parameter values from other observations). The dotted curves show the $\mathcal{C}^{1/2}_{\Omega_b h^2, \Omega_b h^2}$ obtained by inversion of the $\Omega_b h^2$-Q submatrix of $[\alpha]$; this is the error in $\Omega_b h^2$ that could be obtained if all other parameters except the normalization were fixed, either from other observations or by assumption. Realistically, the precision obtained will fall somewhere between these two sets of curves. The results for a mapping experiment which covers only a fraction f_{sky} of the sky can be obtained by replacing $w \to w f_{sky}$ and scaling by $f_{sky}^{-1/2}$ [c.f., Eq. (2)].

The implications of CMB maps for the baryon density depend quite sensitively on the experiment. As long as $\theta_{fwhm} \lesssim 0.5$, the CMB should (with minimal assumptions) at least be able to rule out a baryon-dominated Universe ($\Omega_b \gtrsim 0.3$) and therefore confirm the predictions of BBN. With angular resolutions that approach 0.1° [which might be achievable, for example, with a ground-based interferometry map (Myers 1995) to complement a satellite map], a CMB map would provide limits to the baryon-to-photon ratio that were competitive with BBN. Furthermore, if other parameters can be fixed, the CMB might be able to restrict $\Omega_b h^2$ to a small fraction of the range currently allowed by BBN.

Moreover, the CMB will also provide information on several other parameters (Jungman et al. 1995a,b). Most significantly, the total density Ω can be determined to better than 10% with minimal assumptions and perhaps better than 1%.

Acknowledgments. This work was supported in part by the D.O.E. under contracts DEFG02-92-ER 40699 and DEFG02-85-ER 40231, by the Harvard Society of Fellows, by the NSF under contract ASC 93-18185, and by NASA under contract NAG5-3091 and NAGW-2448, and by NASA under the MAP Mission Concept Study Proposal.

References

Bennett, C. L. et al. 1995, NASA Mission Concept Study

Bouchet, F. R. et al. 1995, astro-ph/9507032

Copi, C. J., et al. 1995, astro-ph/9508029

Crittenden, R. G. & Turok, N. 1995, Phys. Rev. Lett., 75, 2642

Felten, J. E. & Steigman, G. 1995, in Proc. St. Petersburg Gamow Seminar, St. Petersburg, Russia, 12–14 September 1994, A. M. Bykov and A. Chevalier, Sp. Sci Rev. (Dordrecht: Kluwer)

Górski, K. M. et al. 1994, ApJ, 430, L89

Hogan, C. J. & Ruger, A. 1995, this proceedings

S. T. Myers, private communication

Janssen, M. A. et al. 1995, NASA Mission Concept Study

Jungman, G., Kamionkowski, M., Kosowsky, A., & Spergel, D. N. 1995a, astro-ph/9507080, Phys. Rev. Lett., in press

Jungman, G., Kamionkowski, M., Kosowsky, A., & Spergel, D. N. 1995b, in preparation

Knox, L. 1995, Phys. Rev. D, 52, 4307

Interstellar Abundances of the Light Elements

Roger Ferlet

Institut d'Astrophysique de Paris, CNRS, 98bis bvd Arago, 75014 Paris, France

Martin Lemoine

Department of Astronomy & Astrophysics, Enrico Fermi Institute, The University of Chicago, Chicago, IL60637-1433

Abstract. We review the measurements of the interstellar abundances of the light elements D, ^{3}He, ^{6,7}Li, ^{9}Be, and ^{10,11}B, and discuss their impact on primordial nucleosynthesis and the chemical evolution of the Galaxy.

1. Introduction

It is now well accepted that only D, ^{3,4}He, and ^{7}Li are produced in significant amount during primordial nucleosynthesis (BBN, see Schramm, these proceedings). The other light elements ^{6}Li, ^{9}Be, and ^{10,11}B are expected to be mainly produced in spallation reactions of galactic cosmic rays (GCR) all along the galactic evolution (Reeves, 1994 for a review; see however Cassé & Vangioni-Flam, these proceedings).

The number of measurements of the light elements abundances, and the variety of sites in which these have been performed, have considerably increased in the past twenty years. Notwithstanding their meteoritic/solar system abundances, their interstellar abundances were always the first measured, either in the local or in the more distant interstellar medium (ISM). The first measurements of the interstellar D/H ratio were reported in the early seventies. The first successful radio observations of the ^{3}He/H ratio were published in the eighties. The interstellar abundance of ^{7}Li was measured ever since the late sixties. The abundances of ^{6}Li and $^{10+11}$B were first measured in the past few years. As to ^{9}Be, no firm detection of this element in the ISM has been reported as yet.

However, for more than a decade, D and ^{3}He have been the only elements whose interstellar abundances were used to constrain BBN in a direct way. The primordial ^{4}He mass fraction has been traditionally derived from abundances measured in extragalactic H II regions of low metallicities (Pagel, 1995, and references). The interstellar observations of ^{7}Li have been superseded, in this respect, by the detection of ^{7}Li in very metal deficient stars of the galactic halo, usually interpreted as the signature of the primordial abundance of ^{7}Li (Spite & Spite, 1982; Spite, 1995, and references). In effect, abundances measured at low metallicities are less contaminated by the effect of galactic evolution than interstellar abundances, which are representative of the present epoch. Underlying this simple statement is another trivial statement: *interstellar abundances cannot be used, in any case, to constrain only BBN parameters, since the chemical evolution of the Galaxy always appears as a conditional assumption to these constraints.* Simple as it is, it is our opinion that this has been overlooked in the past decades, and that this is still being overlooked.

Hence, to provide a bottomline to this review, the main significance of interstellar abundances is to constrain both BBN and chemical evolution models. Whenever other abundances of a same element, measured at a lower metallicity, are obtained, these latter generally supersede the interstellar abundance in its significance to BBN. Conversely, the interstellar abundance is then used to constrain chemical evolution through the comparison to this latter abundance. In the following, we discuss D, ^{3}He, ^{6,7}Li, and ^{9}Be and B separately. The observations of these light elements give rise to different problems, and cannot, by now, be put together to form a consistent picture of their origin and galactic evolution. We refer the reader to Wilson & Rood (1995) for an extensive review of the light elements interstellar abundances and for further references.

2. Deuterium

2.1. Observations

There are several methods to measure the interstellar abundance of deuterium (see Vidal-Madjar, 1991, Ferlet, 1992). One of them is to observe deuterated molecules such as HD, DCN, *etc*... and to form the ratio of the deuterated molecule column density to its non-deuterated counterpart (H_2, HCN, *etc*....). More than twenty different deuterated species have been identified in the ISM, with abundances relative to the non-deuterated counterpart ranging from 10^{-2} to 10^{-6}. Conversely, this means that fractionation effects are important, and that, as a consequence, this method cannot provide a precise estimate of the true interstellar D/H ratio; rather, this method is used in conjunction with estimates of the interstellar D/H ratio to gather information on the chemistry of the ISM. Another way to derive the D/H ratio comes through radio observations of the hyperfine line of D I at 92cm. The detection of this line is however extremely difficult, and no firm detection has ever been reported (Heiles et al., 1993, Lubovitch, these proceedings). The detection of this line would allow to probe more distant interstellar media than the local medium discussed below; however, because a large column density of D is necessary to provide even a weak spin-flip transition, these observations aim at molecular complexes. As a result, the upper limit derived toward Cas A (Heiles et al., 1993): D/H$\leq$ 2.1 $\times$ 10^{-6} probably simply results from a large fraction of D and H being in molecular form in these clouds. Finally, the only way to derive a reliable estimate of the interstellar D/H ratio is to observe the atomic transitions of D and H of the Lyman series in the far-UV; in absorption in the local ISM against the background continuum of cool or hot stars. These observations have been performed using Copernicus, IUE, and now HST. Both types of target stars present pros and cons.

The main advantage of observing cool stars is that they can be selected in the vicinity of the Sun. This results in low H I column densities, and trivial to nearly trivial lines of sight. In effect, due to the low atomic weight of H I and

Dɪ, to the Dɪ-Hɪ −82km/s isotopic shift, and to the abundance of Hɪ in the local medium, the Dɪ line cannot be detected at Lyman α in the wing of the Hɪ line for Hɪ column densities larger than 10^{19} cm^{-2}. Also, the presence of several interstellar components with different b-values may imply a large error on the Hɪ column density if these components are unresolved. For this reason, deriving the Hɪ column density has always been the limiting factor of accurate D/H ratios measurements. Note that the spectral resolutions of Copernicus and IUE were respectively 15 and 20 km/s, and, as a consequence, a non-trivial line of sight, even in the local ISM, would generally go unresolved. Eventhough HST-GHRS now offers a spectral resolution of 3.5km/s, the thermal width of the Dɪ line in the local ISM is $\simeq$ 8 km/s, so that one has to observe lines of heavier species (thinner lines) to fully use the resolving power of HST. However, as a result of the Hɪ chromospheric emission profile at Lyman α, the interstellar absorption of Nɪ 1200Å (triplet) is not available; yet, Nɪ was shown to be an excellent tracer of Hɪ in the ISM (Ferlet, 1981). Hence, one usually observes the strong lines of Mgɪɪ, Feɪɪ or similar elements, although these elements probe mainly Hɪɪ media and not Hɪ media. Moreover, the chromospheric emission line has to be modeled to set the continuum for the interstellar absorption. Such a procedure necessarily introduces systematic errors. Nevertheless, this method has provided the most precise measurement of the local D/H ratio in the direction of Capella, using HST–GHRS: $(D/H)_{\alpha Aur} = 1.60 \pm 0.09^{+0.05}_{-0.10} \times 10^{-5}$ (Linsky et al., 1993, 1995).

Hot stars are unfortunately located further away from the Sun, so that one always has to face a high Hɪ column density and often a non-trivial line of sight structure. In these cases, Dɪ could not be detected at Lyα, and one has to observe higher order lines, $e.g.$ Lyδ, Lyϵ, hence these measurements have primarily come through Copernicus observations. The stellar continuum is however smooth at the location of the interstellar absorption, and, moreover, the Nɪ triplet is available to probe the velocity structure of the line of sight.

All published D/H ratios are collected in Fig.1, distinguishing hot stars from cool stars observations (see Vidal-Madjar, 1991, Ferlet, 1992 for references). The D/H ratios range from $\sim$ 5. $\times 10^{-6}$ to $\sim$ 4. $\times 10^{-5}$. A large scatter is clearly detected in Fig.1, and represents variation of the D/H ratio in the local ISM, that may be as large as a factor $\simeq$ 4 over scales as small as a few parsecs. The essential question is: do these variations really exist?

Unfortunately, one cannot answer this question and at the same time be perfectly objective. On one hand, note that the Linsky et al. (1995) measurement does not agree with any of the previous D/H measurements toward Capella. Therefore, one cannot say that all these observations and their error bars and perfectly reliable. On the other hand, one could use the observed differences between different measurements toward a same star to get an estimate of the systematics; although this estimate is rough, it does not seem to be able to account the large scatter of Fig.1. Furthermore, note that time variations of the D/H ratio have already been reported toward ϵ Per (Gry et al., 1983), interpreted as due to the ejection of high velocity hydrogen atoms from the star. Finally, to say that the systematics associated with IUE and Copernicus observations are large, that they account for the observed discrepancies, and that only the Linsky et al. (1995) value should be kept, is also very arbitrary. This measurement toward Capella is undoubtedly of an unprecedented quality. The modeling of the chromospheric emission line could be achieved with great preci-

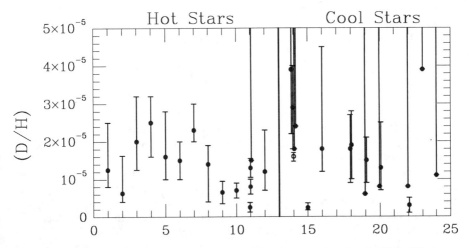

Figure 1. Measurements of the D/H ratio in the local ISM. The left hand-side box collects data obtained toward hot stars, while the right hand-side one collects cool stars observations. The x-axis has no physical significance, and merely labels the differents stars. Data points next to each other, within less than 1 x-axis unit, correspond to the same target star. The open circle represents the Linsky et al. (1995) measurement using HST toward Capella. All other data points come from Copernicus or IUE observations.

sion, observing the binary system at different phases; the line of sight structure was found to be trivial, as only the local cloud, in which the Sun is embedded, was detected. However, this velocity structure was obtained through MgII and FeII observations; hence, it could be that HI media, shifted from the local cloud by less than $\sim 5-8$ km/s went unnoticed in MgII and FeII although they would play an important role in the HI saturated profile. Such systematics were not considered by Linsky et al. (1993, 1995).

To answer to the reliability of the observations shown in Fig.1, one has to re-analyze all these data in a consistent way, looking for possible undetected systematics. Another (complementary) way would be to proceed with further observations of the local D/H ratio using HST as well as Lyman-FUSE, which should be launched in 1998. Finally, note that various explanations to these possible fluctuations of the D/H ratio have been put forward as early as Vidal-Madjar et al. (1978), Bruston et al. (1981): this is a long-standing problem.

2.2. Future prospects

A promising observational strategy is to observe nearby white dwarfs; not only such targets may be selected near to the Sun, circumventing the main disadvantage of hot stars, but they can also be chosen in the high temperature range, so as to provide a smooth stellar profile at Lyα. The NI triplet at 1200Å would be available, allowing an accurate sampling of the line of sight. Such observations have now been conducted using HST toward two white dwarfs: Hz43 (see

Landsmann et al., these proceedings), and G191–B2B (Lemoine et al., 1995; Vidal-Madjar et al., 1996).

In the case of Hz43, the structure of the line of sight seems, at a first glance, to be trivial, *i.e.* consisting of only the local cloud. The D/H ratio, as well as the H<small>I</small> column density, are consistent, in the single-cloud hypothesis, with those of the local cloud, obtained by Linsky et al. (1995). However, due to the relative faintness of this target, the N<small>I</small> triplet was observed at medium resolution only, and other interstellar components cannot be ruled out as of now. Obviously, this target looks very promising, and such observations should be complemented with higher resolution higher signal-to-noise ratio data, to provide a very precise measurement of the D/H ratio.

In the case of G191–B2B, the line of sight was found to comprise at least 2 interstellar components. Data obtained in Cycle 1 at high resolution for Mg<small>II</small> and Fe<small>II</small>, and medium resolution for many other species, including N<small>I</small>, O<small>I</small>, H<small>I</small>, were complemented in Cycle 5 by high resolution observations of N<small>I</small> 1200Å, O<small>I</small> 1302Å, and Lyα. In cycle 1, the following D/H ratios were obtained: $(D/H)_A = 1.0^{+0.4}_{-0.1} \times 10^{-5}$, $(D/H)_B = 1.4^{+0.1}_{-0.3} \times 10^{-5}$, and the B cloud was identified with the local cloud, *i.e.* that observed by Linsky et al. (1995). Analysis of the Cycle 5 data is but preliminary; however, the presence of the A and B clouds has been confirmed in H<small>I</small> media. The profile fitting of the N<small>I</small>, O<small>I</small>, and D<small>I</small> lines yields:

$$\frac{(D/N)_B}{(D/N)_A} = 2.32 \pm 0.45 \qquad \frac{(D/O)_B}{(D/O)_A} = 3.25 \pm 1.75 \qquad \frac{(N/O)_B}{(N/O)_A} = 1.38 \pm 0.79$$

It is important to note that O<small>I</small> is also an excellent tracer of H<small>I</small>, hence of D<small>I</small>. These ratios thus give a hint for possible variations of the D/H ratio between the two detected clouds: although both (N/O) ratios are consistent, the (D/N) and (D/O) ratios are systematically different between both clouds. Cautiousness prevents us from going further as long as the analysis of these data is not complete.

2.3. Conclusion

If the variations of the D/H ratio are illusory, then one could quote as an average: $(D/H)_{ISM} \simeq 1.3 \pm 0.4 \times 10^{-5}$. This rather large error bar arises from a subjective although conservative viewpoint. The significance of this value will be discussed along with the ^{3}He measurements in the next section. If the D/H does vary in the ISM, one has to understand why; until then, *no measurement of the D/H ratio in the ISM should be quoted as reliable.* Moreover, one should expect these variations to be larger in reality than what is observed. The actual value might in fact be very different from what is observed, if these variations are systematic, *i.e.* act in one way only; this in turn would heavily bear on the chemical evolution of deuterium. It appears that the upper bound on Ω_B is obtained from BBN predictions through the interstellar abundance of deuterium: this bound would have to be removed until the variations and their cause are properly understood. There is all hope that the Lyman-FUSE mission will solve these problems (see Friedman, these proceedings). Lyman-FUSE will probe the ISM further than the local medium, it will look for gradients of the deuterium abundance with galactocentric distance and with galactic height in the halo, and it will even probe extragalactic low-redshift objects. These studies, performed with an instrument as suited as FUSE is for them, should greatly clarify the deuterium problem.

3. Helium–3

3.1. Observations

Since the linewidths of ^{3}He and ^{4}He recombination lines are always larger than the isotopic shift, and ^{3}He is so underabundant as compared to ^{4}He (^{3}He/^{4}He$\sim 10^{-4}$), the only reliable way to observe ^{3}He is to detect the hyperfine transition of the ^{3}He$^+$ hydrogen–like ion at 3.46cm. This spin–flip line does not occur for ^{4}He$^+$, hence there is no confusion possible. However, due to the high ionization potential of ^{3}He, this line can only be detected in HII regions and planetary nebulae (PNe). Such measurements are extremely difficult: the line peak is only a few mK, for a continuum whose strength is typically a few tens of K, i.e. 10^4 times stronger. This continuum emission gives rise to baseline features that mimic the ^{3}He$^+$ line, so that sophisticated observational and data analysis methods are necessary to model this baseline. Moreover, whereas the ^{3}He$^+$ line depends on the integrated electron density, the amount of H$^+$ or ^{4}He$^+$, obtained through continuum emission or recombination lines, depend on the integrated squared electron density. Hence one has to model the HII region to derive the ^{3}He/H abundance from the ^{3}He$^+$ observation. Modeling beyond the homogeneous sphere approximation is now underway through high resolution mapping; in any case, it should always increase the value of ^{3}He/H (Balser et al., 1994).

Such a huge observational task was undertaken by the Rood, Bania & Wilson team over the last decade, at the NRAO Green Bank 43m radio-telescope, and yielded ^{3}He abundances in twelve HII regions and a few PNe. The integration time reaches more than 100hrs in a few cases, and more integration is still being planned. The errors on the abundances of ^{3}He have been brought down to $\sim 10\%$. The reader is referred to Bania et al. (1987), Balser et al. (1994), Rood et al. (1995) for a thorough discussion of the observations and the associated uncertainties. They report ^{3}He abundances in HII regions in the range ^{3}He/H$\simeq 1 - 4 \times 10^{-5}$, in the homogeneous sphere approximation. Hence, strong variations of the ^{3}He abundance are detected; no gradient is found with the galacto-centric distance; the highest abundances are found in the Perseus arm, and the lowest at low and high galactocentric distances. However, a negative correlation of the observed abundances with the mass of the HII region seems to be detected. This attests for the existence of some systematic effects associated with the HII regions, that might account for the scatter observed vs galactocentric distance. These are discussed below. Another very important result is the detection of ^{3}He in the planetary nebula NGC3242, at a level of ^{3}He/H$\sim 10^{-3}$ (Rood et al., 1992), i.e. some two orders of magnitude above the HII and pre-solar (^{3}He/H$_\odot \simeq 1.5 \times 10^{-5}$, Geiss, 1993) abundances.

3.2. Interpretation

On the theoretical side, it is expected that D is burned to ^{3}He during the pre-main sequence collapse; ^{3}He could be further synthesized in the p-p chain in low mass stars, $M \lesssim 2 M_\odot$. This nucleosynthetic product should be mixed up to the surface during the first dredge-up, and might be expelled into the ISM through AGB winds (Balser et al., 1994). This is consistent with the formidable ^{3}He abundance in NGC3242. Helium-3 is destroyed, although not entirely, in high mass stars. Altogether, one expects a star generation, averaged over its initial mass function, to be a net producer of ^{3}He (Dearborn, Schramm & Steigman, 1986). One should therefore observe a negative gradient with galactocentric distance, and none is observed. It has recently been suggested that the negative correlation observed *vs* the mass of the HII region could be due to pollution by winds of massive stars (Balser et al., 1994; Olive et al, 1995), due to the fact that D is burned to ^{3}He during the PMS collapse, but ^{3}He is destroyed only in later phases; hence, depending on the number of massive stars and their age, the average wind could be either depleted or enriched in ^{3}He. In this scenario, one should expect the actual interstellar ^{3}He/H abundance, at a given galactocentric distance, to lie between $\simeq 10^{-5}$ and 5×10^{-5}. This effect is however uncertain and poorly understood, and one cannot, at the present time, correct the observed abundances for these systematics.

Nevertheless, there has not been, up to now, a single chemical evolution model able to reproduce the galactic evolution of ^{3}He, which is constrained by the observed abundances of ^{3}He at the pre-solar epoch and in the interstellar medium. Up to recent times, galactic evolution models of D and ^{3}He did not include the production of ^{3}He in low mass stars, and even in these cases, ^{3}He was strongly overproduced with respect to its pre-solar abundance (see *e.g.* Steigman & Tosi, 1992; Vangioni-Flam et al., 1994). The latest works including ^{3}He production from hydrogen burning show that the overproduction cannot be avoided in any scenario of galactic evolution (Galli et al., 1995; Scully et al., 1995). This is embarassing, all the more since the lower bound on Ω_B has been traditionally provided by the constraint $((D+^3He)/H)_p \leq 9. \times 10^{-5}$; this constraint was obtained from a simple modeling of the galactic evolution of D and ^{3}He[1]. Only recently has this bound been superseded by observations of the deuterium abundance in metal-deficient absorbers on quasars lines of sight (Songaila et al., 1994; Carswell et al., 1994; Rudgers & Hogan, these proceedings). This lower bound on Ω_B is now $(D/H)_p \lesssim 2. \times 10^{-4}$, thereby showing that the previous one may have been wrong by a factor 2...

The deuterium problem, which arose from the difficulty of reconciling the interstellar abundance of D with its primordial abundance, as predicted by the observed primordial mass fraction of ^{4}He, through standard chemical evolution (Vidal-Madjar & Gry, 1984), has now moved to embrace both D and ^{3}He isotopes. Indeed, if one wishes to have $(D/H)_p \sim 2. \times 10^{-4}$, then $(^3He/H)_p \sim 2.0 \times 10^{-5}$: one has to burn $\sim 2. \times 10^{-4}$ in deuterium abundance to ^{3}He, without overproducing the pre-solar abundance of ^{3}He, although the primordial abundance of ^{3}He is already higher than this latter! Even then, D has to be destroyed by a factor $\simeq 15$ during the galactic evolution, whereas most models give a destruction factor $\simeq 2 - 3$ only (but see Vangioni-Flam & Cassé, 1995). At the opposite, if one wishes to start with a low primordial deuterium abundance, $\simeq 3. \times 10^{-5}$, to follow the standard prediction of chemical evolution, one has

---

[1] The reason why the galactic evolutions of D and ^{3}He are always considered together is of course due to the burning of D to ^{3}He in stars, and to their primordial origin.

to account for the high primordial mass fraction of ^{4}He, $Y \simeq 0.244$, whereas $Y \sim 0.23$ is observed (but see Sasselov & Goldwirth, 1995). Moreover, even in that case, one still overproduces ^{3}He...

To answer to these problems, some authors have suggested that ^{3}He should not survive the final phases of stellar evolution in low mass stars (Vangioni-Flam et al., 1994); it was furthermore suggested that an extra-mixing during the first dredge-up might destroy ^{3}He, and altogether explain the anomalous ^{12}C/^{13}C, ^{16}O/^{18}O and ^{7}Li/H ratios observed in some giant stars (Deliyannis, 1994; Hogan, 1995). The first numerical simulations of stellar evolution in these phases, including turbulent mixing induced by meridional circulation, show that this might indeed be the case (Charbonnel, 1994, 1995; Wasserburg et al., 1995). This deserves further study, as the issue is crucial.

4. Lithium

The interstellar abundance of ^{7}Li cannot be directly derived from spectroscopic observations, as one always has to apply a correction for the ionization of lithium in the ISM, and for its depletion onto interstellar grains. As a matter of fact, the depletion factor of ^{7}Li is rather derived by identifying the meteoritic abundance to the interstellar abundance. Hence, to derive constraints on the galactic evolution of lithium, it is much more interesting to determine the isotopic ratio ^{7}Li/^{6}Li, which does not suffer from these corrections (see however Steigman, 1995 for a different approach). However, the difficulty now lies in detecting ^{6}Li in the ISM (see Lemoine & Ferlet, 1994).

In effect, the only resonance line of lithium accessible is the fine structure doublet of ^{7}LiI at 6707.761–6707.912 Å, with that of ^{6}LiI shifted by 0.160mÅ toward longer wavelengths. The main component of the fine structure ^{6}Li doublet is thus superimposed on the weaker component of the ^{7}Li doublet, with no hope of resolving the blend, since the separation of these two lines is 0.4 km/s, much less than the typical intrinsic width of the lithium lines, ~ 1.8 km/s in the diffuse ISM. One may then try to detect the weaker component of the ^{6}Li doublet. However, the equivalent width of the ^{7}Li line is no more than a few mÅ on lines of sight already comprising $N(HI) \sim 10^{21}$ cm^{-2}, due to the low interstellar abundance of ^{7}Li, (^{7}Li/H)$\sim 10^{-9}$, and the quasi-total ionization of ^{7}Li to ^{7}LiII. Assuming an isotopic ratio similar to the meteoritic ratio, (^{7}Li/^{6}Li)$_\odot = 12.3$, with an oscillator strengths ratio 2, the equivalent width of the weaker ^{6}Li line should be $\sim 20 - 50\,\mu$Å... Moreover, on such dense lines of sight, several interstellar absorbing clouds are often detected within ~ 10 km/s, whereas the fine structure shift of the LiI lines is 6.8km/s. Due to the complex transition structure and the weakness of the ^{6}Li line, there is no need to say that not to account for all absorbing ^{7}Li components would strongly bias the analysis (see Fig.2a). Therefore, detecting the ^{6}Li absorption requires an extremely high signal-to-noise ratio, typically S/N$\gtrsim$ a few $\times 10^3$, together with a high resolving power, optimally $\lambda/\Delta\lambda \simeq 2 - 3 \times 10^5$ (where the intrinsic width of the lines is resolved).

Ferlet & Dennefeld (1984) obtained data of a very high quality toward ζ Oph, S/N$\simeq$ 4000 per pixel, $\lambda/\Delta\lambda = 10^5$, at the ESO 1.4m CAT + CES, but ^{6}Li was not detected, and the lower limit (^{7}Li/^{6}Li)$\gtrsim 25$ was derived. The first detection of ^{6}Li was reported by Lemoine et al. (1993) in the direction of ρ Oph, in two absorbing clouds of the line of sight, $(^7\text{Li}/^6\text{Li})_A = 11.1\pm2.$, $(^7\text{Li}/^6\text{Li})_B \sim 3$. The data were obtained at the ESO 3.6m Telescope fiber linked to the CES, at S/N$\simeq$ 4000, $\lambda/\Delta\lambda= 10^5$. The ^{6}Li isotope was only marginally detected in the secondary B cloud. In order to derive an accurate velocity structure of the line of sight, KI was simultaneously observed: the KI 7699Å line is $\sim$ 40 times stronger than the LiI line, KI is much heavier than LiI, hence its line is thinner, and finally KI and LiI are known to behave similarly in the ISM (White, 1986). This KI line was observed at $\lambda/\Delta\lambda=10^5$, although the optimal resolving power would be $\lambda/\Delta\lambda \simeq 6 \times 10^5$.

Subsequently, Meyer et al. (1993) reported the detection of ^{6}Li toward ζ Oph and ζ Per, deriving respectively $(^7\text{Li}/^6\text{Li})_{\zeta\, Oph} = 6.8^{+1.4}_{-1.7}$, and $(^7\text{Li}/^6\text{Li})_{\zeta\, Per} = 5.5^{+1.3}_{-1.1}$; the data were obtained at the KPNO 0.9m telescope, at a resolving power $\lambda/\Delta\lambda \simeq 2.5 \times 10^5$, and a signal-to-noise S/N$\simeq$ 2000. Although their high quality data confirm the previous detection in the sense that the interstellar ^{7}Li/^{6}Li ratio seem to have decreased since the formation of the solar system, with a present value $\sim 5-10$, we must stress that their result is biased by an incomplete study of the line of sight structure. In effect, it is clear from their data that the lines of sight toward these stars comprise at least two clouds, that are resolved in LiI; hence one cannot derive a reliable ^{7}Li/^{6}Li ratio using a single-cloud hypothesis. The data of Meyer et al. (1993) are of sufficiently high quality to allow the determination of the ^{7}Li/^{6}Li ratio in both clouds of each line of sight, and such an analysis should be undertaken.

This is furthermore confirmed by our latest observations toward ζ Oph, where two clouds were identified in KI and LiI (Lemoine et al., 1995). The data were obtained at the ESO 3.6m + CES, with S/N$\simeq$ 7500 per pixel, and $\lambda/\Delta\lambda= 10^5$, and two ratios were derived: $(^7\text{Li}/^6\text{Li})_A \simeq 8.6 \pm 2$, $(^7\text{Li}/^6\text{Li})_B \sim 2$. The spectrum is shown together with the profile fitting of the lines in Fig.2a, and the doublet of ^{6}Li, revealed by the residuals of the ^{7}Li absorption, is shown in Fig.2b.

The interpretation is not straighforward, and the reader is referred to Lemoine et al. (1995), Lemoine & Ferlet (1994). Essentially, a ^{7}Li/^{6}Li ratio of the order of $\simeq 9 - 11$, if confirmed to be representative of the ISM, would be a signature of an unknown production of ^{7}Li in the last 4.6 Gyrs. Indeed, it is more or less accepted that there must be an unknown stellar source of ^{7}Li, probably AGB C and S stars, in order to explain the abrupt increase in abundance of ^{7}Li between the halo phase, where $(^7\text{Li}/\text{H})_{Pop\,II} \simeq 10^{-10}$, and the young disk phase, where $(^7\text{Li}/\text{H})_{Pop\,I} \simeq 10^{-9}$; GCR spallation reactions only contribute a weak amount of ^{7}Li to its pre-solar abundance. Using the fact that ^{6}Li is only produced in GCR spallation reactions, one may evaluate the predicted interstellar ^{7}Li/^{6}Li ratio, starting 4.6 Gyrs ago with ^{7}Li/^{6}Li= 12.3, if only GCR produce ^{7}Li: $(^7\text{Li}/^6\text{Li})_0 \sim 5$. Hence if one observes a ratio higher than this latter, there must be some extra ^{7}Li. This argument is due to Reeves (1993). However, one cannot say so until the low "anomalous" $(^7\text{Li}/^6\text{Li})_B$ ratios are accounted for. These may result from a contamination of these B clouds by a strong spallative source of ^{6}Li. These may as well result from the presence of a third absorbing cloud on the line of sight, whose radial velocity would be such that its ^{7}Li absorption would contaminate the ^{6}Li absorption of cloud B. This third cloud would have

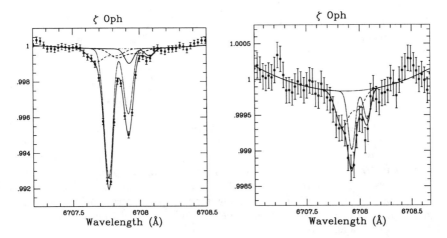

Figure 2. **Left**: Spectrum of the LiI line toward ζ Oph; the solid line shows the total profile, and the dashed lines show the individual ^{7}Li and ^{6}Li components (two absorbing clouds). **Right**: Residuals of the ^{7}Li absorption, revealing the ^{6}Li absorption for both clouds; note the scale in ordinates.

gone unnoticed at the present resolution of our KI observations; further higher resolution KI observations have now been undertaken at the AAT + UHRF, with $\lambda/\Delta\lambda = 6 \times 10^5$, and their analysis is underway.

It appears that these studies do not as yet give a consistent picture of the interstellar ^{7}Li/^{6}Li ratio. This may be due to the extreme difficulty of such observations, but it may also reflect the existence of various processes of creation of both isotopes in the ISM. Notably, in the case of the Vangioni-Flam & Cassé (these proceedings) spallation mechanism, the effective timescale is so short that one would expect to detect anisotropies in the interstellar ^{7}Li/^{6}Li ratio. Independently of these issues, and in the case where there is a unique ^{7}Li/^{6}Li ratio representative of the ISM, deriving this ratio would be highly important as it could confirm or infirm the action of a stellar source of ^{7}Li in the last 4.6 Gyrs. This would help in clarifying the lithium galactic evolution.

5. Beryllium and Boron

Finally, we report briefly on the latest attempts at detecting the beryllium and boron isotopes in the interstellar medium. Deriving the interstellar abundance of beryllium also suffers from an unknown depletion correction; here, however, this depletion factor of beryllium is a key-discriminant of scenarios of grain formation (Snow et al., 1979). Recently, an upper limit of $\delta_{Be} \leq -0.1$ dex has been set on this depletion factor[2]. The observations were conducted using the CFH 3.6m Telescope + Gecko spectrograph, with a resolving power $\lambda/\Delta\lambda = 10^5$, for the

---

[2] $\delta_{Be} \equiv \log_{10}(\mathrm{Be/H}) - \log_{10}(\mathrm{Be/H})_{\odot}$, where $\odot$ denotes the meteoritic abundance.

line of Be II at 3130Å, in the direction of ζ Per, obtaining a signal-to-noise ratio S/N≃ 700 (Lemoine et al., 1996). This line has never been detected, essentially since its equivalent width is not larger than ≃ 60µÅ, and because it falls in the near-UV, where the atmospheric extinction becomes dramatically important. This upper limit suits very well the curve of depletion factors *vs* condensation temperatures, favoring the formation of grains in stellar material, and does not suit the curve of depletion factors *vs* first ionization potential.

As to boron, it has recently been detected in the ISM using HST + GHRS, observing the line B II 1362Å in the direction of ζ Oph, achieving a signal-to-noise ratio S/N≃ 200 with a resolving power $\lambda/\Delta\lambda = 1.7 \times 10^4$ (Federman et al., 1993). The corresponding depletion factor, $\delta_B = -0.4$ fits very well the curve of depletion factor *vs* condensation temperatures. A similar program by the same group is now being conducted on HST + GHRS to measure the isotopic ratio $^{11}B/^{10}B$ on the same line of sight, using higher spectral resolution and higher signal-noise ratio. Such a measurement would certainly be a *première*, and the results are eagerly awaited.

Acknowledgments: we thank A. Vidal-Madjar, E. Vangioni-Flam, and M. Cassé for permanent discussions. M.L. acknowledges support by the NASA, DoE, and NSF, at the University of Chicago, as well as by CNRS.

References

Carswell, R.F., Rauch, M., Weymann, R.J., Cooke, A.J., Webb, J.K.: 1994, MNRAS **268**, L1
Balser, D.S., Bania, T.M., Brockway, C.J., Rood, R.T., Wilson, T.L.: 1994, ApJ **430**, 667
Bania, T.M, Rood, R.T., Wilson, T.L.: 1987, ApJ **323**, 30
Bruston, P., Audouze, J., Vidal-Madjar, A., Laurent, C.: 1981, ApJ **243**, 161
Charbonnel, C.: 1994, AA **282**, 811
Charbonnel, C.: 1995, ApJ **453**, L41
Dearborn, D.S.P., Schramm, D.N., Steigman, G.: 1986, ApJ**302**, 35
Deliyannis, C.P.: 1995, *The Light Elements Abundances*, ed. P. Crane, Springer, p.395
Federman, S.R., Scheffer, Y., Lambert, D.L., Gilliland, R.L.: 1993, ApJ **413**, L51
Ferlet, R.: 1981, AA **98**, L1
Ferlet, R., Dennefeld, M.: 1984, AA **138**, 303
Ferlet, R.: 1992, *IAU Symposium 150*, p.85
Galli, D., Palla, F., Ferrini, F., Penco, U.: 1995, ApJ **443**, 536
Gry, C., Laurent, C., Vidal-Madjar, A.: 1983, AA **124**, 99
Geiss, J.: 1993, *Origin and Evolution of the Elements*, eds. N. Prantzos, M. Cassé, E. Vangioni-Flam, Cambridge, p.89
Heiles, C., McCullough, P.R., Glassgold, A.E.: 1993, ApJ, submitted
Hogan, C.J.: 1995, ApJ **441**, L17
Lemoine, M., Ferlet, R., Vidal-Madjar, A.: 1993, AA **269**, 469

Lemoine, M., Ferlet, R.: 1994, astro-ph/9410055; 1995, *The Light Elements Abundances*, ed. P. Crane, Springer, p.350
Lemoine, M., Ferlet, R., Vidal-Madjar, A.: 1995, AA **298**, 879
Lemoine, M., Vidal-Madjar, A., Bertin, P., Ferlet, R., Gry, C., Lallement, R.: 1995, AA in press
Lemoine, M., Ferlet, R., Vidal-Madjar, A.: 1996, AA in preparation
Linsky, J. et al.: 1993, ApJ **402**, 694
Linsky, J. et al.: 1995, ApJ in press
Meyer, D.M., Hawkins, I., Wright, E.L.: 1993, ApJ **409**, L61
Olive, K.A., Rood, R.T., Schramm, D.N., Truran, J.W., Vangioni-Flam, E.: 1995, ApJ **444**, 680
Pagel, B. E.: 1995, *The Light Elements Abundances*, ed. P. Crane, Springer, p.155
Reeves, H.: 1993, AA **269**, 166
Reeves, H.: 1994, Rev. Mod. Phys. **66**, 193
Rood, R.T., Bania, T.M., Wilson, T.L.: 1992, Nature **355**, 618
Rood, R.T., Bania, T.M., Wilson, T.L., Balser, D.S.: 1995, *The Light Elements Abundances*, ed. P. Crane, Springer, p.201
Sasselov, D., Goldwirth, D.S.: 1995, ApJ **444**, L5
Scully, S.T., Cassé, M., Schramm, D.N., Truran, J.W., Vangioni-Flam, E.: 1995, report UMN-TH-1402/95, astro-ph/9508086
Snow, Th.P.Jr, Weiler, E.J., Oegerle, W.R.: 1979, ApJ 234, 506
Songaila, A., Cowie, L.L., Hogan, C., Rugers, M.: 1994, Nature **368**, 599
Spite, F., Spite, M.: 1982, AA **115**, 357
Spite, F.: 1995, *The Light Elements Abundances*, ed. P. Crane, Springer, p.239
Steigman, G., Tosi, M.: 1992, ApJ **401**, 150
Steigman, G.: 1995, report OSU-TA-4/95, astro-ph/9504048
Tytler, D., Fan, X.M.: 1995, BAAS **26**, 1424
Vangioni-Flam, E., Cassé, M.: 1995, ApJ **441**, 471
Vangioni-Flam, E., Olive, K.A., Prantzos, N.: 1994, ApJ **427**, 618
Vidal-Madjar, A., Laurent, C., Bruston, P., Audouze, J.: 1978, ApJ **223**, 589
Vidal-Madjar, A., Gry, C.: 1984, AA **138**, 285
Vidal-Madjar, A.: 1991, Adv. Space Res. **11**, 97
Vidal-Madjar, A., Lemoine, M., Ferlet, R., Gry, C.: 1996, in preparation
Wasserburg, G.J., Boothroyd, A.J., Sackmann, I.-J.: 1995, ApJ **447**, L37
White, : 1986, ApJ **307**, 777
Wilson, T.L., Rood, R.T.: 1995, ARAA in press

QSO Absorption Lines from Primordially Produced Elements

Edward B. Jenkins

Princeton University Observatory, Princeton, NJ 08544-1001, USA

Abstract. Over the last several years, absorption lines from D and He in very distant gas clouds have been observed in quasar spectra. These results are of use in estimating the density of baryons and uv photons in the universe. This article summarizes the results obtained so far and discusses some technical principles behind the observations.

1. Introduction

In our quest to learn more about the abundances of elements created during the early development of the universe, we strive to observe material that has had the least amount of tampering from other nucleosynthetic processes. One powerful way to work toward this goal is to observe the absorption lines in the spectra of distant quasars. These lines arise from foreground gas clouds or galaxies that have an age of only several Gyr. While heavy elements are known to exist in virtually all such systems, including even those that are responsible for the Lyman-α forest (Cowie et al. 1995; Tytler et al. 1995), their abundances are much lower than those found in our galaxy.[1] This statement applies even to the "damped Lyman-α systems" that are likely to arise from disk galaxies (Lu et al. 1995; Pettini et al. 1995; Wolfe 1995). Thus, while we may never be able to examine absolutely pristine primordial material, we can follow the examples of other abundance studies and relate abundance ratios for certain elements to some measure of the overall amount of heavy element production.

2. Hydrogen and Helium

It comes as no surprise that the most abundant elements of the universe, H and He, are responsible for the most conspicuous features in the spectra of quasars. For a great majority of the gas systems that create the absorption lines, the densities are very low and nearly all of the atoms are maintained in an ionized state by the uv background flux from quasars and galaxies. In very rough terms, it is generally believed that 1 part in 10^4 of the H atoms is neutral, 1 part in 10^6 of the He atoms is neutral, and 1 part in 100 of the He is singly ionized. An unfortunate consequence of this high ionization is that it is

---

[1]An exception seems to be systems that have an absorption redshift nearly equal to the emission redshift of the quasar (Petitjean 1995).

virtually impossible to reconstruct the relative abundances of these two species to any level of accuracy that would be of any use in testing different propositions for Big Bang Nucleosynthesis. About all we can do is to confirm that "Indeed, there's lots of H and He out there, just as one would expect."

Absorption by H is manifested by the familiar Lyman-α forest that has been intensively studied by many investigators since the first systematic investigation by Sargent, et al. (1980). The detection of He absorption is a technically much more difficult task. To observe the most redshifted features, the quasar's emission redshift z(em) must be somewhat greater than 2 to place the strong He II 304Å feature above the Lyman limit absorption by our galaxy (and, for HST, above about 3 to be well within its wavelength coverage). It is not easy to find such a quasar, for it must be bright enough to observe and, at the same time, it must not have a foreground gas cloud with a Lyman limit absorption that obliterates the flux at 304Å in the quasar's rest frame (Picard & Jakobsen 1993). So far, detections of intergalactic He II absorption have been reported for 3 different objects: Q0302-003 (Jakobsen et al. 1994), Q0302-0019 (Tytler et al. 1995), and HS1700+6416 (Davidsen 1995). For the case of Q0302-003, Songaila et al. (1995) have shown that the He absorption is consistent with what one would expect for the clouds responsible for the Lyman-α forest absorption observed from the ground. The optical depths for the He features are about 20 times those of the ones from H, based on ionization equilibria derived from the ratios of certain heavier elements.

In contrast to the very strong absorption by singly ionized helium atoms, the neutral form of this element produces features that are much weaker than the hydrogen features. He I absorption lines were detected in the spectrum of HS1700+6414 by Vogel & Reimers (1995).

In summary, gas clouds showing primordially produced H and He have a conspicuous presence in quasar spectra, but the strong role played by photoionization in setting the strengths of the lines, and the uncertainties therein, preclude our making any meaningful abundance measurements. It is better to make the standard assumption on the He abundance and use the He lines to learn more about the clouds' physical properties (Baron et al. 1989) or the strength and spectral distribution of the intergalactic radiation field (Ostriker & Miralda-Escudé 1990; Miralda-Escudé & Ostriker 1992; Giroux, Fardal, & Shull 1995). With better signal-to-noise ratios and wavelenth resolution, absorption from a smoothly distributed intercloud medium may be detected on top of the contribution from the clouds.

3. Hydrogen and Deuterium

3.1. Measurement Principles

In contrast to the situation with He, a comparison of the deuterium abundance with that of hydrogen is free of the complications introduced by photoionization. Both substances have identical ionization potentials, cross sections, and recombination coefficients. From the standpoint of learning more about Big Bang Nucleosynthesis, deuterium is more useful than helium because its predicted relative abundance changes more dramatically with the parameter η, the present ratio of nucleons to photons.

The spectroscopy of H and D embodies the simplest principles that we are taught in elementary physics: we view transitions from the ground state of a one-electron atom, giving rise to the $1s - np$ Lyman series. The H and D lines are displaced from each other by a small amount (equivalent to a doppler shift of 81 $km\ s^{-1}$), owing to differences in the reduced mass of the electron. The challenge that we face in measuring D/H is the large disparity between the abundances of these two isotopes, of order 10^4 or more. Fortunately, we focus on some very saturated H lines and then see their counterparts from D. Along the lines of sight to quasars, we have at our disposal an enormous range in column densities of individual absorption systems (Petitjean et al. 1993; Hu et al. 1995), and we can select from Lyman series lines whose transition probabilities differ by a factor of 1000.[2]

To measure an abundance, we must have an absorption feature that has a central optical depth τ_0 not very far from unity, otherwise the line is either too weak to be detected or so badly saturated that we can not derive a reliable abundance. Another alternative is to have a line that is so strong that the damping wings can be measured. For this to happen, the wings must completely overpower the effect of weak absorption components at large displacements in velocity.

Table 1. Central Optical Depths for $b = 15\ km\ s^{-1}$

Log N(H)	Lyman-α H	Lyman-α D^a	Lyman-α D^b	Lyman-ξ H	Lyman-ξ D^a	Lyman-ξ D^b	Lyman Limit H
16.0	500	0.10	0.01	0.42			0.063
16.5	1600	**0.32**	0.032	1.3			**0.20**
17.0	5000	**1.0**	0.10	4.2			**0.63**
17.5	1.6×10^4	**3.2**	**0.32**	13.			**2.0**
18.0	5.0×10^4	10.	**1.0**	42.			6.3
18.5	1.6×10^5	32.	**3.2**		0.026		20.
19.0	Damped	100.	10.		0.084		63.
19.5	Damped		32.		**0.26**	0.026	
20.0	Damped				**0.84**	0.084	
20.5	Damped				**2.6**	**0.26**	
21.0	Damped				8.4	**0.84**	
21.5	Damped				26.	**2.6**	

^{a}Under the condition that D/H $= 2 \times 10^{-4}$.
^{b}Under the condition that D/H $= 2 \times 10^{-5}$.

Table 1 presents an outlook on column density ranges where we can operate with different members of the Lyman series. It shows values of τ_0 for lines that

---

[2] A useful tip: the f value of each transition is roughly proportional to its wavelength separation from adjacent members of the series.

have a velocity dispersion $b = 15\ km\ s^{-1}$, a value that is typical for the systems that have values of log $N(\text{H})$ listed in the first column of the table. The table features two lines in the Lyman series: the first is the strongest one, Lyman-α at laboratory wavelength of 1215.7Å (for hydrogen), and the second is the 14th member of the series, Lyman-ξ at 916.4Å. For a reason that will be made clear later in this discussion, there is a special reason for picking Lyman-ξ as a representative of the weakest line. For each line, the table lists values that apply to H and to two representative values of D/H. The last column in the table shows continuum opacities from hydrogen just below the Lyman Limit.

For an observation to be useful, we must be able to measure the column densities of D and H simultaneously. Bold numbers in the table indicate cases where such measurements can be made without appreciable error. For H, one can start with measurements of the Lyman-ξ line at log $N(\text{H}) = 16$, and upon reaching higher column densities move over to measuring the depth of the Lyman Limit absorption when Lyman-ξ and its immediate neighbors become too saturated. There is a gap in the range $18.0 < \log N(\text{H}) < 19.0$ where the amount of hydrogen can not be determined with much reliability. Beyond log $N(\text{H}) = 19$, one can measure the damping wings of Lyman-α and other low members of the series. Depending on the actual value of D/H, deuterium can be measured at values of log $N(\text{H})$ that have bold numbers in the D columns. Intermediate members of the Lyman series can fill in for D column densities that are spanned by the extremes of Lyman-α and Lyman-ξ.

To give some historical perspective on the conditions shown in the table, we note that determinations of D/H toward hot stars in our galaxy with *Copernicus* (see the listing by Friedman, et al. in this conference volume) generally spanned the range $19 < \log N(\text{H}) < 20$, while the measurement of D and H absorption in the Lyman-α line in the HST spectrum of Capella by Linsky, et al. (1993) was at log $N(\text{H}) = 18$. Up to now, measurements of D/H in quasar spectra generally have been in the regime log $N(\text{H}) = 17 \pm 0.5$.

Webb, et al. (1991) investigated the accuracies that might be obtained under realistic observing circumstances. They carried out some computer simulations of spectral data with the noise and resolution (S/N = 15, $R = 30,000$) that one can reasonably expect to find with a 10 hr exposure of a 17.5 magnitude quasar with a 4M telescope, a reasonable standard to apply at the time the article was written.[3] They concluded that under favorable circumstances ($b \approx 25\ km\ s^{-1}$) one could get accuracies of ± several tenths of a dex for D/H. However one aspect of their study that was not realistic was the adoption of a simple gaussian for the velocity profile in each test example, followed by the assumption in the analysis that indeed only this one component was present. This simple velocity behavior is generally not manifested in real absorption systems[4] (Lanzetta 1992; Churchill, Vogt, & Steidel 1995; Lu et al. 1995; Wolfe 1995). As a consequence, ambiguities can arise from the inability to positively identify overlapping components. Also, considerable errors in column densities can

---

[3] Nowadays, with the Keck telescope and other 8–10M class telescopes that are about to come on line, one might expect to do better.

[4] There are occasional exceptions: e.g., the system in front of HS1946+7658 at $z(\text{abs}) = 1.7382$ seems to consist of only one simple component (Lu et al. 1995).

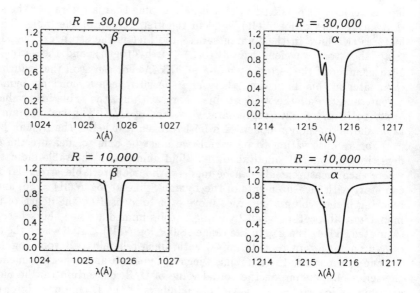

Figure 1. A reconstruction of the Lyman-β and Lyman-α absorption features toward a typical "Lyman Limit" system, with $\log N(\mathrm{H}) = 17.5$ and $b = 15\ km\ s^{-1}$, on the assumption that D/H is only 2×10^{-5}. The D component of Lyman-α is easily measured at a wavelength resolving power of 30,000 (upper panels), but it blends in with the damping wings of the H component when R is reduced to 10,000 (lower panels).

come about when fine-scale velocity structures are present and optical depths are large (Levshakov & Takahara 1995), as is the case with hydrogen in the range where the weakest members of the Lyman series become saturated, but before the damping wings in Lyman-α are reasonably well established.

There is an interesting principle about the study of H and D that is illustrated by the computer simulations shown in Figs. 1 and 2. For pedagogical clarity, these simulations do not include noise or kinematically complex profiles, but they have been smeared by gaussian instrumental profiles of different widths. The central optical depths of H and D in Fig. 1 for Lyman-α were chosen to be exactly the same as those of Lyman-ξ in Fig. 2. Several factors make the determination of D/H easier when there is enough material present to show D in the higher Lyman lines:

1. There are many lines with roughly comparable (but still different) f values to measure, giving the observer more than the usual 1, 2 or 3 lines when Lyman-α, β and perhaps γ are available.

2. At Lyman-ξ, the spacing of the lines is about twice the separation of the D and H lines, making the D lines neatly sandwiched between adjacent H lines. This arrangement resembles the superposition two picket fences that are out of phase with each other and permits D to show a spectacular change in the spectrum, even at low resolution. If D/H is raised to 2×10^{-4},

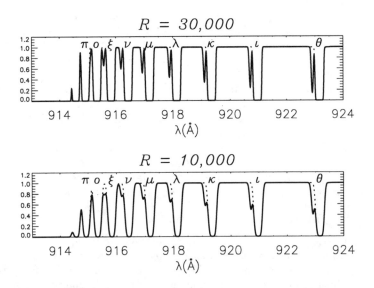

Figure 2. A reconstruction of the higher Lyman series lines seen at two different instrumental resolving powers, as indicated, through a "Damped Lyman-α" system with log $N(\text{H}) = 20.6$ and $b = 15\ km\ s^{-1}$. As in Fig. 1, D/H = 2×10^{-5}.

the average intensities coming through the gaps between the H absorptions of Lyman-ξ, o, and π are reduced to about 1/3 their original intensities. This effect can be seen at resolutions R even poorer than 10,000.

3. The strengths of damping wings from H, which tend to swallow up the D profiles, scale in proportion to $Nf\gamma$ at any particular velocity displacement from the core of the line, instead of Nf for the parts of the profile dominated by doppler spreading. As stated earlier, Nf in the two figures are about the same, but the decay rate γ for Lyman-ξ (summed over all possible downward transitions) is only 0.0025 times that of Lyman-α.

3.2. Actual Measurements

We live in interesting times: measurements of D/H in distant systems are just starting to come in. The first report, indicating D/H = $1.9 - 2.5 \times 10^{-4}$, was published by Songaila, et al. (1994) for a Lyman limit absorption system at $z = 3.32$ in the quasar 0014+813. This was quickly followed by another article (with a more cautiously worded title) reporting different observations of exactly the same system (Carswell et al. 1994). The results of the two studies are essentially consistent with each other. Now, Rugers & Hogan (this conference volume) report on a more detailed analysis of the data taken by Songaila, et al. They conclude that the velocity component analyzed earlier actually can be broken into two components, and, most important, that the D components have velocity dispersions of about $1/\sqrt{2}$ times that of their H counterparts, as one might expect if thermal doppler broadening dominates over the effects of turbulence. The determination of D/H for 0014+813 is significantly higher than ratios found in the local interstellar medium, an effect that is consistent with the notion that D has been consumed by stellar processing in our galaxy [(Steigman & Tosi 1992); for the cosmological implications of these results, see also the articles by Schramm and Hogan in this conference volume.]

Just to keep things a little interesting, a much different finding on D/H at great distances has been reported by Tytler & Fan (1994) [see also Tytler et al. (1995)]. They claimed that D/H = 2×10^{-5} (with a possible systematic error of about 50%) in a system at $z = 3.57$ toward 1937-1009. Unlike the determinations for 0014+813, this result is consistent with the local value of D/H. The conflict between this result[5] and the other one underscores the need for more measurements to be taken for other systems.

3.3. Interlopers: H posing as D?

A question raised by Songaila, et al. (1994) and others (Steigman 1994; Tytler et al. 1995) is the possibility that a wisp of high velocity hydrogen is masquerading as D in a main component. We have no way of knowing for sure[6] that we are not being fooled by this phenomenon, apart from the reassuring indication of a difference in b from an analysis like the one by Rugers & Hogan.

For those who like to figure gambling odds, it is instructive to examine the statistical argument by Songaila et al. that such an interloper could be present, but with a very low probability. They counted the number of lines that could be seen in the same echelle spectral order as that which contained the Lyman-α line being studied. The found 12 lines over a 4300 $km\ s^{-1}$ interval, leading to a mean separation of 360 $km\ s^{-1}$. They then calculated a probability of about 3% that a random forest line would fall within about a 10 $km\ s^{-1}$ wide interval centered on the D line. A flaw in this approach is evident if one notices that 4 other components are identified in the immediate vicinity of the component under consideration, prompting the question, "What is the probability that 4 or more other components could accidentally lie within a velocity span of, say, $\pm 200\ km\ s^{-1}$ centered on the feature?" (Asking this question *a posteriori* is a little unfair of course.) The answer, if one sums the Poisson probabilities, $P(4) + P(5) + P(6) + ...$, gives a similarly low number: 2.5%! The key to this problem is that Songaila, et al. estimated the probability of obtaining a hit from a *completely unrelated* forest component, without considering the added

---

[5] Unfortunately, at the time of writing this article, the author is not aware of any full scale publication of this finding (i.e., one that shows the data and how they were analyzed). Thus, we have no opportunity to form our own opinions about its validity.

[6] One might question whether or not a very strong line from a heavy element in a low ionization stage, such as the 1335Å line of C II or the 1302Å line of O I, might be used to indicate the presence of another component centered near $-81\ km\ s^{-1}$ with respect to the principal one. The answer is probably not - at least not in a very distant system that has a heavy element abundance ratio that is one or two orders of magnitude below solar. For instance, if $\log N(\text{H})$ = 21.5 - the highest value that is generally seen in a damped system - the central optical depth τ_0 of either line should be about 200 if either [C/H] or [O/H] = -1, an optimistically high value for metallicity. Unfortunately, this number multiplied by the largest apparent D/H = 2×10^{-4} gives $\tau_0 = 0.04$, i.e., too small to be seen unless the S/N is extraordinarily high.

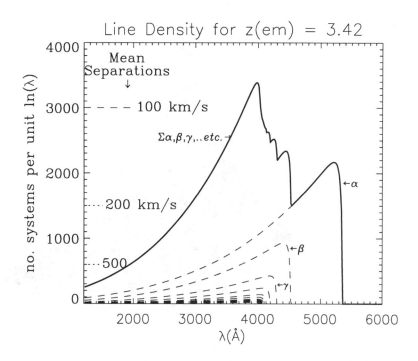

Figure 3. The expected density of forest lines having a central intensity lower than 0.9 times the continuum (i.e., $\tau_0 > 0.1$) as a function of observed wavelength for a quasar at $z(\text{em}) = 3.42$. Higher Lyman lines (β, γ, etc.) of systems at higher redshift pile on top of the Lyman-α lines up to a point (near L20; see Fig. 2) where they blend together. The rounding of the distributions where the series begin is caused by the "proximity effect" (Bajtlik, Duncan, & Ostriker 1988)

complication of satellite lines coming from material physically associated with the gas system being studied. Such components are often seen within one or two hundred $km\ s^{-1}$ of systems with large $N(H)$, although the multiplicity is generally not as bad in low ionization species as with much higher stages of ionization exhibited by such ions as C IV and Si IV.

Returning to the issue of unrelated forest lines, their density increases markedly as a function of $[1 + z(\text{abs})]$. In addition, for a fixed value of $z(\text{abs})$ there may be additional lines from higher series members piled on top of the Lyman-α lines if $[1 + z(\text{em})]$ of the quasar is greater than $\frac{32}{27}[1 + z(\text{abs})]$. This compounding of lines is illustrated in plot of line density vs. λ in Fig. 3 for the spectrum of a quasar at $z(\text{em})$ equal to that of the one observed by Songaila, et al. This diagram may be generalized to include quasars at arbitrary values of $z(\text{em})$ by adding a third dimension, giving the result shown in Fig. 4.

Fig. 4 shows that if one wants to reduce the confusion from random forest lines *unrelated* to the system being considered, the best place to go is in the ultraviolet. Ground based observatories can attain reasonably uncluttered spectra

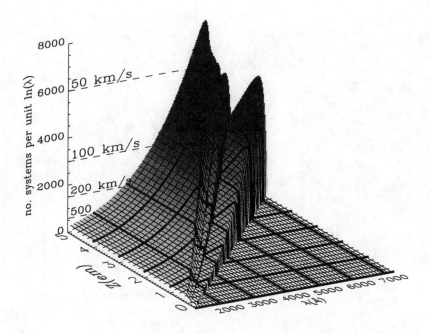

Figure 4. Same as Fig. 3, generalized to include a range of values for z(em).

(mean separation of lines of order 270 $km\ s^{-1}$) at around 4000Å if quasars with z(em) < 2.5 are observed. Another alternative is to study Lyman-α absorption in systems that have z(abs) not much less than z(em), where the proximity effect helps to reduce the unwanted interference, as was the case with the system studied by Songaila, et al (1994) and Carswell, et al. (1994).

4. Closing Remarks

If measurements of He and D were easy, we would have by now a large inventory of results that would give us powerful and broad insights on the density of baryons and uv photons in the universe. Astronomers are off to a good start in meeting this challenge, however. The future looks bright for additional results to build upon (or perhaps modify) the initial findings, as we comission new, large-aperture telescopes on the ground, install a more modern uv spectrograph (STIS) on HST (Woodgate & STIS Team 1992), and launch the Far Ultraviolet Spectroscopic Explorer (Friedman, et al., this conference volume).

References

Bajtlik, S., Duncan, R. C., & Ostriker, J. P. 1988, ApJ, 327, 570

Baron, E., Carswell, R. F., Hogan, C. J., & Weymann, R. J. 1989, ApJ, 337, 609
Carswell, R. F., Rauch, M., Weymann, R. J., Cooke, A. J., & Webb, J. K. 1994, MNRAS, 268, L1
Churchill, C. W., Vogt, S. S., & Steidel, C. C. 1995, in QSO Absorption Lines, ed. G. Meylan (Berlin: Springer), p. 153
Cowie, L. L., Songaila, A., Kim, T.-S., & Hu, E. M. 1995, AJ, in press
Davidsen, A. F. 1995, BAAS, 27, 853
Giroux, M. L., Fardal, M. A., & Shull, J. M. 1995, ApJ, 451, 477
Hu, E. M., Kim, T.-S., Cowie, L. L., & Songaila, A. 1995, AJ, in press
Jakobsen, P., Boksenberg, A., Deharveng, J. M., Greenfield, P., Jedrzejewski, R., & Paresce, F. 1994, Nat, 370, 35
Lanzetta, K. M. 1992, PASP, 104, 835
Levshakov, S. A., & Takahara, F. 1995, preprint
Linsky, J. L. et al. 1993, ApJ, 402, 694
Lu, L. M., Savage, B. D., Tripp, T. M., & Meyer, D. M. 1995, ApJ, 447, 597
Miralda-Escudé, J., & Ostriker, J. P. 1992, ApJ, 392, 15
Ostriker, J. P., & Miralda-Escudé, J. 1990, ApJ, 350, 1
Petitjean, P. 1995, in QSO Absorption Lines, ed. G. Meylan (Berlin: Springer), p. 61
Petitjean, P., Webb, J. K., Rauch, M., Carswell, R. F., & Lanzetta, K. 1993, MNRAS, 262, 499
Pettini, M., King, D. L., Smith, L. J., & Hunstead, R. W. 1995, in QSO Absorption Lines, ed. G. Meylan (Berlin: Springer), p. 71
Picard, A., & Jakobsen, P. 1993, A&A, 276, 331
Sargent, W. L. W., Young, P. J., Boksenberg, A., & Tytler, D. 1980, ApJS, 42, 41
Songaila, A., Hu, E. M., & Cowie, L. L. 1995, Nat, 375, 124
Songaila, A., Cowie, L. L., Hogan, C. J., & Rugers, M. 1994, Nat, 368, 599
Steigman, G. 1994, MNRAS, 269, L53
Steigman, G., & Tosi, M. 1992, ApJ, 401, 150
Tytler, D., & Fan, X.-M. 1994, BAAS, 26, 1424
Tytler, D. et al. 1995, in QSO Absorption Lines, ed. G. Meylan (Berlin: Springer), p. 289
Vogel, S., & Reimers, D. 1995, A&A, 294, 377
Webb, J. K., Carswell, R. F., Irwin, M. J., & Penston, M. V. 1991, MNRAS, 250, 657
Wolfe, A. M. 1995, in QSO Absorption Lines, ed. G. Meylan (Berlin: Springer), p. 13
Woodgate, B., & STIS Team 1992, in Science with the Hubble Space Telescope, ed. P. Benvenuti & E. Schreier (Garching bei München: ESO), p. 525

Primordial D/H from Q0014+813

Martin Rugers and Craig Hogan

University of Washington, Astronomy Department, Box 351580, Seattle, WA 98195

Abstract. This paper presents a new, detailed analysis of the z=3.32 Lyman Limit absorption system in the spectrum of Q0014+813, resulting in a better estimate for the primordial abundance of deuterium along this line of sight: $(D/H) = 1.9 \pm 0.4 \times 10^{-4}$. The cosmological consequences of this value are briefly discussed.

1. Introduction

In Songaila, Cowie, Hogan and Rugers (1994) the finding of an absorption feature, which coincided with the expected wavelength for D, on the blue side of the $z = 3.32$ absorber complex (discussed in Chaffee et al., 1986) in Q0014+813 was reported. The D/H ratio implied by this observation, around 2×10^{-4} is rather higher than that derived from the usual models for Galactic chemical evolution, leading us to question whether this feature is really a Ly-α forest interloper, rather than a D absorption feature. Here we clarify the situation in this one absorber complex by measuring the Doppler width of the D line, which in a purely thermally broadened case will be $(1/\sqrt{2})$ of the Doppler width of the H line. We argue that the interpretation of the lines as deuterium is more plausible than the interloper hypothesis. More details can be found in Rugers and Hogan (1996a).

2. Observation, Reduction, and Fitting

Six exposures for a total of 4 hours were taken of Q0014 at the Keck 10m telescope, November 1993, using the HIRES echelle spectrograph. The data were processed, extracted and corrected for the radial velocity of the observer with respect to the QSO using IRAF. The Voigt profile fitting code VPFIT, developed by Webb and Carswell (see e.g. Webb (1987) and Carswell et al. (1987)), was then used to determine the redshift (z), Doppler parameter (b) and column density (N) by fitting Voigt profiles to the unsmoothed data, convolved with a Gaussian instrument profile, for each absorber component, while making formal error estimates for each parameter from the covariance matrix, based on the reduced χ^2.

Using the results from Songaila, Cowie, Hogan and Rugers as a starting point for the profile fitting, it became clear that a much better fit to the data was obtained using two HI absorbers, with corresponding D absorbers, instead of one. Each of the two D lines at the blue end of the absorber complex has a

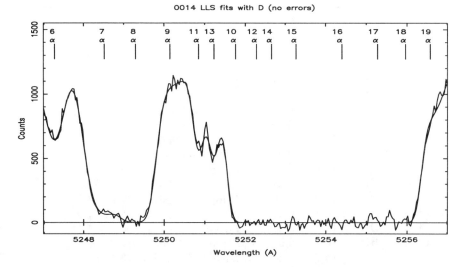

Figure 1. Lyman-α fits to the LLS in Q0014, at z=3.32. The deuterium features are labeled 11 and 13, the corresponding hydrogen features 12 and 14, respectively.

Doppler width consistent with thermal broadening of a DI line corresponding to its HI line. The results of the fits are presented in Figures 1 and 2, and Table 1. Fitting HI lines through Lyman-17, where the lines are optically thin, allows for a well constrained estimate of the HI column density. For the two D candidates we find D/H ratios of $10^{-3.73 \pm 0.12}$ for the $z = 3.320482$ absorber (the bluer of the two), and $10^{-3.72 \pm 0.09}$ for the $z = 3.320790$ absorber. We therefore expect the D/H ratios for these absorbers to lie in the ranges of 1.4 to 2.5×10^{-4} and 1.5 to 2.3×10^{-4}, respectively. Comparing the total D column density from the two D lines with the total column density of the complex, as derived from the Lyman Limit optical depth (between 1.7×10^{17} and 2.4×10^{17}) gives an independent, conservative lower limit of 1.1×10^{-4} for the the D abundance in this system, assuming of course that the D features really are deuterium.

3. Discussion

Significant to our claim that these two lines can indeed be identified as D is the fact that both are narrow, aside from the fact that they lie right at the expected wavelength. The fits for both are consistent with thermal broadening of DI, and are not likely to be due to thermally broadened HI, as the velocity widths are significantly smaller than the thermal broadening of hydrogen at 10^4K, the lowest temperature expected for equilibrium gas under these conditions. Moreover, the ratio of b_H/b_D for each component is consistent with what is predicted thermally: $\sqrt{2}$. Unfortunately, the metallicity of this system is too low for metal lines to be

Table 1. VPFIT results for the Q0014+813 LLS.

Comp.	Species	logN	±	z	±	b	±
1	HI	15.45	0.68	2.310923	0.000015	22.1	4.3
2	HI	13.18	0.09	2.644001	0.000052	25.0	6.5
3	HI	13.22	0.11	2.648422	0.000038	16.8	5.7
4	HI	13.71	0.04	2.649057	0.000025	24.7	3.0
5	HI	12.78	0.34	3.315725	0.000309	34.3	12.4
6	HI	13.24	0.12	3.316360	0.000029	17.6	3.5
7	HI	14.01	0.07	3.317390	0.000050	25.3	2.9
8	HI	14.28	0.05	3.318026	0.000028	19.6	2.2
9	HI	12.99	0.42	3.318726	0.000388	38.0	12.9
10	HI	13.74	0.34	3.320050	0.000064	10.4	5.0
11	DI	13.03	0.10	3.320483	0.000010	7.5	1.2
12	HI	16.76	0.07	3.320482	0.000044	10.1	2.0
13	DI	13.18	0.07	3.320789	0.000012	8.8	1.1
14	HI	16.90	0.06	3.320790	0.000037	12.7	0.8
15	HI	15.22	0.38	3.321300	0.000131	13.1	4.3
16	HI	16.36	0.16	3.322225	0.000041	22.8	2.1
17	HI	15.36	0.19	3.322955	0.000088	17.8	6.0
18	HI	14.44	0.32	3.323513	0.000089	12.0	5.1
19	HI	13.06	0.13	3.324024	0.000078	24.0	6.4

detectable, without which it is extremely hard to verify the double-ness of the main HI absorbers, other than with the improved fit.

The reinforcement of the assertion that D has been detected lies in several factors. One interloping cloud no longer suffices, as two are required, both of which fortuitously have the offest expected for D (the velocities for D and H agree to within the accuracy in both cases). Second, both have to yield the same, false abundance (the estimated D/H is nearly the same for both). Finally, the D lines are unphysically narrow for hydrogen, and agree with what would be expected for DI at the same temperature as HI. Since metal lines are relatively rare, it is unlikely that these two features are due to metal line absorption from a system at a lower redshift. The identification of the features as deuterium makes physical sense and is the natural interpretation.

Using this absorber complex to estimate the primordial deuterium abundance $(D/H)_p = 1.9 \pm 0.4 \times 10^{-4}$, Standard Big Bang Nucleosynthesis gives a baryon to photon ratio η of $1.7 \pm 0.2 \times 10^{-10}$, corresponding to $\Omega_b h^2 = 0.0062 \pm 0.0008$. This value of η is consistent with the Big Bang estimates based on cosmic helium and lithium (Copi et al., 1995, Hata et al., 1995, and contributions by Schramm and Steigman in this volume). A consequence of this high value for D/H is that the baryon density is only two to three times as large as that of known baryons in gas and stars, and that most of the dark matter in galaxies is non-baryonic. The fact that our estimate for D/H is roughly a factor of 10 higher than the value in the ISM of the Milky Way is likely due to processing of the lightest elements in stars (Hogan, this volume).

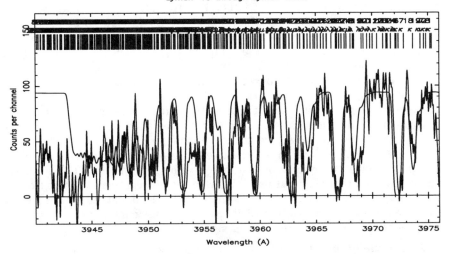

Figure 2. Lyman-10 through the Lyman Limit for the Q0014 absorber.

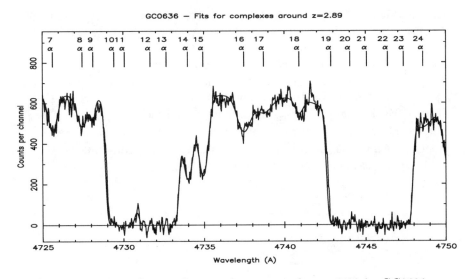

Figure 3. Ly-α fits to the complexes around $z = 2.89$ in GC0636 +680. The D and corresponding H features at $z = 2.89040$ are numbered 9 and 10, at approximately 4728 and 4729.5 Å, respectively.

Deuterium crops up elsewhere in our data, although it is not measured as well as in the Chaffee cloud in Q0014. For example, the LLS at $z = 2.90$ in the spectrum of GC0636+680 (lines numbered 19 through 23 in Figure 3) is not useful for a reliable determination of D/H, as the redshift of the absorber is lower, the velocity structure of the complex is quite intricate, and the column densities of the individual clouds appears high. The presence of D is indicated in several other absorbers, though, as a small absorption dip is present at the expected velocity offset. One example of another possible narrow absorption feature consistent with deuterium is shown in figure 3. Due to the lack of higher-order Lyman series lines, the errors on the columns (and therefore the D/H ratio) in this case are rather large: we deduce a value for D/H of $10^{-3.95 \pm 0.54}$ for the $z = 2.89040$ absorber. We hope to be able to use other such absorbers to obtain a higher confidence level in the estimated value of primordial D/H. It seems likely, however, that the most useful information will continue to come from the few, best, "lucky" examples, such as the Chaffee cloud, rather than from large random statistical samples, both because of the easier interloper rejection and because of the more accurate column measurement.

Acknowledgments. Thanks to Antoinette Songaila and Len Cowie for taking the data, and to Bob Carswell for help in running VPFIT. This work was supported at the University of Washington by NSF grant AST 932 0045.

References

Carswell, R.F., Webb, J.K., Baldwin, J.A., Atwood, B. 1987, ApJ **319**, 709.
Chaffee, F. H., Foltz, C. B., Röser, H.-J., Weymann, R. J. & Latham, D. W. 1985, ApJ, **292**, 362.
Copi, C., Schramm, D., Turner, M. 1995, Science **267**, 192.
Hata, N., Scherrer, R., Steigman, G., Thomas, D., Walker, T., Bludman, S., Langacker, P. 1995, Phys. Rev. Lett., in press.
Rugers, M., Hogan, C. 1996a, ApJ Letters, submitted.
Rugers, M., Hogan, C. 1996b, in preparation.
Songaila, A., Cowie, L., Hogan, C., Rugers, M. 1994, Nature **368**, 599.
Webb, J.K. 1987, *PhD thesis*, Cambridge University.

The Chemical Enrichment History of Damped Lyman-alpha Galaxies

Limin Lu, Wallace L. W. Sargent, & Thomas A. Barlow

Caltech, 105-24, Pasadena, CA 91125 (email: ll@troyte.caltech.edu)

Abstract. Studies of damped Lyα absorption systems in quasar spectra are yielding very interesting results regarding the chemical evolution of these galaxies. We describe some preliminary results from such a program.

1. Introduction

Damped Lyα absorption systems in quasar spectra are generally believed to trace the absorption from interstellar gas in high-redshift galaxies, possibly from the disks or proto-disks of spirals (Wolfe 1988). They can be studied in the redshift range $0 < z < 5$ by combining UV and optical observations. The damped Lyα galaxies are particularly suited for probing the chemical evolution of galaxies over a large fraction of the Hubble time for several reasons: (1) they are relatively common and easy to identify in quasar spectra, so building up a large sample is possible; (2) given their large neutral hydrogen column densities (N(HI)$\sim 10^{20} - 10^{22}$ cm^{-2}), most of the absorbing gas should be neutral so ionization corrections should be minimal (cf. Viegas 1995); (3) the damped Lyα galaxies should be relatively representative of galaxies at high redshifts since they are selected simply because they happen to lie in front of background quasars.

The first systematic investigation of the chemical evolution of damped Lyα galaxies was conducted by Pettini and collaborators (Pettini et al. 1994; 1995), who studied the Zn and Cr abundances in ~ 20 damped Lyα galaxies. The advent of the 10-m Keck telescope allows us to carry out similar investigations in a much more detailed fashion. In this short contribution, we present some preliminary results from such a program. Detailed analysis and discussion may be found in Lu, Sargent, & Barlow (1996; hereafter LSB96).

2. Results

Figure 1 shows the abundance results so far obtained from our Keck program, with the addition of selected measurements from published papers where we believe the effect of line saturation has been treated properly. We also correct (when applicable) the abundance measurements from previous papers for the set of new oscillator strengths compiled by Tripp, Lu, & Savage (1995) so that all the measurements will be on the same footing. References to the data used in constructing figure 1 may be found in LSB96.

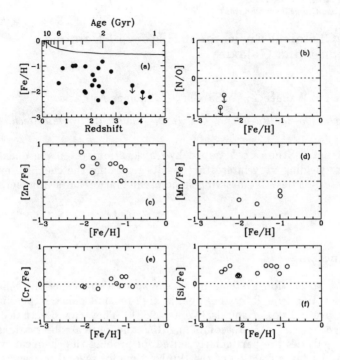

Figure 1. (a) Age-metallicity relation for our sample of damped Lyα galaxies. The conversion from redshift to age is calculated for $q_0 = 0.5$ and $H_0 = 50$. (b)-(f) Abundance ratios of selected elements found in damped Lyα galaxies. The notion [Fe/H] has the meaning [Fe/H]=log(Fe/H)$_{damp}$−log(Fe/H)$_\odot$, and similarly for others. Typical measurement errors of the abundances are 0.1 dex.

2.1. Age-Metallicity Relation

Figure 1(a) shows the age-metallicity relation for our sample of damped Lyα galaxies (filled circles). The solid curve roughly indicates the age-metallicity relation for disk stars in the Galactic solar neighborhood determined by Edvardsson et al. (1993). We note the following:

(1) The damped Lyα galaxies have Fe-metallicities ([Fe/H]) in the range of 1/10 to 1/300 solar, thus representing a population of very young galaxies at least in terms of the degree of chemical enrichment.

(2) The mean metallicity appears to increase with age, providing direct evidence for the buildup of heavy elements in galaxies. It may be significant that all the four galaxies with $z > 3$ have [Fe/H]< -1.7, while at $2 < z < 3$ at least some galaxies have achieved much higher metallicities. This may signal an epoch of rapid star formation in galaxies. We also note that the *intrinsic* trend of increasing metallicity with age would be stronger if Fe is somewhat depleted by dust in these galaxies because the depletion should be the least for the highest redshift galaxies (see section 2.2, however).

(3) Clearly the damped Lyα galaxies have much lower metallicities than the Milky Way disk at any given time in the past. This may bear significantly on the nature of the damped Lyα galaxies. It was suggested initially (cf. Wolfe 1988) that the damped Lyα absorbers may trace disks or proto-disks of high-redshift spirals. But the low metallicities of damped Lyα galaxies cast some doubts on this interpretation. Timmes, Lauroesch, & Truran (1995; also see Timmes 1995, this volume) suggested that the abundance measurements of damped Lyα galaxies are consistent with the chemical enrichment history of the Milky Way disk if the enrichment process in damped Lyα galaxies is delayed by $\sim$3 Gyrs for some reason; this seems to place the Milky Way at a privileged position. On the other hand, the metallicities found for our sample of damped Lyα galaxies are very similar to those found for Galactic halo stars and globular clusters, suggesting the possibility that damped Lyα absorbers may represent a spheroidal component of high-redshift galaxies. This possibility has in fact already been suggested by Lanzetta, Wolfe, & Turnshek (1995) based on considerations of gas consumptions in these galaxies.

2.2. Abundance Ratios and Nucleosynthesis

Panels (b)-(f) of figure 1 show the abundance ratios of various elements in damped Lyα galaxies relative to their corresponding solar ratios. Elemental abundance ratios, in principle, allow one to gain insight of what kind of nucleosynthetic processes may be responsible for the enrichment of the interstellar medium. For example, the well-documented overabundance of even-Z (Z=atomic number) α-group elements relative to the Fe-peak elements in Galactic halo stars is believed to reflect the nucleosynthetic products of massive stars through SN II explosions (cf. Wheeler, Sneden, & Truran 1989). It is interesting that the observed abundance patterns of N/O, Si/Fe, Cr/Fe and Mn/Fe in damped Lyα galaxies are all consistent with measurements in Galactic halo stars (cf. Wheeler et al. 1989). In particular, we note that the observed N/O ratios are not easily explained with dust depletions because N and O are largely unaffected by dust in the Galactic ISM. The observed Mn/Fe ratios are also difficult to explain with dust depletions because in the Galactic ISM dust depletions cause the gas-phase Mn/Fe ratio to be higher than the solar ratio, opposite to what is observed in damped Lyα galaxies. On the other hand, these ratios are easily understood in terms of the odd-even effect (ie, the odd-Z elements generally show underabundances relative to the even-Z elements of same nucleosynthetic origin at low Fe metallicities) and the different nucleosynthetic origins of these elements (cf. Wheeler et al. 1989). These results strongly indicate that we have observed these galaxies during the epoch when SN II are largely responsible for the enrichment of the interstellar medium in these galaxies, while low mass stars have not had enough time to evolve and to dump their nucleosynthetic products into the interstellar medium through mass loss and SN Ia. Thus the chemical enrichment process in these galaxies should not have proceeded more than 1 Gyr when they were observed.

However, the observed Zn/Fe ratio in damped Lyα galaxies is inconsistent with the above nucleosynthesis interpretation. In Galactic stars, Zn/Fe is found to be solar at all metallicities (cf. Wheeler et al. 1989 and references therein). This difference may suggest that, while the observed relative abundance patterns in damped Lyα galaxies are dominated by the effects of nucleosynthesis, there is some dust depletion effect on top of that. The presence of a small amount of dust in damped Lyα galaxies has been claimed from the reddening of the background quasars (cf. Pei, Fall, & Bechtold 1991). On the other hand, recent theoretical studies indicate that Zn can be produced in large quantities in the neutrino driven winds during SN II explosions (Hoffman et al. 1995; see also Woosley 1995, this volume). Since SN II makes little Fe, a Zn/Fe overabundance may be possible in the ejecta of SN II. The puzzle is then why Zn is observed to track Fe abundance in Galactic stars. If indeed the observed Zn/Fe overabundance in damped Lyα galaxies is caused by depletion of Fe onto dust grains, the [Fe/H] measurements in figure 1(a) will underestimate the true Fe-metallicities by ∼ 0.5 dex (on average).

3. Concluding Remarks

Damped Lyα galaxies provide the unprecedented opportunity to directly probe the chemical enrichment history of galaxies over a large fraction of the Hubble time. Some intriguing results have already emerged from the current study. However, many questions remain, eg, why do the damped Lyα galaxies have so low metallicities compared to the past history of the Milky Way disk, and what are the implications? What is the significance of the large scatter in the measured [Fe/H] at any give redshift? How big a role does dust play in modifying the observed abundances and their interpretations? Some of these issues will be addressed in more details in LSB96.

References

Edvardsson, B., et al. 1993, A&A, 275, 101
Hoffman, R.D. et al. 1995, ApJ, submitted
Lanzetta, K.M., Wolfe, A.M., & Turnshek, D.A. 1995, ApJ, 440, 435
Lu, L., Sargent, W.L.W., & Barlow, T.A. 1996, in preparation (LSB96)
Pei, Y., Fall, S.M., & Bechtold, J. 1991, ApJ, 378, 6
Pettini, M. et al. 1994, ApJ, 426, 79
Pettini, M. et al. 1995, in *QSO Absorption Lines*, ed. G. Meylan (Springer-Verlag), 71
Timmes, F.X., Lauroesch, J.R., & Truran, J.W. 1995, ApJ, 451, 468
Tripp, T.M., Lu, L., & Savage, B.D. 1995, ApJS, in press
Viegas, S. M. 1995, MNRAS, 276, 268
Wheeler, J.C., Sneden, C., & Truran, J.W. Jr. 1989, ARA&A, 27, 279
Wolfe, A.M. 1988, in *QSO Absorption Lines: Probing the Universe*, eds. J.C. Blades, D.A. Turnshek, & C.A. Norman (Cambridge Univ Press), 297

Understanding the Deuterium Abundance: Measurements with the FUSE Satellite

Scott Friedman, Warren Moos, and William Oegerle

Center for Astrophysical Sciences, Johns Hopkins University, Baltimore MD 21218

Donald York

Department of Astronomy and Astrophysics, University of Chicago, Chicago, IL 60637

1. Motivation for Deuterium Measurements

Big Bang nucleosynthesis accounts for the existence of the light elements ^{1}H, ^{2}D, ^{3}He, ^{4}He, and ^{7}Li through nuclear reactions that occurred in the first few minutes after the initial explosion. Two properties of deuterium make it a particularly important species to investigate. The first is that the rate at which D was burned into heavier elements depended strongly on the baryon density at that early epoch. The second is that in the subsequent evolution of the universe there have been no significant production mechanisms for D identified, although it can be easily destroyed in stellar interiors. Therefore, estimates of the primordial deuterium abundance relative to hydrogen give a direct measure of the baryon density in the universe, and measurements of the local abundance, which have been contaminated by stellar processing, give a lower limit to the primordial value. Estimates of the primordial D/H ratio have mostly been determined by extrapolating measurements made in the local interstellar medium and using models of the galactic chemical evolution to remove the effects of astration. However, the correction factor is estimated to be in the range 3-5, but could be as large as 10 (Vangioni-Flam & Audouze 1988). Therefore, progress can be made both by measuring D/H in low metallicity environments, which should approximate more closely the true primordial value, and by measuring D/H in many locations in the Galaxy, so that models of astration can be improved.

2. Previous D/H Measurements

Most measurements of the D/H ratio have been made in clouds of gas in the interstellar medium (ISM) using the *Copernicus* satellite, which operated from 1972-1981. The ultraviolet spectrograph, with a resolution of ~20,000, measured the hydrogen Lyman series absorption lines induced by intervening clouds along lines of sight towards UV bright stars. The deuterium lines are displaced by an apparent velocity of 82 km s^{-1} $\approx$ 0.3Å toward the blue, allowing separate estimates of the D and H column densities, from which the ratio may be computed. The fact that *Copernicus* could measure all members of the Lyman series allowed precise determinations of the column densities over a wide range of D/H values because, for deuterium column densities $N(D) > 10^{15}$ cm^{-2}, at least one transition lies on the linear portion of the curve of growth for T > 6000K.

The limited sensitivity of *Copernicus*, however, permitted such measurements only toward about a dozen stars, the most distant of which is less than 1 kpc. Thus, in general only the local ISM was sampled. This material has undergone significant processing and astration, making extrapolation to a primordial value highly model dependent.

Measurements at lower spectral resolution of only the Lyα profiles have also been made using IUE, toward stars closer than 1 kpc (see e.g., Murthy et al. 1990). The results generally were not as precise as those from *Copernicus*, and frequently provided only limits to the D/H ratio.

More recently HST-GHRS has been used to measure material along the lines of sight to Procyon and Capella (Linsky et al. 1995) and the white dwarf G191-B2B (Lemoine et al. 1995), at distances of 3.5, 12, and 48 pc, respectively. These measurements of the Lyα line required high S/N and very high resolution, R $\approx$ 90,000, and are suitable only for short lines of sight. The abundances were determined by carefully fitting the saturated wings of the HI absorption profiles.

Finally, an entirely new environment has recently been sampled using ground based observatories. Songaila et al. (1994) used the Keck telescope to measure the Lyman profiles toward the QSO 0014+813 (z_{em} = 3.42). This sightline traverses a Lyman limit system which, at z_{abs} = 3.32, is sufficiently redshifted to put the absorption lines in the visible range. Exactly the same system was measured by Carswell et al. (1994). Both groups got the same result, D/H = 2.5 $\times$ 10^{-4}, which is about an order of magnitude greater than the values measured in the local ISM. This material has low metallicity (Chaffee et al. 1985) and is expected to more closely reflect the primordial D/H value without the need for major corrections due to astration. Such a high D/H value implies a very low baryon density, Ω_b = $0.005h^{-2}$, where $h = H_0/100$ km s^{-1} Mpc^{-1}, only about a factor of two greater than the total estimated mass contained in luminous stars and gaseous material.

An evidently contradictory value has emerged from a recent measurement (Tytler et al. 1995) using the Keck telescope, of a cloud (z_{abs} = 3.57) along the sightline to QSO 1937-1009 (z_{em} = 3.78). The preliminary result is D/H $\approx$ 2 $\times$ 10^{-5}, about 10 times less than Songaila et al., and much closer to the local values. Adopting this value implies $\Omega_b \approx 0.017h^{-2}$.

The problem with the ground-based measurements of high redshift systems is that it is difficult to be certain that a weak hydrogen cloud at a relative velocity of -80 km s^{-1} is not mistakenly interpreted as the deuterium line. Tytler et al. recognized this possibility in addressing the large discrepancy with the Songaila result, and conclude "Either their absorption line was H rather than D, or primordial nucleosynthesis was dramatically inhomogeneous." The latter possibility argues strongly for making measurements in a wide variety of environments and redshifts. A complete program will require ground-based echelle spectrographs on the largest telescopes for high z material, and HST and FUSE for low z material. The two main requirements are to observe as many Lyman series lines as possible, at least as high as Lyϵ or Lyζ; and to obtain numerous measurements in each of several well defined environments: Galactic disk, Galactic halo, halos at z < 1 and halos at z > 1. Observations at high z or at

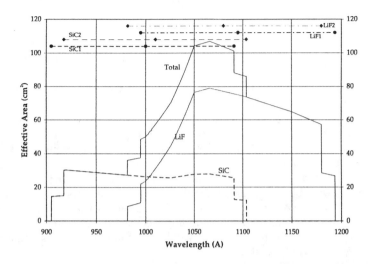

Figure 1. The predicted effective area of the FUSE instrument. Two channels have SiC coated optics and two channels have Al+LiF coated optics, in order to maximize the sensitivity across the full FUSE bandpass. The contributions from these pairs are shown separately and together. The gaps between MCP plates in the detector have been omitted for clarity.

low z in halos offer the chance to relate D/H to heavy element abundances of N and O, which should be anti-correlated with D for consistency with standard views of astration and heavy element production. All low z measurements, on the other hand, offer the chance to define the environment more quantitatively than at high z.

In summary, the situation at high redshift is currently unsettled. Also, within the Galaxy, additional measurements over a much wider range of environments and metallicities will be required to refine the astration corrections.

3. Deuterium Measurements with FUSE

The Far Ultraviolet Spectroscopic Explorer (FUSE) satellite will be the first long duration, high resolution spectroscopy mission since *Copernicus* to cover the Lyman series transitions (excluding Lyα). It consists of a four-channel Rowland spectrograph (Green, Wilkinson, & Friedman 1994) with a resolution of 30,000 and an effective area (Figure 1) that varies between about 30 and 100 cm^2. Its wavelength coverage is 905 – 1195Å, and this entire bandpass is captured simultaneously making for highly efficient, broadband observations.

The photon-counting microchannel plate detectors with delay-line anodes have good sensitivity and very low background levels, giving FUSE a vastly improved faint limit compared to *Copernicus*, which was limited to observations of bright stars within about 1 kpc. In contrast, FUSE will be able to observe stars throughout the Galaxy and, for the first time at this resolution and wavelength, extragalactic objects including many QSOs and AGNs. This will allow probes of more distant material in the Galactic disk, in the halo, and possibly in the halos of other galaxies that fall along the sightlines to bright background sources.

The FUSE Science Team is currently investigating the best candidates to use as background sources for deuterium measurements. For extragalactic sources, which can be used as probes of material in other galaxies and in the halo of the Milky Way, our primary source has been a list of several hundred of the brightest AGNs that have been observed with IUE (Penton & Shull, to be submitted). About 70 such objects have fluxes exceeding 2.5×10^{-14} ergs cm^{-2} s^{-1} Å^{-1} at 1000Å. This limiting flux would give S/N = 20 in each 33mÅ resolution element in about 225,000 seconds.

How many of these sightlines will actually prove to be suitable for deuterium studies is unknown. The intervening clouds must have redshifts $z < 0.28$ or else the most useful DI absorption lines will be shifted out of the FUSE bandpass. The number of Lyman limit systems rapidly decreases at low redshift, but it is expected that there will be at least several with column densities sufficiently great to have measurable deuterium.

For halo studies the requirement is that the clouds have large negative velocities so that the DI lines are shifted away from the absorption lines caused by low velocity galactic hydrogen clouds. Positive velocities are unacceptable unless they are very large, perhaps $V_{LSR} > 200$ km s^{-1}, or else the DI lines will be confused with the galactic HI absorption. The covering factor for H clouds with $|V_{LSR}| > 100$ km s^{-1} down to a limiting column density of $N_H = 7 \times 10^{17}$ cm^{-2}, obtained from 21 cm data, has recently been estimated to be 37% (Murphy, Lockman, & Savage 1995). Over 10% of all extragalactic objects with $f(1000Å) > 2.5 \times 10^{-14}$ ergs cm^{-2} s^{-1} Å^{-1} should meet our requirements. We estimate there are over 150 such sources, some of which are as yet undiscovered but will be found by the time the FUSE mission begins.

Other potential background sources for halo studies include extragalactic HII regions, hot stars in the Magellanic Clouds, and Galactic halo stars. Within the Galaxy a variety of background sources will be used, including hot white dwarfs and early type stars in the plane. Many of the observations, toward both Galactic and extragalactic objects, will be useful for other studies, such as the distribution and dynamics of hot gas, especially OVI, in the Galaxy, and of metal abundances in the interstellar medium.

The vast improvement that FUSE offers will allow a systematic study of astration as a function of environment, which will yield better extrapolations to the primordial D/H value. The extragalactic observations will help tie together the local measurements with the high z results from clouds whose evolutionary histories and metallicities are hard to determine with precision.

4. Status of the FUSE Mission

About one year ago the FUSE project was restructured from a medium-sized Delta-Class Explorer into a Principal Investigator Class mission. The total estimated cost was reduced by about 60%, but the most important science goals and capabilities were retained.

NASA has recently authorized the start of the construction phase of the project. Launch is scheduled for October, 1998. All science planning and mission operations will take place at the Satellite Control Center on the Johns Hopkins University Homewood campus. The mission is designed for a three year lifetime.

A large portion of the observing time will be allocated to a Guest Investigator program, to be administered by the Goddard Space Flight Center. The Call for Proposals for observations will be issued in early 1998. The remaining observing time will be split between Canada and France, our partners who are providing hardware for the instrument, and the FUSE Science Team.

Additional information about the FUSE mission and the science program may be found in the FUSE homepage on the World Wide Web:

$$http://profuse.pha.jhu.edu/fuse.html$$

References

Allen, M.M., Jenkins, E.B., & Snow, T.P. 1992, ApJS, 83, 261

Bruston, P., Audouze, J., Vidal-Madjar, A., and Laurent, C. 1981, ApJ. 243, 161

Carswell et al. 1994, MNRAS, 268, L1

Chaffee, F.H. et al. 1985, ApJ, 292, 362

Green, J.C., Wilkinson, E., & Friedman, S.D. 1994, Proc. SPIE 2283, 12

Lemoine, M. et al. 1995, A&A, in press

Linsky, J.L. et al. 1995, ApJ, 451, 335

Murphy, E.M., Lockman, F.J., & Savage, B.D. 1995, ApJ, 447, 642

Murthy et al. 1990, ApJ, 356, 223

Songaila, A., Cowie, L.L., Hogan, C.J., & Rugers, M. 1994, Nature 368, 599

Tytler, D. et al. 1995, in QSO Absorption Lines : Proceedings of the ESO Workshop, Garching, Germany, ed. G. Meylan: Springer

Vangioni-Flam, E., & Audouze, J. 1988, A&A, 193, 81

The Galactic Center Abundances of Deuterium, Lithium, and Boron

D. A. Lubowich

Department of Physics and Astronomy, Hofstra University, Hempstead, NY 11550

1. Discussion

The abundances of deuterium, lithium, and boron provide important information about big-bang nucleosynthesis, Galactic chemical evolution, stellar evolution, and cosmic-ray spallation reactions. The Galactic Center is the most active and heavily processed region in the Galaxy. Thus the abundances of Li or B which are related to stellar activity should be increased in the Galactic Center, while the abundance of D should be reduced in the Galactic Center unless there are additional sources of D. However D nucleosytheisis mechanisms significantly overproduce Li and B 1000-10000 times unless the D is produced in a region containing only H and He. D, Li, and B have been predicted to be enhanced in the Galactic Center to $D/H = 1 \times 10^{-4}$; $Li/H = 2 \times 10^{-6}$ and $B/H = 2 \times 10^{-6}$ due to AGN activity, cosmic-ray spallation reactions, or gamma-ray induced photodisintegration of C, N, O, or Fe. Based on models of cosmic-ray spallation reactions, Li/H and B/H are proportional to $(O/H)^2$. Any large flux of low-energy cosmic rays will increase the abundances of D, Li, and B by an additional 3-5 times.

Observation of the J = 1-0 and 2-1 lines of DCN telescope in the Galactic Center Sgr A 50 km/s molecular cloud using the NRAO 12m telescope clearly indicates the presence of DCN ($T_r^* = 60$ mK ±mK for the J= 1-0 line) and therefore D (Lubowich, Pasachoff, Balonek, and Tremonti, ApJ, in preparation). When combined with observations of $H^{13}CN$ the Galactic Center $D/H = (2 - 8) \times 10^{-6}$ and is reduced by two to eight times as compared to the ISM $D/H = 1.7 \times 10^{-5}$. Observations of the hyperfine lines of LiI and BI using the NRAO 43M telescope (Lubowich, Turner, and Hobbs, 1996, ApJ, submitted) did not detect either line so that the Galactic Center $N(LiI) < 1.9 \times 10^{16}$ cm^{-2}, Li/H $< 3.9 \times 10^{-8}$, $N(BI) < 2.2 \times 10^{18}$ cm^{-2}, and B/H $< 2.7 \times 10^{-6}$ for the 20 km/s Sgr A molecular cloud with a hydrogen column density $N(H) = 2N(H_2) = 2 \times 10^{-24}$ cm^{-2}. The results for DCN are shown in Figure 1, Figure 2, and Table 1. The Galactic Center abundances of D, Li, and B are shown in Table 2 which includes estimates from nucleosynthesis models (see Reeves, H., 1994, *Rev. Mod. Physics*, 66, 193, for a review of light element nucleosynthesis). There is no evidence of enhanced Li or B in the GC and the D abundance is lower than anywhere else in the Galaxy. However, since astration should significantly reduce the D abundance, the presence of D is *prima facia* evidence that there must be a Galactic source of deuterium.

If the GC Li and B are only produced via cosmic-ray spallation reactions, then the GC time-integrated cosmic-ray flux is less than 40 times the time-integrated cosmic-ray flux in the disk. Thus the Galactic Center has not had an extended period (> 1 Gyr) of AGN activity, a large cosmic-ray flux, nor a large gamma-ray flux and there are no sources of D nucleosynthesis in the Galactic Center. Therefore, the Galactic Center D probably originated in the infall of primordial matter with D/H = 8×10^{-5} implying a baryon density less than the crtitcical density necessary to close the Universe in a flat Einstein de-Sitter Universe.

Acknowledgments. I wish to thank Hofstra University for a Faculty Research and Development Grant.

Table 1. Sgr A 50 km/s Molecular Cloud

Molecule	Transition	Frequency (MHz)	T_r^* (mK)	rms (mK)	Δ V (km/s)
		U72323.9	68	7.1	22.7
		U72355.2	18	7.1	30.6
DCN	1-0	72415.8	61	7.1	30.9
$HC^{13}CCN$	8-7	72475.0	15	7.1	21.9
$HCC^{13}CN$	8-7	72482.2	12	7.1	27.9
$HC^{15}N$	1-0	86055.8	192	27	26.8
SO	2(2)-1(1)	86094.3	226	27	19.8.9
$H^{13}CN$	1-0	86336.4	89	20	20.4
$H^{13}CN$	1-0	86343.1	151	20	28.4
HCN	1-0	88627.0	5408	48	23.3
HCN	1-0	88636.2	4744	48	37.2
HCO+	1-0	89188.5	3070	7.1	25.7
		U 89204.3	532	23	11.7
		U89215.5	250	23	11.7
		U89221.8	130	23	7.8
$HC^{13}CCN$	10-9	90593.9	130	23	8.5
$HCC^{13}CN$	10-9	90602.5	120	23	8.5
HNC	1-0	90660.5	61	2150	20.0
HNC	1-0	90666.7	61	2250	24.7
C_2S	7,7-6,6	90684.3	15	167	12.4
CH_3OH	3(2)-2(2) A	144735.3	50	15	19.2
DCN	2-1	144830.2	39	15	25.2

Table 2. Estimates of Galactic Center Abundances of D, Li, and B

	D/H	Li/H	B/H
Gamma-ray flux (photoerosion)	1×10^{-4}	2×10^{-8}	2×10^{-4}
Cosmic-ray flux to produce GC D/H	1×10^{-6}	2×10^{-6}	2×10^{-8}
100x increase in low-energy CR flux	2×10^{-5}	3×10^{-7}	6×10^{-8}
500x increase in low-energy CR flux	1×10^{-4}	1.5×10^{-6}	3×10^{-7}
AGN activity	7×10^{-4}	3×10^{-7}	3×10^{-7}
Halo spallation model Z=1		3×10^{-7}	3×10^{-8}
Supernova neutrino nucleosynthesis		2×10^{-9}	3×10^{-9}
Cosmic-rays flux to produce [O/H] = .5		3×10^{-8}	6×10^{-9}
Cosmic-ray flux to produce [O/H] = 1		3×10^{-7}	6×10^{-8}
Galactic Center (observations)	$(2-8) \times 10^{-6}$	$< 3.9 \times 10^{-8}$	$< 2.7 \times 10^{-6}$

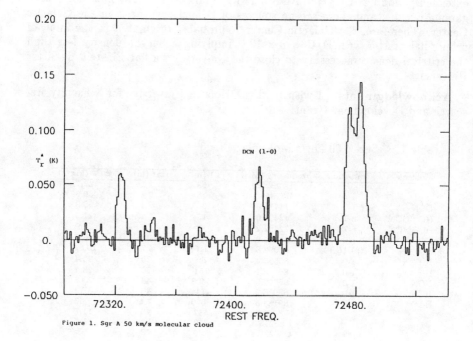

Figure 1. Sgr A 50 km/s molecular cloud

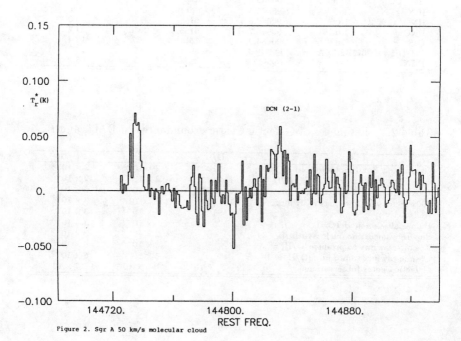

Figure 2. Sgr A 50 km/s molecular cloud

Standard Abundances

Nicolas Grevesse and Arlette Noels
Institut d'Astrophysique, Université de Liège 5, avenue de Cointe, B-4000 Liège, Belgium

A. Jacques Sauval
Observatoire Royal de Belgique, 3, avenue Circulaire, B-1180 Bruxelles, Belgium

Abstract. After a short discussion on the very notion of standard abundances, we present an updated version of the solar chemical composition as derived from the solar photosphere and the meteorites.

1. Introduction

The words Standard Abundances are found everywhere in the astrophysical literature where they now replace Cosmic or Solar Abundances, or Local Galactic Abundances, however keeping the same meaning. Actually this meaning has strong implications which are not always kept in mind by the users.

Heavy elements are produced in different types of stars, evolving with very different lifetimes. A homogeneous increase in Z in the Galaxy would require a similar rate of star formation everywhere, with the same initial mass function and instantaneous recycling of newly-formed elements. It is obvious, both from observations and theory, that there is no single chemical mixture which can be used everywhere in the Galaxy and there is no such idea as a standard chemical composition.

This is however what is generally being done. The solar system chemical composition serves as a standard not only for stars apparently of the same metallicity but also for objects with different Z, the detailed composition being adjusted following the Z's ratio.

The choice of the Sun as a standard is obvious for the following reasons: it is the nearest and best known star, its chemical composition can be derived using various techniques in different outer layers of the star; it has additional reliable indicators like the meteorites and in particular, the CI meteorites, and the planets. It is however questionable whether the Sun is really representative of the mean chemical composition of the Galaxy in the solar environment (see Section 2).

The knowledge of the chemical composition of the different constituents of the Universe is a key data for modelling these objects. The chemical composition is an essential data for testing the nucleosynthesis processes as well as galactic evolution models. It plays also a crucial role in the computation of such important data as the opacities. It is still out of reach to really incorporate in a stellar evolution code, a precise opacity computation taking into account the slightest changes in one or more of the heavy elements, even if the atomic data are now available. So a compromise is adopted which is to compute opacity tables for the solar chemical composition varying only X and Y and adjusting the Z value. This is one of the reasons why the solar chemical composition has become, mostly for pragmatic reasons, the standard chemical composition.

2. Is the Sun a Typical Star?

It has been suggested that the Sun, might be anomalous, i.e. metal rich, and therefore not representative of the local ISM composition at the time of its formation, 4.6 Gyr ago.

The two main reasons are the following:

- The metallicity in the present local ISM region, essentially measured from analyses of the Orion nebula and of nearby B stars, is lower than the value obtained from solar abundances (Gies & Lambert 1992, Cunha & Lambert 1992, Wilson & Rood 1994, Mathis 1996). This is in contradiction with galactic chemical evolution models which predict an increase of metallicity with time. This increase should however vary with the galactocentric position, the highest values being found towards the central regions of the Galaxy.

- The Li abundance measured in the Orion nebula, is very similar to the meteoritic abundance which suggests that it has evolved very little during the last 4.6 Gyr. This observation is also very difficult to reconcile with the theoretical predictions of the variation with time of the Li abundance (Steigman 1993, Cunha et al. 1995).

On the other hand, a very detailed analysis by Edvardsson et al. (1993) of a large number of F and G dwarfs has led to the conclusion that:

- The Sun is indeed a normal star, i.e. it has the same composition as other stars of the same age located at the same galactocentric distance.

- A real dispersion exists among the stars.

3. Standard Abundances from the Sun

The outer layers of the Sun show a very heterogeneous structure in the chromosphere and the corona, overlying a well mixed photosphere where most of the absorption lines are formed, just above the convection zone.

Abundances can be derived from all these regions ranging from the photosphere and sunspots to the chromosphere and corona using classical spectroscopic techniques but also γ-ray spectroscopy. Particle measurements are used to obtain data for the solar wind (SW) and solar energetic particles (SEP).

The most reliable results for most of the chemical elements are without any doubt obtained from analyses of the solar photospheric spectrum for which very high quality data exist from UV to IR. Furthermore, physical processes and physical conditions are rather well known in the photospheric layers. This is not the case for other regions like sunspots, chromosphere or corona, the last two layers being extremely heterogeneous and varying with time. In addition, it is now known that a fractionation process is at work in the solar outer layers: elements with first ionization potential (FIP), lower than about 10 eV are systematically overabundant by a factor of 4.5 in the corona, SW and SEP when compared to photospheric values.

Pioneering works by Payne (1925) and Russell (1929) have shown that the Universe is largely dominated by hydrogen. Russell (1929) succeeded for the first time in deriving the solar abundances of a large number of chemical elements and Russell's mixture has been used for about three decades.

Since then, much progress has been done in the field of solar photospheric abundances. On the one hand, solar photospheric spectra are now available with very high resolution and signal over noise ratio for quite a large range in wavelength, from UV to far IR (for a review, see Kurucz 1995). On the other hand, empirical modelling of the photosphere has now reached a high degree of accuracy (see e.g. Grevesse & Sauval 1994). Last, but not least, accurate atomic data, in particular transition probabilities, have been obtained for quite a number of transitions of solar interest although additional spectroscopic work remains to be done.

The most recent results for solar abundances derived from photospheric spectra are given in Table 1 and comments for a few important elements are to be found in Section 5.

4. Standard Abundances

Other sources of abundances exist in the solar system like planets, comets and meteorites. In the planets, however, elements have either evaporated or fractionated; very few reliable data are available for comets. A very rare class of meteorites, CI meteorites, is known to be representative of the matter from which the solar system formed, 4.6 Gyr ago. Except for the very volatile elements, this class of meteorites has retained all the other elements present in the primitive matter of the solar nebula. It is therefore a very reliable source of standard abundances especially because of the high precision of the measurements: most of the results for CI meteorites are known to within 5 to 10 percent.

In the past decades, there were large and unexplained discrepancies between photospheric and meteoritic results, for quite a large number of elements. These past discrepancies have now gone away, mostly thanks to the increased accuracy of the analyses of the photospheric spectra, essentially due to the use of atomic data of better accuracy. It is now generally accepted that the solar photosphere and the CI meteorites have exactly the same composition, even if some small differences still remain (Anders & Grevesse 1989, Grevesse & Noels 1993a, Palme & Beer 1994).

Since Russell's first analysis, other tables of standard abundances have been published. Most of them are essentially based on a combination of data coming from meteorites as well as from the solar photosphere (Goldschmidt 1937, Suess & Urey 1956, Goldberg et al. 1960, Cameron 1968, 1973, 1982, Ross & Aller 1976, Meyer 1979, 1985, 1989, Anders & Ebihara 1982, Trimble 1975, 1991, Anders & Grevesse 1989, Grevesse et al. 1992, Grevesse & Noels 1993a, Palme & Beer 1994).

We summarize in Table 1 what we believe to be the best values for standard abundances as derived from the photosphere and the meteorites. Abundances are given in the logarithmic scale usually used by astronomers, $A_{e\ell} = \log N_{e\ell}/N_H + 12.0$, where N_i is the abundance by number.

Photospheric abundances are essentially those given in Grevesse et al. (1992) and Grevesse & Noels (1993a), updating results recommended in Anders & Grevesse (1989). Values in parentheses are uncertain and values in brackets are based on other solar or astronomical data. The solar photospheric abundances of S, Sr, La and Ce have been remeasured recently and their values have slightly changed (Delalic et al. 1990, Biémont et al. 1993, Gratton & Sneden 1994).

Meteoritic abundances have been taken from Anders & Grevesse (1989) and Palme & Beer (1994). We took the straight mean between those two tables, the differences being very small for most of the elements. For a few elements however, new measurements led us to adopt revised values. For boron, we took the recent data of Zhai & Shaw (1994) who found a meteoritic abundance of boron about 25 % smaller than the value recommended by Anders & Grevesse (1989). For S, P, Se and Au, we chose the new values obtained by Palme & Beer (1994). The conversion factor from the meteoritic scale, $N_{Si} = 10^6$, to the solar abundance scale, $\log N_H = 12$, has been derived as usual by comparing the solar meteoritic ratio, $R = \log (\text{sol/met})$, for a series of elements which abundances have been accurately measured both in the photosphere and in the meteorites. We now adopt $R = 1.560 \pm 0.013$. The uncertainty in coupling the two scales is thus only of the order of 3 %.

It is still obvious from the results presented in Table 1 that the uncertainties on the photospheric results are much larger than those on the meteorites. Within the uncertainty limits, there is a complete agreement between both sources of solar abundances, except for some elements which photospheric values are still very dubious. We also give in Table 1 the differences between photospheric and meteoritic results.

5. Comments on a Few Elements

5.1. Helium

Helium is a very peculiar element in the sense that its primordial abundance is known with a great accuracy, $Y_p = 0.23 \pm 0.01$ (Y is the usual mass abundance of He; see Wilson & Rood 1994 for detailed references), whereas its solar abundance is unknown. Despite its name and its high abundance, this element is unfortunately undetectable in the photospheric spectrum and in the meteorites. SW and SEP measurements show a very variable but rather low value, with a

Table 1. Element Abundances in the Solar photosphere and in Meteorites

El.	Photosphere	Meteorites	Ph-Met	El.	Photosphere	Meteorites	Ph-Met
01 H	12.00	–	–	42 Mo	1.92 ±0.05	1.97 ±0.02	−0.05
02 He	[10.99 ±0.035]	–	–	44 Ru	1.84 ±0.07	1.83 ±0.04	+0.01
03 Li	1.16 ±0.10	3.31 ±0.04	−2.15	45 Rh	1.12 ±0.12	1.10 ±0.08	+0.02
04 Be	1.15 ±0.10	1.42 ±0.04	−0.27	46 Pd	1.69 ±0.04	1.70 ±0.04	−0.01
05 B	(2.6 ±0.3)	2.79 ±0.05	(−0.19)	47 Ag	(0.94 ±0.25)	1.24 ±0.04	(−0.30)
06 C	8.55 ±0.05	–	–	48 Cd	1.77 ±0.11	1.76 ±0.04	+0.01
07 N	7.97 ±0.07	–	–	49 In	(1.66 ±0.15)	0.82 ±0.04	(+0.84)
08 O	8.87 ±0.07	–	–	50 Sn	2.0 ±(0.3)	2.14 ±0.04	−0.14
09 F	[4.56 ±0.3]	4.48 ±0.06	+0.08	51 Sb	1.0 ±(0.3)	1.03 ±0.04	−0.03
10 Ne	[8.08 ±0.06]	–	–	52 Te	–	2.24 ±0.04	–
11 Na	6.33 ±0.03	6.32 ±0.02	+0.01	53 I	–	1.51 ±0.08	–
12 Mg	7.58 ±0.05	7.58 ±0.01	0.00	54 Xe	–	2.23 ±0.08	–
13 Al	6.47 ±0.07	6.49 ±0.01	−0.02	55 Cs	–	1.13 ±0.02	–
14 Si	7.55 ±0.05	7.56 ±0.01	−0.01	56 Ba	2.13 ±0.05	2.22 ±0.04	−0.09
15 P	5.45 ±(0.04)	5.53 ±0.04	−0.08	57 La	1.17 ±0.07	1.22 ±0.02	−0.05
16 S	7.33 ±0.11	7.20 ±0.04	+0.13	58 Ce	1.58 ±0.09	1.63 ±0.02	−0.05
17 Cl	[5.5 ±0.3]	5.28 ±0.06	0.22	59 Pr	0.71 ±0.08	0.80 ±0.04	−0.09
18 Ar	[6.52 ±0.10]	–	–	60 Nd	1.50 ±0.06	1.49±0.02	+0.01
19 K	5.12 ±0.13	5.13 ±0.02	−0.01	62 Sm	1.01 ± 0.06	0.98 ±0.02	+0.03
20 Ca	6.36 ±0.02	6.35 ±0.01	+0.01	63 Eu	0.51 ±0.08	0.55 ±0.02	−0.04
21 Sc	3.17 ±0.10	3.10 ±0.01	+0.07	64 Gd	1.12 ±0.04	1.09 ±0.02	+0.03
22 Ti	5.02 ±0.06	4.94 ±0.02	+0.08	65 Tb	(−0.1 ±0.3)	0.35 ±0.04	(−0.45)
23 V	4.00 ±0.02	4.02 ±0.02	−0.02	66 Dy	1.14 ±0.08	1.17 ±0.02	−0.03
24 Cr	5.67 ±0.03	5.69 ±0.01	−0.02	67 Ho	(0.26 ±0.16)	0.51 ±0.04	(−0.25)
25 Mn	5.39 ±0.03	5.53 ±0.01	−0.14	68 Er	0.93 ±0.06	0.97 ±0.02	−0.04
26 Fe	7.50 ±0.04	7.50 ±0.01	0.00	69 Tm	(0.00 ±0.15)	0.15 ±0.04	(−0.15)
27 Co	4.92 ±0.04	4.91 ±0.01	+0.01	70 Yb	1.08 ±(0.15)	0.96 ±0.02	+0.12
28 Ni	6.25 ±0.01	6.25 ±0.01	0.00	71 Lu	(0.76 ±0.30)	0.13 ±0.02	(+0.63)
29 Cu	4.21 ±0.04	4.29 ±0.04	−0.08	72 Hf	0.88 ±(0.08)	0.75 ±0.02	+0.13
30 Zn	4.60 ±0.08	4.67 ±0.04	−0.07	73 Ta	–	−0.13 ±0.04	–
31 Ga	2.88 ±(0.10)	3.13 ±0.02	−0.25	74 W	(1.11 ±0.15)	0.69 ±0.03	(+0.42)
32 Ge	3.41 ±0.14	3.63 ±0.04	−0.22	75 Re	–	0.28 ±0.03	–
33 As	–	2.37 ±0.02	–	76 Os	1.45 ±0.10	1.39 ±0.04	+0.06
34 Se	–	3.38 ±0.02	–	77 Ir	1.35 ±(0.10)	1.37 ±0.01	−0.02
35 Br	–	2.63 ±0.04	–	78 Pt	1.8 ±0.3	1.69 ±0.04	+0.11
36 Kr	–	3.23 ±0.07	–	79 Au	(1.01 ±0.15)	0.87 ±0.02	(+0.14)
37 Rb	2.60 ±(0.15)	2.41 ±0.02	+0.19	80 Hg	–	1.17 ±0.08	–
38 Sr	2.97 ±0.07	2.92 ±0.02	+0.05	81 Tl	(0.9 ±0.2)	0.83 ±0.04	(+0.07)
39 Y	2.24 ±0.03	2.23 ±0.02	+0.01	82 Pb	1.95 ±0.08	2.06 ±0.04	−0.11
40 Zr	2.60 ±0.02	2.61 ±0.02	−0.01	83 Bi	–	0.71 ±0.06	–
41 Nb	1.42 ±0.06	1.40 ±0.02	+0.02	90 Th	–	0.09 ±0.02	–
				92 U	(< −0.47)	−0.50 ±0.04	–

ratio N_{He}/N_H of the order of 4 %. Other sources from solar spectroscopy give very uncertain results (see e.g. Laming & Feldman 1994). The giant planets do not help very much as they do show very different and rather small helium contents, whereas the outermost planets are more helium abundant, with large uncertainties however (Grevesse et al. 1992).

The so-called solar helium abundance is therefore derived from calibrations using theoretical stellar evolution models. The most recent calibration, with an adopted value of $Z/X = 0.0244$, from Table 1, leads to a helium mass abundance of $Y = 0.27$. This value is somewhat smaller than the previously found Y value using the same procedure but with an older Z/X ratio of 0.0286 from Anders & Grevesse (1989). In those calibrations, Y is the helium abundance of the nebula from which the solar system formed and the standard theoretical evolutions do not change the Y value in the outer layers during the whole central hydrogen burning phase.

There are now strong indications that this value of Y = 0.27 is too large. On the one hand, the inversion of the observed helioseismic data leads to a value of about 0.23 (Kosovichev et al. 1992). On the other hand, observations from Spacelab recently published (Gabriel et al. 1995), although they are limited to the solar corona, suggest a value of N_{He}/N_H of 0.07 ± 0.011, which means a Y value of about 0.22.

This puzzling difference could be easily explained if helium diffusion has been at work during the 4.6 Gyr of the Sun's evolution. Actually, theoretical evolutions taking helium diffusion into account have shown that a 10 % reduction in the photospheric helium abundance can be expected (Proffitt & Michaud 1991, Christensen-Dalsgaard et al. 1993, Pinsonneault 1995).

5.2. Lithium - Beryllium - Boron

These elements are the only ones for which large differences exist between meteoritic and photospheric abundances. This is easily understandable as those elements are burned at low temperatures; if the convection envelope is deep enough, the surface abundances will be lowered. There are however theoretical uncertainties about the extent of the outer convection zone; here again, the inversion of helioseismic data suggests a convective envelope somewhat deeper than what is found in the models (Christensen-Dalsgaard et al. 1993). Depletion rates to be explained by theoretical models are 140 for Li and 1.9 for Be, the photospheric abundance for B beeing too uncertain to allow giving a depletion rate for B.

5.3. Carbon - Nitrogen - Oxygen

These elements contribute for about 70 % to the metallicity. As they are partly lost in meteorites, the knowledge of their photospheric abundances is of particular interest. A comprehensive discussion is given in Grevesse & Noels (1993a) and Grevesse & Sauval (1994). The uncertainties are still uncomfortably large because of their crucial role in the metallicity. They essentially come from the lack of accuracy in the atomic and molecular data.

5.4. Neon - Argon

The abundances of these two noble gases can only be derived from the coronal spectrum, SW and SEP, which explains the rather large uncertainties given in Table 1. The abundance values quoted in Table 1 are weighted means between SW and SEP values and measurements from impulsive flare spectra (see Grevesse et al. 1992, Grevesse & Noels 1993a).

5.5. Iron

The longstanding puzzling problem of the difference between the photospheric and the meteoritic abundance of iron now seems to be solved. Recent works (Holweger et al. 1990, Holweger et al. 1991, Biémont et al. 1991, Hannaford et al. 1992, Milford et al. 1994, Blackwell et al. 1995a,b, Holweger et al. 1995, O'Mara 1995, Kostik et al. 1996) do show that the abundance derived from Fe II lines nicely agrees with the meteoritic value. These lines are the best indicators of the solar Fe abundance because iron is essentially once ionized in the solar photosphere. Moreover, accurate transition probabilities have recently been determined for some of these lines. A problem still remains with the abundance derived from Fe I lines which shows a dependence on the excitation energy. Low excitation lines lead to a somewhat higher abundance whereas high excitation lines give an abundance in agreement with the meteorites. We do believe this problem has its origin in solar spectroscopy. High excitation Fe I lines are on the whole faint lines in the solar photospheric spectrum whereas low excitation lines are medium strong lines. High excitation lines are also much less sensitive to temperature as well as to possible departures from local thermodynamic equilibrium. Corrections to the low excitation line results, coming from slight temperature modifications, effects of microturbulence, non-LTE effects and collisional broadening effects might explain the difference between the results derived from low excitation and high excitation lines of Fe I.

5.6. Thorium

This element is a radioactive element used as a chonometer for constraining the age of the Galaxy (Butcher 1987). In Table 1, we have not indicated any value for the photospheric abundance of Th for the following reasons. The only line that can be used is a line of Th II at 4019.136 Å. The abundance derived from this line is much larger than the accurately known meteoritic value. Such a discrepancy is unexplainable because Th is a refractory element and its meteoritic abundance is representative of the Th abundance in the original nebula (Anders & Grevesse 1989). It has been shown that the Th II line is blended with a Co I line (Lawler et al. 1990) and also with a V I line (Grevesse & Noels 1993b). As the transition probability of the V I line is still somewhat uncertain, it is impossible to predict its contribution to the Th II line and thus to derive an accurate value for the photospheric abundance of thorium.

6. Conclusions

Much progress has been made during the last two decades in the solar abundance accuracies. They have been due to the availability of high quality spectra covering a large range in wavelength but essentially to definite progress in the accuracy of transition probabilities. The solar photosphere is never *at fault*. Past *errors* have been shown to be due to errors in atomic or molecular data.

The new solar abundances are now in excellent agreement with the meteoritic abundances derived for CI carbonaceous chondrites, the mean difference between photospheric and meteoritic results having vanish to zero. With the results presented in Table 1, i.e. the photospheric data for C, N, O, Ne and Ar, largely lost by the meteorites and the more accurate meteoritic data for the other elements, the classical mass abundances are X = 0.708, Y = 0.275 and Z = 0.017.

Very small differences exist for some elements. They are essentially due to the uncertainties of the photospheric results. Meteoritic data have now reached very high accuracies, e.g. a few percent.

The least well known data are the data for CNO which contribute the most to the metallicity ($\sim$ 70 %) and are largely lost from meteorites. Progress is expected in the near future provided the accuracy of molecular data needed to interpret the best solar indicators of the abundances of C, N and O is improved.

Progress is also to be expected concerning a more realistic description of the heterogeneous outer solar layers through hydrodynamical modelling of the matter motions just above the solar convection zone.

Diffusion seems to be at work in the solar outer layers. The present day solar photospheric He content which we cannot unfortunately measure directly is about 10 % smaller than it was when the Sun was born. Such an effect is not seen in other elements because photospheric and meteoritic abundances are in very good agreement.

Even if the Sun is not a standard or typical star, it is and will remain a unique source of chemical element abundances because it is the best known star to which other stars are compared.

References

Anders, E., & Ebihara, M. 1982, Geochim. Cosmochim. Acta, 46, 2363
Anders, E., & Grevesse, N. 1989, Geochim. Cosmochim. Acta, 53, 197
Biémont, E., Baudoux, M., Kurucz, R. L., Ansbacher, W., & Pinnington, E. H. 1991, A&A, 249, 539
Biémont, E., Quinet, P., & Zeippen, C. J. 1993, A&AS, 102, 435
Blackwell, D. E., Lynas-Gray, A. E., & Smith, G. 1995a, A&A, 296, 217
Blackwell, D. E., Smith, G., & Lynas-Gray, A. E. 1995b, A&A, 303, 575
Butcher, H. R. 1987, Nature, 328, 127
Cameron, A. G. W. 1968, in Origin and Distribution of the Elements, L.H. Ahrens, Pergamon Press, 125
Cameron, A. G. W. 1973, Space Sci.Rev., 15, 121
Cameron, A. G. W. 1982, in Essays in Nuclear Astrophysics, C.A. Barnes, D.D. Clayton & D.N. Schramm, Cambridge University Press, 23
Christensen-Dalsgaard, J., Proffitt, C. R., & Thompson, M. J. 1993, ApJ, 403, L75
Cunha, K., & Lambert, D. L. 1992, ApJ, 399, 586
Cunha, K., Smith, V. V., & Lambert, D. L. 1995, ApJ, 452, 634
Delalic, Z., Erman, P., & Källne, E. 1990, Physica Scripta, 42, 540

Edvardsson, B., Andersen, J., Gustafsson, B., Lambert, D. L., Nissen, P. E., & Tomkin, J. 1993, A&A, 275, 101
Gabriel, A. H., Culhane, J. L., Patchett, B. E., Breevelt, E. R., Lang, J., Parkinson, J. H., Payne, J., & Norman, K. 1995, Advances in Space Research, 15, 63
Gies, D. R., & Lambert, D. L. 1992, ApJ, 387, 673
Goldberg, L., Müller, E. A., & Aller, L. H. 1960, ApJS, 45, 1
Goldschmidt, V. M. 1937, Skr. Nor. Vidensk. Akad. Oslo I. Mat.-Naturv. Kl. No 4
Gratton, R. G., & Sneden, C. 1994, A&A, 287, 927
Grevesse, N., & Noels, A. 1993a, in Origin and Evolution of the Elements, N. Prantzos, E. Vangioni-Flam & M. Cassé, Cambridge University Press, 15
Grevesse, N., & Noels, A. 1993b, Physica Scripta, T 47, 133
Grevesse, N., & Sauval, A. J. 1994, in Molecular in the Stellar Environment, U.G. Jørgensen, Lecture Notes in Physics 428, Springer-Verlag, 196
Grevesse, N., Noels, A., & Sauval, A. J. 1992, in Proceedings of the First SOHO Workshop, ESA SP-348, 305
Hannaford, P., Lowe, R. M., Grevesse, N., & Noels A. 1992, A&A, 259, 301
Holweger, H., Heise, C., & Kock, M. 1990, A&A, 232, 510
Holweger, H., Bard, A., Kock, A, & Kock, M. 1991, A&A, 249, 545
Holweger, H., Kock, M., & Bard, A. 1995, A&A, 296, 233
Kosovichev, A. G., Christensen-Dalsgaard, J., Däppen, W., Dziembowski, W. A., Gough, D. O., & Thompson, M. J. 1992, MNRAS, 259, 536
Kostik, R. I., Shchukina, N. G., & Rutten, R. J. 1996, A&A, in press
Kurucz, R. L. 1995, in Laboratory and Astronomical High Resolution Spectra, A.J. Sauval, R. Blomme & N. Grevesse, ASP Conf. Ser, Vol. 81, 17
Laming, J. M., & Feldman, U. 1994, ApJ, 426, 414
Lawler, J. E., Whaling, W., & Grevesse, N. 1990, Nature, 346, 635
Mathis, J. 1996, this volume
Meyer, J. P. 1979, in Les Eléments et leurs Isotopes dans l'Univers, Institut d'Astrophysique, Université de Liège, 153
Meyer, J. P. 1985, ApJS, 57, 173
Meyer, J. P. 1989, in Cosmic Abundances of Matter, C.J. Waddington, American Institute of Physics, 245
Milford, P. N., O'Mara, B. J., & Ross, J. E. 1994, A&A, 292, 276
O'Mara, B. J. 1995, private communication
Palme, H., & Beer, H. 1994, in Landolt-Börnstein, Group VI, Astron. Astrophys. Vol. 3, Extension and Supplement to Vol. 2, Subvol. a, H.H. Voigt, Springer-Verlag, 196
Payne, C. H. 1925, Stellar Atmospheres, Harvard Obs. Monographs
Pinsonneault, M. 1995, in Stellar Evolution : What Should Be Done, A. Noels, D. Fraipont-Caro, M. Gabriel, N. Grevesse & P. Demarque, Institut d'Astrophysique, Université de Liège

Proffitt, C. R., & Michaud, G. 1991, ApJ, 380, 238
Ross, J. E., & Aller, L. H. 1976, Science, 191, 1223
Russell, H. N. 1929, ApJ, 70, 11
Steigman, G. 1993, ApJ, 413, L73
Suess, H. E. P., & Urey, H. C. 1956, Rev. Mod. Physics, 28, 53
Trimble, V. 1975, Rev. Mod. Physics, 47, 877
Trimble V. 1991, The Astron. Astrophys. Rev., 3, 1
Wilson, T. L., & Rood, R. T. 1994, ARA&A, 32, 1
Zhai, M., & Shaw, D. M. 1994, Meteoritics, 29, 607

Solar Coronal Abundance Anomalies

Jean-Paul Meyer

Service d'Astrophysique, CEA,DSM,DAPNIA,
Centre d'Etudes de Saclay, 91191 Gif-sur-Yvette, France

Abstract. An up-to-date account is given of the observed elemental abundance anomalies in the solar corona and in the heliosphere, as compared to photospheric composition (First Ionization Potential, "FIP" bias, which implies an ion-neutral fractionation in a cool gas). A large variety of sites on the solar surface have now been investigated. Recent observations suggest that the ion-neutral fractionation process and the injection of FIP-biased material into the corona is related to solar activity, and plausibly to the evaporation of cool chromospheric gas at the footpoints of hot coronal loops. This also ties in with the problem of the origin of the slow Solar Wind. Possible links with stellar and galactic cosmic ray physics are briefly mentioned.

1. Introduction

It is now well established that the elemental composition of the solar outer atmosphere and interplanetary medium largely differs from that of the solar photosphere. Heavy elements with First Ionization Potential (FIP) $\lesssim 10$ eV ("low-FIP" elements) tend to be overabundant by factors on the order of ~ 4 relative to those with FIP $\gtrsim 10$ eV ("high-FIP" elements) : there exists a "FIP-bias" (Fig. 1). The status of absolute abundances, relative to H, is not entirely settled (§ 6). Here, I will not cover in detail the entire body of observations which led to this conclusion, for which I refer the reader to earlier reviews by Feldman (1992), Meyer (1991; 1993a or b $\equiv$ "Paper I"), and Saba (1995), in which the relevant references will be found. I will, rather, just give a broad overview of these observations, point out the recent updates, and stress the new viewpoint that the FIP-bias is related to solar activity, and probably to chromospheric evaporation at the footpoints of hot coronal loops. The hurried reader may jump to § 7, where the main observational trends are summarized, and their significance discussed.

2. Basics for any scenario for the FIP-fractionation. Possible sites.

To discuss the composition observations in an appropriate framework, it is preferable to first recall that a FIP-bias inevitably implies an ion-neutral fractionation. This can take place only in a medium in which neutrals and first ions *exist*, *i.e.* certainly in a gas at a temperature $T < 10000$ K, and more precisely around $T \sim 7000$ K, if UV photoionization is not dominant. This implies that the FIP-fractionation takes place in the chromospheric material, possibly in upward jets of chromospheric gas, or simply in the chromospheric temperature plateau around $T \sim 7000$ K. In the chromospheric plateau, indeed, the low-FIP elements are predominantly ionized, while the high-FIP ones are essentially neutral (with a, possibly crucial, caveat for C, which is only ~ 50 % neutral, Paper I). Note also that, in this context, there may be a simple physical reason to the value of ~ 10 eV, right around the Lyα energy, for the borderline between comparatively enhanced low-FIP and unenhanced high-FIP elements : the chromospheric temperature plateau is, indeed believed to result from a thermostating controlled by H excitation to its $n = 2$ level (Athay 1981; Paper I). A FIP-fractionation in specific locations of the photosphere itself cannot be formally ruled out, but seems unlikely. But any scenario of FIP-fractionation at $T > 10000$ K, in the Transition Region or in the corona or in the Solar Wind (SW) acceleration region can be strictly ruled out.

Another essential point : since the *bulk of* the coronal and SW gas are, over large regions of the Sun, affected by the FIP-bias, the *bulk of* the gas transferred from chromosphere into corona and SW in these regions must undergo the ion-neutral fractionation. This imposes strict constraints to models of gas transfer from chromosphere to corona and to SW, *i.e.* to the origin of the coronal and SW gases.

A number of scenarios for the FIP-fractionation have been proposed, which have been reviewed by Hénoux (1995) (see also Marsch et al. 1995; Vauclair 1995). But the observational constraints to models are evolving rapidly. I will, therefore, hardly discuss specific models here. I will, rather, point out the observational constraints to models for FIP-fractionation and gas transfer, and especially recent observations allowing a detailed mapping of the amplitude of the FIP-bias over the solar surface.

I will discuss, in turn, the observations relevant to closed- and to open-field regions on the solar surface. I will, further, distinguish between the compositions of the escaping, SW, material, and of the coronal material on the solar surface itself.

3. Closed-field region FIP-biases

"Closed-field regions" on the solar surface are dominated by closed loops at altitudes up to $\sim 0.20 R_\odot$. They tend to dominate in the equatorial regions, extending up to latitudes which increase with solar activity over the 11-year cycle. Closed-field regions are the regions in which most of the solar activity takes place. They are the source of the *slow*-SW, and composition observations will tell us a lot about the origin of this slow-SW (§ 7.2.2). Solar Energetic Particles (SEP) events, both gradual and impulsive, are all, one way or another, related with solar activity in closed-field regions (§ 3.1 and 3.2.4).

3.1. Closed-field region FIP-biases : The escaping material ; The basic composition pattern

For more than a decade, large scale average compositions of the material escaping the corona have been provided by near Earth, in-situ observations of the slow-SW itself, as well as of *gradual* event SEP's ("gradual-SEP"), which are accelerated out of this very SW throughout the interplanetary medium, and essentially reflect this SW composition. Historically, these observations have been the first to provide decisive evidence for FIP-related deviations relative to photospheric composition in the solar environment, which could obviously not be attributed to the SW or SEP acceleration processes, since these take place at temperatures on the order of 1 MK or more. So, the composition of this escaping material should reflect that of the *source* of the slow-SW, somewhere in the corona (other types of selection effects do not seem important).

Actually, the gradual-SEP data currently yield the most complete determinations of the slow-SW composition : they are rather straightforward [1] and cover most significant elements up to Ni, so that they have allowed an excellent determination of the entire FIP-bias pattern originally present in the slow-SW. The combined slow-SW and gradual-SEP data, presented in Fig. 1, show two well defined "plateaus" for the relative abundances of the low- and the high-FIP heavy elements, the low-FIP elements being enhanced relative to the high-FIP ones by a pretty stable "FIP-bias factor" of $\sim 4.5 \pm 1$, which is *always* present over closed-field regions (Paper I; Geiss et al. 1995; von Steiger et al. 1995; Garrard & Stone 1993; Reames 1995) [2].

He is found depleted by a factor of ~ 2 relative to heavier high-FIP elements (see § 5). The absolute normalization to H of these *relative* heavy element abundances will be discussed in § 6.

3.2. Closed-field region FIP-biases : Mapping specific sites on the solar surface

It is obvious that only detailed observations of specific features on the very surface of the Sun may, in the long run, allow us to pinpoint the sites, hence to determine the mechanisms for the FIP-fractionation. Such observations are currently developing at a rapid pace. We now have a large body of EUV and X-ray spectroscopic observations of a few key low-FIP (*e.g.* Mg, Si, Ca, Fe) and high-FIP (*e.g.* O, Ne) elements, largely thanks to instruments on board rockets and spacecraft such as, *e.g.* SMM, Yohkoh and soon SOHO (ref. in Paper I). Careful reexamination of earlier data, such as the superb *Skylab* spectroheliograms, has also proven extremely powerful : in particular, a group of neighboring Mg VI and Ne VI lines around 400 Å, all formed around ~ 0.4 MK with very similar emissivity profiles, has been very precious in mapping the low-FIP/high-FIP Mg/Ne ratio in specific solar features (*e.g.* Feldman et al. 1987; Feldman 1992; Sheeley 1995a). Additional information is provided by flare γ-ray line observations (Share & Murphy 1995; Ramaty et al. 1995), and by *impulsive* event Solar Energetic Particles accelerated in the immediate flare environment, which get collected near Earth ("impulsive-SEP"; Reames et al. 1994).

Virtually all these data (except γ-ray lines) refer to media with temperatures between ~ 30000 K and ~ 30 MK, hence well *beyond* the ~ 7000 K conceivable

---

[1] All particles are observed, independent of their charge state ; the charge-to-mass dependent biases are well mastered (Meyer 1991; Paper I; Garrard & Stone 1993; Reames 1995).

[2] This two well defined "plateaus" structure contrasts with the results of a number of spectroscopic observations of the corona. This will be discussed in § 5.

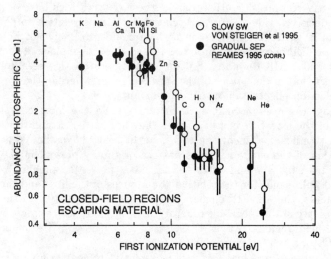

Figure 1. The basic FIP-bias pattern of overabundances relative to photosphere, as observed in the material escaping closed-field regions (§ 3.1). Slow-SW data by von Steiger et al. (1995), and gradual-SEP data by Reames (1995) (roughly corrected for residual charge-to-mass ratio biases, for both heavies and H); the Garrard & Stone (1993) SEP data yield a very similar picture. Normalized to O. The errors on the reference photospheric abundances have been folded into the error bars; they are, in particular, dominant for Ne and Ar.

sites of ion-neutral fractionation, where ions and neutrals coexist. Note also that, due to the higher luminosity of Active Regions and flares, most of the available data have long pertained to these sites; this may have biased our conception of the composition of the entire solar atmosphere.

3.2.1 Orientation : "Quiet Sun" and "Active Sun" (large, hot coronal loops)

Within closed-field regions, it is useful to distinguish the *"Quiet Sun"* and the *"Active Sun"* loops. The coronal magnetic field topology (which controls the gas topology) is essentially controlled by the geometry of the emerging photospheric fields.

When the photospheric field is very disordered on small scales, only small loops can develop above the chromosphere (~ 5000 km), whose footpoints follow the chromospheric supergranulation network. The field appears everywhere *bipolar*, even on small scales. These small loops reach temperatures of up to ~ 0.7 MK only. Since only small amounts of magnetic energy can be stored in each loop, large energy releases cannot take place, and these regions are denoted *"Quiet Sun"*.

By contrast, more ordered regions with consistent photospheric field polarity over significant areas allow larger coronal loops ($\sim 10^4$ to 10^5 km) to develop and connect regions of opposite polarities, with loop temperatures of ~ 1 MK or more. These *unipolar* regions where such ordered fields and associated *large coronal loops* are predominant are denoted *"Active Sun"*, since they are occasionally the site of large magnetic energy releases : the more specific *"Active Regions"* (AR) at ~ 3 MK, and the flares up to ~ 30 MK.

A particularly clear illustration of these two types of regions has been provided by Feldman & Laming (1994). Fig. 2, adapted from their paper, shows conspicuously that *(i)* in the Ne VII line, formed at ~ 0.5 MK, only small, low lying features are visible, essentially footpoints of small loops which trace the chromospheric network, and *(ii)* in the Mg IX line, formed at ~ 1 MK, only larger coronal loop systems are being observed.

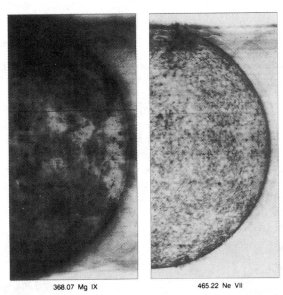

368.07 Mg IX   465.22 Ne VII

Figure 2.   Images of the Sun : *(right)* in a Ne VII line formed at ~ 0.5 MK, showing the cool, low lying loop system tracing the chromospheric network ("Quiet Sun"), and *(left)* in a Mg IX line formed at ~ 1 MK, showing the hotter, large coronal loops ("Active Sun") (§ 3.2.1). The North pole is to the right. While large Mg IX coronal loops are largely absent from the polar coronal hole region, the Ne VII low lying loop system seems similar all over the Sun (§ 4). Adapted from Feldman & Laming (1994).

3.2.2 FIP-bias and local magnetic field topology in closed-field regions. General

Based on the study of a few specific, brighter solar features (erupting prominence, impulsive and gradual flares, AR's), using both a DEM analysis and *Skylab* images in the ~ 400 Å Mg VI and Ne VI lines formed at ~ 0.4 MK, U. Feldman and K. Widing have come to the conclusion that low lying objects, tightly contained by a compact field, have photospheric composition, while larger scale, looser field regions are FIP-biased (ref. in Feldman 1992 and Paper I; Widing & Feldman 1993). This has even suggested that the amplitude of the

FIP-bias might be essentially controlled by the degree of opening of the field, an hypothesis that will be discussed later (§ 7.2.3).

More generally, as first noted by N. Sheeley in the 1970's and published in 1995 (Sheeley 1995a), the *Skylab* images show a very remarkable and intriguing behavior of the $\sim$ 400 Å Mg VI and Ne VI lines over the entire Sun ! Whenever the Mg VI/Ne VI line ratio is observed in a region of locally *bipolar* field, it is found comparatively *low*, consistent with a photospheric composition ; bipolar field regions imply small scale loops. By contrast, whenever a locally *unipolar* field is being observed, Mg VI/Ne VI is comparatively *high*, FIP-enhanced relative to photospheric composition and roughly consistent with SW composition ; the observed unipolar field regions are cooler footpoints of large, hot coronal loop. This is illustrated in Fig. 3, adapted from Sheeley (1995a).

So, all over the solar surface, bipolar field features at $\sim$ 0.4 MK have kept a non-FIP-biased, photospheric composition. What is the nature of this observed non-FIP-biased material ? Part of it is newly emerging, heated up photospheric material, associated with newly emerging field in AR's, which have, at least for the time being, kept their photospheric composition (Sheeley 1995a; Paper I). But photospheric Mg VI/Ne VI ratios are also systematically observed in bipolar field features *not associated with any rise of gas* (*e.g.* chromospheric flare ribbons, compact flares, erupting prominences). Actually, whenever Quiet Sun, chromospheric network associated gas happens to be lit up in Mg VI and Ne VI lines, its composition is photospheric (Sheeley 1995b). This suggests that the entire Quiet Sun material, well above the $\sim$ 7000 K lower chromosphere, may have kept its photospheric composition.

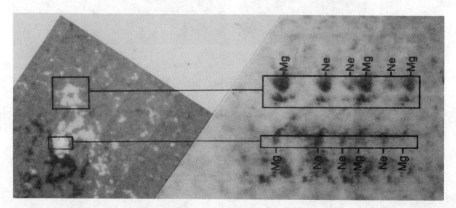

Figure 3. A conspicuous example of the correlation between the magnetic field topology and the $\sim$ 400 Å Mg VI and Ne VI line ratios in the *Skylab* spectroheliograms (§ 3.2.2). Left : magnetogram ; black and white are opposite polarities. Right : the same features in Mg VI and Ne VI lines, dispersed horizontally. As compared to the Ne lines, the Mg lines are much more intense in the upper, magnetically unipolar feature, which is the footpoint of a large coronal loop (SW-like composition), than in the lower, small scale magnetically bipolar feature (near photospheric composition). Adapted from Sheeley (1995a).

3.2.3 Average FIP-biases over Quiet Sun and larger coronal loops

Solar limb *Skylab* images of Quiet Sun regions, taken in Mg VI and Ne VI lines at 0.4 MK which are predominantly emitted by low lying, cool loops, have first shown that the average Quiet Sun is, at most, only weakly FIP-biased, by a factor of $\lesssim 2$ (Feldman & Widing 1993).

A very precious new piece of information has been recently provided by a global study of the entire disk by Laming et al. (1995) ("the Sun as a star", actually performed with the purpose of interpreting observations of *stellar* coronae, § 7.3). Its, at first strange, outcome is that the entire Sun appears roughly non-FIP-biased when observed with lines forming below ~ 0.9 MK, but does appear FIP-biased when observed with lines forming above ~ 0.9 MK. On average on the disk, lines at $T < 0.9$ MK originate mainly in the small Quiet Sun loops attached to the chromospheric network (§ 3.2.1.).

This observation, together with those of the preceding section, implies that *the Quiet Sun, chromospheric network associated, $T \lesssim 0.7$ MK small loop system has a roughly photospheric composition* ; only the larger true "coronal" loops above $T \sim 1$ MK are FIP-biased. The implications of these findings will be discussed in § 7.2.1.

3.2.4 FIP-biases in full-fledged Active Regions and flares

Due to the higher luminosity of these sites, the vast majority of the EUV and X-ray spectroscopic observations have actually dealt with AR's or flares. They are discussed in detail in Feldman (1992) and in Paper I, where the earlier references will be found ; for more recent work, see Saba & Strong (1993), Sterling et al. (1993), Fludra et al. (1993), Widing & Feldman (1993, 1995), Waljeski et al. (1994), McKenzie & Feldman (1994), Phillips et al. (1994, 1995), Athay (1994), Monsignori Fossi et al. (1994), Schmelz (1995), Fludra & Schmelz (1995), Sheeley (1995a), Antonucci & Martin (1995). The problems involved in determining these abundances have been further discussed in depth by Mason & Monsignori Fossi (1994), Mason (1995), and Saba (1995).

As discussed in § 3.2.2, essentially photospheric compositions are found in low-lying, compact AR's or flares, while large AR's, both quiescent and flaring, typically show the same low-FIP element enhancements, by factors of ~ 4.5, as the escaping slow-SW material (Feldman 1992; Paper I; Sheeley 1995a). Impulsive-SEP data and γ-ray line data, both relevant to the flaring AR environment, yield the same low-FIP element enhancements (Reames et al. 1994; Ramaty et al 1995). But significantly higher enhancements are virtually *never* found in closed-field AR's.

A time evolution of the composition is actually seen within each developing AR. Each AR starts with emerging small, compact loops with photospheric composition, with evolve into large coronal loops with SW-type composition over time scales of $\sim$ a day (Fig. 4). Further, in the center of well formed AR's, newly emerging loops with photospheric composition are observed at low altitude, under the canopy formed by the large loops with FIP-biased composition (Sheeley 1995a; Paper I). Note that this suggests that the observed correlation of the FIP-bias with the opening of the field (§ 3.2.2) may, in the case of AR's, just reflect a time evolution. The implications of these findings will be discussed in § 7.2.1 and 7.2.3.

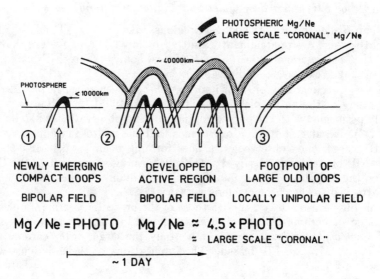

Figure 4. Time evolution of the composition of an Active Region, as discussed in § 3.2.4. From Paper I.

3.2.5 FIP-biases over Sunspots

The situation of the FIP-bias in sunspots, while not yet entirely clarified, may become a source of precious clues (Feldman 1992; Paper I; Athay 1994; Sheeley 1995a). Sunspots are located in very active regions of the Sun. Over many of them, only FIP-biased material is being observed. But in the environment of some sunspots, apparently those which are in contact with a magnetically neutral line (Sheeley 1995a), photospheric-type material is being observed in the umbra, in regions within the dominant sunspot field polarity, but near the neutral line. By contrast, FIP-biased material is found further out, in the penumbra and in neighboring plages (see, however, Athay 1994), where the field is associated with previously emerged flux, and/or connected with neighboring coronal loops (Sheeley 1995a) ; according to Athay (1994), this is particularly the case in regions of opposite polarity. The possible implications of these findings will be discussed in § 7.2.4.

4. Open-field region FIP-biases

In *"Open-field regions"* on the solar surface, the field opens up directly into the interplanetary medium, right above the low lying, chromospheric network associated loop system. This network-associated loop system is about the same as in Quiet Sun areas (§ 3.2.1; Fig. 2). But there are essentially no large, hot coronal loops at higher altitudes. The freely escaping coronal gas is comparatively cool, hence less luminous. These regions, therefore denoted *"Coronal Holes"*, dominate in particular over polar regions. They are the source of the *fast*-SW. That the fast-SW originates in a cooler corona is evidenced by its observed cooler

O^{7+}/O^{6+} freezing temperature, which lies around ~ 1.2 MK, as compared to ~ 1.6 MK in the slow-SW (*e.g.* Geiss et al. 1995).

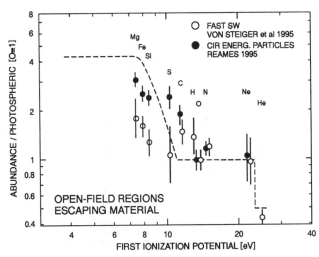

Figure 5. The FIP-bias pattern observed in the material escaping open-field regions (§ 4.1), plotted as in Fig. 1. Fast-SW data by von Steiger et al. (1995), and CIR energetic particle data by Reames (1995) (H and He are not plotted, since they are contaminated by pick-up ions). The closed-field region pattern has been sketched as a dashed line for comparison.

4.1. Open-field region FIP-biases : The escaping material

Earlier indications that the fast-SW is much less FIP-biased than the slow one (Paper I) have been recently substantiated by the beautiful Ulysses spacecraft data, as illustrated in Fig. 5 (Geiss et al. 1995; von Steiger et al. 1995). The typical coronal hole SW FIP-biases seem to lie around a factor of ~ 1.5 only.

An impressive correlation is actually found between the SW low-FIP/ high-FIP Mg/O ratio and the SW speed and, even more tightly, with the O^{7+}/O^{6+} freezing temperature, as shown in Fig. 6. This suggests some correlation between the physical conditions in the $\sim 10^4$ K chromosphere, where the FIP-fractionation is bound to take place, and in the $\sim 10^6$ K corona (Geiss et al. 1995).

Further, energetic particles *accelerated out of the fast-SW* in Corotational Interaction Regions (CIR) some ~ 3 a.u. away from the Sun indicate a significant FIP-bias by a factor of ~ 2.5, but again definitely lower than that found in material escaping from closed-field regions (Fig. 5) (Paper I; Reames 1995 and ref. therein).

In both the fast-SW and the CIR energetic particle data, C (FIP = 11.3 eV) is enhanced relative to O(FIP= 13.6 eV), *i.e.* C seems to behave as a low-FIP element, in sharp contrast with the very solid data relevant to closed-field regions (gradual- and impulsive-SEP's, slow-SW; Fig. 1) (see § 5).

In spite of the weaker general FIP-bias, He in the fast-SW is at least as deficient, relative to heavier high-FIP elements, as in the material escaping closed-field regions (Fig. 5 vs. Fig. 1) (see § 5).

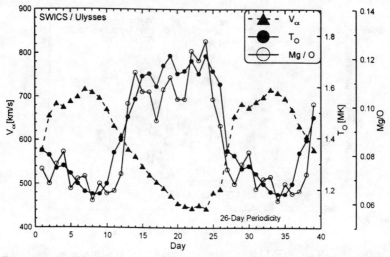

Figure 6. The impressive correlation in time of the SW low-FIP/ high-FIP Mg/O ratio with the SW speed and, even more tightly, with the coronal O^{7+}/O^{6+} freezing temperature (§ 4.1). From Geiss et al. (1995).

4.2. Open-field region FIP-biases : Mapping specific sites on the solar surface

4.2.1 Average open-field region FIP-bias near Sun

Solar limb *Skylab* Mg VI-Ne VI images of coronal hole regions yield FIP-bias factors of $\sim$ 2 to 2.5, which should be largely relevant to the low lying, chromospheric network associated, cool loop system (§ 3.2.1; Feldman & Widing 1993). This, comparatively weak, but significant, FIP-bias is comparable to those observed in the material escaping coronal holes. The significance of this similarity is not obvious (§ 7.2.3).

4.2.2 Polar plumes and other very open, diverging field features

Within coronal holes, radial structures of $\sim$ 4-fold denser material, but lower bulk radial velocity, are present : the *"Polar Plumes"*, which appear mainly, but not only, in the polar holes. Polar plumes appear associated with active, EUV-enhanced, small bipoles in the network at their base. They may result from the reconnection of the active bipole field with with a neighboring unipolar, open flux concentration of the ambient coronal hole field (Widing & Feldman 1992; Wang & Sheeley 1995 and ref. therein). Note that polar plumes are believed to contribute only a minor fraction to the total coronal hole SW flux.

In sharp contrast with the weak bulk coronal hole FIP-biases, very large low-FIP element enhancements, up to factors of $\sim$ 10 have been observed in polar plumes ! These are significantly larger than those observed in closed-field regions (§ 3). Further, even larger FIP-biases, with enhancement factors of ~ 15 have been found in other very open, diverging field regions outside coronal holes, such as active regions connected with a widely open field feature (Paper I ; Widing & Feldman 1992, 1993) !

5. More on the shape of the FIP-bias pattern

Very-low-FIP elements. The "nominal" FIP-bias pattern, as yielded by the material escaping closed-field regions (slow-SW, gradual-SEP's) shows two well defined abundance plateaus for all heavy elements with FIP's between ~ 4 and 8 eV and ~ 11 and 22 eV, respectively (§ 3.1; Fig. 1). In particular the SEP data show *no* significant enhancement of very-low-FIP K, Na, Al, Ca (FIP ~ 4 to 6 eV) relative to low-FIP Mg, Si, Fe (~ 7.5 to 8 eV) (less than a factor of ~ 1.2 ; Garrard & Stone 1993; Reames 1995).

This contrasts with a number of spectroscopic observations of the corona, which find extra very-low-FIP element enhancements by factors of up to ~ 2.5 relative to Mg, Si, Fe (*e.g.* Feldman 1993 and ref. therein; McKenzie & Feldman 1994; Monsignori Fossi et al. 1994; Fludra & Schmelz 1995; Antonucci & Martin 1995). If real, such an extra enhancement of very-low-FIP elements would suggest an ion-neutral fractionation taking place, not only in a ~ 7000 K chromospheric-plateau-type gas, but also in cooler gases with temperatures extending down to say, the temperature minimum at ~ 4000 K (or even cooler in a sunspot environment). In such cooler gases, Mg, Si and Fe could be partly neutral, and thus behave, to some degree, as high-FIP elements.

Carbon. In the coronal hole escaping material, C is observed to behave as a low-FIP element (von Steiger et al. 1995; Reames 1995; § 4.1; Fig. 5). This suggests that the FIP-fractionation takes place at a slightly higher temperature in coronal holes than in closed-field regions.

Neon. As discussed in Paper I, the photospheric Ne/O ratio $\sim 0.16 \pm 0.05$ is reliably determined, both from very rich non-solar HII region and hot star data, and from solar EUV and X-ray observations of compact features with photospheric Mg/O ratio (Feldman 1992; Paper I; Widing & Feldman 1995). Very similar values of Ne/O ~ 0.14 to 0.20 are found in *all* the escaping material (Fig. 1 and 5; § 3.1 and 4.1), and in most EUV and X-ray FIP-biased coronal material observations (Feldman 1992; Paper I; Waljeski et al. 1994; Widing & Feldman 1995; and ref. below), indicating that the high-FIP element plateau is, indeed, very close to flat up to Ne. Recent γ-ray line data on low altitude, FIP-biased material in a number of flares systematically yield somewhat higher Ne/O ratios ~ 0.25 (Ramaty et al. 1995) ; it is not certain that this difference is beyond the errors on the various types of determinations.

But, on some occasions, Ne/O has been found much higher, ~ 0.35, in coronal material ! This seems to happen in non-flaring as well as in flaring coronal features, and especially in post-flare loops (Paper I; Saba & Strong 1993; Fludra & Schmelz 1995; Schmelz 1995). This can in no case be understood in terms of a FIP-dependent fractionation, *i.e.* of an ion-neutral separation in a medium whose ionization is controlled thermally or by lower energy UV photons. A tentative explanation by Shemi (Paper I), in terms of an X-ray photoionization of the site of ion-neutral fractionation, might possibly explain such high ratios, as long as they are only occasional (Share & Murphy 1995) ; but it applies only to flaring sites.

Helium. In the material escaping both closed- and open-field regions (gradual-SEP's, slow- and fast-SW), He is found definitely depleted by a factor of ~ 2 relative to heavier high-FIP elements (§ 3.1 and 4.1; Fig. 1 and 5). This He depletion should originate at low altitude, plausibly in the chromospheric FIP-fractionation environment, *not* in the SW acceleration region. This is indicated, both by the stability of the He/H ratio in coronal holes (Geiss 1982), and by the low He/C,O ratios observed in impulsive-SEP's, which sample material at coronal, not SW, altitudes (Reames et al. 1994; Paper I). Any model for the He depletion will have to account for the factor of ~ 2 depletion of He relative to Ne (while the FIP of He, 24.5 eV, is close to that of Ne, 21.5 eV, which itself is not depleted relative to O, with a FIP of 13.6 eV only) [3], and for the large depletion of He in the fast-SW, in spite of comparatively weak general FIP-bias in this medium (Fig. 5 vs. Fig. 1).

Note that a recent tentative determination of He/H in the corona itself, based on the observation of scattered Lyα light, has not yielded such a large depletion of He relative to H (Gabriel et al. 1995).

6. The absolute abundances, relative to Hydrogen

Anchoring the heavy element FIP-biased pattern to H, in order to get absolute abundances, is not an easy task. As discussed in Meyer (1991) and Paper I, the first hints in the early 1980's suggested photospheric absolute abundances of the low-FIP heavy elements, and a depletion of high-FIP ones. Two years ago, most data seemed to converge on the opposite conclusion : an absolute enhancement of low-FIP heavy elements, and photospheric abundances for the high-FIP ones ; (this is actually a simpler situation, since H itself is a high-FIP element, and is roughly neutral or ionized together with the heavier high-FIP elements). To-day, the situation seems, again, somewhat confused. The data are presented in Fig. 7.

In the escaping material, all indicators point to an absolute enhancement of low-FIP elements, *i.e.* that H behaves roughly like heavier high-FIP elements. These indicators have now become quite reliable. The direct slow-SW data show a well defined, *slightly* high H/O ratio $\sim 1.6 \times$ photospheric (von Steiger et al. 1995). If anything, this ratio can represent only an *upper* limit to the H/O ratio in the coronal source gas ; it seems, indeed, unlikely that the H/O ratio be decreased in the SW acceleration process, since it is H that drags along heavy elements in this process. In gradual-SEP's, there were difficulties in interpreting the observed H/heavy-element ratios in terms of source medium composition, due to the large difference in charge-to-mass ratio between H and heavies ($Q/A = 1$

---

[3] If the reference photospheric abundance of Ne actually turns out to lie near the upper bound of its error bar, this difficulty is alleviated, but not entirely removed (Fig. 1 and 5).

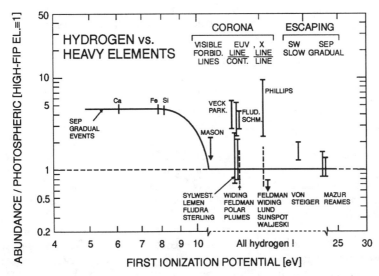

Figure 7. Anchoring the heavy element FIP pattern to Hydrogen. Most data are referenced in Paper I, and the more recent ones in § 6. The gradual-SEP data have been roughly corrected for charge-to-mass ratio biases.

for H and ≤ 0.5 for heavies), which affect their respective acceleration efficiencies. These difficulties have now been overcome, and it is now understood that, at lower energies, the SEP abundances approach those in their source medium (Mazur et al. 1992). On this basis, H/heavies ratios have been derived from SEP data by Mazur et al. (1993) and Reames (1995), which should be relevant to their source medium, the slow-SW. They also indicate roughly photospheric absolute abundances of high-FIP elements.

The truly coronal measurements yield a more confusing picture. Because H has no lines in the corona, most spectroscopic determinations consist of delicate line-to-continuum studies. Two years ago, all recent spectroscopic studies seemed to indicate that H behaves like heavier high-FIP elements, in agreement with the escaping material data (Paper I and ref. therein). While Sterling et al. (1993) and Waljeski et al. (1994) have confirmed this trend, several recent line-to-continuum (Fludra et al. 1993; Fludra & Schmelz 1995) and line-to-line (Phillips et al. 1994, 1995) [4] studies have indicated, on the contrary, an absolute depletion of high-FIP elements. Some of the problems involved in the various visible, EUV and X-ray spectroscopic determinations of absolute coronal abundances have been recently discussed in depth in a paper by Saba (1995), to which we refer the reader.

So, the situation is not entirely settled. My own personal prejudice is that the evidence from the escaping material in-situ analysis is more straightforward

---

[4]Phillips et al. have compared the Fe line intensities within a flare in the corona (collisionally excited Fe XXV resonance line) and in the underlying photosphere (Kα and Kβ fluorescence lines due to the X-ray irradiation by the same flare).

than the conflicting evidence from spectroscopy, so that it will probably turn out that we have an absolute enhancement of low-FIP elements.

7. Summary, discussion, and outlook

7.1. Summary of the observations

The elemental composition of large fractions of the solar outer atmosphere is affected by the "FIP-bias" described in § 1 (Fig. 1). This bias must result from an ion-neutral fractionation, which can take place only in a gas at $T < 10000$ K, most plausibly chromospheric gas (§ 2). The "step" of the FIP-pattern, around ~ 10 eV may be related to the thermostating of the ~ 7000 K chromospheric temperature plateau at the Lyα energy. Most relevant composition observations refer to much hotter media, well above the region where the fractionation can take place (except for the γ-ray data).

Essentially, FIP-biases by a rather constant factor of ~ 4.5 are observed in all materials associated with the large hot coronal loops at $T \gtrsim 1$ MK present in the closed-field regions, and associated with more or less intense coronal activity (§ 3.2.1; Fig. 2) : ordinary large, hot coronal loops, non-compact AR's and flares [5], and escaping slow-SW [6] (Fig. 1).

Otherwise, outside sites of large scale coronal activity, only weakly or non-FIP-biased material is observed : *(i)* in compact active loops and newly emerging material in AR's [7], *(ii)* everywhere in the network-associated low lying, cooler loop system at $T \lesssim 0.7$ MK (§ 3.2.1), both in the Quiet Sun areas within closed-field regions [8] and at the base of coronal holes [9], and *(iii)* in the fast-SW escaping coronal holes [10] (Fig. 5).

Sunspots yield a contrasted picture (§ 3.2.5).

In one, rare, type of situation, much larger FIP-biases, reaching factors of ~ 15, are observed : when an active loop is associated with an open field configuration. This has been observed in two types of sites : some AR's with associated diverging field in the equatorial regions, and polar plumes within coronal holes [11] ; note that, in polar plumes, the underlying active loop tends to be quite compact.

---

[5] EUV and X-ray spectroscopy (composition – local field topology correlation, § 3.2.2 ; averaged disk and limb studies, § 3.2.3 ; AR and flare studies, § 3.2.4 ; sunspot studies, § 3.2.5), γ-ray line data, and impulsive-SEP data (§ 3.2.4).

[6] Slow-SW, observed directly, and via gradual-SEP's (§ 3.1).

[7] EUV and X-ray spectroscopy (composition – local field topology correlation, § 3.2.2 ; AR and flare studies, § 3.2.4 ; sunspot studies, § 3.2.5).

[8] EUV and X-ray spectroscopy (composition – local field topology correlation, § 3.2.2 ; averaged disk and limb studies, § 3.2.3).

[9] EUV limb spectroscopy (§ 4.2.1).

[10] Fast-SW, observed directly, and via CIR energetic particles (§ 4.1).

[11] EUV spectroscopy (§ 4.2.2).

Relative to H, we *probably* have an absolute enhancement of the low-FIP elements, with roughly photospheric abundances for the high-FIP ones. This is clearly indicated by the escaping material data ; but there is still conflicting evidence among the coronal spectroscopic studies (§ 6; Fig. 7).

In *all* escaping materials, He is found depleted by a factor of ~ 2, relative to the heavier high-FIP elements (§ 5; Figs. 1 and 5). While the low-FIP element plateau is definitely flat in the material escaping closed-field regions (Fig. 1), a number of spectroscopic studies suggest an excess of very-low-FIP elements (FIP ~ 4 to 6 eV) relative to low-FIP Mg, Si, Fe (~ 8 eV) in the corona (§ 5). C seems to behave as a low-FIP element in the material escaping in coronal holes (§ 5; Fig. 5). Occasionally, Ne is observed enhanced relative to O, in quiescent as well as flaring AR's (§ 5). The possible implications of these various observations are briefly considered in § 5.

7.2. Discussion

7.2.1 FIP-bias and solar activity

The results on the fine scale correlation between FIP-bias and local magnetic field topology (§ 3.2.2), together with and the average limb and disk studies (§ 3.2.3), have shown that the Quiet Sun, chromospheric network associated, $T \lesssim 0.7$ MK, low lying loop system has a roughly photospheric composition ; only the larger true "coronal" loops above $T \sim 1$ MK are FIP-biased. Since these cooler $T \lesssim 0.7$ MK loops are situated above the ~ 7000 K region of conceivable FIP-fractionation, we can conclude that *no significant FIP-fractionation takes place in the Quiet Sun environment !* This also implies that *the Quiet Sun loop system cannot be the source of the hotter,* ~ 1.5 *MK coronal material and of the slow-SW*. The material must remain confined in the low lying Quiet Sun loop system (including spicule material, unless a specific fractionation takes place in the spicules themselves).

The observed time evolution of the AR composition (§ 3.2.4; Fig. 4), implies an influx of FIP-biased gas into the AR loops, since it is impossible to have an ion-neutral fractionation within a ~ 3 MK AR loop ! A gas influx from reconnecting ambient, *previously FIP-biased* coronal loops is not plausible, since AR loops are denser and hotter than ordinary coronal loops. So, the new, FIP-biased gas can come only from chromospheric evaporation at the loop footpoints. But we know that no FIP-fractionation takes place in the *quiet* chromosphere. So, the observed time evolution seems to require *a coupled evaporation and ion-neutral fractionation in the chromosphere below the AR loop footpoints*. One interesting scenario along this line has been tentatively developed by Antiochos (1994) [12].

That the γ-ray line data, pertaining to very low altitude, dense, chromospheric gas at the footpoints of a flaring loop, yield FIP-biased gas compositions (Ramaty et al. 1995) confirms that the FIP-fractionation is, indeed, taking place

---

[12] Antiochos' (1994) scenario, based on a selective *influx of ions* into the evaporating chromospheric gas, may have a problem in accounting for the lack of enhancement of C in the slow-SW and gradual-SEP's, since C tends to be ~ 50 % ionized in the chromosphere (see Paper I).

at very low altitudes. The same is true of some observations of very low ionization states in the neighborhood of sunspots (Feldman 1992; Athay 1994).

So, the AR environment is certainly a site in which FIP-fractionation takes place, and in which newly FIP-biased gas is being injected into the corona. It seems likely that the less active, ordinary large coronal loops at $T \gtrsim 1$ MK get their FIP-biased gas by the same evaporation-fractionation process as AR loops ; reconnection with AR loops might also play a role.

FIP-fractionation seems related with chromospheric evaporation. Apart from density considerations, one reason why it takes place in the $T > 1$ MK large, coronal loops, and not in the $T < 0.7$ MK network-associated low lying loops, could be that the evaporation rate goes as $T^{3.5}$ (Antiochos 1994 and ref. therein), and that a sufficiently large evaporation rate might be required for the FIP-fractionation to be effective. Note, however, that the amplitude of the FIP-bias attained *saturates* in closed loop systems, whatever the loop temperature beyond $T \sim 1$ MK : ordinary coronal loops around $T \sim 1.5$ MK, quiescent AR loops at $T \sim 3$ MK, and flaring loops with $T \sim 20$ MK, all show the same FIP-bias factor of ~ 4.5 as the escaping slow-SW material ! This saturation certainly requires an explanation. It seems significantly broken only when local activity, and presumably evaporation, is associated with a widely open field structure (§ 7.2.3 below).

7.2.2 Origin of the slow Solar Wind

This has a bearing on the question of the origin of the *slow*-SW. The striking composition similarity between slow-SW and large coronal loops strongly suggests that *the slow-SW material originates in large coronal loop gas, not in lower altitude material*. This idea actually fits well with the current picture of the slow-SW dynamics and topology. Quite generally, the SW speed is anti-correlated with the degree of divergence of the field along which it escapes. Slow-SW is associated with a rapid divergence, *i.e.* large expansion factors of the field away from the Sun (*e.g.* Wang & Sheeley 1990, 1994; Wang 1994 and ref. therein). This suggests an origin of the slow-SW in narrow open-field channels, first tightly squeezed between the upper coronal loop systems, and fanning out abruptly at higher altitudes [13]. At the squeezing points, reconnection, with its associated transfer of gas, is very likely to take place (Sheeley 1995a,b).

7.2.3 Role of the magnetic field topology ?

We now address the question : to which extent does the degree of opening of the field play a crucial role in controlling the FIP-bias ?

The observation of near-photospheric compositions in compact loops, of SW-type FIP-biases (factors of ~ 4.5) in larger coronal loops (§ 3.2.2), and of much larger FIP-biases (factors of ~ 10 to 15) in polar plumes and other very open field structures (§ 4.2.2), have first suggested that the key parameter controlling the amplitude of the FIP-bias was the field geometry : the more open

---

[13] Slow-SW actually originates mainly in at the boundaries of large, especially polar, coronal holes (solar minimum), and above small, isolated holes (solar maximum) (Wang & Sheeley 1994; Wang 1994).

the field, the larger the FIP-bias (*e.g.* Feldman 1992; Widing & Feldman 1992, 1993).

The clear evidence that the FIP-bias is weak in the bulk coronal holes, the prototype of open-field regions (Fig. 5; § 4.1), together with the interpretation of the AR data in terms of a time evolution (Fig. 4; § 3.2.4), have led to question this views (Paper I; Sheeley 1995a). The entire issue, actually, needs to be reexamined. Sunspot studies may also yield valuable clues (§ 7.2.4). In any case, several parameters can be correlated, and finding out which is the key causal factor is not straightforward.

Compact field sites, which are always low-lying, can, indeed, be associated with *(i)* higher densities, which make any diffusive ion-neutral separation more difficult, and/or *(ii)* lower temperatures, which reduces chromospheric evaporation (in the Quiet Sun and coronal hole network), and/or *(iii)* short time scales for FIP-bias build up (in newly emerging material and impulsive flares). So, there are a number of reasons why compact field features may keep a near-photospheric composition.

Regarding the weak FIP-bias found in the fast coronal hole SW, its significance depends upon the source of this material. One may be impressed by its similarity with the comparable weak bias found in the underlying network-associated cool loop system (§ 4.1, 4.2.1), and consider a link between the two reservoirs. We have seen, however, that in closed-field regions the upper coronal/SW material does not originate in the underlying network-associated cool loop system, which must be pretty leakproof. Most likely, the same is true in coronal holes, which have a very similar network-associated loop system (§ 4; Fig. 2). The weak FIP-bias in the fast-SW seems, therefore, rather related to the upper coronal open-field structure in coronal holes, as opposed to the closed hot coronal loops that dominate elsewhere. Then, the essential point may be that, in coronal holes, the upflow of material is driven by plain pressure gradient, rather than by thermal conductivity driven evaporation as in the hot coronal loop system. This difference is essential, if the FIP-fractionation is, indeed, related to the evaporation process, as the AR observations suggest. So, in this particular case, it seems that an open field yields a lower FIP-bias.

In the case of polar plumes and of AR's with an associated diverging field (§ 4.2.2), by contrast, the association between an active feature (even if compact, in the case of plumes) and a wide field opening seems to specifically produce extra-high FIP-biases.

7.2.4 The sunspot puzzle

The presence of photospheric-type material above some sunspots (§ 3.2.5) could turn out to be very instructive. It has been tentatively interpreted *(i)* in terms of the observed "low-FIP" elements Si or Fe remaining neutral in the very cool sunspot chromosphere (Paper I), *(ii)* in terms of the presence of newly emerging material near the magnetically neutral line, towards the center of the AR in which the sunspot is located (Sheeley 1995a), and *(iii)* in terms of a fractionation mechanism depending on the field intensity (Antiochos 1994), or on the field polarity or on the gas pressure (Athay 1994).

7.3. Outlook : The stellar and the galactic cosmic ray connection

The stellar connection. Investigations of the composition of later-type *stellar* coronae are now starting. Studies based on the EUVE spacecraft data find a FIP-effect in the coronae of some stars, not in others (Drake et al. 1995, 1996 and ref. therein; Brickhouse 1996). Laming et al.'s (1995) analysis of the entire Sun disk (§ 3.2.3) has been performed with the purpose of interpreting such stellar data ; it suggests, *e.g.* , that the observable FIP-bias in a stellar coronae may depend on the formation temperature of the investigated lines, and on the degree of activity of the star. Studies based on the ASCA spacecraft data, more directed towards active, binary systems, tend to find a general underabundance of heavy elements relative to H, and no ordering in terms of FIP (*e.g.* Singh et al. 1996 and ref. therein).

The galactic cosmic ray connection ? As summarized in the Introduction of Meyer (1993a), where the relevant references will be found, a FIP-bias extremely similar to that found in the solar corona has been (actually first !) found in galactic cosmic rays (GCR), which have typically $\sim$ GeV energies. This very similarity, together with the difficulty in finding in the ISM a cool gas-phase not depleted of its refractory elements as a possible source for the GCR particles, led to the suggestion that the GCR material originates in the coronae of later-type stars possessing a neutral chromosphere like the Sun, and presumably affected by the same FIP-bias as the solar corona. However, while they can "freeze" the coronal composition in a distinct population of MeV energetic particles, later-type stars are energetically unable to accelerate the particles to the GeV range. Only supernova (SN) shock waves can do the job (even though the SN environment seems unable to provide a source material with an appropriate composition !). The following two stage scenario was therefore proposed : later-type stars inject MeV energetic particles with a FIP-biased composition, which get later reaccelerated to GeV energies by neighboring SN shock waves. Star formation regions, in which massive SN explode in the immediate vicinity of many young, very active, later-type stars, seemed a tantalizing environment for such a scenario.

However, an apparent abundance correlation with FIP can mimic an actual correlation with volatility ! As a general rule, indeed, low-FIP elements tend to condense in refractory compounds, and high-FIP ones in volatile ones (if they condense at all !) : an apparent excess of low-FIP elements can therefore mimic an actual excess of those elements that are condensed in dust grains in the ISM. The ambiguity can be removed only by considering the few, exceptional, volatile, though low-FIP, elements. These elements (*e.g.* Na, Ge, Pb) are not the easiest to observe, but the three of them now appear to have a comparatively low abundance, suggesting that volatility, rather than FIP, is the relevant parameter. If this is confirmed, it will imply a preferential acceleration of the interstellar grain material, as compared to the gas, into GCR's. Plausibly, this could take place directly in SN shock waves, in which grains also get destroyed.

Acknowledgments. Inspiring, cheerful discussions with Neil Sheeley have been essential for this paper ! N. Sheeley also allowed me to quote some of his recent results, prior to publication. I also owe a lot to Spiro Antiochos, Uri Feldman, Martin Laming, Ron Murphy, Reuven Ramaty, Don Reames, Julia

Saba, Gerry Share, Alfonse Sterling, Rudi von Steiger, and Ken Widing for very profitable discussions regarding various aspects of this paper.

References

Antiochos, S.K. 1994, Adv. Space Res., 14, (4), 139
Antonucci, E., & Martin, R. 1995, ApJ, 451, 402
Athay, R.G. 1981, ApJ, 250, 709
Athay, R.G. 1994, ApJ, 423, 516
Brickhouse, N.S. 1996, in : Astrophysics in the EUV, IAU Colloq. No. 152,
    S. Bowyer & R.F. Malina eds., (Kluwer), in press
Drake, J.J., Laming, J.M., & Widing, K.G. 1995, ApJ, 443, 393
Drake, J.J., Laming, J.M., & Widing, K.G. 1996, in : Astrophysics in the EUV,
    IAU Colloq. No. 152, S. Bowyer & R.F. Malina eds., (Kluwer), in press
Feldman, U. 1992, Physica Scripta, 46, 202
Feldman, U. 1993, ApJ, 411, 896
Feldman, U., & Laming, J.M. 1994, ApJ, 434, 370
Feldman, U., Purcell, J.D., & Drohne, B.C. 1987, *An Atlas of EUV
    Spectroheliograms from 170 to 625 Å*, NRL Rep. 90-4100 & 91-4100
Feldman, U., & Widing, K.G. 1993, ApJ, 414, 381
Fludra, A., Culhane, J.L., Bentley, R.D., Doschek, G.A., Hiei, E., Phillips,
    K.J.H, Sterling, A., & Watenabe, T., 1993, Adv. Space Res., 13, (9), 395
Fludra, A., & Schmelz, J.T. 1995, ApJ, 447, 936
Gabriel A.H., Culhane, J.L., Patchett, B.E., Breeveld, E.R., Lang, J., Parkinson,
    J.H., Payne, J., & Norman, K. 1995, Adv. Space Res., 15, (7), 63
Garrard, T.L., & Stone, E.C. 1993, 23rd Intern. Cosmic Ray Conf.,
    Calgary, 3, 384
Geiss, J. 1982, Space Sci. Rev., 33, 201
Geiss, J., Gloeckler, G., & von Steiger, R. 1995, Space Sci. Rev., 72, 49
Hénoux, J.C. 1995, Adv. Space Res., 15, (7), 23
Laming, J.M., Drake, J.J., & Widing, K.G. 1995, ApJ, 443, 416
Marsch, E., von Steiger, R., & Bochsler, P. 1995, A&A, 301, 261
Mason, H.E. 1995, Adv. Space Res., 15, (7), 53
Mason, H.E., & Monsignori Fossi, B.C. 1994, A&A Rev., 6, 123
Mazur, J.E., Mason, G.M., Klecker, B., & McGuire, R.E. 1992, ApJ, 401, 398
Mazur, J.E., Mason, G.M., Klecker, B., & McGuire, R.E. 1993, ApJ, 404, 810
McKenzie, D.L., & Feldman, U. 1994, ApJ, 420, 892
Meyer, J.P. 1991, Adv. Space Res., 11, (1), 269
Meyer, J.P. 1993a, in : Origin and Evolution of the Elements,
    N. Prantzos, E. Vangioni-Flam & M. Cassé eds.,
    (Cambridge Univ. Press), p. 26 [Paper I]
Meyer, J.P. 1993b, Adv. Space Res., 13, (9), 377 [Paper I]

Monsignori Fossi, B.C., Landini, M., Thomas, R.J., & Neupert, W.M. 1994, Adv. Space Res., 14, (4), 163

Phillips, K.J.H., Pike, C.D., Lang, J., Watanabe, T., & Takahashi, M. 1994, ApJ, 435, 888

Phillips, K.J.H., Pike, C.D., Lang, J., Zarro, D.M., Fludra, A., Watanabe, T., & Takahashi, M. 1995, Adv. Space Res., 15, (7), 33

Ramaty, R., Mandzhavidze, N., Kozlovsky, B., & Murphy, R.J. 1995, ApJL, in press

Reames, D.V. 1995, Adv. Space Res., 15, (7), 41

Reames, D.V., Meyer, J.P., & von Rosenvinge, T.T. 1994, ApJS, 90, 649

Saba, J.L.R. 1995, Adv. Space Res., 15, (7), 13

Saba, J.L.R., & Strong, K.T. 1993, Adv. Space Res., 13, (9), 391

Schmelz, J.T. 1995, Adv. Space Res., 15, (7), 77

Share, G.H., & Murphy, R.J. 1995, ApJ, 452, 933

Sheeley, N.R. 1995a, ApJ, 440, 884

Sheeley, N.R. 1995b, private communication

Singh, K.P., White, N.E., & Drake, S.A. 1996, ApJ, in press (Jan 10, 1996)

Sterling, A.C., Doschek, G.A., & Feldman, U. 1993, ApJ, 404, 394

Vauclair, S. 1995, A&A, in press

von Steiger, R., Wimmer Schweingruber, R.F., Geiss, J., & Gloeckler, G. 1995, Adv. Space Res., 15, (7), 3

Waljeski, K., Moses, D., Dere, K.P., Saba, J.L.R., Strong, K.T., Webb, D.F., & Zarro, D.M. 1994, ApJ, 429, 909

Wang, Y.M. 1994, ApJL, 437, L67

Wang, Y.M., & Sheeley, N.R. 1990, ApJ, 355, 726

Wang, Y.M., & Sheeley, N.R. 1994, J. Geophys. Res., 99, 6597

Wang, Y.M., & Sheeley, N.R. 1995, ApJ, 452, 457

Widing, K.G., & Feldman, U. 1992, ApJ, 392, 715

Widing, K.G., & Feldman, U. 1993, ApJ, 416, 392

Widing, K.G., & Feldman, U. 1995, ApJ, 442, 446

Isotopic Abundances in Stars as Inferred From the Study of Presolar Grains in Meteorites

Ernst Zinner

McDonnell Center for the Space Sciences and the Physics Department, Washington University, St. Louis, MO 63130, USA

Abstract. Presolar grains in primitive meteorites formed in stellar atmospheres. Their isotopic compositions reflect those of their sources and thus provide new constraints on nucleosynthesis and stellar evolution. Diamond, silicon carbide, graphite, aluminum oxide and silicon nitride have been identified to date. Stellar sources include oxygen- and carbon-rich red giant stars as well as supernovae.

1. Introduction

By the 1950s it had been conclusively established that all chemical elements from carbon on up are produced in stars, and the seminal papers by Burbidge et al. (1957) and Cameron (1957) provided a theoretical framework for their stellar nucleosynthesis. According to these authors, the elements are produced by different nuclear processes with very different isotopic compositions, depending on the specific stellar source. The solar system is thought to have formed from materials contributed by many different stars. The fact that the isotopic composition of solar system materials, even primitive meteorites, which represent samples of the oldest solar system objects, was found to be very uniform has been explained by the extremely thorough mixing of the source material (Cameron 1962). The average composition of the solar system does therefore not give any information about the contribution from individual stars.

The full range of isotopic compositions of the stellar materials that formed the solar system was not realized until a few years ago when preserved stardust was discovered in meteorites and individual stellar grains could be isolated and studied in detail in the laboratory (Anders & Zinner 1993; Ott 1993). These grains have isotopic compositions completely different from those of solar system materials and are believed to have condensed in stellar outflows and supernova ejecta and thus to reflect the elemental and isotopic composition of their stellar sources. The range of their isotopic compositions not only dwarfs that observed in solar system objects but by far exceeds the range obtained by spectrosopic observations in stars. For example, Fig. 1 compares the oxygen isotopic ratios measured in stars with those measured in individual grains of stardust from meteorites. The study of these grains thus provides new information on the isotopic composition of stars by extending measurements to types of stars and to elements that cannot be studied astronomically and, as a consequence, sets new constraints on theories of nucleosynthesis and stellar evolution.

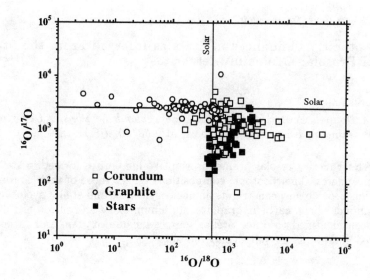

Figure 1. Oxygen isotopic ratios of presolar graphite and corundum grains are compared to those measured spectroscopically in stars. Grain data are from Nittler et al. (1994), Amari et al. (1995) and unpublished data by S. Amari and L. Nittler. Star data are from Harris & Lambert (1984), Harris et al. (1987), Smith & Lambert (1990), and Kahane et al. (1992).

Although the presence of stardust in meteorites had already been indicated in the sixties by the presence of "exotic", i.e., isotopically anomalous, noble gas components of Ne and Xe (Reynolds & Turner 1964; Black & Pepin 1969; Srinivasan & Anders 1978), it took more than twenty years before the carriers of these noble gases were identified as diamond, silicon carbide and graphite (Lewis et al. 1987; Bernatowicz et al. 1987; Amari et al. 1990). Two additional types of stardust, aluminum oxide (corundum) and silicon nitride were found by isotopic measurements of individual grains in the ion microprobe (Hutcheon et al. 1994; Nittler et al. 1994; Nittler et al. 1995b). Furthermore, silicon carbide (SiC) and graphite contain tiny subgrains of Ti, Zr and Mo carbides that were identified within their parent grains by transmission electron microscopy (Bernatowicz et al. 1991; Bernatowicz, Amari, & Lewis 1992; Bernatowicz, Amari, & Lewis 1994).

2. Isotopic Analyses

The physical-chemical separation methods developed at the University of Chicago (Amari, Lewis, & Anders 1994) provide almost pure samples of diamond, SiC and graphite. The isotopic compositions of "bulk samples", i.e. collections of large numbers of grains, of these three types have been obtained for the noble gases (see Anders & Zinner 1993, references therein, as well as Amari, Lewis, &

Anders 1995; Lewis, Amari, & Anders 1994), of diamonds isotopic analyses have also been made for the elements C, N, Sr and Ba (Russell, Arden, & Pillinger 1991; Lewis, Huss, & Lugmair 1991), and of SiC for Sr, Ba, Nd, Sm, and Dy (Ott & Begemann 1990; Podosek et al. 1994; Prombo et al. 1993; Richter, Ott, & Begemann 1992; Richter, Ott, & Begemann 1993; Richter, Ott, & Begemann 1994; Zinner, Amari, & Lewis 1991). The ion microprobe plays a crucial role for the analysis of meteoritic stardust because this instrument makes it possible to measure the isotopic compositions of individual grains down to sizes of less than 1μm. In this paper I shall concentrate on single grain measurements. Presolar diamonds have a size of only $\sim$2nm and are too small for single grain analysis. This review will not deal with them and the reader is referred to the review by Anders and Zinner (1993) and subsequent papers by Huss and Lewis (1994a;1994b; 1995) for details and information on noble gas studies.

Although most SiC grains are <0.5 μm in diameter, some grains are as large as 20 μm and allow isotopic measurements not only of the major elements, but also of trace elements. Isotopic analyses on single grains include the elements C, Si, N, Mg, Ca (Zinner, Tang, & Anders 1989; Zinner et al. 1991; Stone et al. 1991; Ireland, Zinner, & Amari 1991; Amari et al. 1992; Virag et al. 1992; Alexander 1993; Hoppe et al. 1993; Hoppe et al. 1994; Nittler et al. 1995b). Stellar graphite grains are round and have diameters ranging from $\sim$0.7 μm to 20 μm. Single grains have been analyzed for the isotopic compositions of C, N, O, Mg, Si, K, Ca, and Ti (Amari et al. 1990, 1993; Amari, Zinner, & Lewis 1995a, 1995b, 1995c; Hoppe et al. 1995; Zinner et al. 1995a, 1995b). Corundum grains identified to date range from $\sim$0.5 μm to $\sim$3 μm and had their O, Mg and Ti isotopes measured (Hutcheon et al. 1994; Huss et al. 1994; Huss, Fahey, & Wasserburg 1994; Nittler et al. 1994, 1995). Finally, only a few silicon nitride grains, $\leq$ 1μm in size, have been identified and their N, Si, C and Mg isotopes measured (Nittler et al. 1995b).

For ion probe analysis, grains from meteoritic separates are spread out on a metal foil. A given selected grain is bombarded with a finely focused ion beam (O$^-$ or Cs$^+$) and positive and negative secondary ions are analyzed according to their mass in a mass spectrometer (see Zinner et al. 1989 for a more detailed description). While essentially all SiC and graphite grains isolated from primitive meteorites are isotopically anomalous, identifying them as being of stellar origin, most corundum grains found in meteorites formed in the solar system and only a small fraction are of presolar origin. Isotopic ion imaging (Nittler et al. 1994) has become important for the identification of rare populations of isotopically peculiar grains, specifically of stellar corundum and silicon nitride grains, and for the selection of rare subtypes of SiC (e.g., those of type X, see Amari et al. 1992; Nittler et al. 1995b) from the mainstream of SiC. This technique allows simultaneous isotopic measurements of typically 20 grains and, in automated operation, the analysis of up to several thousand grains in a day (Nittler et al. 1994).

3. Range of Isotopic Compositions

The range of isotopic compositions measured in single presolar grains is shown in Figs. 1-5. The range in $^{18}O/^{16}O$ ratios spans more than four orders of magnitude

(Fig. 1). While most graphite grains have large excesses in ^{18}O, most corundum grains have ^{18}O depletions, and it is obvious that these two types of grains must have come from very different stellar sources. The spread in ^{12}C/^{13}C ratios is also almost four orders of magnitude, and the distribution of ^{12}C/^{13}C is quite different for SiC and graphite (Fig. 2). Most of the SiC grains ("mainstream") have a distribution of the C-isotopic ratio similar to that observed in carbon stars, and an origin from such stars is also indicated by other pieces of evidence. Approximately 1% of the SiC grains (named type X grains) differ not only in their C-, but also in their N-, Si-, and Al-isotopic ratios from the mainstream grains (Figs. 3-5).

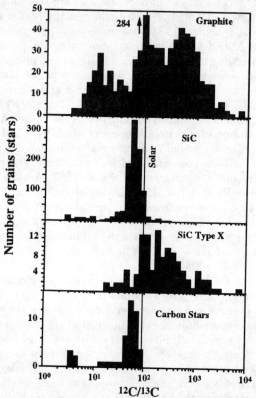

Figure 2. The C-isotopic compositions of presolar graphite and silicon carbide span a much wider range than those observed in carbon stars. Grain data are from Hoppe et al. (1993, 1994, 1995), Alexander (1993), Amari et al. (1995), Nittler et al. (1995b), and unpublished data by S. Amari and P. Hoppe. Carbon star data are from Dominy & Wallerstein (1987).

It should be noted that the numbers of mainstream and type X grains plotted in Figs. 2 and 3 do not correspond to their original abundances, because X grains have been preferentially selected by ion imaging for more detailed studies. The small deviations of the N-isotopic ratios from the solar ratio in

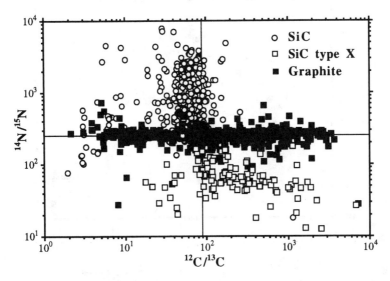

Figure 3. The presolar grain types SiC, the subtype X and graphite have distinct C- and N-isotopic ratios. Data are from Hoppe *et al.* (1993, 1994, 1995), Amari *et al.* (1995), Nittler *et al.* (1995b), and unpublished data by S. Amari.

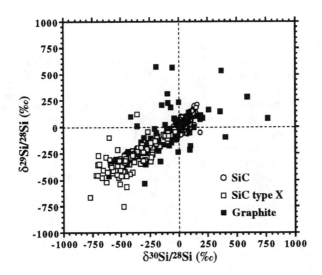

Figure 4. Also in their Si-isotopic compositions SiC, subtype X and graphite grains are distinct from one another. Data are from Virag *et al.* (1992), Alexander (1993), Hoppe *et al.* (1993, 1994), Amari *et al.* (1995), Nittler *et al.* (1995b), and unpublished data by S. Amari.

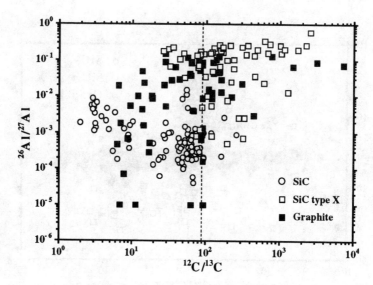

Figure 5. Ratios of $^{26}Al/^{27}Al$ inferred from ^{26}Mg excesses are lowest in mainstream SiC and highest in SiC of type X. Whereas mainstream SiC grains are believed to come from carbon stars, graphite grains with high $^{26}Al/^{27}Al$ ratios and SiC grains of type X most likely have a supernova origin. Data are from Zinner et al. (1991), Hoppe et al. (1994, 1995), Amari et al. (1995), Nittler et al. (1995b), and unpublished data by S. Amari.

graphite grains is surprising in view of the large range of their C-isotopic ratios (Fig. 3), but might be the result of isotopic equilibration with solar system nitrogen, either in the solar nebula or in the laboratory (Hoppe et al. 1992, 1995; Zinner et al. 1995b).

The dichotomy between mainstream and type X grains is also shown in their Si-isotopic ratios (Fig. 4): most mainstream grains have excesses in ^{29}Si and ^{30}Si relative to ^{28}Si (Stone et al. 1991; Alexander 1993; Hoppe et al. 1994), whereas X grains have large depletions in these two isotopes (Amari et al. 1992; Nittler et al. 1995b). There is a much larger scatter in the Si-isotopic ratios of selected graphite grains with high enough Si concentrations for isotopic analysis (Amari et al. 1995). Many SiC and graphite grains show large excesses in ^{26}Mg (some grains have essentially monoisotopic ^{26}Mg), without doubt the result of the decay of short-lived ^{26}Al ($t_{1/2} = 7.1 \times 10^5$ years) (Zinner et al. 1991). Inferred $^{26}Al/^{27}Al$ ratios range up to 0.01 in mainstream SiC, up to 0.1 in graphite and almost up to one in SiC grains of type X (Fig. 5).

4. Stellar Sources

Based on their isotopic compositions, two important stellar sources can be identified for meteoritic presolar grains: red giant stars of low to medium mass during late stages of their evolution when they experience substantial mass loss through stellar winds, and supernovae, massive stars that explode at the end of their evolution.

4.1. Red Giant Stars

Silicon Carbide. Most SiC grains are believed to originate from red giants, specifically from thermally pulsing AGB stars because (Zinner 1995): 1) the distribution of $^{12}C/^{13}C$ ratios in single SiC grains is similar to that in carbon stars (Fig. 2) ; 2) AGB stars are the main contributors of carbonaceous dust to the interstellar medium (Whittet 1992); 3) AGB stars have dusty envelopes and show the 11.2 μm emission feature characteristic of SiC (Little-Marenin 1986) and 4) AGB stars are believed to be the main source of s-process elements, and "bulk samples" (collections of many grains) of meteoritic SiC carry the s-process signature in the isotopic compositions of Kr, Xe, Ba, Nd and Sm (Anders & Zinner 1993).

Most isotopic compositions measured in single SiC grains generally agree with a carbon star origin. The mainstream grains have isotopically heavy to moderately light C and light N (Fig. 3). Such compositions can be explained by CNO nucleosynthesis during core H-burning and mixing of the nucleosynthetic products into the star's envelope at the end of H-burning (first dredge-up) (Bressan et al. 1993; El Eid 1994), and subsequent dredge-up of ^{12}C produced by He-burning during the thermally-pulsing AGB phase (Boothroyd & Sackmann 1988a, 1988b; Lattanzio 1989).

The C- and N-isotopic compositions of grains with $^{12}C/^{13}C$ ratios smaller than ~20 do not agree with theoretical models of the first and third dredge-up. The problem of low $^{12}C/^{13}C$, not only in J-stars, has been recognized for some time (e.g., Smith & Lambert 1990; Sneden 1991). Hot bottom burning (HBB), H-burning at the bottom of the convective envelope of AGB stars, does not solve this problem because this process is believed to prevent carbon-star formation (e.g., Boothroyd, Sackmann, & Ahern 1993; Frost & Lattanzio 1995). Furthermore, HBB fails to produce the low $^{14}N/^{15}N$ ratios observed in SiC grains with low $^{12}C/^{13}C$ ratios (Boothroyd, Sackmann, & Wasserburg 1994). Failure to produce low $^{14}N/^{15}N$ ratios excludes also models of deep mixing that would produce low $^{12}C/^{13}C$ ratios in carbon stars (Charbonnel 1995; Wasserburg, Boothroyd, & Sackmann 1995). The C- and N-isotopic composition of these grains remains unexplained and thus a challenge to nuclear astrophysicists.

The spread in the Si- and Ti-isotopic compositions of single mainstream SiC grains (Hoppe et al. 1994) cannot be explained by nucleosynthesis taking place in a single AGB star and indicates contributions from several stars with different metallicities (Alexander 1993; Gallino et al. 1994).

Corundum Most presolar corundum grains in meteorites have been located by isotopic imaging of the $^{16}O/^{18}O$ ratio in the ion microprobe. These grains show a large range in their O-isotopic ratios (Fig. 6) and $^{26}Al/^{27}Al$ ratios (Nittler et al. 1994, 1995; Huss et al. 1994). While core H-burning and subsequent dredge-up affects the O-isotopic composition of the surface of stars, the higher temperatures needed for the synthesis of ^{26}Al is only achieved during H shell burning. Some grains lack any evidence for ^{26}Al and must have formed in red giant atmospheres before the second or third dredge-up, most have ^{26}Al and apparently formed in AGB stars during the thermally pulsing phase or in intermediate-mass stars that experienced second dredge-up or hot bottom burning.

The O-isotopic compositions of corundum grains with excesses in ^{17}O and moderate depletions in ^{18}O relative to the solar system can be explained by mixing into the envelope (first dredge-up) of material processed during core H-burning in the star's interior where ^{17}O is produced and ^{18}O is destroyed in the CNO cycle. However, similar to the distribution of Si-isotopic ratios in mainstream SiC grains, the spread in O-isotopic compositions found in meteoritic corundum grains cannot be accounted for by a single star and can only be explained by the assumption that different stars with different masses (resulting in variations in the ^{16}O/^{17}O ratio) and different initial isotopic compositions (variations in both ^{16}O/^{17}O and ^{16}O/^{18}O, see Fig. 6) contributed corundum grains to the solar system (Boothroyd et al. 1994). The initial O-isotopic compositions in turn can be related to stellar metallicities via a chemical evolution model (Timmes, Woosley, & Weaver 1995) and the bold line with arrows in Fig. 6 indicates the range in metallicities relative to that of the sun observed for stars of a given epoch (Edvardssen et al. 1993).

The high ^{16}O/^{18}O ratios found in a fraction of the corundum grains (Fig. 6) cannot be explained by variations in metallicity of the original stars. Some of them most likely come from AGB stars of intermediate mass in which hot bottom burning, which destroys ^{18}O very efficiently, took place (Boothroyd, Sackmann, & Wasserburg 1995). However, HBB is believed to occur only in stars with masses of more than 4-5 M$_\odot$ (Boothroyd et al. 1995; Frost & Lattanzio 1995), and in such stars the ^{16}O/^{17}O ratio is expected to be smaller than ~1000 (Boothroyd et al. 1994, 1995). In order to explain the O-isotopic compositions of grains with higher ^{16}O/^{17}O ratios and large ^{18}O depletions, Wasserburg, Boothroyd, & Sackmann (1995) invoked deep mixing, a phenomenon termed "cool bottom processing". In their model some extra mixing takes place between the bottom of the convective envelope and the H-burning shell and cycles ^{18}O to regions of high enough temperature that it is destroyed. It has already been mentioned above that such deep mixing (see also Charbonnel 1995) can produce small ^{12}C/^{13}C ratios and thus can explain the observation of anomalously low ^{12}C/^{13}C ratios in low-mass stars but not the N-isotopic ratios in SiC grains.

A few corundum grains have both ^{17}O and ^{18}O excesses (low ^{16}O/^{18}O). One possiblity is that these grains are from stars with higher than solar metallicities. The most extreme of these data points would require a metallicity Z of between 0.04 and 0.06, a little higher than the range expected in stars at or before the formation of the solar system. An alternative is that the ^{18}O excesses are produced by the first few thermal pulses of AGB stars (Mowlavi 1995). The question is only whether this ^{18}O can be dredged up since most models do not show any significant dredge-up during the first few pulses.

In summary, red giant stars with different masses and different metallicities must have contributed circumstellar corundum grains to the solar system. Some

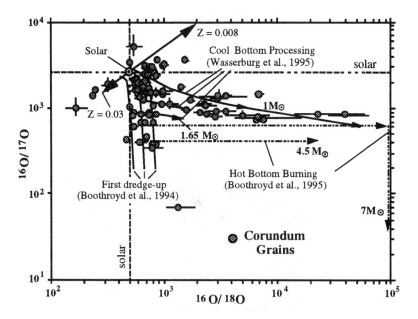

Figure 6. The O-isotopic compositions of circumstellar corundum grains give evidence for contributions from red giants with different mass and metallicity as well as from AGB stars with hot bottom burning and cool bottom processing. Data are from Nittler et al. (1994, 1995), Huss et al. (1994), and unpublished data by G. Huss and L. Nittler.

of these grains show evidence for hot bottom burning, others for cool bottom processing. Some grains might have ^{18}O excesses because of dredge-up of this isotope during the earliest thermal pulses of AGB stars, but an origin from stars with unusually high metallicity cannot be excluded as an alternative explanation.

4.2. Supernovae

Three types of circumstellar grains in meteorites are believed to come from supernovae (Clayton 1975a, 1975b): low density graphite grains (Amari et al. 1995), SiC grains of the rare type X (Amari et al. 1992), and even rarer silicon nitride grains (Nittler et al. 1995b). The C-, N-, O-, and Al-isotopic compositions of single grains from the low density graphite fraction KE3 (1.65 to 1.72 g cm^{-3}) from Murchison (Amari et al. 1995) are shown in Fig. 7, the isotopic compositions of SiC grains of type X in Figs. 2-5. The isotopic compositions of silicon nitride are indistinguishable from those of X grains (Nittler et al. 1995b). Most KE3 grains have ^{15}N excesses and large ^{18}O excesses (up to a factor of 200, corresponding to ^{16}O/^{18}O = ~2.5), and ^{13}C as well as ^{12}C excesses and large ^{26}Al/^{27}Al ratios. Most grains also have ^{28}Si excesses, but some have ^{29}Si and some both ^{29}Si and ^{30}Si excesses. SiC grains of type X constitute only about 1% of all presolar SiC in meteorites. They are characterized by ^{15}N excesses (Fig. 3), ^{28}Si excesses (Fig. 4) and high ^{26}Al/^{27}Al ratios (Fig. 5). In all three signatures they are similar to, but more extreme, than low density graphite grains.

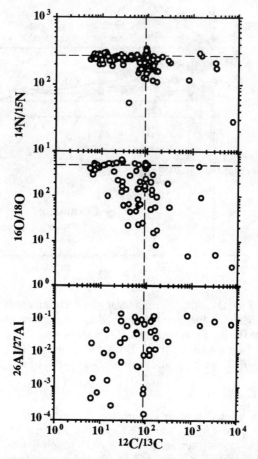

Figure 7. The isotopic compositions of low-density graphite grains are characterized by large ^{15}N and ^{18}O excesses and high ^{26}Al/^{27}Al ratios. These signatures, together with the Si-isotopic data (Amari et al. 1995), indicate a supernova origin (Zinner et al. 1995a). Data are from Amari et al. (1995), and unpublished data by S. Amari.

The ^{18}O excesses in the latter grains indicate a massive star origin (Amari et al. 1995). This isotope is produced by partial He-burning from abundant ^{14}N generated in the CNO cycle during previous H-burning in all stars, but only in massive stars can it come to the surface where grains are formed. The surface of Wolf-Rayet stars is expected to be enriched in ^{18}O during the transition from the WN to the WC phase (Langer 1991; Meynet & Arnould 1994), but the ^{15}N and ^{28}Si excesses cannot be explained by such stars. That leaves Type II supernovae as the most likely candidates for the sources, not only of the low density graphite grains, but also the type X and silicon nitride grains. Oxygen-18 is expected to be found in the He-burning zone of massive stars at the end of their evolution just before their explosion as supernovae (Woosley & Weaver 1995) and ^{15}N is produced in the same zone by explosive reactions. Aluminum-26, on the other hand, is expected to be produced in the overlying He/N shell during H-burning.

If the presence of large amounts of ^{26}Al (product of the He/N zone) and of large excesses of ^{18}O and ^{15}N (products of the He/C zone) in the same grains already indicates mixing between different SN layers, much more extensive mixing is required to explain the large ^{28}Si excesses in type X SiC grains and most low density graphite grains. This isotope is produced by O-burning in interior layers (Woosley & Weaver 1995). Mixing of SN layers can fairly successfully reproduce most isotopic compositions of low density graphite grains (Zinner et al. 1995a) and it remains to be seen whether it can also account for the isotopic compositions of type X SiC grains. Violent mixing has indeed been observed during supernova explosions (Shigeyama & Nomoto 1990; Colgan et al. 1994), and macroscopic mixing is obtained in postexplosion hydrodynamic models (Müller, Fryxell, & Arnett 1991; Herant & Benz 1992; Herant & Woosley 1994).

The most convincing evidence for a SN origin of low density graphite and SiC of type X are large ^{44}Ca excesses in these grains (Amari et al. 1992, 1995; Nittler et al. 1995b; Hoppe et al. 1995). While ^{44}Ca excesses can be produced by neutron capture, the expected accompanying, even larger, relative excesses in ^{42}Ca and ^{43}Ca are not observed. Furthermore, the size of the ^{44}Ca excesses (^{44}Ca/^{40}Ca ratios range up to 138 solar in grain KFA1-302 - see Fig. 8) leaves little doubt that they are due to the decay of ^{44}Ti ($t_{1/2}$ = 58 years). This short-lived isotope is produced by Si- or statistical equilibrium burning in the ^{28}Si-rich zone and the even more interior zone. Grains with ^{44}Ca excesses have also ^{28}Si excesses (i.e., smaller than solar ^{30}Si/^{28}Si ratios) (Fig. 8) and both isotopes have to be mixed into C-rich zones during the supernova explosion in order to be present in carbonaceous grains.

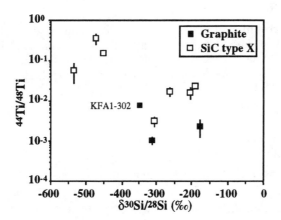

Figure 8. Ratios of ^{44}Ti/^{48}Ti inferred from ^{44}Ca excesses are found in SiC and graphite grains with ^{28}Si excesses. The correlated presence of ^{44}Ti and ^{28}Si, both of which are produced in interior layers of supernovae, is evidence for a SN origin of these grains. Data are from Amari et al. (1995), Nittler et al. (1995a), and Hoppe et al. (1995).

Another indication of a SN origin are ^{41}K excesses found in low density graphite grains (Amari et al. 1995a). These excesses are most likely from the decay of ^{41}Ca ($t_{1/2} = 1.05 \times 10^5$ years), with inferred ^{41}Ca/^{40}Ca ratios ranging up to 0.016. This is much higher than the ratio expected for the envelope of AGB stars (Wasserburg et al. 1995), but in agreement with models of Type II supernovae (Woosley & Weaver 1995).

Although there is overwhelming evidence that certain presolar grains are supernova condensates, in detail there are discrepancies between the isotopic compositions of low density graphite and especially of SiC X grains, and existing supernova models. Problems are the high ^{26}Al/^{27}Al ratios associated with large ^{15}N excesses in the same grains (^{26}Al and ^{15}N are believed to be produced in different zones of supernovae; Woosley & Weaver 1995), and the distribution of Si isotopic compositions in X grains (Nittler et al. 1995b). Grain data thus provide information not available before and set new constraints on models of nucleosynthesis in massive stars.

4.3. Other Stars

While contributions from other stellar sources with C>O, Wolf Rayet stars during the WN and WC phases and novae, cannot be excluded, direct evidence for these sources is much weaker than for AGB stars and supernovae. Graphite grains of higher densities with high ^{12}C/^{13}C ratios could be from WR stars (Hoppe et al. 1995). Such an origin is also in agreement with Kr isotopic data (Amari et al. 1995). One graphite grain with a ^{12}C/^{13}C ratio of 3.8 (Hoppe et al. 1995) is extremely rich in ^{22}Ne-E(L) without accompanying ^{4}He and has a ^{20}Ne/^{22}Ne ratio of only 0.0067 (Nichols et al. 1994), much lower than that expected from He burning (Amari et al. 1995). In this grain ^{22}Ne must have been from ^{22}Na decay and a nova origin is not unlikely (Clayton & Hoyle 1976). More studies are, however, necessary in order to establish what fraction of meteoritic stardust comes from these types of stellar sources.

Acknowledgments. I thank S. Amari, P. Hoppe and L. Nittler for providing unpublished data. I enjoyed many fruitful conversations with S. Amari, R. Gallino, P. Hoppe, G. Huss, L. Nittler, C. Travaglio and S. Woosley. L. Nittler and L. Trower provided much welcome help with LaTeX. This work was supported by NASA.

References

Alexander, C. M., O'D. 1993, Geochim. Cosmochim. Acta, 57, 2869.
Amari, S., Anders, E., Virag, A., & Zinner, E. 1990, Nature, 345, 238.
Amari, S., Hoppe, P., Zinner, E., & Lewis, R. S. 1992, ApJ, 394, L43.
Amari, S., Hoppe, P., Zinner, E., & Lewis, R. S. 1993, Nature, 365, 806.
Amari, S., Lewis, R. S., & Anders, E. 1994, Geochim. Cosmochim. Acta, 58, 459.
Amari, S., Lewis, R. S., & Anders, E. 1995, Geochim. Cosmochim. Acta, 53, 1411.
Amari, S., Zinner, E., & Lewis, R. S. 1995a, Meteoritics, 30, 480.

Amari, S., Zinner, E., & Lewis, R. S. 1995b, in Nuclei in the Cosmos III, ed. M. Busso, R. Gallino and C. M. Raiteri (New York: AIP), p. 581.
Amari, S., Zinner, E., & Lewis, R. S. 1995c, ApJ, 447, L147.
Amari, S., Zinner, E., Lewis, R. S., & Woosley, S. 1995, Lunar Planet. Sci. XXVI, 37.
Anders, E., & Zinner, E. 1993, Meteoritics, 28, 490.
Bernatowicz, T. J., Amari, S., & Lewis, R. S. 1992, Lunar Planet. Sci. XXIII, 91.
Bernatowicz, T. J., Amari, S., & Lewis, R. S. 1994, Lunar Planet. Sci. XXV, 103.
Bernatowicz, T. J., Amari, S., Zinner, E. K., & Lewis, R. S. 1991, ApJ, 373, L73.
Bernatowicz, T. J., Fraundorf, G., Tang, M., Anders, E., Wopenka, B., Zinner, E., & Fraundorf, P. 1987, Nature, 330, 728.
Black, D. C., & Pepin, R. O. 1969, Earth Planet. Sci. Lett., 6, 395.
Boothroyd, A. I., & Sackmann, I.-J. 1988a, ApJ, 328, 653.
Boothroyd, A. I., & Sackmann, I.-J. 1988b, ApJ, 328, 671.
Boothroyd, A. I., Sackmann, I.-J., & Ahern, S. C. 1993, ApJ, 416, 762.
Boothroyd, A. I., Sackmann, I.-J., & Wasserburg, G. J. 1994, ApJ, 430, L77.
Boothroyd, A. I., Sackmann, I.-J., & Wasserburg, G. J. 1995, ApJ, 442, L21.
Bressan, A., Fagotto, F., Bertelli, G., & Chiose, C. 1993, ApJS, 100, 647.
Burbidge, E. M., Burbidge, G. R., Fowler, W. A., & Hoyle, F. 1957, Rev. Mod. Phys., 29, 547.
Cameron, A. G. W. 1962, Icarus, 1, 13.
Charbonnel, C. 1995, ApJ, 453, L41.
Clayton, D. D. 1975a, ApJ, 199, 765.
Clayton, D. D. 1975b, Nature, 257, 36.
Clayton, D. D., & Hoyle, F. 1976, ApJ, 203, 490.
Colgan, S. W. J., Haas, M. R., Frickson, E. F., & Lord, S. D. 1994, ApJ, 427, 874.
Dominy, J. F., & Wallerstein, G. 1987, ApJ, 317, 810.
Edvardssen, B., Anderson, J., Gustaffson, B., Lambert, D. L., Nissen, P. E., & Tomkin, J. 1993, A&A, 275, 101.
El Eid, M. 1994, A&A, 285, 915.
Frost, C. A., & Lattanzio, J. C. 1995, 32nd Liège Int. Astroph. Coll., in press.
Gallino, R., Raiteri, C. M., Busso, M., & Matteucci, F. 1994, ApJ, 430, 858.
Harris, M. J., & Lambert, D. L. 1984, ApJ, 285, 674.
Harris, M. J., Lambert, D. L., Hinkle, K. H., Gustafsson, B., & Eriksson, K. 1987, ApJ, 316, 294.
Herant, M., & Benz, W. 1992, ApJ, 387, 294.
Herant, M., & Woosley, S. E. 1994, ApJ, 425, 814.
Hoppe, P., Amari, S., Zinner, E., Ireland, T., & Lewis, R. S. 1994, ApJ, 430, 870.

Hoppe, P., Amari, S., Zinner, E., & Lewis, R. S. 1992, Meteoritics, 27, 235.
Hoppe, P., Amari, S., Zinner, E., & Lewis, R. S. 1995, Geochim. Cosmochim. Acta, 59, 4029.
Hoppe, P., Geiss, J., Bühler, F., Neuenschwander, J., Amari, S., & Lewis, R. S. 1993, Geochim. Cosmochim. Acta, 57, 4059.
Hoppe, P., Strebel, R., Eberhardt, P., Amari, S., & Lewis, R. S. 1995, Science, submitted.
Huss, G. R., Fahey, A. J., Gallino, R., & Wasserburg, G. J. 1994, ApJ, 430, L81.
Huss, G. R., Fahey, A. J., & Wasserburg, G. J. 1994, Meteoritics, 29, 475.
Huss, G. R., & Lewis, R. S. 1994a, Meteoritics, 29, 791.
Huss, G. R., & Lewis, R. S. 1994b, Meteoritics, 29, 811.
Huss, G. R., & Lewis, R. S. 1995, Geochim. Cosmochim. Acta, 59, 115.
Hutcheon, I. D., Huss, G. R., Fahey, A. J., & Wasserburg, G. J. 1994, ApJ, 425, L97.
Ireland, T. R., Zinner, E. K., & Amari, S. 1991, ApJ, 376, L53.
Kahane, C., Cernicharo, J., Gomez-Gonzalez, J., & Guelin, M. 1992, A&A, 256, 235.
Langer, N. 1991, A&A, 248, 531.
Lattanzio, J. C. 1989, ApJ, 344, L25.
Lewis, R. S., Amari, S., & Anders, E. 1994, Geochim. Cosmochim. Acta, 58, 471.
Lewis, R. S., Huss, G. R., & Lugmair, G. 1991, Lunar Planet. Sci. XXII, 807.
Lewis, R. S., Tang, M., Wacker, J. F., Anders, E., & Steel, E. 1987, Nature, 326, 160.
Little-Marenin, I. R. 1986, ApJ, 307, L15.
Meynet, G., & Arnould, M. 1994, in Proceedings of the European Workshop on Heavy Element Nucleosynthesis, ed. E. Somorjai and Z. Fülöp (Debrecen: Institute of Nuclear Research of the Hung. Ac. of Science), p. 52.
Mowlavi, N. 1995, private communication..
Müller, E., Fryxell, B., & Arnett, W. D. 1991, A&A, 251, 505.
Nichols, R. H., Jr., Kehm, K., Brazzle, R., Amari, S., Hohenberg, C. M., & Lewis, R. S. 1994, Meteoritics, 29, 510.
Nittler, L., Alexander, C., Gao, X., Walker, R., & Zinner, E. 1995, in Nuclei in the Cosmos III, ed. M. Busso, R. Gallino and C. M. Raiteri (New York: AIP), p. 585.
Nittler, L. R., Alexander, C. M. O., Gao, X., Walker, R. M., & Zinner, E. K. 1994, Nature, 370, 443.
Nittler, L. R., Amari, S., Lewis, R. S., Walker, R. M., & Zinner, E. 1995a, Meteoritics, 30, 557.
Nittler, L. R., et al. 1995b, ApJ, 453, L25.
Ott, U. 1993, Nature, 364, 25.
Ott, U., & Begemann, F. 1990, ApJ, 353, L57.
Podosek, F. A., Prombo, C. A., Amari, S., & Lewis, R. S. 1994, ApJ, in press.

Prombo, C. A., Podosek, F. A., Amari, S., & Lewis, R. S. 1993, ApJ, 410, 393.
Reynolds, J. H., & Turner, G. 1964, J. Geophys. Res., 69, 3263.
Richter, S., Ott, U., & Begemann, F. 1992, Lunar Planet. Sci. XXIII, 1147.
Richter, S., Ott, U., & Begemann, F. 1993, in Nuclei in the Cosmos, ed. F. Käppeler and K. Wisshak (Philadelphia: IOP Publishing), p. 127.
Richter, S., Ott, U., & Begemann, F. 1994, in Proc. of the European Workshop on Heavy Element Nucleosynthesis, ed. E. Somorjai and Z. Fülöp (Debrecen, Hungary: p. 44.
Russell, S. S., Arden, J. W., & Pillinger, C. T. 1991, Science, 254, 1188.
Shigeyama, T., & Nomoto, K. 1990, ApJ 360, 242.
Smith, V. V., & Lambert, D. L. 1990, ApJ Suppl., 72, 387.
Sneden, C. 1991, in Evolution of Stars: The Photospheric Abundance Connection, ed. G. Michaud and A. Tutukov (Netherlands: IAU), p. 235.
Srinivasan, B., & Anders, E. 1978, Science, 201, 51.
Stone, J., Hutcheon, I. D., Epstein, S., & Wasserburg, G. J. 1991, Earth Planet. Sci. Lett., 107, 570.
Timmes, F. X., Woosley, S. E., & Weaver, T. A. 1995, ApJ Suppl., 98, 617.
Virag, A., Wopenka, B., Amari, S., Zinner, E., Anders, E., & Lewis, R. S. 1992, Geochim. Cosmochim. Acta, 56, 1715.
Wasserburg, G. J., Boothroyd, A. I., & Sackmann, I.-J. 1995, ApJ, 447, L37.
Wasserburg, G. J., Gallino, R., Busso, M., Goswami, J. N., & Raiteri, C. M. 1995, ApJ, 440, L101.
Woosley, S. E., & Weaver, T. A. 1995, ApJ Suppl., 101, 181.
Zinner, E. 1995, in Nuclei in the Cosmos III, ed. M. Busso, R. Gallino and C. M. Raiteri (New York: AIP), p. 567.
Zinner, E., Amari, S., Anders, E., & Lewis, R. S. 1991, Nature, 349, 51.
Zinner, E., Amari, S., & Lewis, R. S. 1991, ApJ, 382, L47.
Zinner, E., Amari, S., Travaglio, C., Gallino, R., Busso, M., & Woosley, S. 1995a, Lunar Planet. Sci. XXVI, 1561.
Zinner, E., Amari, S., Wopenka, B., & Lewis, R. S. 1995b, Meteoritics, 30, 209.
Zinner, E., Tang, M., & Anders, E. 1989, Geochim. Cosmochim. Acta, 53, 3273.

Radioisotope Production in the Early Solar Nebula by Local High Energy Plasma Winds

Martin S. Spergel

Department of Natural Sciences, York College of CUNY, Jamaica, NY 11451

Abstract. Radioisotopes in the Early Solar Nebula (ESN) are seen to be produced locally by the ESN's ambient stellar winds. Recent observational data supports arguments against transport of radioisotopes into the ESN from nearby molecular clouds or by input from Asympotic Giant Branch (AGB) stars. Insertion by supernova explosions are seen to be implausible. Production by the nucleosynthesis induced by the high energy components of the ESN's winds and enhanced by electrically shielded targets will offer the best production model. This production limited to the inner solar system because of rapid attenuation. This limitation on ^{26}Al can be used to solve the difficulty of its usually predicted overabundance. Cosmogenus by Galatic cosmic rays (GCR) will also be a factor in radionuclide production but for the outer less dense regions of the ESN.

1. Introduction

There has been a long standing interest in determining the sources of production of ^{26}Al (Clayton etal 1977, Clayton 1994, Wasserburg & Papanastassiou 1982, Cameron 1992) along with other radioisotopes and their remnants found in our solar system. The time and site of the production of radioisotopes are important for the purpose of dating evolutionary events in the early solar nebula. Radioisotope production had been believed to have been at sites of AGB stars (Cameron 1992), supernova (Truran & Cameron 1978) or in molecular clouds (Clayton etal 1977, Clayton 1994, Ramaty 1996).

However, recently additional doubts have arisen on these extrasolar sources of radioisotope production. The low observational estimates by Kastner & Myers (1994) of the probability of encounters between mass-losing evolved stars and molecular clouds for the production of ^{26}Al argue strongly against such a non local site for the locally observed ^{26}Al. Furthermore, the observed low production for the production of ^{26}Al from AGB (Prantzos 1993) is seen strongly to rule out production at AGB sites.

Spergel (1995) in exploring the transmission of supernova cosmic rays and solar high energy particles (SHEP) within the ESN, showed rapid attenuation of supernova cosmic rays within the inner ESN. Considering possible ranges of intensities of supernova cosmic rays and SHEP the calculation (Spergel 1995) ruled out supernova cosmic rays as an important source of extrasolar particles for even local cosmogenic production.

I believe that the sources, for the presence of radioisotopes, will be within the solar nebula. Clayton and Lin (1995), as this paper was in preparation, have also reconsidered the standard arguments for the production of ^{26}Al and decided for production in the ESN. They (Clayton and Jin 1995) suggest production in an inner solar sytem shell region at a quiescent phase of the ESN because of limitations of ionization loss. It is suggested here and below in this extended abstract that the source of the incident particles be the ESN stellar-T Tauri-like ionized and neutral winds.

2. Local Production by T Tauri-like winds

High energy plasma winds and associated neutral winds (due to charge exchange scattering of the stellar winds with the ambient nebula gases) are frequenty observed within the present solar heliosphere (Simpson etal 1995, Keppler etal 1995). High energy winds are also seen with T Tauri-like stars, presumed precursors to solar-like stars. Such stellar winds in the emerging sun will provide in the ESN a source of solar high energy particles (SHEP) which can interact with such insitu targets such as ^{26}Mg to produce the ^{26}Al.

It is suggested, based on the expected presence in the ESN of stellar winds of 300-700 km/s (which would yield temperatures up to $\sim 3 \times 10^7$ K) that nucleosythesis or cosmogenous (frequently the term is used to describe this process when it develops outside a stellar burning region) will develop within the nebula cloud to produce the necessary radionuclides. Observational evidence exists, as seen below, which support the presence of such energetic particles and nucleosynthesis at these energies.

The presence of new cross section data (Castellani 1995), indicate that an important but little consider process for the production of radioisotopes will operate at these energies. Cross sections for nucleosynthesis of protons or alpha particles (Rolfs and Rodney 1988) will be enhanced, in their interaction with neutral atomic or molecular particles over those seen in the interaction with ionized nuclei. Such enhancement have been recently observed (Castellani 1995) in low energy interaction on electrically shield targets. There (Castellani 1995) it was also suggested, that in stellar convective zones, electron clouds of the plasma shield may also shield bare target nuclei.

We further suggest that neutral stellar winds induced by charge exchange scattering in the nebula will provide additional candidates for the enhanced cross sections of the neutral particles. Measured values of low energy proton scattered on atomic and molecular targets indicated that fusion cross sections are enlarged and elastic cross sections are reduced, therefore simple extrapolation of accelerator data can lead to an underestimate in the relevant excitation cross sections in lower energy proton induced production.

3. Discussion

A further increase in radionuclide production will be seen in linking: this enhanced proton nucleosynthesis from interactions with the shielded atomic or molecular nuclei; with the increased particle intensities for the lower energy particles available in the solar wind. Particularly then, ^{26}Al levels would come from enhanced interactions in the inner ESN by the collisions of the intense T Tauri-like plasma winds with the atoms and molecules of the ESN. Interactions like ^{26}Mg (p,n) ^{26}Al in this neutral electrical setting may provide the needed selective production and overcome some of the astrophysics earlier objections to production in the ESN.

One can expect to find high production rates from the plasma winds in the inner ESN with strong effects from galatic cosmic rays (GCR) as suggested by Clayton and Lin (1995) to be see in the outer less dense regions of the ESN. Subsequent mixing stages may cloud these production regions.

Future directions will examine the details of the gas and dust interaction with the SHEP. The interface between the GCR and the T Tauri wind region should be explored with attention paid to the effort of that the ^{26}Mg in the SHEP or GCR plays (Ramaty etal 1996).

Possible tests for this model would come from examing the ^{26}Mg /n capture nuclei ratios in meteorites. Particularly of importance are the n capture isotopes produced selectively by low energy neutrons. These n capture nuclei should be more abundant in the inner solar system where solar secondary neutrons are present. One would expect then to find lower ratios if they were formed in the inner ESN and large ratios if the ^{26}Al endproduct was formed in the outer ESN.

Acknowledgments. Supported under NIH-MARC grant #443789

References

Castellani V., Fiorentini G., Ricci B. and Straniero O. 1995, The fate of Li and Be in stars and in the Laboratory, INFNFE-04-95 (Preprint).
Clayton D. D., Dwek E. and Woosley S. E. 1977, ApJ, 214, 300-315.
Clayon D. D. 1994, Nature, 368, 222-224.
Clayton,D.D. and Lin, L. 1995, ApJ, 451,L87-91
Kastner J.J. and Myers P.C. 1994, ApJ, 421, 605-614
Keppler E., Franz M., Korth A., Preuss M. K., Blake J. B., Seidel R., Quenby J. J. and Witte M. 1995, Science, 268, 1013-1016.
Prantzos N. 1993, ApJ, 405, L55-L58
Ramaty,R. Kozlovsky, B. and Lingenfelter, R.E. 1996, ApJ, submitted
Rolfs C. E. And Rodney W. S. 1988, Cauldrons in the Cosmos, in Nuclear Astrophysics, (U.of C. Press), 165-168.
Simpson, J. A., Anglin J. D., Bothmer V., Connell J. J., Ferrando P., Heber B., Kunow H., Lopate C., Marsden R. G., McKibben 1995, Science, 268, 1019-1023.
Spergel M. S. 1995, Astrophysics and Sp.Sci, 223, 187.
Truran J.W. and Cameron A.G.W. 1978, ApJ, 219, 226-229

Lithium in Stars

J.A. Thorburn[1]

Yerkes Observatory, University of Chicago, P.O. Box 258, Williams Bay, WI 53191

Abstract. The observational status of lithium in Population I and II stars is assessed. Implications of the observations for the primordial lithium abundance, Galactic chemical evolution, and stellar physics are discussed.

1. Introduction

Abundances of lithium in stars provide crucial information on three different classes of astrophysical phenomena: internal stellar physics, Galactic chemical enrichment, and elemental production in the Big-Bang. Because deuterium, helium, and lithium are the only three elements produced in detectable amounts through standard Big-Bang nucleosynthesis (BBN), their primordial abundances constitute a powerful validity test of the simplest Big-Bang model. Standard BBN predicts a unique function relating each elemental abundance to η, the baryon-to-photon ratio. If observations of primordial abundances correspond to an overlapping range of η, BBN models may then be utilized to constrain the physical conditions in the very early universe and global parameters such as Ω_B, the density of normal matter in the universe.

Lithium can also be produced by cosmic ray reactions in the interstellar medium, by novae, by stellar flares, by thermally pulsing asymptotic giant branch stars, and by the neutrino process near Type II supernovae. The original lithium produced during the Big-Bang may be overwhelmed by subsequent Galactic chemical evolution. In addition, lithium is destroyed in stellar interiors at a temperature of 2.5×10^6 K, much cooler than the onset of hydrogen burning. Only the outermost 1–3% by mass of a typical star retains an appreciable amount of lithium. As gas is cycled through stellar generations, the original lithium is progressively removed from the interstellar medium through astration. Studying the history of Galactic lithium abundances exposes parameters of keen interest to Galactic chemical evolution: the initial mass function of stars throughout Galactic history, the efficiency of gas recycling and mixing in the interstellar medium, the chemical yields from supernovae, the spectrum of cosmic rays now and in the distant past, and the rate of mass infall from the Galactic halo.

---

[1] Hubble Fellow

Because lithium is a fragile element which is easily burned in stellar interiors, the evolution in surface lithium abundance can be used as a diagnostic of internal mixing. Any process which induces an exchange of material between the stellar surface and the depths where lithium is absent causes the surface abundance of lithium to decline. By tracking the surface lithium abundance histories of stars during the pre-main sequence (pre-MS) and main sequence (MS) evolution, we are able to probe the depth of mixing in stars and its dependence on mass, age, composition, rotational state, and rotational history. Observations of surface lithium abundances in stars reveal internal properties which might otherwise remain hidden.

For these reasons, lithium abundances of stars form a complex dataset, in whose interpretation cosmology, Galactic chemical evolution, and stellar physics are woven seamlessly together. Understanding the implications of lithium observations for any one of these fields entails knowledge of all three. This review will briefly summarize the observations of lithium in stars and explore some of their implications for the three classes of astrophysical phenomena.

2. Lithium in Population I Stars

The realization that Sun-like stars substantially deplete their surface lithium dates to the discovery of the "classical lithium problem": a two order-of-magnitude disparity in the solar photospheric lithium abundance ($N(Li) = \log(Li/H) + 12 = 1.1$) relative to the meteoritic value ($N(Li) = 3.3$; e.g. Anders & Grevesse 1989). Standard models of the Sun—which do not include rotation, convective overshoot, diffusion, or mass-loss—predict very little decline in surface lithium abundance during the first 4.5 Gyr of the Sun's life (Pinsonneault, Kawaler, & Demarque 1990). Assuming that the Sun is typical for its age and spectral type, additional physics is probably needed to explain the depth of mixing—as inferred from the lithium deficiency—characteristic of G stars. Multitudinous explanations for lithium depletion of the solar photosphere are extant in the literature. Of these, the dominant categories include convective overshoot (Straus, Blake, & Schramm 1976; D'Antona & Mazzitelli 1984; VandenBerg & Poll 1989; Ahrens, Stix, & Thorn 1992), mass-loss (Hobbs, Iben, & Pilachowski 1989; Boothroyd, Sackmann, & Fowler 1991; Swenson & Faulkner 1992), increased interior opacities (Swenson, Stringfellow, & Faulkner 1990), and rotationally induced mixing (Vauclair et al. 1978; Baglin, Morel, & Schatzman 1985; Pinsonneault et al. 1990; Charbonnel et al. 1994; Chaboyer, Demarque, & Pinsonneault 1995). As a test of these hypotheses, many investigations of lithium in Sun-like members of star clusters have been pursued. Open clusters form a valuable test site for lithium depletion mechanisms because they have well determined ages, and they consist of many coeval stars with identical initial composition. Through studies of lithium in many open clusters, the dependence of depletion on stellar age and mass can be tested directly. The degree to which lithium depletion is controlled by parameters other than the three fundamental atmospheric parameters (age, composition, and mass) can also be estimated by comparing lithium abundance of otherwise indistiguishable stars; if depletion is not affected by a fourth parameter, then stars of identical mass, age, and initial composition should experience the same amount of lithium depletion.

Studies of lithium in open clusters have demonstrated that Population I stars of roughly solar mass deplete lithium progressively during their lives, but with moderate variation between otherwise indistinguishable stars. For example, stars in the Pleiades cluster (age 70 Myr) show lithium abundances N(Li) ≈ 3.3 in F stars with a gradual decrease in N(Li) among G and K members (Soderblom et al. 1993). A dispersion in N(Li) among stars of similar temperature is also evident; this observation requires either a high degree of non-coevality in the Pleiades cluster or significant lithium depletion from a mixing mechanism not controlled by the fundamental atmospheric parameters. In contrast, F and G members in the Hyades cluster (age 700 Myr) show a complex dependence of N(Li) upon effective temperature: among middle F stars, N(Li) is sharply lower than at cooler or warmer temperatures, a phenomenon known as the "F gap" (Boesgaard & Tripicco 1986). N(Li) measurements for late F stars are distributed about a nearly constant N(Li), but with significantly greater scatter than can be attributed to observational error. In the G and K stars, N(Li) declines rapidly, and less scatter is seen (Thorburn et al. 1993). The difference in N(Li) between the Pleiades and Hyades is negligible at 6400 K, but it exceeds an order of magnitude in the region of the "F gap" and in the late G stars. Similar comparisons to the clusters NGC 752 (age 1.7 Gyr) and M67 (age 4 Gyr) show a continuing pattern of decreasing N(Li) with MS age among stars of solar effective temperature (e.g. Hobbs & Pilachowski 1988).

The observed dependence of N(Li) upon MS age in Sun-like stars is incompatible with convection or convective overshoot being the dominant cause of lithium depletion. Convective removal of lithium is effective when the convection zone is at maximum size and the bottom of the convection zone is either hotter than 2.5×10^6 K or extends beneath the lowest level where lithium survives: the former is most apt to occur during the pre-MS (when a star is fully convective) and the latter will be important upon evolution from the MS. Any alteration of convection theoretically capable of enhancing lithium depletion on the MS, such as convective overshoot, is far more efficient during the pre-MS and it will have an overwhelming effect at that time. Because the observations of lithium in open clusters suggest predominantly MS lithium depletion, non-convective mixing probably plays the dominant role in depleting lithium from Sun-like stars.

The scatter in N(Li) among otherwise similar stars in the Pleiades and Hyades shows the importance of a factor other than the fundamental stellar parameters to lithium depletion. Several pieces of evidence suggest that rotation is the key. In the Hyades, the coincidence in temperature range between the "F gap," the break in the Kraft rotational curve (Kraft 1965), and the presence of a "Ca II gap" (Böhm-Vitense 1995) strongly implies a connection between spin-down and lithium depletion in F type stars. Fast rotation is associated with high lithium abundance in the young Pleiades G and K stars (Soderblom et al. 1993) and in pre-MS stars (Martín et al. 1994b). Short period, tidally-locked binaries in the Hyades and in M67 show higher lithium abundances than do single stars or widely separated binaries of comparable spectral type (Thorburn et al. 1993; Deliyannis et al. 1995): rotational history may be an important factor in lithium depletion. Although rotation itself appears to be related to depletion, the observational evidence clearly contradicts the theory that meridional circulation itself causes lithium depletion in young stars; if it did, high rotational velocity would be anticorrelated with lithium abundance rather than positively correlated. Similarly, if mixing produced by the loss of angular momentum (Pinsonneault et al. 1990) explained the observational data, slowly rotating stars in the Pleiades must necessarily have lost more angular momentum than fast rotators of the same mass. Unless the departures from coevality are substantial in the Pleiades, explaining the lithium data through angular momentum loss may require modification of Skumanich (1972) law for stellar spin-down: $v \propto t^{-0.5}$. Thus, although the observational evidence points to rotation as an important contributor to lithium depletion in Population I stars, the true nature of its role in this process is not fully understood. Despite a wealth of observational data, a comprehensive solution to the "classical lithium problem" in Population I stars remains elusive.

3. Lithium in Population II stars

The surprising discovery that lithium is retained in the atmospheres of the Galaxy's oldest stars dates was announced by Spite & Spite (1982). Prior to this time, it was believed that lithium would not be observable in halo stars, due to their great age and small mass: Population I stars of comparable mass and age do not show lithium on their surface. Even more remarkable was the recognition that halo stars of effective temperature 5700–6300 K show a lithium abundance N(Li) = 2.2 ± 0.2—a factor-of-ten lower than the solar meteoritic value—with virtually no scatter, despite the wide range in composition, diverse Galactic orbits, and a spread of ages. This close agreement in lithium abundances among halo stars near the MS turnoff is qualitatively very different from the large dispersion in lithium abundances for Population I stars, and it was interpreted as evidence that lithium depletion in warm halo stars is very ineffective, perhaps even inoperative. Given the great age, relatively uncontaminated elemental composition, and widely removed birth places of the "Spite plateau" stars, the coincidence in their surface lithium abundances appeared fantastically unlikely unless these stars started with and preserve the relic lithium produced in BBN. A cosmological interpretation of halo star lithium is supported by predictions of standard stellar models, which entail an undetectably small amount of lithium depletion in iron-poor stars (Deliyannis, Demarque, & Kawaler 1990). The fortuitous agreement between the Spite plateau lithium fraction and expectations from standard BBN (e.g. Smith, Kawano, & Malaney 1993), lent further credence to a primordial interpretation of halo star lithium.

However, subsequent investigations of halo star lithium abundances pose difficulties for the Spites' original hypothesis. Several severely lithium deficient stars with effective temperatures in the plateau interval have been discovered (Hobbs, Welty, & Thorburn 1991; Thorburn 1992; Spite et al. 1994). Although lithium-free halo stars make up only 5% of all Population II main sequence turnoff stars with lithium observations, their extreme lithium deficiency is not easily accommodated in the simplest interpretations of the Spite plateau. Furthermore, the existence of even a few lithium deficient halo stars hints that lithium depletion can occur in halo stars, by mechanisms which are as yet unidentified. If a few stars are affected strongly by lithium depletion, can we exclude the possibility that all halo stars are affected by depletion?

Several recent studies have suggested that halo stars *should* deplete lithium substantially during their MS lifetimes, if our knowledge of Population I stars is applied straightforwardly to Population II. One suggested means of depleting lithium substantially in halo stars is through angular momentum loss (Pinsonneault, Deliyannis, & Demarque 1992). As in disk stars, the spin-down of a halo star is expected to produce shear in the internal rotation which is unstable to vertical mixing. Depletion of lithium by an order-of-magnitude or more is the predicted result for halo stars, assuming a distribution of initial angular momenta for halo stars similar to that for disk stars: the initial lithium abundances of halo stars could be similar to those observed in young open clusters. This stellar model also predicts a nearly flat lithium plateau, where N(Li) has little dependence on effective temperature in the range 5700–6300 K. If a slow mixing mechanism of this type operates in halo stars, it might require a higher primordial lithium abundance than is consistent with standard BBN.

Other mechanisms which produce moderate lithium depletion include microscopic diffusion (Proffit & Michaud 1991), a stellar wind (Vauclair & Charbonnel 1994; Swenson 1995), and other mass loss scenarios (Dearborn, Schramm, & Hobbs 1992). In the case of microscopic diffusion, the observational evidence contradicts model predictions: no decline in N(Li) is observed among the hottest Spite plateau stars. Therefore, microscopic diffusion can probably be eliminated as a significant cause of lithium depletion in halo stars (Ryan et al. 1996). In the case of each mass loss scenario, an appreciable amount of lithium depletion only occurs if the rate of mass loss exceeds the Solar wind loss rate by two orders of magnitude. Because little observational evidence exists for any mass loss in halo main sequence turnoff stars, lithium depletion through these processes is regarded as speculative at present.

The rarer, more fragile isotope ^{6}Li has been detected in several halo stars near the main sequence turn-off (Smith, Lambert, & Nissen 1993; Hobbs & Thorburn 1994). This discovery could potentially eliminate all lithium depletion scenarios which invoke substantial nuclear burning or mass-loss in halo stars unless the initial ^{6}Li abundance was extremely high; in both cases, even a slight amount of ^{7}Li depletion would lead to total elimination of ^{6}Li from the star's surface. Even depletion mechanisms which call for slow mixing, with a corresponding gentler differential depletion of ^{6}Li relative to ^{7}Li (and lithium relative to beryllium), are constrained by the presence of ^{6}Li in halo stars unless the initial ^{6}Li abundance was very high. Paradoxically, although the degree of lithium depletion may be limited by observations of ^{6}Li, the mere existence of ^{6}Li in these stars is antithetical to a direct identity between halo star surface lithium abundances and the lithium fraction produced in standard BBN: standard BBN does not produce measureable amounts of ^{6}Li (e.g. Thomas et al. 1993). Rather, the presence of ^{6}Li in halo stars demands either a high primordial ^{6}Li abundance with moderate lithium depletion, minor Galactic contribution through cosmic ray spallation (e.g. Prantzos, Casse, & Vangioni-Flam 1993) to a low primordial lithium abundance, or *in situ* generation of ^{6}Li on the stellar surfaces (Lambert 1995; Deliyannis & Malaney 1995). Weak-to-moderate lithium production in early Galactic history has also been invoked to explain a mild dependence of

N(Li) upon [Fe/H] in the most metal-poor plateau stars (Thorburn 1994; Ryan et al. 1996). Alternatively, the observed correlation between N(Li) and [Fe/H] in very metal-poor stars may simply reflect systematic error in the model stellar atmospheres used to infer lithium abundance from line strengths (Kurucz 1995). However, abundances of the elements beryllium and boron—which are also produced by cosmic ray spallation—have been observed to scale with metallicity in halo stars (Boesgaard & King 1993; Duncan et al. 1996). If the production of beryllium and boron in metal-poor stars is due to cosmic ray nucleosynthesis, commensurate ^{6}Li and ^{7}Li production is expected. A viable explanation for ^{6}Li in halo stars must be referenced to the observations of beryllium and boron in the same stars (e.g. Steigman et al. 1993).

Recent suggestions of a high primordial deuterium abundance (e.g. Songaila et al. 1994) may also complicate our understanding of evolution in the other fragile light elements. If the low primordial lithium/no lithium depletion scenario is correct, Galactic lithium production is needed to increase the cosmic lithium abundance abruptly from N(Li) = 2.2, characteristic in halo stars, to N(Li) = 3.3, characteristic of young, undepleted disk stars. The two most promising means of producing lithium at the needed time and in the needed amount are neutrino nucleosynthesis in Type II supernovae, by which many odd-Z nuclei including ^{7}Li are produced (Woosley et al. 1990), and the ^{7}Be transport mechanism in thermally-pulsing asymptotic giant branch stars (Cameron & Fowler 1971; Sackmann & Boothroyd 1992). Although either of these mechanisms is thought to be capable of producing sufficient lithium to reproduce the observational data (Brown 1992; Timmes, Woosley, & Weaver 1995), our understanding of evolution in lithium abundances, as well as boron and beryllium, may be challenged by suggestions that the true primordial deuterium abundance—and thus the cosmic astration factor of deuterium—is an order-of-magnitude higher than previously thought. Because astration factors of lithium, beryllium, and boron are essentially identical to that for deuterium, studies of Galactic light element evolution will have to be revised to compensate for this effect. Although studies of lithium are not strongly affected, investigations of beryllium and boron abundances may underestimate the true slope of Be/H and B/H as a function of Fe/H or O/H and misidentify the production mechanisms of beryllium and boron as a result of this systematic effect. For example, the apparent transition in log(Be/H) vs. [Fe/H] noted between disk-like and halo-like metallicities (Deliyannis et al. 1996) could be a signature of astration rather than a change in beryllium production mechanism. Achieving a mutually consistent interpretation of all light element abundances is a mandatory but largely unappreciated step towards understanding abundance evolution in any one light element.

Observations of lithium abundances in cool, short-period, tidally-locked Pop. II binaries show a similar pattern of lithium overabundances as do Population I examples (Balachandran et al. 1993; Ryan & Deliyannis 1995). These data imply that lithium depletion depends upon rotation similarly between Population I and cool Population II stars: cool halo stars probably are affected by rotational lithium depletion. However, cool halo stars are known to be depleted in lithium, given their low abundances compared to the plateau. The consequences of the cool star observations for the warmer stars, where depletion has played a lesser role, are not obvious. At present, no tidally-locked binary in the plateau temperature interval has been observed with an anomalously high lithium abundance.

The strongest argument for rotational lithium depletion in Spite plateau stars can be made from lithium abservations of slightly evolved subgiants in the very metal-poor globular cluster M92 (Deliyannis, Boesgaard, & King 1995). Similarly to the studies of the Pleiades and Hyades young open clusters, the investigation of M92 reported significantly different lithium abundances for three stars with almost indistiguishable colors. The full range of lithium abundance among the three stars was roughly a factor-of-two, with temperatures near the cool edge of the Spite plateau. As in the case of the Hyades and Pleiades, this observation can be interpreted as evidence that lithium abundances in globular cluster stars depend upon stellar properties other than age, initial composition, and mass. Rotation-related lithium depletion is one of the simplest explanations.

However, the actual ramifications of globular cluster lithium observations for field stars of the Galactic halo are unclear. Globular clusters provide a far more crowded stellar environment than the halo field, with far greater opportunity for interactions between stars. One possible means by which close stellar encounters may affect lithium abundances of stars is suggested by the known presence of millisecond pulsars in globulars. Low mass X-ray binaries, where a neutron star is associated with a late type main sequence companion, are associated with strong, localized lithium production (Martín et al. 1994a), as inferred from the peculiarly high lithium abundances of their secondaries. Apart from the possibility of stellar encounters, the logical leap from globular clusters—as well as Pop. I open clusters—to halo field stars is hampered by uncertainties in our knowledge of Population II stellar spin-down. Can the distribution of initial angular momenta for halo stars safely be postulated from observations of young disk stars? Does the rotation of the kinematic system have any bearing on the angular momenta of its constituent stars? In the case of globular clusters with identical [Fe/H], differences in giant branch CNO abundances have long been attributed to differences in stellar rotation. Complementary differences in lithium depletion among globular cluster stars in the Spite plateau temperature interval might also be expected for this reason. The logical leap from lithium abundances in globular clusters to field halo stars then depends on the answers to several questions. Are field stars derived from dissociated globular clusters? If so, observations of many globular clusters might serve to illuminate the factors contributing to lithium depletion in halo stars. Does the non-rotating nature of the Galactic halo field imply a different set of initial angular momenta and thus a different amount of lithium depletion than the more compact, sometimes rotating globular clusters? If so, then does the weaker rotation of the halo field and globular clusters suggest lower initial angular momenta (and less lithium depletion) for halo stars than are characteristic of the Galactic disk? Thus, although the observations of lithium in M92 subgiants are provocative and may ultimately require modifications to the common interpretation of the Spite plateau, interpretation of the data is far from unique. The effect of rotational lithium depletion on the halo field Spite plateau is not conclusively established by the globular cluster lithium observations.

4. Conclusions

The study of stellar lithium abundances is a complex endeavor in which three separate fields of astrophysics—stellar physics, Galactic chemical evolution, and Big-Bang nucleosynthesis—converge and become inextricably intertwined. Lithium abundances of Sun-like Population I stars contain a wealth of information about the mechanisms and depth of stellar mixing, which are otherwise unobservable and unknowable in most stars. Observational evidence suggests that the simplest stellar models, which do not include rotation, mass-loss, or microscopic diffusion, are inadequate to account for lithium abundances of the Sun and comparable main sequence stars. Evidence of the integral role of rotation in lithium depletion continues to mount: tidally-locked binaries show higher lithium abundances than single or widely separated counterparts in coeval samples, and fast rotators show higher lithium abundances than slow rotators among pre-MS and young MS stars. However, although some predictions of specific models are consistent with the data—for example, mixing due to angular momentum loss predicts high lithium abundances in tidally-locked binaries—an entirely self-consistent explanation of all existing data is not yet forthcoming.

The interpretation of Population II lithium abundances is complicated by the necessity of calibrating models of stellar physics to the Sun, a star whose lithium depletion has not been explained satisfactorily. Taken at face value, the lithium abundances of field halo stars appear to indicate very little lithium depletion of and minor Galactic contribution to a primordial lithium fraction $N(Li) = 2.2 \pm 0.2$ which is compatible with standard BBN. However, this interpretation is not unique, nor is it secured by the detection of the more fragile isotope ^6Li in warm halo stars. Several scenarios are possible which would entail severe lithium depletion in most halo stars and a corresponding need for a high primordial ^7Li and ^6Li abundance. Recent observations of a lithium abundance dispersion in the stars of M92 constitute the most compelling evidence of moderate-to-severe rotation-related lithium depletion in "Spite plateau" stars. Resolving these two fundamentally opposed interpretations will require greater knowledge of stellar physics and of the unobservable properties of halo stars, such as the distribution of initial angular momenta and angular momentum loss law in early main sequence evolution, both of which bear sensitively on the theoretical estimates of lithium depletion. Finally, our understanding of lithium abundances in halo stars must be placed in a context which self-consistently accounts for chemical evolution of beryllium and boron, two other elements which are destroyed in stars but are also produced in the early Galaxy, as well as of lighter elements such as deuterium and ^3He. At present, this all-inclusive account of lithium in stars remains an aspiration, not a reality.

Acknowledgments. I heartily thank Doug Duncan for delivering this talk, during my sudden illness. His willingness to substitute for me with almost no advance warning went beyond the call of duty.

This work has received support from NASA through grant HF-1064.01-94A awarded by the Space Telescope Science Institute which is operated by the Association of Universities for Research in Astronomy, Inc., for NASA under contract NAS5-26555.

References

Ahrens, B., Stix, M., & Thorn, M. 1992, A&A, 264, 673
Anders, E., & Grevesse, N. 1989, Geochim. Cosmochim. Acta, 53, 197
Baglin, A., Morel, P.J., & Schatzman, E. 1985, A&A, 149, 309
Balachandran, S., Carney, B.W., Fry, A.M., Fullton, L.K., & Peterson, R.C. 1993, ApJ, 413, 368
Boesgaard, A.M., & King, J.R. 1993, AJ, 106, 2309
Boesgaard, A.M., & Tripicco, M.J. 1986, ApJ, 302, L49
Böhm-Vitense, E. 1995, A&A, 297, L25
Boothroyd, A.I., Sackmann, I.-J., & Fowler, W.A. 1991, ApJ, 377, 318
Brown, L.E. 1992, ApJ, 389, 251
Cameron, A.G.W., & Fowler, W.A. 1971, ApJ, 164, 111
Chaboyer, B., Demarque, P., & Pinsonneault, M.H. 1995, ApJ, 441, 865
Charbonnel, C., Vauclair, S., Maeder, A., Meynet, G., & Schaller, G. 1994, A&A, 283, 155
D'Antona, F., & Mazzitelli, I. 1984, A&A, 138, 431
Dearborn, D.S.P., Schramm, D.N., & Hobbs, L.M. 1992, ApJ, 394, L61
Deliyannis, C.P., Boesgaard, A.M., & King, J.R. 1995, ApJ, 452, L13
Deliyannis, C.P., Boesgaard, A.M., King, J.R., & Duncan, D.K. 1996, AJ, in press
Deliyannis, C.P., Demarque, P., & Kawaler, S.D. 1990, ApJS, 73, 21
Deliyannis, C.P., King, J.R., Boesgaard, A.M., & Ryan, S.G. 1995, ApJ, 434, L81
Deliyannis, C.P., & Malaney, R.A. 1995, ApJ, 453, 810
Duncan, D.K., Primas, F., Boesgaard, A.M., Deliyannis, C.P., Hobbs, L.M., King, J., Ryan, S., & Rebull, L.M. 1996, this volume
Hobbs, L.M., Iben, I., Jr., & Pilachowski, C. 1989, ApJ, 347, 817
Hobbs, L.M., & Pilachowski, C. 1988, ApJ, 334, 734
Hobbs, L.M., & Thorburn, J.A. 1994, ApJ, 428, L25
Hobbs, L.M., Welty, D.E., & Thorburn, J.A. 1991, ApJ, 373, L47
Kraft, R.P. 1965, ApJ, 142, 681
Kurucz, R.L. 1995, ApJ, 452, 102
Lambert, D.L. 1995, A&A, 301, 478
Martín, E.L., Rebolo, R., Casares, J., & Charles, P.A. 1994a, ApJ, 435, 791
Martín, E.L., Rebolo, R., Magazzù, A., & Pavlenko, Y.V. 1994b, A&A, 282, 518
Pinsonneault, M.H., Deliyannis, C.P., & Demarque, P. 1992, ApJS, 78, 179
Pinsonneault, M.H., Kawaler, S.D., & Demarque, P. 1990, ApJS, 74, 501
Prantzos, N., Casse, M., & Vangioni-Flam, E. 1993, ApJ, 403, 630
Proffit, C.R., & Michaud, G. 1991, ApJ, 371, 584
Ryan, S.G., Beers, T.C., Deliyannis, C.P., & Thorburn, J.A. 1996, ApJ, Feb. 20 issue

Ryan, S., & Deliyannis, C.P. 1995, ApJ, 453, 819
Sackmann, I.-J., & Boothroyd, A.I. 1992, ApJ, 392, L71
Skumanich, A. 1972, ApJ, 171, 565
Smith, M.S., Kawano, L.H., & Malaney, R.A. 1993, ApJS, 85, 219
Smith, V.V., Lambert, D.L., & Nissen, P.E. 1993, ApJ, 408, 262
Soderblom, D.R., Jones, B.F., Balachandran, S., Stauffer, J.R., Duncan, D.K., Fedele, S.B., & Hudon, J.D. 1993, AJ, 106, 1059
Songaila, A., Cowie, L.L., Hogan, C.J., & Rugers, M. 1994, Nature, 368, 599
Spite, M., Molaro, P., François, P., & Spite, F. 1993, A&A, 271, L1
Spite, F., & Spite, M. 1982, A&A, 115, 357
Steigman, G., Fields, B.D., Olive, K.A., Schramm, D.N., & Walker, T.P. 1993, ApJ, 415, L35
Straus, J.M., Blake, J.B., & Schramm, D.N. 1976, ApJ, 204, 481
Swenson, F. 1995, ApJ, 438, L87
Swenson, F.J., & Faulkner, J. 1992, ApJ, 395, 654
Swenson, F.J., Stringfellow, G.S., & Faulkner, J. 1990, ApJ, 348, L33
Thomas, D., Schramm, D.N., Olive, K.A., & Fields, B.D. 1993, ApJ, 406, 569
Thorburn, J.A. 1992, ApJ, 399, L83
Thorburn, J.A. 1994, ApJ, 421, 318
Thorburn, J.A., Hobbs, L.M., Deliyannis, C.P., & Pinsonneault, M.H. 1993, ApJ, 415, 150
VandenBerg, D.A., & Poll, H.E. 1989, AJ, 98, 1451
Vauclair, S., & Charbonnel, C. 1995, A&A, 295, 715
Vauclair, S., Vauclair, G., Schatzman, E., & Michaud, G. 1978, ApJ, 223, 567
Woosley, S.E., Hartmann, D.H., Hoffman, R.D., & Haxton, W.C. 1990, ApJ, 356, 272

Be Abundances in the Alpha Centauri System

F. Primas

Department of Astronomy and Astrophysics, University of Chicago, Chicago, IL 60637

D.K. Duncan

Department of Astronomy and Astrophysics, University of Chicago, Chicago, IL 60637

R.C. Peterson

Lick Observatory, University of California, Board of Study, and Department of Astronomy and Astrophysics, University of Chicago, Chicago, IL 60637

J.A. Thorburn

Yerkes Observatory, University of Chicago, Williams Bay, WI

Abstract. High signal-to-noise *Hubble Space Telescope* GHRS spectra of α Cen A (spectral type G2 V) and α Cen B (spectral type K1 V) have been analyzed in the Be II λ3130 spectral region, for the purpose of making a precise determination of light element destruction in stars whose masses are slightly greater than and slightly less than the Sun's.

A detailed spectrum synthesis has been made using a version of the Kurucz line list which has been carefully tested by comparison to the Sun and to metal-poor stars, including one which appears to be Be-free. Using a model atmosphere with [M/H] = +0.10, our analysis gives [Be/H] = -0.04 ± 0.12 dex for α Cen A and -0.54 ± 0.24 dex for α Cen B. Using an atmospheric model of solar metallicity, we find [Be/H] = 0.06 ± 0.12 dex and -0.34 ± 0.24 dex for α Cen A and B respectively. Either analysis fits our data equally well.

Since α Cen A and B both show large destruction of Li (which is destroyed at a lower temperature than Be), the lack of Be destruction in α Cen A and moderate destruction in α Cen B should allow rather accurate determination of the depth of mixing once models have been computed.

1. Introduction

Light element abundances can be used as a fundamental test of models of the outer structure of solar-type stars, since these elements undergo nuclear destruction by (p,α) reactions at very low temperatures. Li, Be, and B are destroyed in progressively deeper layers in the outer regions of stars. Thus their surface abundances are the observable consequences of internal processes. When combined with determinations of the other two light elements, Be abundances provide strong constraints on stellar models. The α Cen system is excellent for study since has a well-determined orbit and parallax, and both stars have very well known physical parameters, including $M(A)=1.1M_\odot$ and $M(B)=0.9M_\odot$. Soderblom and Dravins (1984) have accurately determined the Li abundances, finding destruction of almost a factor of 100 in α Cen A, and > 100 in B, compared to protostellar values.

2. Data

Data was obtained using the Echelle-B grating of the Goddard High Resolution Spectrograph of the Hubble Space Telescope on July 3, 1993. Nominal echelle resolution of 90,000 was degraded to approximately 35,000 due to the aberrated telescope point spread function observed through the Large Science Aperture. S/N was carefully tested and shown to be equal to the photon statistics limit of ≥ 100 per 0.008Å pixel.

3. Abundance analysis

3.1. Be Linelist

Our spectrum syntheses use the version of SYNTHE distributed by Kurucz (1993) on CD-ROM # 18, modified to run on UNIX SPARC stations by Steve Allen (UC Santa Cruz). The model atmospheres are those which included blanketing by nearly 60 million atomic and molecular features.

Our beginning line list included only atomic and molecular lines with laboratory-measured wavelengths. (Duncan et al. 1996 discuss why Kurucz-predicted lines are good for model atmosphere computation but not detailed spectrum synthesis). The great similarity between the Sun and α Cen A allowed refinement of the line list. We adjusted downward the gf values of the few lines which were off, and added a few atomic and molecular transitions (taken from Nave et al. 1994) of known wavelength but without laboratory-determined gf values until a good match was achieved first of all in the Sun, then α Cen A and B, and then spectra of the metal-poor stars HD106516 and HD140283 (Molaro et al. 1995). None of these changes affect the Be determination.

3.2. Spectrum Synthesis

Chmielewski et al. (1992) report a complete summary of all the most recent determinations of the stellar parameters for both α Cen A and B. We started from the Kurucz' grid models (T_{eff}, $\log g$ and [M/H]) 5750, 4.5, 0.00 for α Cen A and 5250, 4.5, 0.00 for α Cen B, and tested adjusting all three parameters. The α Cen A system is found to be slightly more metal rich than the Sun by most authors. We can fit our data equally well models of [M/H] = 0.00 or +0.10 but with slightly different Be abundances.

The Be II line is a doublet, and in carefully fitting both components we found evidence of slight blending of the 3131.066Å line. Trial of a number of possible candidates resulted in a best guess of Fe I, at 3131.043 Å, as the blend. This is mainly constrained by HD106516, a star of metallicity [M/H] = -1.00 but apparently with no Be. This blend affects the derived Be abundances by about 0.05 dex, smaller than the errors given below.

4. Uncertainties

Our best-determined Be abundances and the effects of changes due to uncertainties in stellar T_{eff}, $\log g$, metallicity, and continuum location are given in Table 1, and a sample synthesis fit in Fig. 1. A change of ±250 K in the temperature of α Cen A corresponds to a 25% variation in the calculated flux. Since absolute fluxes from HST are known to ~10%, this provides an extra temperature constraint ±100 K. Similar consideration gives ±65 K for the temperature of α Cen B. Placement of the continuum is a significant source of error. This was estimated by choosing different regions for normalization to the models we could still (barely) fit the Be feature and continuum with vertical shifts of 3-4%. We consider this a conservative estimate of this error source. One systematic effect was quantified by running a synthesis including the entire Kurucz list of predicted lines. Although the wavelength of any one of these lines can be off by several Å, the hundreds of weak lines this list adds shows the statistical effect on continuum placement.

Table 1. Be Abundances and Uncertainties.

	$[\text{Be/H}]_{\alpha\ CenA}$	$[\text{Be/H}]_{\alpha\ CenB}$
Abundances:		
$[M/H]=+0.10$	-0.04 ± 0.12	-0.54 ± 0.24
$[M/H]= 0.00$	0.06 ± 0.12	-0.34 ± 0.24
Uncertainties:		
$T_{eff}\pm100$ K	±0.04 dex	±0.04 dex
$\log g\pm0.03$ dex	±0.018 dex	±0.024 dex
$[M/H]\pm0.10$ dex	±0.10 dex	±0.20 dex
Continuum ±3.5%	±0.05 dex	±0.10 dex
Model ± predicted lines	±0.03 dex	±0.06 dex
Total uncertainty	±0.12 dex	±0.24 dex

5. Discussion

Li, Be, and B burn at progressively lower temperatures (i.e. 2.5, 3.5 and 5 $\times 10^6$ K respectively). Li is strongly depleted in the sun, and standard stellar models cannot reproduce this since the bottom of the surface convection zone is predicted not to reach the burning temperature. Many additional mixing mechanisms have therefore been suggested including mass loss, diffusion, rotationally-driven mixing, gravitational waves, and others. α Cen A and B both show large destruction of Li. The lack of Be destruction in α Cen A and moderate destruction in α Cen B allow rather accurate determination of the depth of mixing, to a temperature intermediate between 2.5 and 3.5 $\times 10^6$ K in

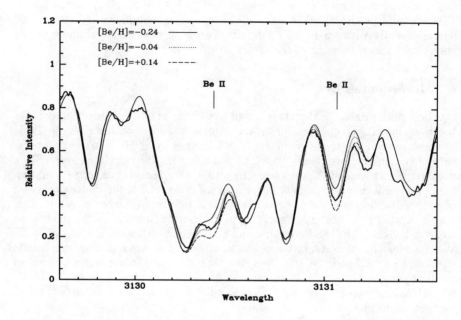

Figure 1. Spectrum synthesis around the Be II doublet for α Cen A

the case of α Cen A, and to very close to 3.5 ×10^6 K in the case of α Cen B. Strict limits to possible main sequence mass loss are also implied. More detail on the testing of such models will be presented in Primas et al. 1996.

References

Chmielewski, Y., Friel, E., Cayrel de Strobel, G., Bentolila, C. 1992, A&A, 263, 219

Kurucz, R.L. 1993, CD-ROM #13, #18

Molaro, P., Bonifacio, P., Castelli, F., Pasquini, L., Primas, F., in The light element abundances, P.Crane, Springer-Verlag, 415

Nave, G., Fuhr, J.R., Learner, R.C.M., Thorne, A.P., Brault, J.W. 1994, ApJS, 94, 221

Primas, F., Duncan, D.K., Peterson, R., Thorburn, J., Deliyannis, C.P., 1996, ApJ, in preparation

Soderblom, D.R., Dravins, D. 1984, A&A, 140, 427

Acknowledgments. This research was based on observations obtained with the NASA/ESA Hubble Space Telescope through the Space Telescope Science Institute, which is operated by the Association of Universities for Research in Astronomy, Inc., under NASA contract NAS5-26555.

The Evolution of Boron in the Galaxy

D.K. Duncan, F. Primas, K.A. Coble, L.M. Rebull

University of Chicago, Department of Astronomy and Astrophysics

A.M. Boesgaard

University of Hawaii Institute for Astronomy

Constantine P. Deliyannis

Yale University, Department of Astronomy

L.M. Hobbs

University of Chicago, Yerkes Observatory

J.R. King

University of Texas, Department of Astronomy

S. Ryan

Anglo-Australian Observatory

Abstract. The Goddard High Resolution Spectrograph (GHRS) of the Hubble Space Telescope (HST) has been used to obtain spectra of the boron 2500 Å region in eight stars ranging from [Fe/H] = -0.3 to -2.96, including the most metal-poor star ever observed for B. Spectrum synthesis using latest Kurucz model atmospheres has been used to determine [B/H] for each star, and particular attention paid to the errors of each point, to permit judgement of the goodness-of-fit of models of galactic chemical evolution.

A straight line of slope 0.98 gives an excellent fit to all available data. There is no indication of a change in slope between halo and disk metallicities. The B/Be ratio is typically 10. This data supports models of light element production by cosmic ray spallation of C,N,O nuclei onto protons and He nuclei, probably in the vicinity of massive supernovae in star-forming regions. It *does not support* the long-held view of light element production occuring in the general ISM from the collisions of high-energy protons onto C,N,O nuclei.

1. Introduction

The evolution of Li, Be, and B is a powerful descriminant between different models of the chemical and dynamical evolution of the Galaxy. Light element production depends on the intensity and shape of the cosmic ray spectrum, which in turn depends on the supernova and massive star formation rates. It also could depend on the rise of the (progenitor) CNO abundances and the decline of the gas mass fraction, which is affected by rates of infall of fresh (unprocessed) material and outflow, e.g. by supernova heating. The present observations were sought to constrain these models.

2. Data

The Goddard High-Resolution Spectrograph (GHRS) of the Hubble Space Telescope (HST) was used with the G270M grating to obtain spectra of resolution 26,000 in the BI region (2500 Å) of seven stars ranging in metallicity from [Fe/H] = -0.3 to -2.85. Typical S/N achieved was 25 per pixel (0.026Å), or 50 per diode. To this was added the data for BD -13°3442 ([Fe/H]=-2.96, the subject of a separate investigation), that of the Sun, and three stars observed by Duncan, Lambert, and Lemke (1992).

3. Analysis and Errors

Boron abundances were determined via spectrum synthesis using the Synthe program distributed by Kurucz (1993) on CD-ROM, modified to run on Unix SparcStations by Steve Allen (UC Santa Cruz). We used the linelist of Duncan *et al.* 1996, which consists almost entirely of laboratory-measured lines, and which fits both the Hyades giants and metal-poor stars.

Considerable care was spent in determining the error for each of the B determinations. Sources of random error we considered included stellar effective temperature, metallicity, continuum placement, and photon statistics in the points defining the line itself. In more metal-poor stars the continuum is easier to define, but the B line is weaker and less certain. At disk metallicity, continuum errors are larger but make less difference since the line is deep. Suggestions of systematic NLTE effects which would increase B abundances for all the stars are beyond the scope of the present investigation. The NLTE suggestion will be tested observationally in 1995 by HST observations of BI and BII.

A typical spectrum synthesis fit is shown in Figure 1. Errors for HD 76932 (one of the more metal-rich stars) and BD 26°3578 (one of the more metal-poor) are given as examples in the table.

4. Discussion

Figure 2 immediately shows the main result of this investigation: [B] is very linear with [Fe/H] over both disk and halo metallicities. A least squares fit to this data yields a slope of 0.98 and reduced chi square of 0.66, indicating an excellent fit.

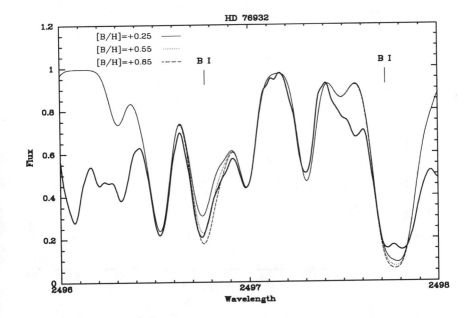

Figure 1. Sensitivity of model to boron abundances. Best-fit model with B abundances changed by ±0.3 dex.

Uncertainty	Effect on B abundances
HD 76932	
±100 K in T_{eff}	± 0.18 dex
±0.10 dex in [Fe/H]	± 0.15 dex
±1.5% in the placement of the continuum	± 0.02 dex
photon statistics	± 0.18 dex
uncertainties in $\log g$ and microturbulence (ξ)	insignificant
Net error :	± 0.29 dex.
BD 26°3578	
±125 K in T_{eff}	± 0.15 dex
±0.10 dex in [Fe/H]	± 0.15 dex
±1% in the placement of the continuum	± 0.10 dex
photon statistics	± 0.10 dex
uncertainties in $\log g$ and microturbulence (ξ)	insignificant
Net error :	± 0.25 dex.

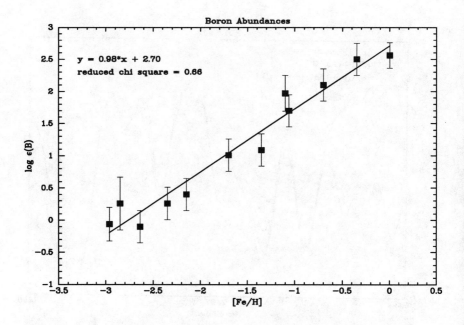

Figure 2. Variation of B abundance with [Fe/H]. Least-squares linear fit : $\log \epsilon(B) = 0.98[\mathrm{Fe/H}]+2.70$ with a reduced chi square $\chi^2_\nu = 0.66$.

Least-squares fitting with two line segments joined at any intermediate metallicity worsens the fit.

Figure 3 shows Be and Li abundances for the same stars. The least squares fit of Figure 3, reduced 10X, gives a good fit to the Be data. The two stars which show reduced Li abundances are the two coolest stars in the sample, and therefore might be expected to have deeper convection zones and thus destroyed some of their Li.

These data are now being used to test models of galactic chemical evolution. A preliminary conclusion is that the B/Be ratio strongly supports cosmic ray (CR) spallation as the source of the B and Be observed; no evidence is found for direct contribution of B from supernovae (Woosely et al. 1990). The slope of nearly exactly 1 supports models in which B and Be are produced by spallation of C,N,O onto protons. This was originally suggested by Duncan, Lambert, and Lemke (1992), and has been modelled in detail by Casse, Lehoucq, and Vangioni-Flam (1995) and Ramaty, Kozlovsky, and Lingenfelter (1995). Casse et al. and Ramaty et al. find that winds from massive stars in star-forming regions and massive star supenovae produce a flux of C and O which through collisions with protons and He nuclei can reproduce both the magnitude and slope of B production seen in Figure 2. They further identify the recent detection of gamma ray line emission from the Orion Nebula as direct evidence for this process.

These models differ from the long-accepted picture (Reeves, Fowler, and Hoyle, 1970) that spallation of CR protons onto C,N,O nuclei in the general interstellar medium is the primary source of light element production. The new

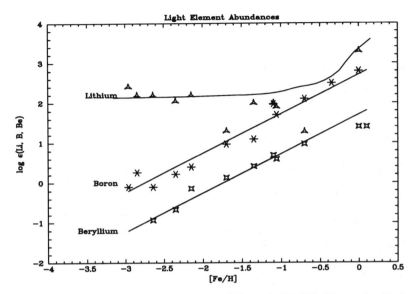

Figure 3. Variation of Li, Be, and B with [Fe/H]. Fit to the B data is that of Figure 2; fit to Be data is B fit − 1 dex. Fit to Li data from Duncan et al. (1992).

models also predict a higher $^{11}B/^{10}B$ ratio, and seem to solve the long-standing problem of isotopic composition seen for B in cosmic rays.

References

Casse, M., Lehoucq, R., and Vangioni-Flam, E. 1995, Nature, 373, 318.
Duncan, D., Peterson, R., Thorburn, J., Pinsonneault, M., and Deliyannis, C. 1996, ApJ, in press.
Duncan, D.K., Lambert, D.L., and Lemke, M. 1992, ApJ, 401, 584.
Kurucz, R.L. 1993, CDROM # 18.
Ramaty, R., Kozlovsky, B., and Lingenfelter, R., 1995, ApJ, in press.
Woosley, S., Hartmann, D., Hoffman, R., and Haxton, W. 1990, ApJ, 356, 272.

Acknowledgments. This research was based on observations obtained with the NASA/ESA Hubble Space Telescope through the Space Telescope Science Institute, which is operated by the Association of Universities for Research in Astronomy, Inc., under NASA contract NAS5-26555.

Boron abundance of BD-13°3442

L.M. Rebull, D.K. Duncan
University of Chicago, Department of Astronomy and Astrophysics

A.M. Boesgaard
University of Hawaii Institute for Astronomy

Constantine P. Deliyannis
Yale University, Department of Astronomy

L.M. Hobbs
University of Chicago, Yerkes Observatory

J.R. King
University of Texas, Department of Astronomy

S. Ryan
Anglo-Australian Observatory

Abstract. The Goddard High Resolution Spectrograph (GHRS) of the Hubble Space Telescope (HST) has been used to obtain a spectrum of the boron 2500 Å region of BD-13°3442. At a metallicity of [Fe/H]= −2.96, this is the most metal-poor star ever observed for B, and 30 hrs of spacecraft time (8.25 hours exposure time) resulted in a definite detection. Spectrum synthesis using the latest Kurucz model atmospheres yields an abundance of $\log \epsilon$ (B)= −0.06±0.25. This value fits a linear extrapolation of the B vs. [Fe/H] relation found for 11 halo and disk stars of less extreme metallicity (Duncan *et al.* 1996b), arguing against multiple sources of B such as an inhomogeneous Big Bang or direct production in supernovae. The slope of such a fit is almost exactly 1, which could be evidence of light element production by cosmic ray spallation of C,N,O nuclei onto protons, rather than the reverse, an alteration of the theory of light element production which has been accepted for many years.

1. Introduction

The evolution of Li, Be, and B is a powerful descriminant between different models of the chemical and dynamical evolution of the Galaxy. The theory of light element production which has been accepted for many years (e.g. Reeves and Meyer 1978) suggests production by cosmic ray (CR) spallation of protons onto CNO nuclei in the interstellar medium. Thus the light element abundances depend on the intensity and shape of the cosmic ray spectrum, which in turn depends on the supernova and massive star formation rates. They also could depend on the rise of the (progenitor) CNO abundances and the decline of the gas mass fraction, rates of infall and outflow, etc. The present observations were sought to constrain these models.

2. Data

The Goddard High-Resolution Spectrograph (GHRS) of the Hubble Space Telescope (HST) was used with the G270M grating to obtain a spectrum in the B I region (2500 Å). Data was collected for a total of 8.25 hours exposure time (30 hours spacecraft time) in six separate "visits" over several months from December 1994 to June 1995. The S/N of the final spectrum was 100, limited only by photon statistics; resolution was 24,000 (0.025 Å/pixel).

3. Analysis

Boron abundances were determined via spectrum synthesis using the Synthe program distributed by Kurucz (1993) on CD-ROM, modified to run on Unix SparcStations by Steve Allen (UC Santa Cruz). We used the linelist of Duncan et al. (1996a), which consists almost entirely of laboratory-measured lines, and which fits both the Hyades giants and metal-poor stars.

Considerable care was spent in determining the error for each of the B determinations. Sources of random error we considered included temperature, metallicity, continuum placement, and photon statistics in the points defining the line itself. In more metal-poor stars the continuum is easier to define, but the B line is weaker. At disk metalicity, continuum errors are larger but make less difference since the line is deep. Suggestions of NLTE effects which would increase B abundances will be tested observationally in 1995 by HST observations of BI and BII.

Figure 1 shows a comparison of BD-13°3442, BD3°740, and HD 140283, the three most metal-poor stars observed in the B region. It is clear that BD-13°3442 is the most metal-poor star of the three. (BD3°740 was observed by Duncan et al. 1996b; HD 140283 was observed by Duncan, Lambert, and Lemke 1992.)

Published values of stellar parameters for BD-13°3442 are [Fe/H] = -2.9 to -3.14, $\log g$ = 3.6 to 3.8, and T_{eff}= 6100 to 6300. Our final best-fit was achieved with a model of [Fe/H] = -2.96, $\log g$ = 3.0, and T_{eff}= 6100.

Considerable time was spent in determining errors; the results of this analysis appear in the table. The spectrum synthesis fit is shown in Figure 2.

4. Discussion

When the data of BD-13°3442 is combined with that of Duncan et al. (1996b and this volume), a straight line of slope 0.98 provides an excellent fit to the data.

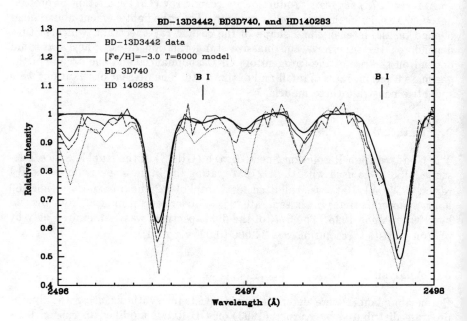

Figure 1. Comparison of BD-13°3442 ([Fe/H]=−2.96), BD-3°740 ([Fe/H]=−2.85), and HD 140283 ([Fe/H]=−2.64) in the boron region. BD-13°3442 is clearly the most metal-poor and HD 140283 is clearly the most metal-rich.

Uncertainty	Effect on B abundances
BD-13°3442	
±125 K in T_{eff}	± 0.1 dex
±0.15 dex in [Fe/H]	± 0.2 dex
±0.5% in the placement of the continuum	± 0.1 dex
Photon statistics	± 0.1 dex
uncertainties in $\log g$ and microturbulence (ξ)	insignificant
Net error :	± 0.26 dex.

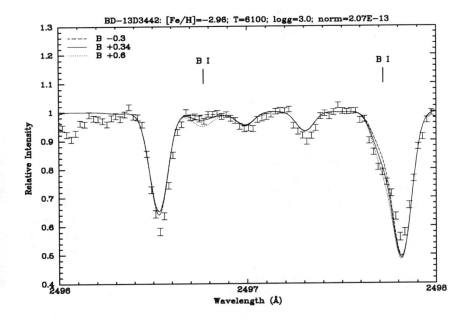

Figure 2. The best-fit value for [B/H] was +0.34; boron abundances also shown here are [B/H]= −0.3 and [B/H]= +0.6.

This may indicate a mechanism in which B and Be are produced by spallation of C,N, and O onto protons either near supernovae, as suggested by Duncan, Lambert, and Lemke (1992), or in winds from massive stars in star-forming regions (Casse, Lehoucq, and Vangioni-Flam, 1995), rather than primarily from CR proton spallation onto C,N, and O nuclei in the general interstellar medium. More detailed comparison with such models is being carried out.

References

Casse, M., Lehoucq, R., and Vangioni-Flam, E. 1995, Nature, 373, 318.
Duncan, D.K., Lambert, D.L., and Lemke, M. 1992, ApJ, 401, 584.
Duncan, D., et al., 1996b, in preparation
Duncan, D., Peterson, R., Thorburn, J., Pinsonneault, M., and Deliyannis, C. 1996a, ApJ, in press.
Kurucz, R.L. 1993, CDROM # 18.
Reeves, H. and Meyer, J.-P. 1978, ApJ, 226, 613.

Acknowledgments. This research was based on observations obtained with the NASA/ESA Hubble Space Telescope through the Space Telescope Science Institute, which is operated by the Association of Universities for Research in Astronomy, Inc., under NASA contract NAS5-26555.

Carbon and Oxygen Nucleosynthesis in the Galaxy: Problems and Prospects

Suchitra C. Balachandran [1]

Laboratory for Astrophysics, National Air and Space Museum, MRC 321, Smithsonian Institution, Washington DC 20560

Abstract. The abundances of the CNO elements are of vital importance in various astrophysical contexts. This paper reviews our current understanding of their evolution as a function of iron. Since stars of different masses produce different ratios of CNO to Fe, these trends are particularly useful in our understanding of Galactic chemical evolution. The CNO abundances are not reliably measured, especially at low metallicities, because of the few available optical features and the problems associated with their use. Improved and reliable measurements of carbon and oxygen are becoming available with the advent of high resolution infrared spectrographs which allow the vibrational-rotational transitions of CO and OH to be examined.

1. Introduction

Being the most abundant triad after hydrogen and helium, the importance of carbon, nitrogen and oxygen is widely recognized. For example, the CNO elements form an important opacity source in the stellar interior and play a vital role in energy generation via the CNO cycle. Furthermore, since oxygen and carbon are overproduced with respect to iron in massive and intermediate-mass stars compared to the solar ratio, the evolution of these ratios as a function of metallicity forms the basis for testing models of Galactic chemical evolution. In particular, because the models are now sophisticated enough to make accurate predictions with small uncertainties (e.g., Timmes et al. 1995), it is important to be able to provide high quality observational data for comparison.

This is not an exhaustive review. It builds upon the review on the evolution of the elements by Wheeler, Sneden and Truran (1989, hereafter WST) which contains the status of CNO measurements at that time, and deals solely with trying to piece together the evolution of these elements as a function of metallicity. Oxygen abundances have since been subject to a considerable amount of debate and controversy and those details are summarized below. While some of the recent measurements of carbon abundances in very metal-poor dwarfs are unreliable because they are derived from high-excitation lines, some information may be gleaned from trends in the higher metallicity disk stars. The present re-

---

[1] Visiting Astronomer at the Infrared Telescope Facility which is operated by by the University of Hawaii under contract to the National Aeronautics and Space Administration

view discusses the sometimes conflicting observational results, summarizes their limitations, investigates the causes of the conflicts and explains why improved results may soon become available.

2. The Status of Oxygen

2.1. Background

Since oxygen is produced mainly in massive stars, its abundance at low metallicities provides information about the earliest epochs of nucleosynthesis. The decline in [O/Fe] as [Fe/H] increases towards solar values is interpreted as due to the growing nucleosynthetic importance of Type Ia supernovae which preferentially produce iron over oxygen. The trend of [O/Fe] with [Fe/H] is hence of particular interest because it provides a measure of the timescale for the initial appearance of SN Ia.

Although oxygen is the most abundant of the CNO triad, it has few atomic transitisions and its abundance has been measured from a small number of optically accessible lines, of which the most commonly used are the O I triplet lines at 7775 Å. The modest overabundance of O with respect to Fe, [O/Fe] $\sim$ +0.5 at [Fe/H] $\sim$ -1.0, was well-known from previous O I-based studies (see WST for a summary). However, results from the O I triplet caused a mild furor when Abia and Rebolo (1989) extended the examination to lower metallicities and showed that [O/Fe] continued to increase with decreasing metallicity, reaching values as high as [O/Fe] $\sim$ +1.0 to +1.5 at [Fe/H] < -2.0. These values were in stark contrast to abundances obtained from the analysis of metal-poor giants which indicated that [O/Fe] levelled off at $\sim$ +0.3 to +0.5 for [Fe/H] < -1.0.

Two principal causes have been targeted to explain the observed oxygen abundance differences between the metal-poor dwarfs and giants. The available optical transitions conspire to make it nearly impossible to examine the same oxygen features in both dwarfs and giants. The high excitation potential (9.15 eV) O I triplet lines are undetectably weak in the cool atmospheres of the giants and the [O I] features used to derive abundances in giants are gravity-sensitive and weak in dwarfs. The O I triplet forms deep in the stellar photosphere near the convective boundary and non-LTE or other effects may, conceivably, result in spuriously high abundances. Alternatively, although the standard stellar models predict that the oxygen abundance in a giant is not altered by the dredge-up of ON-cycled material until later stages of evolution, non-standard mixing resulting in a lower oxygen abundance is clearly evidenced in some globular cluster giants (e.g., Kraft et al 1993). There is therefore general consensus that the pristine oxygen abundance is most reliably determined from the dwarfs. The high oxygen abundances measured in dwarfs from the O I triplet lines have a significant effect on ages derived from theoretical isochrones; globular clusters are several billion years younger if oxygen-enhanced mixtures are used (Vandenberg 1992). A flurry of activity therefore followed Abia and Rebolo's study, and these were aimed primarily at finding alternative features to determine oxygen abundances in dwarfs and at understanding the formation of the O I triplet.

2.2. Recent Results

From very high resolution and high S/N spectra, Spite & Spite (1991) and Nissen & Edvardsson (1992) measured the weak [O I] feature in dwarfs. The latter study demonstrated that [O/Fe] increased from the 0.0 to +0.5 as [Fe/H] decreased from 0.0 to -0.8. At slightly lower metallicities, the former study claimed that the measured equivalent width of the [O I] feature was smaller than would be expected from the O I-based abundance. At [Fe/H] = -1.4 and -1.6, Spite & Spite (1991) obtained [O/Fe] = +0.48 and +0.59 respectively, in contrast to published O I-based ratios which were about 0.5 dex higher. However, Balachandran & Carney (1996) have shown that, at least at these intermediate metallicities, a large part of the difference between [O I]- and O I-based ratios is due to differences in the adopted stellar parameters, analysis variations and, surprisingly, in the metallicities adopted or derived. Reanalyzing published equivalent widths for HD 103095 they obtained [Fe/H] = -1.22 ± 0.04, [O/Fe] = +0.42 ± 0.11 from O I and [O/Fe] = +0.33 ± 0.12 from [O I]. The oxygen abundance ratios thus agree within the errors, demonstrating the requirement for a consistent analysis. It should be noted that the [O I] equivalent widths measured in metal-poor dwarfs are extremely small ($\sim$ 2 to 3 mÅ). It is thus conceivable that [O I]-based results may only be upper limits due to the inclusion of minute unrecognized blends. Of course, the small equivalent widths preclude all efforts to extend such observations to even lower metallicities where the differences between [O I]- and O I-based abundances may be larger. Spite and Spite's results were sufficient, however, to cast some doubt on the O I-based [O/Fe] trend. Confirmation that standard abundance analyses of high excitation lines produce anomalously high abundances came from Tomkin et al's (1992) finding that the high-excitation (7.5 eV) C I lines yielded high carbon abundances in metal-poor stars compared to results from other carbon indicators. However, neither Tomkin et al. nor the detailed studies of Kiselman (1991; 1993) were able to account for these spurious abundances on the basis of non-LTE effects. Kiselman suggested that non-homogeneous, three-dimensional model atmospheres which include granulation may be required to understand the formation of the high-excitation lines but, at present, we are far from being able to account for this anomaly quantitatively.

An alternative to such painstaking measurements of the [O I] feature was attempted by Bessell et al. (1991) and more recently by Nissen et al. (1994) who measured the electronic A-X transistions of the OH molecule. These features lie in the near-UV around 3150 Å and thus pose their own set of problems. Ground-based data acquisition is a challenge because of the low efficiency of most spectrograph optics at these short wavelengths and the proximity of this region to the atmospheric cut-off. Therefore, while these two studies extended to metallcities lower than the [O I] studies, they were based on spectra with S/N ratios of $\sim$ 20 to 30. The more exhaustive analysis of Nissen et al. (1994) resulted in [O/Fe] abundances which are in between the dwarf and giant values; [O/Fe] was found to increase from +0.4 to +0.8 as [Fe/H] decreased from -1.5 to -3.0. Balachandran and Carney (1996) have pointed out that there may be a systematic uncertainty in this result which arises from the analysis. Nissen et al's results are based on spectral syntheses which were performed by comparison to the solar spectrum. Their initial fitting of the solar spectrum required the reduction of the OH oscillator strengths by far larger values than the errors in the experimental or theoretical transistion probabilities. These large changes in the oscillator strengths may have been necessitated because theoretical predictions do not match the observed solar fluxes (Kurucz 1992; Bell et al. 1994). Since the observed and predicted stellar fluxes are more closely matched at lower metallicities, the slope in Nissen et al's result may be suspect.

In the final conclusion, the large O I-based abundances are clearly under suspicion, but alternative optical indicators are hard-pressed to delineate the trend of [O/Fe] versus [Fe/H].

3. The Status of Carbon

Both carbon and nitrogen are overproduced relative to oxygen and iron in intermediate-mass stars which go through the AGB phase and become planetary nebulae. Since nitrogen abundances are notoriously difficult to measure, and, at least to first order, the carbon and nitrogen abundances are expected to provide redundant information, carbon alone is discussed here.

Most of our current knowledge about the trend of [C/Fe] with metallicity is still based on the three studies summarized by WST: Laird (1985); Carbon (1987); and Tomkin et al. (1986). All three studies derived carbon abundances from the spectra of dwarfs using CH lines in the G-band; the first two are larger surveys based on low resolution spectra and the third is based on a smaller sample of high resolution data. Collating the data from the three studies after considering the details of each analysis, WST issued a plea for improved measurements and drew the tentative conclusion that C/Fe was roughly solar until [Fe/H] $\sim$ -1.5 and climbed to [C/Fe] = +0.5 as [Fe/H] fell below -1.5.

For two principal reasons, the results from these studies provide at best a qualitative estimate of the evolution of carbon abundances and do not allow a direct comparison with theoretical predictions. The CH G-band features are in a heavily blended portion of the spectrum. Therefore, in addition to the usual uncertainties associated with abundance determinations, spectral synthesis, especially of low-resolution data, is prone to error due to incomplete line lists or uncertainties in continuum placement, for example. Furthermore, there have been previous assertions that CH-based carbon abundances do not agree with values derived from C_2 Brown (1987) and CO (see Brown et al. 1990). Secondly, and perhaps more importantly, the oxygen abundance is unknown in a large fraction of the examined stars, with each study adopting an acceptable (but different) [O/Fe] ratio. This shortcoming, which has been recognized by all the authors and pointed out by WST, has a significant effect upon the derived carbon abundance as demonstrated by Carbon et al. (1987), because the CH line strengths in cool stars are directly dependent upon the amount of carbon tied up in CO which is, of course, dependent upon the oxygen abundance.

Although several important questions remain largely unanswered, some additional insight may be gained, and some further questions raised, from the few carbon abundance studies of disk dwarfs carried out since the WST review.

- Is the carbon abundance constant between 0.0 > [Fe/H] > -1.5?

Recent studies have used high-excitation (8.6 eV) C I lines (Tomkin et al. 1995) and the [C I] line (Andersson and Edvardsson 1994) to determine carbon abundances in disk samples as a function of metallicity. The C I strengths, especially in the warmer stars, are not as susceptible to changes in the input oxygen abundance and all the stars in the [C I] study have previously measured oxygen abundances. While questions may be raised about the suitability of using high-excitation C I lines which have been demonstrated to produce anomalously high abundances at low metallicities, it could be argued that the disk metallicities are not low enough to result in significant errors and indeed, at these metallicities, O I-based oxygen abundances agree with those from other indicators. Tomkin et al. (1995) futher contend that there is good agreement between the abundances derived from C I and [C I]. However, there are only 12 stars in common and most of these are are within ± 0.15 dex of the solar metallicity; only upper limits were placed on most of the lower metallicity stars by the [C I] study. Nevertheless, the possibility that the C I-based results may produce spurious trends with metallicity should not be dismissed.

The essential finding from the two studies is that [C/Fe] increases steadily with decreasing metallicity, going from roughly [C/Fe] ~ -0.2 at [Fe/H] ~ +0.2 to [C/Fe] ~ +0.3 at [Fe/H] ~ -0.8. Although WST deduced that the [C/Fe] ratio is flat in this metallicity range, Tomkin et al. (1995) suggest that the errors in the CH-based studies are large enough to accomodate the slope observed in the disk dwarfs. It is curious, however, that none of the three CH-based studies obtained [C/Fe] ratios as large as the +0.4 at [Fe/H] = -1.0; the low-resolution studies obtained [C/Fe] ~ 0.0 and high-resolution study obtained [C/Fe] ~ -0.3. Tomkin et al.'s (1986) high resolution study contains a large sample of warm stars whose CH-based abundances should not be adversely affected by their adopted choice of a uniform [O/Fe] = 0.0. One may conservatively assume that stars above the solar temperature are unaffected by significant CO formation since the CO lines are very weak in the solar spectrum. The subset of these warm stars produces abundances consistent with the entire sample for -1.0 < [Fe/H] < 0.0. Until this discrepancy between the CH and C I abundances is resolved, the absolute carbon abundances should be treated with some caution.

- Is there a scatter in [C/Fe] at a given metallicity?

WST cautioned that the observed scatter of ± 0.2 dex may only be an artifact of the analyses. However, the disk studies show a similar scatter throughout their metallicity range. Given the insensitivity of these warm stars to the input oxygen abundance and the relative simplicity of abundance determinations from unblended C I lines, the internal errors in the disk sample are probably smaller, suggesting that the observed star-to-star scatter is real. It is interesting to note that the scatter in the carbon abundances is larger than that in seen in oxygen abundances over the same metallicity range (Nissen and Edvardsson 1992).

- Does the [C/Fe] ratio increase sharply at metallicities below [Fe/H] < -1.5?

Galactic chemical evolution models (e.g., Timmes et al. 1995) do not predict a sharp rise in carbon abundances at low metallicities. An observational finding to the contrary is therefore important. WST were dubious about the reality of this increase though it is clearly evident in the combined CH-based data. Although Tomkin et al.'s (1986) high-resolution study suggests a sharp break, the subset of hot stars ($T_{eff} > 5800$ K) which should be unaffected by significant CO formation, suggests rather a steady increase in [C/Fe] with decreasing metallicity. However the scatter at all metallicities is large. Since that review, Tomkin et al's (1992) study provides the only additional set of carbon abundances at low metallicites. While these abundances are based on the high-excitation C I lines and are therefore unreliable at low metallicities, some facts may be gleaned from the C/O ratio. Tomkin et al. (1992) pointed out that while both the carbon and oxygen abundances derived from high excitation C I and O I lines were anomalously high at low metallicities, the C/O ratio was independent of temperature, suggesting that the errors in the two abundance determinations cancelled out in the ratio. They found that C/O was essentially constant from solar metallicity to [Fe/H] = -2.5 at [C/O] = -0.57 ± 0.12. *If this C/O ratio is indeed accurate, a sharp increase in the carbon abundance at [Fe/H] < -1.5 would require a corresponding sharp increase in the oxygen abundance at the same metallicity.* No such finding has been made from the existing oxygen abundance data.

We are left with several alternatives. Since the C I lines have a lower excitation potential of 7.5 eV compared to the 9.15 eV O I lines, carbon and oxygen abundances may be altered differently as a function of metallicity (by this as yet indentified process) and the flat C/O ratio may be incorrect. Carbon may be rising sharply at [Fe/H] < -1.5 or follow the gradual trend seen in the disk dwarfs. Similarly oxygen may either rise initially and flatten out at the value seen in the giants or rise slowly and continuously. These alternatives have different implications on Galactic chemical evolution and both carbon and oxygen abundances require closer scrutiny at low metallicities.

4. The Infrared Alternative

The final verdict on the trend of carbon and oxygen with metallicity is presently far from conclusive and it appears unlikely that much progress can be made by using the available optical features. However, the recent advent of high resolution infrared spectrographs now allows us to explore a new avenue that holds promising advantages. The infrared spectrum, rich in molecular lines, contains the CO and OH vibrational-rotational molecular bands. CO is particularly useful for the derivation of carbon abundances because it is the primary site where carbon resides in cool stars. The OH features are unique as well. They are seen in cool stars ($T_{eff} < 5000$ K) and are temperature but not gravity sensitive. Thus, for the first time, oxygen abundances can be compared consistently in dwarfs and giants using the same features. Details about the use of OH lines are given in Lambert et al. (1984), Smith & Lambert (1985), and Balachandran & Carney (1996).

So far, the only metal-poor dwarf in which carbon and oxygen abundance analyses have been carried out using these infrared features is HD 103095, better known as Groombridge 1830, the bright halo dwarf with [Fe/H] ~ -1.2 (Balachandran & Carney 1996). They obtained [O/Fe] = + 0.29 ± 0.05 and [C/Fe] = -0.32 ± 0.05. In a reanalysis of other atomic oxygen equivalent widths from the literature, they found that the OH-based oxygen abundance was consistent with those obtained from the atomic indicators. It is worth reiterating that previously debated oxygen abundance differences, at least from solar metallicity to this moderately-poor value, may be due to differences in analysis techniques.

In the context of previous discussions, two points should be noted from these results, though with the strong caveat that there is, as yet, only a single data point from the infrared data. Firstly, the carbon abundance of [C/Fe] = -0.32 is far lower than the value that would be predicted if the disk dwarf trend is extrapolated; that extrapolation would result in [C/Fe] ~ +0.6. The measured carbon abundance of HD 103095 is well outside the ± 0.2 dex scatter about the mean trend seen in the disk dwarfs. The CO-based abundance is however consistent with mean CH-based value at this metallicity. Secondly, the C/O ratio derived from the infrared molecular features is identical to the mean value obtained by Tomkin et al. (1992) from the high-excitation atomic features. Abundances from additional stars will soon provide the basis for more detailed comparisons.

The infrared molecular features have their limitations in that they are only useful in cool stars. Also, current high-resolution infrared spectrographs are only able to observe bright dwarfs (K > 10) and the small array sizes result in modest wavelength coverage. However, the CO and OH lines are clean and well-separated, and lines free of telluric contamination can be located. Improving technology will reduce the readout noise in the infrared arrays and, together with larger telescopes, will allow more extensive samples to be compiled in time. Larger-sized array detectors and cross-dispersed echelle spectrographs are already being planned, and these will allow larger wavelength coverage in a single exposure. It may then possible to look for features predicted by the Galactic chemical models, such as the bump in the carbon abundances when contributions from intermediate-mass stars become significant but SN Ia have not yet appeared.

Acknowledgments. This review was written during a visit to the Centro de Fisica Nuclear at the University of Lisbon. The author is pleased to acknowledge the hospitality received. The author's work on stellar abundances is supported by the NSF grant AST 93-14851.

References

Abia, C. & Rebolo, R. 1989, ApJ, 347, 186
Andersson, H., & Edvardsson, B. 1994, A&A, 290, 590
Balachandran, S. C., & Carney, B. W., 1996, AJ, in press
Bell, R. A., Paltaglou, G., &Tripicco, M. J., MNRAS, 268, 771
Bessell, M. S., Sutherland, R. S., & Ruan, K. 1991, ApJ, 383, L71
Brown, J. A. 1987, ApJ, 317, 701

Brown, J. A., Wallerstein, G., & Oke, J. B. 1990, AJ, 100, 1561
Carbon, D. F., Barbuy, B., Kraft, R. P., Friel, E. D., & Suntzeff, N. B., PASP, 99, 335
Kraft, R. P., Sneden, C., Langer, G. E., & Shetrone, M. D. 1993, AJ, 107, 1773
Kiselman, D. 1991, A&A, 245, L9
Kiselman, D. 1993, A&A, 275, 269
Kurucz, R. L. 1992, Rev. Mex. Astron. Astrof. 23, 181
Laird, J. B. 1985, ApJ, 289, 556
Lambert, D. L., Brown, J. A., Hinkle, K. H., & Johnson, H. R. 1984, ApJ, 284, 223
Nissen, P. E., & Edvardsson, B. 1992, A&A, 261, 255
Nissen, P. E., Gustafsson, B., Edvardsson, B., & Gilmore, G. 1994, A&A, 285, 400
Smith, V. V. & Lambert, D. L. 1985, ApJ, 297, 326
Spite, M., & Spite, F. 1991, A&A, 252, 689
Timmes, F. X., Woosley, S. E., &Weaver, T. A. 1995, ApJS, 98, 617
Tomkin, J., Sneden, C., & Lambert, D. L. 1986, ApJ, 302, 415
Tomkin, J., Lemke, M., Lambert, D. L., & Sneden, C. 1992, AJ, 104, 1568
Tomkin, J., Woolf, V. M., Lambert, D. L., & Lemke, M. 1995, AJ, 109, 2204
Vandenberg, D. A. 1992, ApJ, 391, 685
Wheeler, C., Sneden, C. & Truran, J. W., Jr. 1989, ARA&A, 27, 279 (WST)

Carbon Stars and Elemental Abundances

W. K. Rose

Department of Astronomy, University of Maryland, College Park, MD 20742, e-mail: wrose@astro.umd.edu

Abstract. In this paper calculations related to the formation of late-type carbon stars, non-LTE radiative transfer computations interpreting mm wavelength CO observations and calculations of infrared continuum radiation emitted from dust particles in the stellar winds of late-type stars are discussed. The predicted relationship between s-process nucleosynthesis and carbon star formation is also described.

Because the CO molecule is very stable in the atmospheres of asymptotic giants most carbon is in the form of CO if the C/O abundance ratio is normal (i.e., < 1) whereas most oxygen exists as CO if the above ratio is reversed. Infrared observations of M giants show that their infrared spectra are dominated by H_2O bands (Woolf, Schwarzschild, & Rose 1964) whereas the infrared spectra of late-type carbon stars contain C_2 and other carbon molecules.

As is well known helium burning synthesizes ^{12}C from 3α reactions, ^{16}O from the $^{12}C(\alpha,\gamma)^{16}O$ reaction and a smaller amount of ^{20}Ne from the reaction $^{16}O(\alpha,\gamma)^{20}Ne$. The latter reaction is inhibited (i.e., blocked) because parity conservation in nuclear reactions implies that the $J^\pi = 2^-$ and $J^\pi = 4^-$ levels of ^{20}Ne do not produce resonances (Rolfs & Rodney 1988). During a helium shell flash in an asymptotic giant branch star the temperature becomes sufficiently high that much more ^{12}C than ^{16}O is synthesized because the 3α reactions are more temperature sensitive than the $^{12}C(\alpha,\gamma)^{16}O$ reaction. Stellar interior model calculations (Rose & Smith 1970, 1972, Iben & Renzini 1983) show that helium shell flashes cause carbon-rich mass in the core to be dredged to the surface layers. It is widely accepted that quasi-periodic generation of helium shell flashes leads to inversion of the C/O abundance ratio and therefore to the production of late-type carbon stars.

Most s-process nucleosynthesis is believed to occur in red giants as a consequence of helium shell flashes and therefore concurrent with the formation of carbon stars. During the peak burning phases the temperatures within the helium shells are $T \sim 3 \times 10^8 K$ and neutrons can be formed by the $^{22}Ne(\alpha,n)^{25}Mg$ and $^{13}C(\alpha,n)^{16}O$ reactions. Neutron production via the $^{22}Ne(\alpha,n)^{25}Mg$ reaction takes place in the following manner. ^{14}N whose abundance has been increased by the CNO cycle within the overlying hydrogen burning shell is mixed into the helium burning shell where ^{22}Ne is synthesized by the reactions $^{14}N(\alpha,\gamma)^{18}F(e^+,\nu)^{18}O(\alpha,\gamma)^{22}Ne$. Neutron production via the $^{13}C(\alpha,n)^{16}O$ reaction occurs if protons which synthesize ^{13}C from ^{12}C are mixed into the helium burning shell during a thermonuclear runaway. Stellar interior calculations referenced above show that the entropy discontinuity at the boundary layer separating the core and hydrogen shell formally prevents convection from penetrating into the hydrogen burning shell. However, the convective zone produced by helium shell flashes extends to within less than one pressure scale height of the entropy barrier discussed above. Therefore, it is plausible to assume that a small fraction of a scale height of protons are mixed into the helium-carbon-rich core where they rapidly synthesize ^{13}C from ^{12}C and generate neutrons via the ^{13}C$(\alpha, n)^{16}$O reaction. Convective overshoot and shear turbulence are two possible physical mechanisms that may cause mixing at the core-hydrogen shell interface. The observed very high abundances of CNO in nova shells require that mixing between the cores and hydrogen-rich envelopes of white dwarfs responsible for nova outbursts must occur.

Cowan & Rose (1977) investigated the possibility that some r-process nucleosynthesis might occur via the ^{13}C$(\alpha, n)^{16}$O reaction in some red giants. They concluded that neutron production was insufficient for actinide isotopes such as ^{232}Th, ^{235}U, and ^{238}U to be synthesized. However, some r-process nucleosynthesis might take place in red giants undergoing helium shell flashes. Intense neutron fluxes from the ^{13}C$(\alpha, n)^{16}$O reaction can occur only if the number density of protons is sufficiently low that its rate is much faster than the ^{13}C$(p, \alpha)^{14}$N reaction rate. In previously published non-LTE radiative transfer calculations (Rose 1984, 1989, 1995) mm wavelength ^{12}CO and ^{13}CO line profiles from late-type carbon stars were discussed. The computations include simultaneous solutions of the radiative transfer equation for CO emission lines, the equations of statistical equilibrium and an energy equation. In the first of the above papers ^{12}CO line profiles were predicted for a range of assumed physical input parameters, namely mass loss rate, ^{12}CO/H$_2$ abundance ratio and stellar luminosity. In addition, a detailed comparison was made between calculations and observations of IRC 10216. The assumed distance to IRC 10216 in Rose (1984) was 250 pc, and therefore its predicted luminosity would be lowered in an obvious manner if the actual distance is less. These calculations indicate that distances greater than 250 pc are unlikely. Independent computations by a number of researchers give very similar theoretical predictions for this object. In Rose (1989, 1995) calculations of ^{13}C/^{12}C abundance ratios and ^{13}CO line profiles were described. In the first of the above papers it was concluded that all 12 carbon stars observed by Wannier & Sahai (1987) have ^{13}C/^{12}C abundance ratios $\leq 1/50$ and most probably have lower values (i.e., $\lesssim 1/90$). The estimated ^{13}C/^{12}C abundance ratio from infrared CO measurements given by Jaschek & Jaschek (1993) are $1/40 - 1/70$. The disagreement in predicted ^{13}C/^{12}C abundance ratios is not as large as it might seem because my calculated ^{13}CO line strengths vary rather weakly with ^{13}C abundances for ^{13}C/^{12}C ratios $\lesssim 1/50$. However, my calculations do favor lower ^{13}C/^{12}C ratios for the late-type carbon stars observed by Wannier & Sahai (1987) than those listed by Jaschek & Jaschek (1993). Hydrogen deficient RCrB stars, which may be formed after the asymptotic giant branch, are examples of stars known to be both carbon-rich and to have very low ^{13}C/^{12}C abundance ratios. Improved measurements of ^{13}CO line profiles in late-type carbon stars will hopefully resolve this issue.

Although supernovae and novae contribute some dust to the interstellar medium most dust formation occurs in winds from late-type carbon and oxygen-rich stars. In previous publications (Rose 1987, 1989) models for infrared continuum radiation from evolved stars that are undergoing rapid mass loss were calculated. These models show that their observed far infrared spectra imply a dust absorption coefficient Q_λ which varies as $1/\lambda$. These results are true for both carbon and oxygen-rich red giants. They show that the dust particles embedded in outflowing gas from late-type carbon stars are amorphous rather than graphitic. Similar results have been independently obtained by Sopka et al. (1985) and Whittet (1989).

The observed mid-infrared ($\sim 10\mu$m) peaks in the continuum radiation from the Milky Way and M31 have been attributed to very small interstellar dust particles such as PAHs and/or infrared emission from evolved stars that are undergoing rapid mass loss (Rose 1990). It is now widely believed that a significant amount of dust present in the solar nebula was accreted onto macroscopic bodies without undergoing vaporization and recondensation (Whittet 1992). The infrared spectra of comets show evidence for both silicate and carbonaceous dust.

References

Cowan, J. J., & Rose, W. K. 1977, ApJ, 212, 149
Iben, I., & Renzini, A. 1983, ARA&A, 21, 271
Jaschek, C., & Jaschek, M. 1993, The Classification of Stars (Cambridge: Cambridge Univ. Press)
Rolfs, C. E., & Rodney, W. S. 1988, Cauldrons of the Cosmos, (Chicago: Univ. Chicago Press)
Rose, W. K. 1984, ApJ, 285, 237
───────── . 1987, ApJ, 312, 284
───────── . 1989, in AIP Conference Proceedings No. 183, ed. C. H. Waddington (New York: AIP)
───────── . 1990, in IAU Symposium No. 139, eds. S. Bowyer & C. Leinert (Dordrecht: Kluwer)
───────── . 1995, in The Analysis of Emission Lines, ed. R.E. Williams (Cambridge: Cambridge Univ. Press), in press
Rose, W. K., & Smith, R. L. 1970, ApJ, 159, 903
───────── . 1972, ApJ, 173, 385
Sopka, R. J., et al. 1985, ApJ, 294, 242
Wannier, P. G., & Sahai, R. 1987, ApJ, 319, 367
Whittet, D. C. B. 1989, in IAU Symposium No. 135, Interstellar Dust, eds. L. J. Allamandola & A. G. G. M. Tielen (Dordrecht: Kluwer), p. 455
───────── . 1992, Dust in the Galactic Environment (London: IOP Publ. Ltd.)
Woolf, N. J., Schwarzschild, M., & Rose, W. K. 1964, ApJ, 140, 833

Spectroscopic Constraints on the Helium Abundance in Globular Cluster Stars

W.B. Landsman

Hughes STX, Code 681, NASA/GSFC, Greenbelt, MD 20771

A.P.S. Crotts

Columbia University, Dept. of Astronomy, New York, NY 10027

I. Hubeny and T. Lanz

USRA, Code 681, NASA/GSFC, Greenbelt, MD 20771

R.W. O'Connell and J. Whitney

U. of Virginia, Astronomy Dept., Charlottesville, VA 22903

T.P. Stecher

Code 680, NASA/GSFC, Greenbelt, MD 20771

Abstract.
 The globular cluster ω Centauri has a rich population of UV-bright stars, most of which were discovered on a 1620 Å image of the cluster obtained in 1990 with the Ultraviolet Imaging Telescope (UIT). We have obtained ground-based and IUE low-dispersion spectroscopy of seven UV-bright stars in ω Cen, in order to study the variation of photospheric helium abundances across the UV-bright region of the HR diagram. All seven stars are radial velocity members of the cluster and show helium lines in their spectra. The three least luminous target stars have $T_{eff} \sim$ 20,000 K and show photospheric helium depletions characteristic of stars on the hot horizontal branch. The fourth target star (UIT-644) also has $T_{eff} \sim$ 20,000 K but with a large overabundance of helium and nitrogen, perhaps indicating a prior ejection of its hydrogen-rich envelope. The remaining three target stars have sdO spectra with strong Balmer and He II lines, but none from He I. These three stars have an approximately solar helium abundance and are promising spectroscopic probes of the cluster helium abundance.

1. Introduction

The helium abundance, Y, in globular cluster stars is of interest in cosmology for two reasons. First, the helium abundance in these old low-metallicity stars can provide a useful upper limit to the primordial abundance, and second, the isochrone ages of the globular clusters depend upon the assumed helium content (Shi 1995). Most globular cluster stars are too cool to show helium in their spectra, and the hot horizontal branch (HB) stars often show greatly depleted helium abundances due to gravitational settling of helium in their high gravity atmospheres (e.g. Heber et al. 1986). Thus, the determination of the helium content in globular clusters has usually relied on indirect methods, such as number counts of stars in different evolutionary phases (the "R method", Caputo et al. 1987) or the width of the horizontal branch (Dorman et al. 1989).

The color-magnitude diagrams of globular clusters often show one or two UV-bright stars, which are at least a magnitude more luminous than the zero-age horizontal branch (ZAHB). The most luminous UV-bright stars are thought to be post-asymptotic giant branch (AGB) stars evolving at constant luminosity across the HR diagram toward their final white dwarf stage, while the less luminous UV-bright stars are probably directly evolved from the hot horizontal branch. The UV-bright stars are promising spectroscopic probes of the cluster helium abundance for three reasons. First, because they are lower gravity than stars near the ZAHB, they are less likely to have suffered gravitational settling of photospheric helium. Second, unlike the long-lived ($\sim 10^8$ years) HB stars, the rapid evolutionary time scale ($\sim 10^5$ years) of post-AGB stars may be insufficient to allow stratification of photospheric helium to occur, even if a relevant physical process (e.g. diffusion, radiative levitation) is operating. Finally, the UV-bright stars are easier spectroscopic targets than the hot HB stars in any particular cluster, simply due to their higher luminosity. Note that a small enhancement of the surface helium abundance in UV-bright stars over the main-sequence value ($\Delta Y \sim 0.02$) is expected due to the first dredge-up on the red giant branch (Sweigart et al. 1989). From high-resolution optical spectra, Conlon et al. (1994) find an approximately solar helium abundance ($Y = 0.28$) in the UV-bright star Barnard 29 in M13 (but also see Adelman et al. 1994).

The globular cluster ω Centauri has an especially rich population of UV-bright stars due to the extreme luminosity of the cluster, and the presence of a populous hot HB. A 1620 Å image of the cluster obtained with the Ultraviolet Imaging Telescope (UIT) in 1990 revealed more than two dozen hot stars at least one magnitude brighter than the ZAHB (Whitney et al. 1994). In February 1995, we obtained low-dispersion (3.1 Å) CTIO 4m spectra of seven of the brightest stars in the Whitney et al. catalog. All seven stars are radial velocity members of the cluster and show helium lines in their spectra. We have also obtained IUE low-dispersion spectra for 6 of the 7 target stars. The optical spectra are analyzed with NLTE hydrogen-helium model atmospheres to estimate the effective temperatures, gravities, and helium abundances.

2. Results

Table 1 summarizes the results of our observations. The first column gives the star ID number from Whitney et al., the second column gives an alternate name from Dickens (1988) if available, and the third column gives the 1620 Å UIT magnitude, m162, as tabulated in Whitney et al. (1994). (Note that UIT-644 and UIT-1275 are two UV-bright stars in the core of ω Cen, labeled as UIT-1 and UIT-2, respectively, by Landsman et al. 1992.) The next three columns give the approximate temperatures, gravities, and helium abundances as derived from a preliminary model atmosphere analysis. The final column gives the bolometric luminosity, which is derived from T_{eff} and m162 using the distance (5100 pc) and reddening (E(B–V) = 0.15) of ω Cen.

Table 1. UV-Bright Stars in ω Cen observed at CTIO

ID	Name	m162	T_{eff} (K)	log g	Y	log L/L$_\odot$
697	Dk 3089	12.98	20,000	4.0	0.065	2.07
829	UIT-829	13.52	20,000	4.0	0.065	1.98
1275	UIT-1275	12.09	22,500	4.0	0.13	2.46
644	UIT-644	10.46	20,000	3.5	0.68	2.86
197	ROA 5342	12.01	60,000	4.5	0.26 – 0.43	3.1
330	Dk 3873	12.63	70,000	5.0	0.26 – 0.43	2.8
151	UIT-151	12.63	70,000	5.0	0.26 – 0.43	2.8

The seven target stars can be discussed in the following three groups.

1. The three stars (UIT-1275, Dk-3089, and UIT-829) with the lowest luminosity also have the lowest helium abundance. These stars have B-type spectra with Balmer lines and He I lines, but none from He II. Evolutionary tracks suggest that these stars are evolved directly off of the ZAHB, and are similar to the low-gravity field HB stars discussed by Saffer et al. (1995). As these stars evolve off of the ZAHB toward lower gravity, they apparently retain some depletion of photospheric helium.

2. UIT-644 also has a B-type spectrum but with much stronger helium lines and with strong nitrogen lines. The luminosity of UIT-644 is sufficiently high that it probably is on a post-AGB track (c.f. Landsman et al. 1992). UIT-644 may have ejected its hydrogen rich envelope, revealing processed material with a high helium and nitrogen abundance.

3. ROA 5342, UIT-151, and Dk 3873 have sdO type spectra with strong Balmer and He II lines, but none from He I. The absence of He I indicates that $T_{eff} > 60,000$ K, making these the hottest stars ever found in a globular cluster. However, there is a fairly large ($\pm 10,000$ K) uncertainty in the T_{eff} determination, because no metal lines are detected in the optical spectra (nor are any expected at our S/N and resolution), and because the slope of the UV continuum is not a useful indicator of T_{eff} at these high temperatures. The helium abundance is approximately solar, so that these stars might be useful probes of the cluster helium abundance. More precise determinations of the helium abundance will be possible once T_{eff} is better determined, for example, by modeling the ionization balance of metals in moderate resolution UV spectra. Note that these sdO stars in ω Cen differ from the hot sdO stars in the field, which are generally helium-rich (e.g. Drilling & Beers 1995)

3. Discussion

There are two important requirements to be met before spectroscopic measurements of hot stars will be useful for constraining the cluster helium abundance. The first is to demonstrate that the photospheric helium abundance in any particular star has not been modified by gravitational settling or other surface effects. In particular, it would be gratifying to find agreement in derived Y values from UV-bright stars with different atmospheric parameters. The results here suggest that the best spectroscopic probes are high-luminosity ($\log L/L_\odot > 2.5$) UV-bright stars on post-AGB tracks.

The second requirement is that the photospheric helium abundance measurements have sufficient precision to critically test the existing indirect helium abundance determinations. For the sdO stars in ω Cen, the accuracy of the helium abundance determinations are currently limited both by the S/N of the optical spectra, and by the lack of moderate resolution UV spectra to better constrain the effective temperatures. FOS spectra of UIT-151 are currently scheduled for early 1996, and should considerably improve the helium abundance measurement presented here.

Acknowledgments. Funding for the UIT project is supported through the Spacelab Office at NASA Headquarters under Project number 440-51.

References

Adelman, S.J., Aikman, G.C.L, Hayes, D.S., Phillip, A.G.D., & Sweigart,A.V. 1994, A&A, 282, 134
Caputo, F., Martinez Roger, C., & Paez. E. 1987, A&A, 183, 228
Conlon, E.S., Dufton, P.L., & Keenan, F.P. 1994, A&A, 290, 897
Dickens, R.J., Brodie, I.R., Bingham, E.A., & Caldwell, S.P. 1988, Rutherford Appleton Laboratory, RAL 88-04
Dorman, B., VandenBerg, D.A. & Laskaraides. P.G. 1989, ApJ, 343, 750
Drilling, J.S., & Beers, T.C. 1995, ApJ, 446, L27
Heber, U., Kudritzki, R.P., Caloi, V., Castellani, V., Danziger, J., & Gilmozzi, R. 1986, A&A, 162, 171
Landsman, W.B. et al. 1992, ApJ, 395, L21
Saffer, R.A., Bergeron, P., Koester, D., & Liebert, J. 1994, ApJ, 432, 351
Shi, X. 1995, ApJ, 446, 637
Sweigart, A.V., Greggio, L., & Renzini, A. 1989, ApJS, 69, 911
Whitney, J.H. et al. 1994, AJ, 108, 1350

Selected Elemental Abundances in Five Oxygen-Poor Stars in ω Centauri

D. Zucker, G. Wallerstein, and J. A. Brown

Astronomy Department, University of Washington, Seattle, WA 98195

Abstract. We have derived abundances or upper limits for Fe, O, Na, Al, Sc, and Eu in five oxygen-poor red giants in Omega Centauri, a globular cluster which displays a metallicity range of some 1.5 dex. All of the stars show large Al excesses and slight Sc excesses, and all but one reveal an excess of Na. We find either solar or slightly deficient Eu abundances in these stars, in agreement with other recent results. We have also used spectral synthesis to derive $^{12}C/^{13}C$ ratios which indicate significant (but not complete) CN cycling of surface material and which are similar to ratios obtained for oxygen-rich stars in Omega Centauri and in other globular clusters. We argue that, based on current reaction rates, even if CN cycling were complete it would not be possible for these stars to create internally the Al excesses observed.

1. Introduction

Omega Centauri, our galaxy's most massive and luminous globular cluster, shows a metallicity range of approximately 1.5 dex. The cluster's ellipticity can be explained by rotation (Meylan 1986) and its metallicity distribution is not bimodal (Butler, Dickens and Epps 1978), suggesting that ω Cen was not formed by mergers but rather is a closed system that has retained it s own stellar ejecta. We can therefore trace the enrichment of various species in the cluster by measuring elemental abundances relative to iron, and compare the dependence of the abundance of a species in ω Cen with its behavior in the Galaxy as a whole (Wheeler, Sneden and Truran 1989), as well as with models of Galactic enrichment (Timmes, Woosley and Weaver 1995).

One of the most interesting elements to study is oxygen, which typically exhibits an enhanced abundance in metal-poor field stars (Wheeler, Sneden and Truran 1989) and globular clusters. Paltoglou and Norris (1989) obtained the puzzling result that 5 of 15 ω Cen giants in their sample were substantially deficient in oxygen. Norris and Da Costa (1995b) found 13 of an overlapping set of 38 red giants in ω Cen to be oxygen-poor and have shown in detail (1995a) that there is a clear correlation of enhanced sodium and aluminum with depletion of oxygen (also indicated by Paltoglou and Norris 1989), similar to the oxygen-sodium anticorrelation in M13 (Kraft et al. 1992).

2. Observations and Analysis

We obtained two spectroscopic data sets in April and May of 1994 at the Cerro Tololo Inter-American Observatory. The first set was taken with the 1.5m telescope and the Bench-Mounted Echelle at a resolution of $\sim 20,000$. Typically five half-hour or hour exposures were combined to yield spectra with S/N of ~ 50–60. Two stars, ROA 150 and 144, were observed with the 4m echelle and the same camera at a resolution of $\sim 35,000$, yielding S/N of ~ 60–80. All data were reduced and equivalent widths measured using IRAF tasks. For our target stars we used model atmosphere parameters published in Norris and Da Costa (1995b), generally collected from previous photometric work. Our analysis was performed by the methods previously described in Brown and Wallerstein (1989, 1992).

3. Results

In Table 1 we show our derived abundances and compare them with the results of Paltoglou and Norris (1989), and Norris and Da Costa (1995b). The agreement among analyses is good; such agreement between different authors is a much better indication of the reliability of the results than is a statistical analysis in a single paper. We find that, near $[Fe/H] = -1.5$, substantial deficits in oxygen are generally accompanied by excesses in $[Na/Fe]$ of about +0.4 and excesses in $[Al/Fe]$ of $\sim$ +1.0. Scandium in some of the stars appears to be slightly enhanced, although $[Sc/Fe]$ is not more than +0.1–+0.2. The trends indicated are fully consistent with the results of Paltoglou and Norris (1989) and of Norris and Da Costa (1995a,1995b).

We also looked for europium, for which Smith et al. (1995) found a surprising deficiency in oxygen-rich ω Cen stars, relative to field stars of similar metallicities. Unfortunately the $\lambda 6645$ line of Eu II was located near some bad pixels in the 1.5m data, forcing us to search for other lines. In the two stars for which we obtained 4m data, the $\lambda 6645$ line was measurable. As shown in Table 1, we see no evidence for an excess of europium. Norris and Da Costa (1995a,1995b) found either small negative values or upper limits for $[Eu/Fe]$; we believe that our results are in general agreement.

It is well-known that both low-metallicity field giants and low-mass open cluster giants have a much lower $^{12}C/^{13}C$ ratio than is predicted by standard stellar evolution models (Sneden, Pilachowski and VandenBerg 1986; Gilroy 1989). Rather than a ratio near 20 the ratio in globular cluster red giants is typically between 4 and 8 (Brown and Wallerstein 1989; Bell, Briley and Smith 1990). Since the ratio of reaction rates with protons is close to 3.5 (independent of temperature) an observed ratio of 3.5 would indicate that virtually all the material on the surface of the star has been mixed to depths at which the $^{12}C(p,\gamma)^{13}N(\beta^+,\nu)^{13}C$ reaction sequence has been active. Similarly, a ratio of 7 would indicate that approximately half the surface material has been processed in this way. For our five target stars we used the $\Delta V = 2$ bands of the red CN system to derive the $^{12}C/^{13}C$ ratios, which we give in Table 2. The oxygen-poor stars can be characterized by a ratio of 5 ± 1, hence roughly 2/3 of the surface material has experienced substantial CN cycling. As indicated in Table 2, the oxygen-poor stars in ω Cen are not significantly different from the oxygen-rich stars in the same cluster or those in M22, M4, and 47 Tuc. The carbon isotope ratios therefore do not support the suggestion that oxygen-po or stars are more deeply mixed than oxygen-rich stars.

Table 1.   Abundances of Iron and Other Elements Relative to Iron

Star (ROA)	[Fe/H]			[O/Fe]			[Na/Fe]		
	a	b	c	a	b	c	a	b	c
42*	−1.6	−1.7	−1.6	< −0.7	< −0.5	< −0.35	+0.2	+0.4	+0.3
100*	−1.4	−1.5	−1.4	−0.5	< −0.4	< −0.4	+0.6	+0.65	+0.65
139*	−1.4	−1.45	−1.5	< −0.9	−0.05	< −0.3	−0.1	+0.2	+0.8
144†	−1.4	−1.7	−1.7	< −0.9	< −0.3	< −0.2	+0.2	+0.5	< +0.4
150*	−1.2	−1.25	−1.3	...	−0.5	−1.15	+0.2	+0.6	+0.55
150†	−1.4	−1.25	−1.3	< −0.35	−0.5	−0.15	+0.35	+0.6	+0.55

	[Al/Fe]			[Sc/Fe]			[Eu/Fe]		
	a	b	c	a	b	c	a	b	c
42*	+1.1	+0.9	...	+0.1	+0.1	+0.4	< 0.0	...	...
100*	+1.7	+1.15	...	+0.3	0.0	+0.25	0.0	...	...
139*	+0.9	+0.7	...	+0.2	+0.1	+0.2	< 0.0	...	...
144†	+1.1	+0.95	...	0.0	+0.1	+0.3	−0.1	...	...
150*	+1.0	+1.1	...	−0.1	+0.15	+0.3	...	...	...
150†	+1.4	+1.1	...	−0.1	+0.15	+0.3	< +0.05	...	...

^aFrom this work; * indicates data from CTIO 1.5m, † data from CTIO 4m
^bFrom Norris and Da Costa (1995b)
^cFrom Paltoglou and Norris (1989)

4. Discussion

The enhancement of ^{23}Na in oxygen-poor stars has been explained by Langer, Hoffmann and Sneden (1993) as due to proton capture by ^{22}Ne or three-proton-capture by ^{20}Ne. The processed material would then have been carried to the surface by co nvection. The analogous process of proton capture by ^{25}Mg and ^{26}Mg to produce ^{27}Al is much more difficult to justify (three proton captures by ^{24}Mg is not viable because of its small cross-section for proton capture). The initial abundances of the heavy isotopes of Mg have been found to be low in two metal-poor dwarfs (Tomkin and Lambert 1980; Lambert and McWilliam 1986) and in metal-poor field giants (Barbuy 1985). Thus it does not seem possible to significantly enhance the surface ^{27}Al through this reaction chain, especially since the carbon isotope ratios we have observed indicate that no more than 2/3 of the surface material could have been processed. Unless this difficulty can be overcome, the possibility that excess aluminum and perhaps sodium were produced in an earlier generation of stars cannot be excluded.

Table 2. Carbon Isotope Ratios for O-poor Stars in ω Cen Compared with O-rich Stars

Star (ω Cen)	$^{12}C/^{13}C$	Cluster	$^{12}C/^{13}C$
42*	5 ± 1	ω Cen^a	4 − 6
100*	4 ± 1	M22^{b,d}	4 − 5
139*	5 ± 1	M4^{b,e}	3 − 8
150*	5 ± 1	47 Tuc^{b,c}	4 − 7
150†	6 ± 0.5		
144†	6 ± 0.5		

*Data from CTIO 1.5m
†Data from CTIO 4m
^{a}From Brown and Wallerstein (1993)
^{b}From Brown, Wallerstein, and Oke (1990)
^{c}Bell, Briley and Smith (1990) found a range of 3 − 8.
^{d}Smith and Suntzeff (1989) found a range of 3 − 10 from the CO bands in 4 stars and one star that may be a post-AGB star with a ratio of ≥ 40.
^{e}Smith and Suntzeff found a range of 4 − 10 from 6 stars.

References

Barbuy, B. 1985, A&A, 151, 189
Bell, R. A., Briley, M. M. and Smith, G. H. 1990, AJ, 100, 187
Brown, J. A. and Wallerstein, G. 1989, AJ, 98, 1643
Brown, J. A. and Wallerstein, G. 1992, AJ, 104, 1818
Brown, J. A. and Wallerstein, G. 1993, AJ, 106, 133
Brown, J. A., Wallerstein, G., Cunha, K. and Smith, V. V. 1991, A&A, 249, L13
Brown, J. A., Wallerstein, G. and Oke, J. B. 1990, AJ, 100, 1561
Butler, D., Dickens, R. J., and Epps, E. 1978, ApJ, 225, 148
Kraft, R. P., Sneden, C., Langer, G. E. and Prosser, C. F. 1992, AJ, 104, 645
Langer, G. E., Hoffmann, R. and Sneden, C. 1993, PASP, 105, 301
Meylan, G. 1986, A&A, 184, 144
Norris, J. E. and Da Costa, G. S. 1995, ApJ, 441, L81 (1995a)
Norris, J. E. and da Costa, G. S. 1995, ApJ, 447, 680 (1995b)
Paltoglou, G. and Norris, J. E. 1989, ApJ, 336, 185
Smith, V. V., Lambert, D. L. and Cunha, K. 1995, AJ, (submitted)
Smith, V. V. and Suntzeff, N. B. 1989, AJ, 97, 1697
Sneden, C., Pilachowski, C. A. and VandenBerg, D. 1986, ApJ, 311, 826
Sneden, C. 1973, PhD Thesis, University of Texas at Austin
Timmes, F. X., Woosley, S. E. and Weaver, T. A. 1995, ApJS, 98, 617
Tomkin, J. and Lambert, D. L. 1980, ApJ, 235, 925
Wheeler, J. C., Sneden, C. and Truran, J. W., Jr. 1989, ARA&A, 27, 279

Cosmic Abundances
ASP Conference Series, Vol. 99, 1996
Stephen S. Holt and George Sonneborn (eds.)

Abundance Anomalies in Globular Cluster Red Giant Stars. I. Synthesis of the Elements Na and Al

Robert M. Cavallo and Roger A. Bell

Dept. of Astronomy, University of Maryland, College Park, MD 20742

Allen V. Sweigart

Code 681, NASA/GSFC, Greenbelt, MD 20771

1. Introduction

A basic problem in the field of cosmic abundances is understanding the production of the elements within globular cluster (GC) stars. Recent reviews (e.g. Briley et al. 1994, Kraft 1994) have examined in detail the discrepancies between what canonical stellar evolution sequences predict and what is actually observed. These discrepancies raise a fundamental question: are the abundance anomalies the result of primordial contamination by an earlier generation of stars or do GC stars process these elements in thier interiors and mix them to the surface in the course of normal evolution?

The observed correlations of sodium and aluminum with nitrogen and anticorrelations with oxygen have stirred debate as to the origins of these intermediate mass elements. It was thought that these elements would have to come from external sources, e.g., Cottrell and Da Costa (1981) suggest that perhaps 5 - 10 $M_\odot$ stars shed their processed envelopes into the cluster environment, thereby contaminating the newly formed low mass stars. However, Denisenkov and Denisenkova (1990) showed that Na and Al can be formed via proton captures on ^{22}Ne. In addition, Langer, Hoffman, and Sneden (1993; hereafter LHS93), citing a lack of initial ^{22}Ne to account for all of the observed increase in ^{23}Na, hint that ^{23}Na is actually in a cycle with ^{20}Ne. LHS93 also discuss two methods of producing ^{27}Al. The first is a simple proton capture on ^{26}Mg. The second method starts with ^{25}Mg$(p,\gamma)^{26}$Al. ^{26}Al has a half-life of 7.5 x 10^5 years; therefore, it can either proton capture and beta decay or beta decay and proton capture, resulting in ^{27}Al in each case.

We present new results of combining red giant branch (RGB) sequences with a nuclear reaction network with the aim of following the production of Ne, Na, Mg, and Al in the hydrogen shell burning region (the " H shell").

2. The RGB Sequences and the Network

Our sequences begin on the lower RGB where the H shell has burned through the hydrogen discontinuity left behind after the first dredge-up (Sweigart & Mengel 1979). At this point mixing is no longer hindered by a mean molecular weight gradient. The sequences end at the tip of the RGB with the onset of the helium flash. Table 1 describes the properties of the sequences. To save space, we present detailed information only for sequence 1. Similar results hold true for the other sequences except that the extent of the nuclearly processed region above the H shell decreases with higher metallicity.

Table 1. Properties of RGB Sequences.

Sequence #	Sequence Name	Mass ($M_\odot$)	Z
1	m795.z14	0.795	0.0001
2	m80.z44	0.80	0.0004
3	m875.z43	0.875	0.004
4	m15.z18652	1.5	0.01865

The nuclear reaction network was kindly provided to us by Dr. W. David Arnett of the University of Arizona. It consists of 110 nuclei from ^{12}C to ^{49}Ca plus ^{1}H, ^{4}He, and free neutrons. The input abundances are scaled according to solar values (Anders & Grevesse 1989) and contain all isotopes from ^{19}F to ^{44}Ca which constitute more than 1% of their respective elements. We obtain initial CNO abundances from the RGB sequences. The output of the code yields isotopic profiles around the H shell at different points along the RGB.

3. Results

Figures 1a and 1b show the Ne, Na, Mg, and Al profiles for sequence 1 at the point on the RGB where mixing can begin. In figure 1a, the initial rise in ^{23}Na as one goes in towards the shell is due to the proton captures on ^{22}Ne. However, it is clearly seen that the majority of ^{23}Na comes from ^{20}Ne, confirming the results of LHS93. The right hand ordinate describes the ratio of the rate of production of ^{21}Na from ^{20}Ne (the slowest proton capture rate in the cycle) to the rate of the ^{23}Na$(p,\alpha)^{20}$Ne reaction. Above the shell, the production of ^{23}Na is favored.

Figure 1b shows the production of ^{27}Al from ^{26}Mg and ^{25}Mg (through ^{26}Al) as suggested by LHS93. However, within the H shell, the reaction ^{27}Al$(p,\alpha)^{24}$Mg occurs causing a build up of ^{24}Mg as seen in figure 1a. ^{24}Mg undergoes a series of proton captures to create ^{27}Al, but at this point on the RGB, the reaction rate ^{27}Al$(p,\alpha)^{24}$Mg favors the production of ^{24}Mg.

Figures 2a and 2b show the same profiles for sequence 1 at the onset of the helium flash. Notice this time that a significant amount of ^{23}Na is siphoned into ^{24}Mg within the H shell. However, by this point on the RGB, the reaction rates favor the production of ^{27}Al as seen in figure 2b.

4. Discussion

Based on the results from sequence 1 alone, we can see the existence of two cycles in the H shell, the first being the Ne-Na cycle and the second being the Mg-Al cycle. These are connected via the reaction ^{23}Na$(p,\gamma)^{24}$Mg which occurs about once in every 100 Ne-Na cycles for sequence 1. These cycles are depicted graphically in figure 3. Timmermann et al. (1988) and Champagne et al. (1988) do not favor the existence of a Mg-Al cycle because the (p,α) reaction is slower than previously thought (Caughlin & Fowler 1988). However, at the

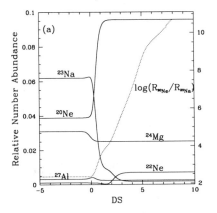
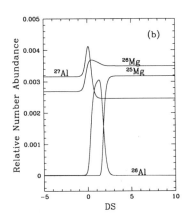

Figure 1. Neon, sodium, magnesium, and aluminum profiles around the H shell region of sequence 1 at the point on the RGB where mixing can begin. R_{20Ne}/R_{23Na} is the ratio of the reaction rates $^{20}Ne(p,\gamma)^{21}Na$ to $^{23}Na(p,\alpha)^{20}Ne$. The abscissa DS is the distance in mass from the center of the H shell in units of the shell's thickness

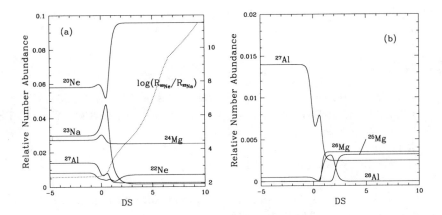

Figure 2. Neon, sodium, magnesium, and aluminum profiles around the H shell region of sequence 1 at the tip of the RGB. The relationships between Ne & Na and Mg & Al are clearly visible here and in figures 2a and 2b

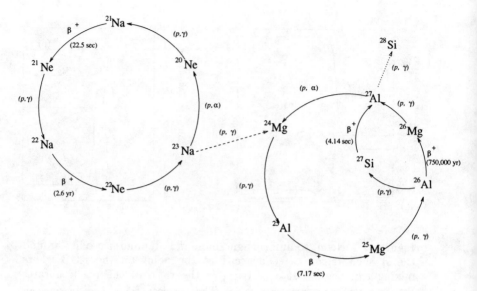

Figure 3. A graphic depiction of the Ne-Na and Mg-Al cycles. The reaction types are marked in parentheses. The half-lifes are given parenthetically for β decays.

temperatures of interest, neither paper can completely exclude the existence of the cycle. These results will be discussed more fully in a future publication.

Acknowledgments. R.M.C. was supported by a NASA Graduate Student Fellowship. A.V.S. was partly supported by NASA RTOP 188-41-51-03.

References

Anders, E. & Grevesse, N. 1989, Geochim. Cosmochim. Acta, 53, 197
Briley, M. M., Bell, R. A., Hesser, J. A., & Smith, G. H. 1994, Can. J. Phys., 72, 772
Caughlin, G. R. & Fowler, W. A. 1988 ADANDT, 40, 283
Champagne, A. E. 1988, Nucl Phys, A487, 433
Cottrell, P. L. & Da Costa, G. S. 1981, ApJ, 245, L79
Denisenkov, P. A. & Denisenkova, S. N. 1990, Sov. Astron. Lett., 16, 275
Kraft, R. P. 1994, PASP, 106, 553
Langer, G. E., Hoffman, R., & Sneden, C. 1993, PASP, 105, 301
Sweigart, A. V. & Mengel, J. G. 1979, ApJ, 229, 624
Timmermann, R. et al 1988, Nucl Phys, A477, 105

Cosmic Abundances
ASP Conference Series, Vol. 99, 1996
Stephen S. Holt and George Sonneborn (eds.)

Chemical composition of supergiants in the Magellanic Clouds

V. Hill

DASGAL, Observatoire de Paris-Meudon, 92105 Meudon Cedex

1. Introduction

As a contribution to the study of the chemical history of the Magellanic Clouds (MCs), we have undertaken a systematical study of a sample of F and K supergiant stars reachable with high resolution spectroscopy and analysed with homogeneous methods and codes. We report here some of the most recent results concerning a sample of six K stars in the SMC (Hill et al., in preparation), compared with previous results (Hill et al. 1995, Spite et al. 1989, hereafter SBS89).

2. Choice of the samples and analysis

Within each sample, the stars were chosen with similar *(B-V)* colors; on the other hand, their were chosen to be widely spread out over the galaxy. In the LMC where the dynamics are dominated by a wide disk-like rotation, the extension of the sampling is moreover insured by choosing a whole range of radial velocities (V_r). The structure of the SMC is somewhat more clumpy. The HI gas is splitted into four kinematic components (Martin et al. 1989), and the two main ones are refered to as VH (high velocity) and VL (low velocity). All our stars but one (PMMR23, belonging to VL) belong to the VH component. The resulting samples consisted of nine F supergiants in the LMC ($6200K \leq T_{eff} \leq 7500K$, $0.1 \leq \log g \leq 0.7$) and six K supergiants in the SMC ($4000K \leq T_{eff} \leq 4300K$, $-0.7 \leq \log g \leq 0.3$).

The model atmospheres that we used to derive the abundances were interpolated, for the cooler stars (K supergiants) in Plez's (1992) grid of models, and for the hotter stars (F supergiants) in Bikmaev's grid (1993 private communication), both adapted from the MARCS code. The effective parameters of the stars (T_{eff}, $\log g$, V_t) were determined by an iterative process described in Hill et al. (in preparation).

3. Homogeneity

Unlike previous studies of supergiant stars in the MCs (Luck & Lambert 1992, and Russell & Bessell 1989), we found a surprisingly uniform iron content among the sample stars : the standard deviation for each sample ($\leq 0.11 dex$) is of the order of the expected intrinsec uncertainty (dues to temperature errors mainly). The mean iron abundances found in each Cloud were, for the LMC (sample of nine F supergiants) $\overline{[Fe/H]}=-0.27\pm0.07$ dex and for the SMC (sample of six

K supergiants) $\overline{[Fe/H]}$=–0.74±0.1 dex; adding the three SMC F supergiants (from Spite et al. 1989) yields $\overline{[Fe/H]}$=–0.70±0.11 dex. This uniformity holds for most other elements (e.g. $[O/H]_{LMC}$=-0.35±0.08 dex), adding strenght to the evidence of homogeneity of the chemical content over the sample.

In the LMC, the range of V_r of the stars insures that they belong to various parts of the galaxy, and that the mixing of the ISM over the galaxy must be efficient to give birth to stars with such similar metal content. In the SMC, all our stars but one belong to the VH component : the homogeneity found therefore only concerns this component. However, the star PMMR23 belonging to the VL group does not depart significantly from the sample ($[Fe/H]$=–0.80 dex).

4. Carbon and Nitrogen

Table 1. CNO abundances in the MCs.

	[Fe/H]	ϵ(C)	ϵ(N)	ϵ(O)	[C+N/H]	[C/Fe]	[O/Fe]
LMC F stars	-0.27 ±0.07	8.09	8.60^a	8.56	0.06^a	-0.19 ±0.12	-0.15 ±0.08
SMC K stars	-0.74 ±0.10	7.56	7.31	7.92	-0.88	-0.28 ±0.04	-0.21 ±0.08
SMC F stars	-0.62	7.85	8.12^a	8.14	-0.30^a	-0.15	-0.16
LMC H II regions^b	—	7.90	6.97	8.43	-0.70	—	—
SMC H II regions^b	—	7.16	6.46	8.02	-1.41	—	—
Sun^c	0.00	8.55	7.97	8.91	0.00	0.00	0.00

^{a}Nitrogen abundances are from LL92 for stars not belonging to our sample
^{b}from Dufour, 1984
^{c}from Grevesse & Noels, 1993

In Table 1 we report the mean carbon, nitrogen and oxygen abundances for our stars together with the values for the Magellanic Clouds' H II regions from Dufour (1984). Convective mixing brings CNO-processed material to the outer atmospheric layers during stellar evolution along the red giant branch. Such process may lead to carbon deficiencies accompanied by nitrogen enhancements (the effect on oxygen is negligible). C+N abundance (which should be free of mixing effects) is found to be overdeficient relative to iron ($[(C + N)/H] < [Fe/H]$) in the SMC K supergiants by around the same factor as found in our Galaxy between the Sun and the solar neighbourhood young objects such as supergiants (Luck & Lambert, 1985) or main sequence B stars in Orion (Cuhna & Lambert, 1994). In F stars, we did not derive the nitrogen abundance for our stars so that the values given in Table 1 are the mean values for stars not belonging to our sample, taken from LL92. The high nitrogen abundance may be a systematic effect due to the lines used in their analysis.

The H II regions C+N abundances in the Galaxy are consistent with that of these young objects, whereas the H II regions in both Clouds show a strong overdeficiency compared to supergiants, due to the low carbon abundance in the H II regions. In the stars, the carbon to iron abundance is lower than solar in both Clouds, but higher than the H II regions' values by ≈0.2 and ≈0.5 dex respectively in the LMC and SMC. It has been argued that the H II regions value, derived from highly ionised species may be underestimated. But Garnett

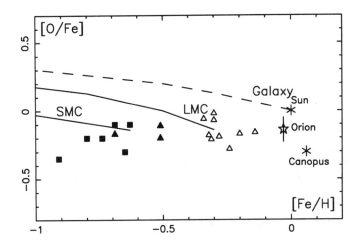

Figure 1. [O/Fe] versus [Fe/H]. For the **Galaxy**: Sun and Canopus (asterisks) and a mean for Orion B stars from Cunha & Lambert (1994) (open star), where the vertical line shows the r.m.s. For the **SMC**: K supergiants (filled square; this work), F supergiants (filled triangles; SBS89). For the **LMC**: F supergiants (open triangles; Hill et al. 1995a). The chemical evolution models by RD92 are shown for the Galaxy (dashed line), the LMC and the SMC (solid lines).

et al. (1995) found from an HST study using more reliable lines that C/O in giant H II regions in metal-poor dwarf galaxies are lower than expected; could this be an indication that the Clouds are indeed depleted in carbon ?

5. Oxygen

Fig.1 shows the oxygen to iron ratios for SMC, LMC and Galactic stars, together with the prediction of chemical evolution models from Russell & Dopita (1992). Oxygen is depleted by around 0.2 dex ($\overline{[O/Fe]}_{LMC}$ =−0.15 ± 0.08 dex and $\overline{[O/Fe]}_{SMC}$ =−0.21 ± 0.08 dex) in both the LMC and the SMC (with respect to the Sun). The oxygen abundances of H II regions in both Clouds are consistent with these values. This low oxygen to iron ratio may be explained by a different chemical evolution in the Clouds from that of the Galaxy. Russell & Dopita (1992) constructed models that fit these observations, by using a different slope of the IMF (more high-mass stars in the Clouds than in the Galaxy, and therefore more Type II Supernovae). But other α elements do not seem to be systematically depleted in the Clouds as oxygen is : $\overline{[\alpha/Fe]}_{LMC} \approx$ +0.17 ±0.10 dex (as a reference, Canopus gives [α/Fe] =+0.12 dex) and $\overline{[\alpha/Fe]}_{SMC} \approx$ -0.04 ±0.11 dex (Kstars), where α is the mean of Si, Ca and Ti.

Another drawback to the chemical evolution explanation of the oxygen underabundance lays in the fact that in our own Galaxy, the solar neighbourhood H II regions and supergiants show also a depletion of around 0.2 to 0.3 dex relative to the Sun (and also relative to the average zero metalicity G-dwarfs), and so do B stars and H II regions in Orion. Compared to our local young population, the Clouds are therefore not significantly overdepleted in oxygen.

6. Heavy r and s process elements

The lighter elements Y and Zr appear to be normal in both Clouds, whereas the elements with $Z \geq 56$ (La, Eu, and also to a lesser extent Ce and Nd), seem to be enhanced by around 0.3 dex in the LMC and 0.5 dex in the SMC. The r-process seem therefore to be very important in the Clouds; in terms of chemical evolution, this would ask for a dominating supply of elements produced by supernovae Type II of moderate mass (heavy -r process elements are believed to be mostly produced in moderate mass SN II).

7. Conclusion

The homogeneity found for the abundances of the supergiants suggests a very efficient mixing in the Clouds, or a surprisingly similar evolution all over the galaxy. The stellar and H II regions oxygen abundances are in good agreement; the O/Fe ratios are lower than the solar value, but compatible with the abundance of the young objects of the stellar neigbourhood. On the other hand, the carbon and nitrogen (C+N) abundances are found to be higher in the supergiants than in the H II regions. The enhancement of the very heavy elements is confirmed, suggesting a high proportion of moderate mass SN II in the Clouds.

References

Cunha, K., Lambert, D.L., 1994, ApJ, 426, 170
Dufour, R.J., 1984, in Structure and evolution of the Magellanic Clouds, IAU Symp. 108, S. Van den Bergh & K. de Boers, Dordrecht: Reidel, 353
Garnett, D.R., Skillman, E.D., Dufour, R.J., Peimbert, M., Torres-Peimbert, S., Terlevich, R., Terlevich, E., Shields, G.A., 1995, ApJ, 443, 64
Grevesse, N., Noels, A., 1993, in: La Formation des Elements Chimiques, B. Hauck, S. Paltani & D. Raboud, (Lausanne, Switzerland), 205
Hill, V., Andrievsky, S., & Spite, M., 1995, A&A, 293, 347
Luck, R.E., Lambert ,D.L., 1985, ApJ, 298, 782
Luck, R.E., Lambert, D.L., 1992, ApJS, 79, 303
Martin, N., Maurice, E., Lequeux, J., 1989, A&A, 215, 219
Plez, B., Brett, J.M., Nordlund, A., 1992, A&A, 256, 551
Russell, S.C., Bessell, M.S., 1989, ApJS, 70, 865
Russell, S.C., Dopita, M.A., 1992, ApJ, 384, 508
Spite, M., Barbuy, B., Spite, F., 1989, A&A, 222, 35 (SBS89)

X-Ray Measurements of Coronal Abundances

Stephen A. Drake

Universities Space Research Association and Code 660.2, Goddard Space Flight Center, Greenbelt, MD 20771

Abstract. The availability of high throughput, moderate-resolution X-ray spectroscopes, such as those currently operating on the Japanese/US *ASCA* observatory, has enabled the measurements of the elemental abundances in stellar coronae. I discuss the assumptions inherent in such abundance analyses, and the potential sources and magnitudes of uncertainties associated with them. The results to date show that almost all stars whose X-ray spectra have been studied in detail show a general deficiency of heavy elements in their coronae relative to the solar photospheric values by factors of $\sim 2 - 5$. This is somewhat unexpected since most of the stars studied are believed to be normal Population I stars, and no such systematic effect has been observed in the solar corona. Independent analyses of high-resolution extreme-ultraviolet spectra of a number of stars have generally found similar results, indicating that the apparent metal-deficiency is almost certainly not an artifact. I discuss the reliability of the individual abundances for those elements which have strong lines in the *ASCA* spectral range.

1. Introduction

It is impossible to cover every aspect of the subject of stellar coronal abundances in the length constraints of the present review, so I will primarily limit it to the discussion of the abundances inferred from **X-ray spectroscopy of late-type, 'coronal' stars**. There have recently been published some excellent reviews and research papers on the topic of the **solar** coronal and energetic particle abundances, e.g., Meyer (1991), Reames (1992), Feldman (1992), Laming *et al.* (1995), Fludra & Schmelz (1995), Meyer (1996), and thus I will only briefly touch upon these results. Likewise, the discussion of stellar coronal abundances as determined by observations in the **extreme-ultraviolet (EUV)** using the *Extreme Ultraviolet Explorer (EUVE)* and *Solar Heliospheric Observatory (SOHO)* spectrometers will be limited, and readers should refer for more details to the recent articles by Mason & Monsignori Fossi (1994), Schrijver *et al.* (1995), and Drake *et al.* 1995. These exclusions essentially mean that this review will concentrate on the interpretation of the soft X-ray spectra of stars of spectral types F through M that have been obtained by the *Advanced Satellite for Cosmology and Astrophysics (ASCA)* since its launch on February 20, 1993. Furthermore, since analyses of many of the spectra that *ASCA* has obtained have in many cases not yet been published, in my comparisons of theoretically predicted versus observed spectra I will perforce mostly use *ASCA* data that I and my collaborators have obtained as part of several of our own Guest Investigator programs.

The structure of this review is as follows: in §2, I describe the theoretical concept of a coronal plasma, and the assumptions that are incorporated when calculating the X-ray spectrum emitted by such a plasma; in §3, I give some examples of the *ASCA* spectra that have obtained of cool stars; in §4, I summarize the coronal elemental abundances that such analyses have yielded and address the issue of how uncertain are these determinations; and, in §5, I summarize my conclusions and present some suggestions as for promising avenues for future studies in this area.

2. Coronal Plasma Models

Coronal plasma models have been extensively used, initially to model the solar coronal spectrum (hence the name), subsequently to model the spectra of other types of astrophysical plasmas, e.g., supernovae remnants, stellar coronae, shocked regions in the winds of hot stars, etc., that have physical conditions more or less similar to the solar corona. Good recent reviews of such models have been given by Mewe (1991) and Raymond & Brickhouse (1995). For present purposes, I will concentrate on (a) a discussion of the 'built-in' assumptions typically incorporated in these models, (b) an examination of potential sources of errors in these models, and (c) how such model spectra are compared with observed spectra of astrophysical plasmas so as to infer physical parameters such as the elemental abundances.

A coronal plasma model is an 'ideal' construct in that real plasmas, while they may in many ways approximate it, will always violate some or all of its defining requirements, viz: the plasma (i) is collisionally ionized and radiatively cooled, (ii) is in ionization equilibrium, (iii) is optically thin at all energies to its radiation, (iv) has electron and ion components with Maxwellian energy distributions with the same temperature, and (v) is not affected by any external radiation through processes like photoionization. Conditions close to these are often found in low-density ($N_e \leq 10^{10}$ cm^{-3}), high-temperature ($T_e \sim 10^{6-8}$K) plasmas such as the solar corona. Given these assumptions and assuming that the atomic physics for line and continuum emission is accurately described, for any specified T_e, N_e, and adopted set of elemental abundances, the emissivity per unit volume of the model plasma as a function of energy, or model spectrum, can be calculated. The most commonly used coronal plasma codes nowadays appear to be the Mewe & Kaastra (or MEKA), Raymond & Smith, and Landini & Monsignori Fossi codes. Since there are thousands of lines of astrophysically abundant elements in the ultraviolet, EUV, and soft X-ray spectral regions that contribute, together with continuum processes such as free-free, bound-free, and two-photon emission, these coronal emission models require large amounts of atomic data.

What are the primary sources of uncertainty in these coronal models? The major internal uncertainties are probably (i) the particular ionization equilibrium that has been adopted, and (ii) error and omissions in the tabulated lines, their predicted energies and collsion strengths. This latter problem is particularly acute in the Fe L-shell region of 0.5 to 1.5 keV (8 to 24 Å) which contains a large number of lines, principally the $n = 3$ and $n = 4$ to $n = 2$ 'L-shell' line complexes from a variety of ionization stages of Fe (Fe XVII - XIV), the analogous Ni L-shell complexes, as well as the resonance lines of the He- and H-like ions of N, O, Ne, and Mg. Liedahl et al. (1995) have discussed the deficiencies of the Fe L-shell atomic data that were included in pre-1995 versions of the standard coronal plasma models, and recent versions of these codes (e.g., the MEKAL code of Mewe, Kaastra, & Liedahl 1995) have incorporated these updated to the atomic data, but further improvements are still clearly necessary, e.g., to the Ni L-shell lines' atomic data.

The major external uncertainty in using such coronal models to fit observed spectra is that of ascertaining whether the physical conditions in the astrophysical plasma match those assumed in the model. For example, most stellar coronae are probably not in an isothermal, constant-density, optically thin, steady state, but have both temperature and density variations, at least some strong resonance lines that are optically thick, and can exhibit variability on timescales as short as minutes (during flares). There is also some evidence from studies of the solar transition region and corona (Laming et al. 1995) that there may be elemental abundance stratification as a function of temperature. Another possible complication is the presence of 'non-thermal' continuum emission that might contaminate the otherwise thermal coronal emission.

The effects of all these departures from 'ideal' coronal models need to be examined on a case-by-case basis, but some general and reasonably secure conclusions can be stated. Firstly, when modeling observed soft X-ray spectra of stellar coronae with sufficient signal-to-noise, isothermal models are almost invariably unacceptable, and at least two discrete temperature components are required. However, as first noted by Craig & Brown (1976), low- and moderate-resolution soft x-ray spectra are rather insensitive to the finer details of the temperature structure of the emitting plasma, and in general the inversion procedure to find the emission measure EM $= \int N_e^2 dV$ as a function of temperature (the so-called differential emission measure or DEM) is ill-defined. Thus, for example, all *ASCA* spectra of stars that have been modeled to date can typically be well-fit by models with either 2, 3, or 4 discrete-temperature (hereafter 2-T, 3-T, or 4-T) components, or by more complex functional representations such as high-order polynomials. Other effects such as those due to high densities and optical depth, while they can affect specific lines and line ratios quite strongly, are probably negligible, at least at the spectral resolution of present soft X-ray spectrometers (see Mewe 1991 and Singh et al. 1996 for a more detailed discussion). Finally, nonthermal continua and departures from equilibrium conditions are probably not significant in most stellar coronae, with the possible exception of during the impulsive phases of flares: since most stellar coronal spectra are accumulated over periods of hours to days, these effects should normally be negligible.

The standard way in which the spectra of astrophysical plasmas are compared with those predicted by coronal plasma models is to use 'global' fitting procedures such as those that are incorporated into the XSPEC and SPEX spectral analysis software packages. An initial model is adopted, with an assumed temperature structure and set of elemental abundances, together with an estimated interstellar hydrogen column density n_H. Thus, if one uses a 2-T model with abundances assumed to be the same as those in the solar photosphere, this means that there are 5 free parameters (the temperature and emission measure of each component and n_H) in such a model. The values of the free parameters are varied until, in the context of this particular model, the best fit of the calculated with the observed spectrum is obtained. Failure of such 'solar-abundance' models to fit the observed spectra then presumably would imply that the elemental abundances prevailing in stellar coronae are not in the same proportions as in the solar photosphere. Since much of the solar coronal plasma also does not appear to have the same abundances as are found in the solar photosphere, it is perhaps not too surprising that, as discussed below, stellar coronae also appear to have peculiar elemental abundances

3. *ASCA* Spectroscopy of Stellar Coronae: The Data and the Modeling

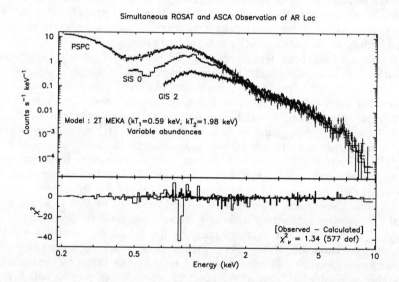

Figure 1. The simultaneous *ASCA* SIS 0 and GIS 2, and *ROSAT* PSPC spectra of AR Lac are plotted as crosses and the best-fit, 2-T variable-abundance MEKA model as the histograms in the top box, while the resultant contributions to the χ^2 statistic are shown in the lower box, all as a function of energy (from Singh *et al.* 1996). The hotter component has an emission measure 1.5 times that of the cooler one. The best-fit abundances for this model are shown in Fig. 3.

The *ASCA* Observatory has been described in some detail by Tanaka *et al.* (1994), so here I will only very briefly summarize its salient characteristics. It comprises four imaging X-ray telescope (Serlemitsos *et al.* 1995) at the foci of which are two Solid State Imaging Spectrometers, SIS0 and SIS1 (Burke *et al.* 1991), and two GAs Imaging Spectrometers, GIS2 and GIS3 (Ohashi *et al.* 1991), respectively. The energy coverages of the two types of detectors are quite similar (0.5 – 10 keV for the SIS's compared to 0.8 – 10 keV for the GIS's), but the energy resolution of the SIS's (2% FWHM at 5.9 keV decreasing to 6% at 1.0 keV) is a factor of 3 – 4 better than that of the GIS's. Because this latter property makes them more sensitive to line emission, I will mostly limit my discussion to SIS spectra. The effective areas of the 4 detectors are similar ($\sim 100 - 150$ cm^2) over the central part (1.5 – 5.0 keV) of their energy ranges. In our group's experience, an accumulated SIS total counts of $\geq 10^4$ is required in order to obtain good constraints on the properties of a coronal-type X-ray source.

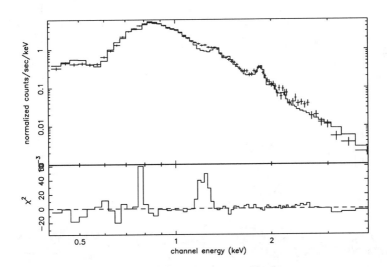

Figure 2. The same as in Fig. 1, but showing the *ASCA* SIS 0 spectrum of Capella and the best-fit 2-T variable-abundance MEKAL model (i.e., the MEKA model with the Fe L-shell atomic data updated), from Drake *et al.* (1996). Notice the poor fit of the model in the vicinity of the Ne X Lyman$-\beta$ line at 1.2 keV. The temperatures of the two components are 0.55 and 0.9 keV with the hotter component having an emission measure 0.47 times that of the cooler one. The best-fit abundances for this model are shown in Fig. 4.

Some examples of *ASCA* spectra, showing both the actual data as well as the best-fit model spectra, of the RS CVn binary AR Lac (chosen as an example of a relatively hard coronal source), and the nearby active binary system Capella (a fairly soft coronal source), are shown in Figures 1 and 2, respectively. The particular best-fit models that I show here are 2-T models in which the elemental abundances were allowed to vary from their solar photospheric values (but the abundance patterns of the two components were assumed to be identical). For the solar photospheric reference abundances, I have adopted the values given by Anders & Grevesse 1989: notice that more recent estimates of the Fe abundance in the solar photosphere (e.g., Grevesse et al. 1992) are about 0.15 dex smaller than the Anders & Grevesse Fe value of log $N_{Fe} = 7.67$. I show the elemental abundances as determined from these coronal spectra as a function of their first ionization potentials (FIP) in Fig. 3 (AR Lac) and Fig. 4 (Capella).

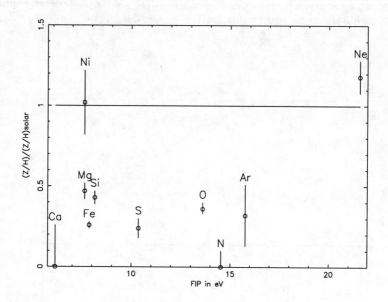

Figure 3. The elemental abundances which give the best fit to the *ASCA* SIS 0, GIS 2, and *ROSAT* PSPC spectra of AR Lac are plotted as a function of first ionization potential. The error bars shown are the formal-fit errors only and do not include the possible systematic errors that are discussed in the text.

Notice that, except for the elements Ni and Ne in AR Lac, all elements appear to have lower abundances than in the solar photosphere (hereafter referred to as subsolar abundances). Typical depletions range from factors of 2 to ≥ 5. The error bars shown indicate the formal-fit errors and mostly reflect the errors due to counting statistics. In AR Lac there is no evidence for any trend of the depletion factor with FIP, whereas in Capella there is some evidence that elements with FIP values greater than 10 eV are more depleted than those with low-FIP values. This low- versus high-FIP effect is seen in the solar corona where the ratio of their respective depletions is about 4 (cf. Feldman 1992), and it is generally (but not unanimously: see Fludra & Schmelz 1995) believed that it is the abundances of the low-FIP elements that are enhanced relative to their photospheric values. The results shown here for Capella are more consistent with the alternate hypothesis that it is the high-FIP elements that are depleted

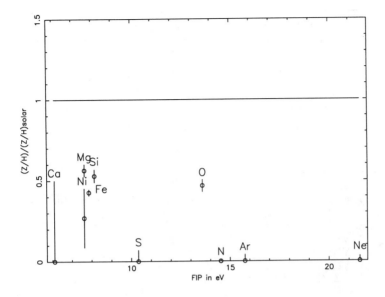

Figure 4. The elemental abundances and their formal-fit errors, for the best-fit model to the *ASCA* SIS 0 spectrum of Capella, are plotted as a function of first ionization potential.

in the solar corona, but, as noted below, the systematic uncertainties in the derived abundances of 2 of the 4 high-FIP elements (Ne and N) are large, so that this is not a strong conclusion.

The spectrum of AR Lac is very similar to the *ASCA* spectra of other very active binaries, e.g., Algol and RZ Cas, and the very active single K dwarf Speedy Mic, and the temperatures and abundances inferred for these other stars are likewise similar to those inferred for AR Lac. Likewise, the *ASCA* spectra of the intermediate-activity giant β Cet and dwarf star π^1 UMA are similar to that of Capella, and again similar temperature structures and abundance patterns are inferred. This suggests that it may be the activity level of a corona, as determined by the X-ray luminosity L_x or surface flux S_x, for example, that influences the elemental abundances that prevail in the coronal structures: the RS CVn and Algol binaries with $L_x \sim 10^{31}$ erg s^{-1} thus show one pattern, while moderate activity dwarf and giant stars with $L_x \sim 10^{29-30}$ erg s^{-1} show a different pattern. To this date, no *ASCA* observation has been made of a star with a low-activity corona similar to that of the sun ($L_x \leq 10^{28}$ erg s^{-1}) in order to see whether a solar-like FIP-effect is evident in its soft X-ray spectrum.

Mewe et al. (1996) have analyzed simultaneous *ASCA* and *EUVE* spectra of the very active, single K dwarf AB Dor using the SPEX spectral analysis package and (a) confirmed that the coronal plasma of this active star has an essentially bimodal distribution and (b) inferred coronal abundances that are depleted relative to their solar values by factors of 2 − 3 for the best-determined elements, except for Mg (0.86 times solar) and Ne (1.5 times solar). The average depletion of ~ 0.5 that Mewe *et al.* infer for AB Dor is not as extreme as the ~ 0.25 value that our group has inferred for the very similar star Speedy Mic. Given the differences in the analysis techniques and datasets it remains to be seen whether or not this difference is significant.

4. *ASCA* Spectroscopy of Stellar Coronae: The Elemental Abundances

Table 1. Lines of He- and H-like Ions of Abundant Elements in 0.4 – 10 keV Range.

Ion	Energy keV	Wavelength Å	Formation Temperature ($\log T$ in K)
N VI	0.43	28.8	6.2
N VII	0.50	24.8	6.3
O VII	0.57	21.6	6.3
O VIII	0.65	19.0	6.5
Ne IX	0.92	13.4	6.6
Ne X	1.02	12.1	6.8
Mg XI	1.35	9.17	6.8
Mg XII	1.47	8.42	7.0
Si XIII	1.86	6.65	7.0
Si XIV	2.01	6.18	7.2
S XV	2.46	5.04	7.2
S XVI	2.62	4.73	7.4
Ar XVII	3.14	3.95	7.35
Ar XVIII	3.32	3.73	7.6
Ca XIX	3.90	3.18	7.45
Ca XX	4.10	3.02	7.75
Fe XXV	6.70	1.85	7.85
Fe XXVI	6.97	1.78	8.2
Ni XXVII	7.80	1.59	7.95
Ni XXVIII	8.05	1.54	8.3

How Reliable are *ASCA*-Derived Coronal Abundances? The answer to this question is a function of (i) how good of a representation of the real corona the best-fit model is, (ii) which element is being referred to, and (iii) how well-exposed the particular *ASCA* spectrum under consideration is in the spectral region(s) containing the strongest line(s) of this element. Some concerns were raised in the 'early days' of *ASCA* analysis as to whether the very simplified temperature structures (usually 2-T or 3-T models) generally adopted were systematically biasing the values of the elemental abundances that were being inferred. Our group and others (e.g., Singh *et al.* 1995, 1996; Kaastra *et al.* 1996) have extensively tested this possibility by fitting the same observed *ASCA* spectra with both 2-T, power-law DEM, and high-order polynomial DEM models and the general conclusion is that the same pattern of elemental abundances (at least within the combined statistical errors) is recovered by all those models which can acceptably fit the data. What about the individual elements which have strong lines in the soft X-ray region and hence measurable abundances? I have listed the energies and corresponding wavelengths of the 9 elements which have the resonance line complexes of their He- and H-like ions in the *ASCA* spectral range in Table 1. It is in essence the fit to these lines that yields the relative elemental abundances, while it is the line-to-continuum ratio that determines their normalization relative to hydrogen. The errors quoted in the tables giving the results of analyses of *ASCA* spectra (e.g., White et al. 1994, Drake et al. 1994, Singh et al. 1995, 1996) and shown here in Figs. 3 and 4 are typically $0.05 - 0.2$ dex and are the formal fit errors, and thus should be regarded as lower limits to the 'real' errors. The relative magnitudes of the errors in the individual abundance parameters, for instance, are indicative of which elemental abundances are best and which are least well-determined, and are fairly robust. I now discuss the accuracy of the abundances on an element-by-element basis.

Fe. The Fe abundance is determined by the contrast of the L-shell 'hump' at 1 keV compared to the higher-energy continuum in soft sources like Capella and β Ceti, and by the additional ratio of the He-like Fe XXV K feature at 6.7 keV to the continuum in the harder spectra of active binary stars. The Fe abundances inferred using the K line are generally consistent with those obtained from the L complex in those stars whose spectra are hard enough that the K line is detected, according to both analyses of coronal spectra (e.g., Singh et al. 1996) and of the X-ray spectra of clusters of galaxies (Mushotzky et al. 1996). This suggests that any systematic uncertainties in the inferred Fe abundances due to the uncertainties in the atomic physics of L-shell region are smaller than the typical statistical uncertainties for Fe (i.e., ≤ 0.1 dex). The reality of the low Fe abundances inferred from *ASCA* spectra of coronal stars is supported by *EUVE* spectra of cool stars (e.g., Schrijver et al. 1995) which also show lower than expected contrast of Fe lines to the EUV continuum than is predicted by solar-abundance models.

Si. The Si abundance is predominantly determined from the Si XIII/XIV complex at 1.9 keV: this feature is strong, and should not suffer from significant contamination from other species, and the atomic physics of these H- and He-like ions should be fairly accurately known. The major external uncertainty in the Si abundance is due to the presence of Si in the SIS spectrometer itself. Mushotzky et al. (1996) have studied this problem, and conclude that uncertainty in the SIS calibration due to this is quite small, and that the resultant uncertainty in nay inferred Si abundance is ≤ 0.1 dex.

S. The S abundance is mostly based on the observed strength of the He- and H-like S XV/XVI complex at 2.5 keV: this feature is predicted to be quite strong in solar-abundance coronal plasmas of the temperatures that prevail in active stellar coronae, but is often weaker than this in actual spectra, hence the inferred S underabundances.

Mg. The Mg abundance is based on the observed strength of the the He- and H-like Mg XI/XII complex at 1.4 keV: this feature is also predicted to be one of the most distinguishable ones present in coronal plasmas, and indeed usually is. It does suffer from confusion due to lines of other species, primarily Ni and Fe lines, particularly for plasmas with significant material at temperatures ≥ 1 keV. Mushotzky et al. (1996) discuss in more detail the confusion in this spectral region, uncertainties in the detector response, and the resultant uncertainties in the inferred Mg abundances: they conclude that the latter are probably greater than the typical statistical uncertainties found for this element, i.e., ≥ 0.2 dex.

O. The O abundance is mostly derived from the observed strength of the O VIII line at 0.65 keV: this feature is not seen as an individual line in *ASCA* SIS spectra but is blended with the low-energy edge of the Fe L shell complex which is predicted for these temperatures to be at ~ 0.73 keV. Again, as Mushotzky et al. (1996) note, the presence of O in the SIS detector itself, as well as the position of this feature close to the low-energy edge of the SIS's spectra range, do certainly add a systematic uncertainty to any inferred O abundances, but, as for Si, they estimate that it is relatively small (≤ 0.1 dex).

Ca. The Ca abundance is mostly based on the He-like and H-like Ca XIX/XX feature at 4.0 keV (a relatively unconfused region) that is observable in high signal-to-noise *ASCA* spectra of the harder coronal sources. The uncertainties in the Ca abundance are thus believed to be dominated by the statistical errors.

Ar. The Ar abundance is primarily based on the He-like and H-like Ar XVII/XVIII complex that is predicted to be at 3.3 keV. Similarly to the Ca line, the Ar line is in a relatively unconfused spectral region, and the uncertainties in the inferred Ar abundance should be predominantly statistical.

Ne. The stongest predicted lines of Ne are in the K-shell complex at 1.0 keV which lies right in the middle of the Fe L complex. Despite this, Mushotzky et al. (1996) present evidence based on simulations that Ne abundances inferred from *ASCA* spectra do not seem to exhibit any significant systematic errors. However, a recent analysis of the *ASCA* SIS spectra of the relatively soft coronal sources Capella and β Cet (Drake et al. 1996) has found that the best-fit MEKAL models to these spectra have great difficulty in fitting this region of the spectrum, and, indeed, prefer that the Ne abundance be vanishingly small, suggesting that there may be significant systematic problems in the inferred Ne abundances for at least some coronal sources. Thus, Ne abundances measured by *ASCA* should be treated with some caution, in my opinion.

N. The abundance of N in *ASCA* spectra is primarily inferred from the presence (or absence) of a line feature due to N VII Ly α at 0.50 keV, which is predicted to be strong in solar-abundance plasmas (and whose atomic physics is known very accurately). To my knowledge, no SIS spectrum of any star has shown evidence for this feature, implying either a universal absence of N in stellar coronae, or that the effective area of the SIS has been substantially overestimated at this energy, which lies close to the low-energy cut-off and hence in a spectral region where the effective area is rapidly decreasing towards shorter energies. An *ASCA* Calibration Workshop held in December 1995 has, in fact, concluded that the current SIS calibration below 0.6 keV is unreliable.

Ni. The Ni L-shell complex sunstantially overlaps in energy the Fe L-shell complex. The Ni K-shell line complex lies at a very high energy (8.1 keV) and has never, to my knowledge been detected in any coronal source. The uncertainties in the atomic physics of the Ni L-shell lines are probably large, and I hope that Liedahl and his collaborators will do the same type of thorough review of these Ni lines as they did for their Fe L-shell counterparts. Until such a study is done, any Ni abundance measured by *ASCA* should be taken with several grains of salt: I estimate that the present systematic uncertainties for Ni are ≥ 0.3 dex.

5. Conclusions

The abundances of stellar coronae measured by *ASCA* indicate that the heavy element to hydrogen ratios in these plasmas is are (at face value) only about one-third of their solar photospheric values. The best-determined abundances of elements like Fe, Si, and S are estimated to have total (statistical plus systematic) errors of 0.1-0.2 dex that rival the accuracy at which photospheric abundances can be measured from the optical spectra of stars, while the errors for some of the other elements such as Mg, Ne, and Ni are somewhat larger. Various explanations have been offered to explain these findings:

(i) that the abundances are really subsolar, either due to their photospheric abundances being subsolar and their coronae reflecting this, or to the operation of a differentiation mechanism in their coronae. There is in fact a long-standing finding (Naftilan & Drake 1977) which may support the first alternative, recently confirmed with better data and a larger sample (e.g. Randich *et al.* 1994), that the active stars in many RS CVn binary systems exhibit underabundances in their photospheric Fe abundances of factors of several to 10 compared to the solar value. However, since single stars like AB Dor which have solar photospheric abundances also exhibit subsolar coronal abundances, a different explanation would have to found for these stars.

(ii) that the weakness of the lines relative to the continuum is some kind of artifact unrelated to the elemental abundances; e.g., there is an extra non-thermal continuum in the soft X-ray region, or that the lines are supressed due to radiative transfer effects, or that the apparent continuum is actually dominated by a cloud of line emission from thousands of weak lines that have not been included in the line lists used by the plasma codes. None of these suggestions seem likely in my opinion: the fact that a similar result is inferred from analyses of EUV spectra of stars, e.g., Schrijver *et al.* (1995), Stern *et al.* (1995), Mewe *et al.* (1995) means that any such proposed mechanism must operate with about the same magnitude in both the X-ray and EUV spectral regions, which seems somewhat implausible.

Where should we go from here? In my opinion, we should use the remaining *ASCA* observing cycles to try and obtain high S/N (≥ 100) spectra of a wider range of coronal stars, particularly concentrating on stars that are (a) less pathological in their activity levels than the RS CVn and Algol binaries that dominated the early *ASCA* observations of stars, and (b) have well-determined photospheric abundances. Joint observations with *EUVE* and/or observations of stars for which *EUVE* spectra have been or will be obtained should prove particularly invaluable.

Acknowledgments. I gratefully acknowledge my collaborators in the analyses of the *ASCA* spectra of cool stars, particularly K. P. Singh, Nicholas White, Rolf Mewe, Jelle Kaastra, and Theodore Simon. I also have benefited from discussions with Richard Mushotzky and Stephen White.

References

Anders, E., & Grevesse, N., 1989, Geochimica et Cosmochimica Acta, 53, 197
Burke, B. E. *et al.* 1991, IEEE Trans., ED-38, 1069

Craig, J. D., & Brown, J. C. 1976, A&A, 49, 239
Drake, J. J., Laming, J. M., & Widing, K. G. 1995, in Astrophysics in the Extreme Ultraviolet: IAU Colloquium No. 152, S. Bowyer & R. F. Malina, Cambridge: Cambridge Univ. Press, in press
Drake, S. A., Singh, K. P., White, N. E., & Simon, T. 1994, ApJ, ApJ, 436, L87
Drake, S.A., Singh, K.P., White, N.E., Mewe, R., & Kaastra, J. S. 1996, in Cool Stars, Stellar Systems, and the Sun: Ninth Cambridge Workshop, R. Pallavicini & A. K. Dupree, San Francisco: ASP, in press
Feldman, U. 1992, Phys. Scripta, 46, 202
Fludra, A., & Schmelz, J. T. 1995, ApJ, 447, 936
Grevesse, N., Noels, A., & Sauval, A. J. 1992, in Proceedings of the First SOHO Workshop: ESA SP-348, Noordwijk: ESA Publications Division, ESTEC, 305
Kaastra, J. S. et al. 1996, A&A, submitted
Laming, J. M., Drake, J. J., & Widing, K. G. 1995, ApJ, 443, 416
Liedahl, D. A., Osterheld, A. L., & Goldstein, W. H. 1995, ApJ, 438, L115
Mason, H. E., & Monsignori Fossi, B. C. 1994, A&AR, 6, 123
Mewe, R. 1991, A&AR, 3, 127
Mewe, R., Kaastra, J. S., & Liedahl, D. A. 1995, Legacy, No. 6., 16
Mewe, R., Kaastra, J. S., White, S. M., & Pallavicini, R. 1996, A&A, submitted
Meyer, J.-P. 1991, Adv. Space Res., 11, (1)269
Meyer, J.-P. 1996, in this volume
Mushotzky, R. F. et al. 1996, ApJ, in press
Naftilan, S. A., & Drake, S. A. 1977, ApJ, 216, 508
Ohashi, T. et al. 1991, Proc. SPIE, 1549, 9
Randich, S., Giampapa, M. S., & Pallavicini, R. 1994, A&A, 283, 893
Raymond, J. C., & Brickhouse, N. S. 1995, in Laboratory, Solar, and Stellar, Diffuse Astrophysical and Extragalactic Plasmas, R. Wilson, T. W. Hartquist, & A. J. Willis, Dordrecht: Kluwer, in press
Reames, D. V. 1992, Proceedings of the First SOHO Workshop: ESA SP-348, Noordwijk: ESA Publications Division, ESTEC, 315
Schrijver, C. J., Mewe, R., van den Oord, G. H. J., & Kaastra, J. S. 1995, A&A, 302, 438
Serlemitsos, P. J., et al. 1995, PASJ, 47, 105
Singh, K. P., Drake, S. A., & White, N. E. 1995, ApJ, 445, 840
Singh, K. P., White, N. E., & Drake, S. A. 1996, ApJ, in press
Stern, R. A., Lemen, J. R., Schmitt, J. H. M. M., & Pye, J. P. 1995, ApJ, 444, L45
Tanaka, Y, Inoue, H., & Holt, S. S. 1994, PASJ, 46, L37
White, N.E. et al. 1994, PASJ, 46, L97

ASCA Measurements of Coronal Elemental Abundances in an Active K0 Dwarf Star

K. P. Singh [1], S. A. Drake [2], and N. E. White

Code 660.2, Laboratory for High Energy Astrophysics, NASA/GSFC, Greenbelt, MD 20771

Abstract.
We present *ASCA* X-ray spectroscopic observations of a rapidly rotating K0 type dwarf star, HD 197890, with a period of ~0.3 days. Our moderately deep X-ray observation covering two rotation periods of the star has provided us with an opportunity to estimate the elemental abundances prevailing in its hot (10^6 - $10^{7.5}$K) coronal plasma. The derived abundances are compared with the elemental abundances of the solar photosphere and corona, and with similar measurements of other single late-type stars, and of active stellar binaries containing late-type companions.

1. Introduction

X-ray emission from late spectral type stars is believed to originate in $10^{6.0}$ − $-10^{7.5}$K coronal plasma confined to magnetic loop structures. As the stellar rotation speed increases, the dynamo generated magnetic flux and hence the luminosity and temperature of the plasma increases (Pallavicini et al. 1981). X-ray spectroscopy of active late type stars with *ASCA* is finding that the coronal abundances in these stars are quite different from the solar photospheric abundances (e.g., Antunes et al. 1994; White et al. 1994). These studies have included single giants like β Cet, subgiants in RS CVn type or Algol-type binaries, and a number of active dwarf stars, e.g., the flaring stellar dMe binary system YY Gem (Gotthelf et al. 1994), the G 1.5 V star π^1 UMa (Drake et al. 1994), and the K1 V star AB Dor (White et al. 1995). Here, we present new results based on a preliminary analysis of *ASCA* SIS observations of an active (single) K dwarf star, HD 197890.

Since the detection of HD 197890 (SAO 212437) as a strong and variable EUV source in the *ROSAT* All Sky Survey with the Wide Field Camera (Pounds et al. 1993), it has been found to be one of the most rapidly rotating single stars discovered so far (Anders et al. 1993 − A93; Bromage et al. 1992 − B92). Having a period of only 0.3 days it was nicknamed 'Speedy Mic' by B92. The

---

[1] NRC Senior Research Associate, on leave from Tata Institute of Fundamental Research, Bombay, India

[2] Universities Space Research Association

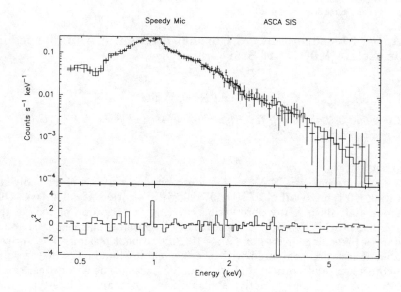

Figure 1. ASCA SIS spectrum of Speedy Mic and the best fit 2T non-solar abundance model shown as a histogram (top panel). The residual χ^2 from the fit are shown in the bottom panel.

optical and UV characteristics of Speedy Mic based on the photometric and spectroscopic studies (A93, B92, Jeffries 1993, Robinson et al. 1994) can be summarized as: (i) m_V=9.3 and B–V = 0.82, (ii) K0 V spectral type, (iii) a photometric period of 7.0±1.0 h (Rossby number = 0.015), (iv) very broad Ca II K emission line, (v) strong Mg II, C II and C IV emission lines in the UV, (vi) anomalously high lithium abundance (equivalent width of 630 mÅ), (vii) $v\sin i$ value of 170±20 km s^{-1}, (viii) radius R $\sin i$ = 0.98 ± 0.14 R$_\odot$, (ix) v_{rad} = -6.5 km s^{-1}, and (x) $F_{bol} = 6.5 \times 10^{-9}$ ergs cm^{-2} s^{-1}. In addition, time resolved Hα spectroscopy by Jeffries (1993) has shown the presence of transient absorption features, similar to those found for K dwarf AB Dor that may be due to co-rotating clouds of cool material ($N_H \sim 10^{20}$ cm^{-2}) (A93; Jeffries 1993) Based on its very high rotation speed and high lithium abundance A93 suggest that HD 197890 is a young pre-main sequence (PMS) object similar to the well studied K0 dwarf AB Dor. High lithium abundance in similar stars has also been suggested to result from the post main sequence evolution (Fekel & Balachandran 1993). Its distance has been estimated to be 40 pc.

2. Observations

Speedy Mic was observed with *ASCA* (Tanaka et al. 1994) from 1995 April 20, 23:25 UT to April 21, 12:35 UT, covering two rotation periods. A useful exposure time of ~22870 s was realized. X-ray emission from Speedy Mic was detected with a count rate of 0.15 counts s^{-1} with the SIS CCD detectors onboard $ASCA$. The source was found to be steady for most of the observation except near the end when it increased in its intensity by ~50% and stayed at that level for about 2000s. No evidence of rotational modulation of the X-ray emission was found. The average X-ray flux observed in the 0.5 – 10 keV energy band was 3.9×10^{-12} ergs cm^{-2} s^{-1} which implies an X-ray luminosity of 7.5×10^{29} ergs s^{-1}.

3. Analysis and Results

The X-ray spectral data extracted from the SIS0 and SIS1 detectors following a standard procedure, and combined for analysis are shown in Figure 1. The spectrum shows little evidence for discrete lines such as the resonance lines of the He- and H-like ions of Mg, Si, and S at 1.4, 1.9, and 2.5 keV, respectively, but is dominated by the broad and essentially unresolved hump of the Fe L-shell complex centered at ~ 0.9 KeV together with a fairly featureless higher-energy 'continuum' emission that is detected up to energies ~ 4 keV. The SIS spectra were analyzed using the XSPEC version 9.0 software. We used the latest version of the coronal plasma emission code known as MEKAL (named after MEwe, KAastra, and Liedahl 1995) which incorporates the Liedahl et al. (1995) updates to the Fe L line strengths in the older version (Kaastra & Mewe 1993). We have tried (a) single-temperature plasma model, (b) a model consisting of two or more discrete plasma components at different temperatures, and (c) a continuous emission measure plasma model parameterized by the exponential of a sum of terms of a 6th-order Chebyshev polynomial in the log EM–log T plane (see Lemen et al. 1989, Singh et al. 1996) (CCP model). The elemental abundances in the plasma for all of these different models could be varied with respect to the solar photospheric values taken from Feldman (1992). The results based on this analyses are given in Table 1. Single temperature models are found to be unacceptable. Two temperature and CP6 models, with solar or non-solar photospheric abundances, were found to be acceptable. The non-solar models with coronal abundances as given in Table 1 are *preferred* over the solar abundance models, however. The best fit 2T non-solar abundance model is shown plotted as a histogram in Fig. 1.

4. Conclusions

The present analysis of X-ray spectra favours lower than solar photospheric abundances in the corona of Speedy Mic, a trend already seen in coronal X-ray spectra of a number of active stars e.g., YY Gem, Algol, AR Lac etc. The pattern of coronal abundances for different elements is close to that found in YY Gem (Gotthelf et al. 1994), and Speedy Mic shows a significant hard component (kT = 2.5 keV) in its X-ray spectrum similar to that also found for YY Gem. The presence of such a hard component, which is absent from the spectrum of less active ($L_x/L_{bol} \sim 10^{-5}$) stars such as π^1UMa, β Cet, and Capella, appears to be a signature of stars, like Speedy Mic, YY Gem, and RS CVn and Algol binaries, whose coronae are exceptionally active, with $L_x/L_{bol} \sim 10^{-3}$.

Table 1.
Results from analysis of ASCA SIS Spectrum

Emission Measure Model	Abundances (relative to Solar phot.)	(kT, log EM) (keV, cm^{-3})	N_H (range) 10^{20} cm^{-2}	χ^2_ν	Deg. of Freedom
1T	Solar	(1.15,52.54)	0	8.3	92
1T	$Z_{all} = 0.08$	(0.80,53.26)	7.8	2.0	91
2T	Solar	(0.69,52.15) (2.5,52.55)	0	1.1	90
2T	$Z_{all} = 0.44$	(0.70,52.47) (2.1,52.65)	0.7 (0–2)	0.9	89
2T	O=0.65,Ne=0.8 Mg=0.14, Si=0.25 S=0.10, Fe=0.27	(0.66,52.62) (2.5,52.60)	2.8 (0–4)	0.6	83
CCP	Solar	–	2.0 (0–3.5)	1.0	87
CCP	O=0.52,Ne=0.75 Mg=0.22, Si=0.4 S=0.10, Fe=0.56	–	1.7 (0–3.0)	0.8	80

Acknowledgments. We wish to thank the entire *ASCA* team for making this observation possible.

References

Anders, G.J., et al. 1993, MNRAS, 265, 541 (A93)

Antunes, A., Nagase, F., & White, N.E. 1994, ApJ, 436, L83

Bromage, G.E., et al. 1992, ASP Conf. Ser. Vol. 26, *Cool Stars, Stellar Systems, and the Sun*, eds. Giampapa, M.S. & Bookbinder, J.A. (B92).

Drake, S.A., Singh, K.P., White, N.E., & Simon, T. 1994, ApJ, 436, L87

Fekel, F.C. & Balachandran, S., 1993, ApJ, 403, 708

Feldman, U. 1992, Phys. Scripta, 46, 202

Gotthelf, E.V., Jalota, L., Mukai, K., White, N.E. 1994, ApJ, 436, L91

Jeffries, R.D. 1993, MNRAS, 262, 369

Kaastra, J.S., & Mewe, R. 1993, Legacy, 3, 16

Lemen, J.R., Mewe, R., Schrijver, C.J., & Fludra, A. 1989, ApJ, 341, 474

Liedahl, D.A., Osterheld, A.L., & Goldstein, W.H., 1995, ApJ, 438, L115

Mewe, R., Kaastra, J.S., & Liedahl, D.A. 1995, Legacy, 6, 16

Pallavicini, R. et al. 1981, ApJ, 248, 279

Pounds, K.A., et al. 1993, MNRAS, 260, 77

Robinson, R.D., et al. 1994, MNRAS, 267, 918

Singh, K.P., White, N.E., & Drake, S.A. 1996, ApJ, 456, 000.

Tanaka, Y., et al., 1994, PASJ, 46, L37

White, N.E., et al. 1994, PASJ, 46, L97

Cosmic Abundances
ASP Conference Series, Vol. 99, 1996
Stephen S. Holt and George Sonneborn (eds.)

The r-, s-, and p-Processes

Bradley S. Meyer, Jason S. Brown, Ning Luo

Department of Physics and Astronomy, Clemson University, Clemson, SC 29634-1911

Abstract. We discuss the three major processes for synthesis of the heavy (mass number $A \gtrsim 70$) nuclei. For each process, we discuss the mechanism in some detail. We also summarize some of the current thinking on where these processes occur.

1. Introduction

This paper deals with the origin of the ~200 naturally-occurring nuclides with mass number A greater than 70. We will explain briefly the formation processes for these isotopes, and then consider where in Nature these processes might have occurred. The science of heavy-element formation has had a distinguished ~40 year history, so no one paper can cover all relevant aspects, but we will attempt to isolate out the key points. Those readers who seek a deeper discussion may wish to turn to Meyer (1994) and references therein.

Because of the large coulomb barriers possessed by the heavy nuclei, their charged-particle-capture production is not efficient. They are instead produced by neutron capture. The lasting contribution of Burbidge et al. (1957) and Cameron (1957) is the recognition that two distinct neutron-capture processes, occurring on different timescales, were each responsible for roughly half of the solar system's supply of heavy nuclei. The 35 low-abundance proton-rich nuclei shielded from neutron-capture production were thought to be made by proton capture. We now believe disintegration reactions are primarily responsible for these nuclei.

Burbidge et al. named the two neutron-capture processes the r- and s-processes. The process responsible for the shielded proton-rich nuclei was called the p-process. These epithets have stuck with time because they so usefully summarize our knowledge of how the heavy elements formed. Their solar system distributions are shown in Figure 1. We discuss each process separately.

2. The r-process

The r-process is the *rapid* process of neutron capture nucleosynthesis. By rapid it is meant that neutron capture reactions occur more rapidly than nuclear β^--decay reactions. The consequence is that the nuclear system is able to establish a nearly exact equilibrium between (n,γ) (i.e. neutron capture) reactions and (γ,n) (i.e. neutron disintegration) reactions before nuclei increase their charge by β^--decay. In $(n,\gamma)-(\gamma,n)$ equilibrium, an abundance maximum occurs

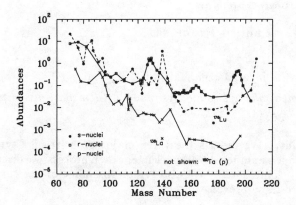

Figure 1. The solar-system abundances of r-nuclei, s-nuclei, and p-nuclei, relative to Si=10^6. Only isotopes for which 90% or more of the inferred production comes from a single process are shown. The data are from Anders & Grevesse (1989) and Käppeler et al. (1989).

at some neutron number N for each proton number Z, and the locus of all of these maxima in the neutron number-proton number plane is the r-process path (see Figure 2). This path is typically ~15 or more neutrons away from the valley of β-stability, although its exact location depends on the instantaneous neutron-number density and temperature. Once the bath of neutrons disappears, however, the nuclei decay back to the stability valley.

Clearly the r-process requires a huge abundance of neutrons. A typical neutron-number density during the r-process is $n_n \sim 10^{26}$ cm^{-3}. This is ~ 200 grams of neutrons per cm^3! An alternative, and more useful, way of looking at this is to note that the r-process assembles actinides like ^{238}U from seed nuclei with mass number A$\approx$ 70 − 100 by neutron capture. The r-process thus needs of order 100 free neutrons per seed nucleus.

Figure 2 shows how the abundance pattern characteristic of the r-process builds up. The r-process path intersects nuclei with closed neutron shells at the magic neutron numbers $N = 28, 50, 82, 124$, and 184. The path "kinks" upward at these neutron numbers because of the added stability of nuclei with closed shells. The key point here is that the more neutron rich nuclei are with respect to the β-stable nuclei, the greater the asymmetry between their numbers of neutrons and protons and, consequently, the larger their β^--decay energies and a fortiori the larger their β^--decay rates. Thus, the upward kink in the r-process path leads to nuclei with smaller β^--decay rates compared to the rest of the nuclei on the r-process path. This slows the flow to higher Z and leads to an abundance build up at the closed neutron shells in the same way that a sudden closing one of several freeway lanes usually leads to a traffic jam.

Now we consider how to get a large abundance of free neutrons. The r-process typically needs ~100 free neutrons per seed nucleus. What requirements this imposes on the degree of neutron richness depends on the entropy per baryon of the matter. Low-entropy matter at the beginning of the r-process consists of seed nuclei and free neutrons. A convenient measure of the neutron richness of

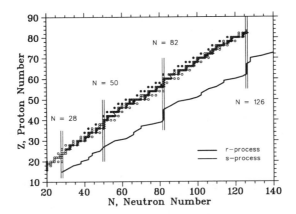

Figure 2. The r-process and s-process paths in the neutron number-proton number plane. The circles show naturally-occurring isotopes. The filled circles show the position of the p-nuclei in this plane. The r-process path is for a temperature $T_9 = 1.53$ and a neutron number density $n_n = 1.67 \times 10^{26}$ cm^{-3}. The unique (no branchings) s-process path is used for the calculations shown in Figure 6

a system is Y_e, which is the electron-to-baryon ratio or, by charge neutrality, the proton-to-baryon ratio. (For us, a baryon is simply a neutron or a proton.) For symmetric matter, $Y_e = 0.5$, while for pure neutron matter, $Y_e = 0$. For the low-entropy material, if we take a typical seed nucleus like ^{78}Ni, we need $Y_e = 28/(78 + 100) = 0.16$ to get 100 free neutrons per seed. This is quite neutron rich.

In high-entropy matter, there is a large abundance of ^{4}He nuclei, which are inert with respect to neutron capture. In this case, the degree of neutron richness needed is less because most protons are locked up in ^{4}He. Typically one finds at the beginning of a high-entropy r-process that the mass fraction X_α is ~ 0.4, the seed nucleus abundance per baryon is ≈ 0.003, and the seed nucleus is something like ^{100}Kr (Z=36). Since the abundance per baryon of ^{4}He is $X_\alpha/4 = 0.1$, the Y_e necessary for 100 neutrons per seed ($Y_n = 100 \times Y_{seed}$) is $Y_e = (2 \times 0.1 + 36 \times 0.003)/(4 \times 0.1 + 100 \times 0.003 + 0.3) = 0.31$. This is considerably less neutron rich than in the low-entropy case, and we currently believe that the solar system's supply of r-process matter was produced in expansions of high-entropy matter.

In such expansions of high-entropy matter, the material begins as free neutrons and protons at high temperature. As the matter expands, it cools, and the neutrons and protons assemble into ^{4}He nuclei. Further cooling allows the ^{4}He nuclei to assemble into seed nuclei. However, because the assembly of seed nuclei depends on the three-body reactions sequences ^{4}He + ^{4}He + ^{4}He $\to$ ^{12}C and ^{4}He + ^{4}He + n $\to$ ^{9}Be followed by ^{9}Be + ^{4}He $\to$ ^{12}C + n, the degree of assembly and the resulting neutron/seed ratio depends sensitively on the entropy per baryon.

To explore this we made a series of 160 r-process calculations for initial Y_e ranging from 0.2 to 0.5 and entropy per baryon ranging from 50 to 500 in

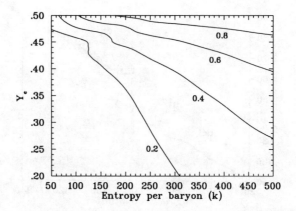

Figure 3. The final mass fraction in ^{4}He nuclei in adiabatic expansions of matter beginning at high temperature and the indicated initial Y_e. The expansion timescale $\tau = 0.1$ s [see equation (1)].

units of Boltzmann's constant. The expansions were truly adiabatic in that the entropy per baryon was constant throughout the expansion. This necessitated solving the detailed integral equations relevant to the equation of state of an ideal $e^+ - e^-$ plasma. This was done at each timestep via Gauss-Legendre quadrature subroutines written by Brown (1996). Also included in the entropy were the baryons and photons. The network is that described in Meyer (1995), but no neutrino-nucleus interactions were included. The calculations began at time $t = 0$ with $T_9 = 10$, where $T_9 = T/10^9$K. The initial density ρ_0 was computed from the entropy by assuming the matter was in nuclear statistical equilibrium. We then allowed the matter to expand adiabatically with

$$\rho(t) = \frac{\rho_0}{(1 + t/2\tau)^2}, \tag{1}$$

which models matter expanding spherically at constant velocity. The expansion timescale τ was taken to be 0.1 s. The calculations proceeded until all reactions froze out.

The final fraction of the mass in ^{4}He nuclei is shown in Figure 3. Here we see the strong sensitivity to entropy. For given initial Y_e, larger entropy leads to a higher mass of leftover ^{4}He due to less assembly into seed nuclei. Note the dependence on initial Y_e also. A lower initial Y_e leads to more assembly of ^{4}He nuclei because of the greater efficacy of the ^{4}He + ^{4}He + n → ^{9}Be followed by ^{9}Be + ^{4}He → ^{12}C + n reaction sequence.

Figure 4 shows the ratio of free neutrons to seed nuclei at $T_9 = 2.5$, the point at which the actual r-process phase of the expansion begins. These contours show that the higher the entropy, the less neutron rich the system needs to be to have a given free neutron/seed ratio at $T_9 = 2.5$.

It is incorrect to think that all r-process matter experienced the same set of initial Y_e, entropy S, and expansion timescale τ. Some superposition of differing components occurred in matter ejected during the r-process events. The exact distribution of ejected matter is a function of the dynamics of the event,

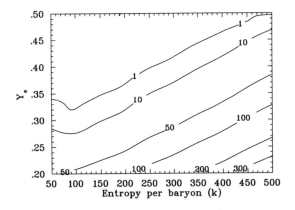

Figure 4. The ratio of the abundance of free neutrons to the abundance of seed nuclei at $T_9 = 2.5$ in adiabatic expansions of matter beginning at high temperature and the indicated initial Y_e. The expansion timescale $\tau = 0.1$ s [see equation (1)].

and we require detailed knowledge of the event to get that distribution. We can, however, invert the problem and provide interesting constraints on these dynamics from the r-process. We see the solar system's r-process distribution in Figure 1. The famous A=130 and A=195 peaks are readily apparent. Because this distribution results from a superposition of components, some component(s) must have had a final distribution with relative heights of the 130 and 195 peaks at least in their solar system ratio of $Y(195)/Y(130) = 0.3$. Because the actual calculated peaks do not always fall exactly at 130 and 195 due to variations in the nuclear dynamics during expansion, we took the maximum abundance in the range $A = 120 - 140$ to be peak 1 and that in $A = 190 - 210$ to be peak 2. These ranges were generous enough that the two abundance peaks always fell in them. Figure 5 shows the locus of (initial Y_e, S) points for which $Y(Peak2)/Y(Peak1) = 0.3$. Only for (initial Y_e, S) points below this line will $Y(Peak2)/Y(Peak1)$ be greater than 0.3. If we assume that an expansion timescale of $\tau = 0.1$ s is appropriate, r-process events must eject some matter with an initial Y_e, S pair below the line in Figure 5. Of course the location of the line also depends on the expansion timescale, and we are exploring this with further calculations.

Where in Nature might the conditions we derived occur? Woosley & Hoffman (1992) first called attention to the high-entropy winds blowing from protoneutron stars. Meyer et al. (1992) showed indeed that this site could give rise to an r-process. Subsequent work by Takahashi et al. (1994) and Woosley et al. (1994) has clarified the picture in terms of realistic supernova models. It is not yet entirely clear that present supernova models really do give rise to conditions appropriate for a successful r-process. Moreover, convection in the early stages of the explosion further clouds the picture (e.g. Herant et al. 1992, Burrows et al. 1995). Also, neutrino-nucleus interactions can hinder the r-process (Meyer 1995). Nevertheless, high-entropy winds in supernovae appear to be the most

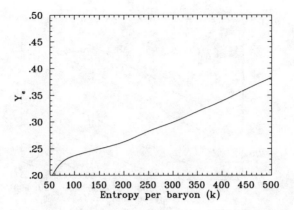

Figure 5. The locus of (initial Y_e, S) points for which $Y(Peak2)/Y(Peak1) = 0.3$ in adiabatic expansions with expansion timescale $\tau = 0.1$ s. A "successful" r-process needs to have some component with (initial Y_e, S) lying below this line.

promising avenue to follow. The constraints we derived above may help in this endeavor.

3. The s-process

The s-process is the *slow* neutron-capture process. Here the neutron-capture reactions are slower than the β^--decays. This means that the s-process path always stays near the valley of β^--stability because, if a nucleus captures a neutron and becomes β^--unstable, it decays before it captures another neutron. Figure 2 shows an s-process path.

If we assume that neutron capture is always slower than β^--decay, and that the temperature and density remain constant, we have the so-called classical s-process model. In this model, the rate of change of the abundance Y_A of the species with mass number A is given by

$$\frac{dY_A}{dt} = n_n <\sigma v>_{A-1} Y_{A-1} - n_n <\sigma v>_A Y_A, \qquad (2)$$

where n_n is the neutron number density of neutrons and $<\sigma v>_A$ is the Maxwell-Boltzmann-weighted average of the neutron-capture cross section times the relative speed v of a neutron and the target species. Equation (2) shows that the rate of change of species with mass number A is due simply to capture in from $A - 1$ and capture out of A. This holds as long as there is no branching of the s-process path. Branching occurs when a β^--decay timescale is long enough that a neutron capture can occur before β^--decay so that the s-process path splits or branches into two separate paths. The two separate paths usually merge again within a few mass numbers. We neglect any branching for the present discussion; however, since neutron-capture rates and some β^--decay rates depend on the temperature and density, analysis of branchings provides crucial constraints on s-process conditions (see e.g. Käppeler et al. 1989).

If we now define the cross sections σ_A to be $\sigma_A = <\sigma v>_A / v_T$, where v_T is the thermal speed of the neutrons, equation (2) becomes

$$\frac{dY_A}{d\tau} = \sigma_{A-1} Y_{A-1} - \sigma_A Y_A, \qquad (3)$$

where the neutron exposure τ is given by

$$\tau = \int_0^t n_n v_T dt'. \qquad (4)$$

τ is the appropriate evolutionary parameter for the classical s-process. It has units of cm^{-2} or mb^{-1}, where 1 millibarn (mb) = 10^{-27} cm^2.

A key point of equation (3) is that if a steady s-process flow is achieved, then $dY_A/d\tau \to 0$ and $\sigma_A Y_A \to constant$. The fact that the solar system's s-process abundances times their neutron-capture cross sections is approximately constant over a large range in A gives important clues about how the s-process occurred.

We have made a series of classical s-process calculations using the unique (no branchings!) s-process path shown in Figure 2. The cross sections we used are from Bao & Käppeler (1987) and the β^--decay rates are from Tuli (1990). The calculations were for $\tau = 0.06$ mb^{-1}, $\tau = 0.3$ mb^{-1}, and $\tau = 1.45$ mb^{-1}. The results are shown in Figure 6. For the calculations given by the solid curves, the initial seeds were a solar distribution. For the calculations given by the various dashed curves, the initial seed was ^{56}Fe only. There are several interesting points in this figure. First, there is a large difference between the solar seeds and ^{56}Fe only seeds for small τ. For example, for $\tau = 0.3$ mb^{-1}, there is little material with $A > 90$ due to the fact that the ^{56}Fe nuclei were only able to neutron capture at most ~ 40 neutrons, primarily because of the small cross sections for the $N = 50$ nuclei. For solar seeds, however, seed nuclei with A up to 209 existed and there is a fairly large abundance of large A nuclei even for small τ. For τ as large as 1.45 mb^{-1}, however, the solid and dashed curves are nearly the same. This is because ^{56}Fe is so abundant that when enough neutron captures can occur, it dominates the resulting s-process distribution. These points are of interest because the first classical s-process studies considered only initial ^{56}Fe seeds.

Another point regarding figure 6 is the degree to which steady flow is achieved. Obviously for large τ, steady flow holds over large ranges in A. Note for $\tau = 1.45$ mb^{-1} that there is a break in the roughly constant σY curves near $A = 140$. This is the mass number at which the s-process path intersects a closed neutron shell, $N = 82$. The closed shell nuclei have relatively small cross sections, so there is a damming of the flow. It therefore takes longer for the nuclei past the closed shell to achieve steady flow with those nuclei in front of the closed shell. For smaller τ's, there is not the same degree of steadiness in the flows because not enough neutron captures have occurred. Note, however, that when the seeds are a solar distribution, approximate steady flow can be achieved for $\tau = 0.3$ mb^{-1} for $A > 100$. These high mass nuclei have large neutron-capture cross sections (except at closed shells), so they do have enough neutron captures to reach approximate steady flow.

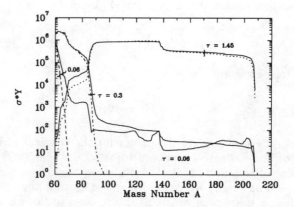

Figure 6. σY curves for classical s-process models for the indicated values of neutron exposure (in mb^{-1}). Solid curves are for calculations using an initial solar seed distribution. The various dashed curves show results for an ^{56}Fe-only seed distribution.

The key point recognized early in s-process studies was that a single neutron exposure τ did not characterize the solar s-process abundance distribution. Seeger et al. (1965) found that a linear superposition of s-processes weighted by an exponential distribution $\rho(\tau) \propto exp(-\tau/\tau_0)$ was needed. Clayton & Rassbach (1967) and Clayton & Ward (1974) then showed that three distinct exponential distributions were necessary. One component with $\tau_0 \approx 0.06$ mb^{-1}, the weak component, was needed to explain the s-nuclei with $A \lesssim 90$. This component was presumably produced in core helium burning in massive stars (e.g. Käppeler et al. 1994). A component with $\tau_0 \approx 0.3$ mb^{-1}, the main component, gives rise to the s-nuclei with $A \gtrsim 90$. This component is likely to come from low-mass AGB stars (e.g. Käppeler et al. 1990). Finally, a strong component with $\tau_0 \approx 7.0$ mb^{-1} may be required to explain the abundances of the $A = 204 - 209$ s-nuclei. The origin of this component is still not clear.

We must finally note the exciting discovery of presolar grains of silicon carbide (SiC) isolated from the Murray and Murchison meteorites (Srinivasan & Anders 1978, Tang & Anders 1988). These grains show a xenon isotopic composition that is predominantly s-process in character. These grains possibly formed in the outflow from AGB stars. At any rate, they confirm the notion that the r-nuclei and s-nuclei formed in distinct processes.

4. The p-process

The r- and s-processes are unable to synthesize the 35 rare proton-rich isotopes we call the p-nuclei. Some other mechanism than a sequence of neutron captures and β^--decays must be responsible. This mechanism is the p-process.

The p-process was originally thought to be a proton-capture process (Burbidge et al. 1957). Nuclei exposed to high temperatures and proton fluxes might capture protons to produce the p-nuclei. In order to produce the heaviest p-nuclei this way, however, an extremely high density must prevail for a long time. For example, even with a mass density in protons as large as 10^6 g/cm^{-3} at a temperature $T_9 = 1$, the timescale to assemble the heaviest p-nuclei is well over 10^4 seconds (see figure 9 of Meyer 1994). It is extremely difficult to imagine any astrophysical environment having such conditions. Moreover, the timescale becomes even longer for a lower density of protons. The timescale for proton capture drops dramatically to $\sim 10^{-4}$ seconds for $T_9 = 3$ and a proton mass density of 10^6 g/cm^{-3}, but disintegration reactions, which tear the nuclei apart, dominate capture reactions for these temperatures. Proton capture is unlikely to have been responsible for the synthesis of the majority of the p-nuclei.

The answer to the question of formation of the p-nuclei was first clearly laid out by Woosley & Howard (1978). Instead of calling on proton-capture reactions to assemble the p-nuclei, these researchers noted that *disintegration* reactions would work. In particular, if an initial distribution of r- and s-nuclei is exposed to high temperature for a long enough period of time, the nuclei all melt and become iron-group nuclei. If, on the other hand, the melting is quenched before it is complete, a distribution of proton-rich nuclei results! The actual reaction sequence Woosley & Howard found was a series of (γ, n) reactions followed by a (γ, p) and (γ, α) cascade down the proton-rich side of the valley of β-stability.

This gamma-process, as Woosley & Howard called it because of the dominance of the photodisintegration reactions, was envisioned to occur in the O/Ne-rich zone of a massive star undergoing a Type II supernova explosion. As the supernova shock drives through this zone, the matter is heated to $T_9 = 2 - 3$. The zone quickly expands and cools, however, thereby quenching the melting of the nuclei. The supernova shock heats matter less as it moves out in stellar radius, so the innermost parts of the O/Ne zone are heated more than the outermost parts. The innermost parts of the zone thus experience more melting and make the lighter p-nuclei. By contrast, the outermost parts of the zone experience the least melting; they make the heavier p-nuclei.

Though the gamma-process model is nearly 20 years old, it is still the most plausible model for synthesis of the p-nuclei. It has now been studied in detail with realistic stellar evolution and supernova models (see Rayet et al. 1995). The new calculations seem to be showing, however, that the gamma-process in Type II supernovae may only be responsible for $\sim 1/4$ the solar system's supply of the p-isotopes. The remaining component may have come from a similar gamma-process occurring on the very outer skin of Type Ia supernovae (Howard et al. 1991).

The major failing of the gamma-process model is its inability to produce ^{92}Mo, ^{94}Mo, ^{96}Ru, and ^{98}Ru in their solar proportions. A glance at figure 1 shows why this must be. Most of the p-isotopes are $\sim 100\times$ less abundant than their r- and s-process counterparts. These p-isotopes could have been produced if $\sim 1\%$ of all r- and s-nuclei experienced a gamma-process. This would fail to produce ^{92,94}Mo and ^{96,98}Ru by a factor of ~ 100, however, because the light p-nuclei are nearly as abundant as their r- and s-nuclei counterparts. Some other mechanism is probably responsible for producing these light p-nuclei.

A promising site for producing ^{92}Mo is in the α-process near the mass cut of Type II supernovae (Fuller & Meyer 1995, Hoffman et al. 1996). This model does not seem to explain the production of the other three underproduced light p-isotopes. Hencheck et al. (1995) explored synthesis of these nuclei in Thorne-Żytkow objects (TZOs). Co-production of the four Mo and Ru isotopes is possible. The question here is whether there have been enough (or any!) TZOs to have made the solar system's supply of the light p-isotopes.

In summary, most of the p-process isotopes were probably made in gamma-processes occurring in Type II and Ia supernovae. ^{92}Mo very likely was synthesized in the α-process occurring near the mass cut in Type II supernova events. The synthesis of the other three isotopes is still up in the air!

References

Anders, E., & Grevesse, N. 1989, Geochim. Cosmochim. Acta, 53, 197
Bao, Z. Y., & Käppeler, F. 1987, At. Data Nucl. Data Tables, 36, 411
Brown, J. S. 1996, Masters thesis, Clemson University, in preparation
Burbidge, E. M., Burbidge, G. R., Fowler, W. A., & Hoyle, F. 1957, Rev. Mod. Phys., 29, 547
Burrows, A., Hayes, J., & Fryxell, B. A. 1995, ApJ, 450, 830
Cameron, A. G. W. 1957, Chalk River Report CRL-41, Atomic Energy Can. Ltd.
Clayton, D. D., & Rassbach, M. E. 1967, ApJ, 148, 69
Clayton, D. D., & Ward, R. A. 1974, ApJ, 193, 397
Fuller, G. M., & Meyer, B. S. 1995, ApJ, 453, 792
Hencheck, M., Boyd, R. N., & Meyer, B. S. 1995, ApJ, submitted
Herant, M., Benz, W., & Colgate, S. A. 1992, ApJ, 395, 642
Hoffman, R. D., Woosley, S. E., Fuller, G. M., & Meyer, B. S. 1996, ApJ, in press
Howard, W. M., Meyer, B. S., & Woosley, S. E. 1991, ApJ, 375, L5
Käppeler, F., Beer, H., & Wisshak, K. 1989, Rep. Prog. Phys, 52, 945
Käppeler, F., Gallino, R., Busso, M., Picchio, G., & Raiteri, C. M. 1990, ApJ, 354, 630
Käppeler, F. et al. 1994, ApJ, 437, 396
Meyer, B. S., Mathews, G. J., Howard, W. M., Woosley, S. E., & Hoffman, R. D. 1992, ApJ, 399, 656
Meyer, B. S. 1994, ARA&A, 32, 153
Meyer, B. S. 1995, ApJ, 449, L55
Rayet, M., Arnould, M., Hashimoto, M., Prantzos, N., & Nomoto, K. 1995, A&A, 298, 517
Seeger, P. A., Fowler, W. A., & Clayton D. D. 1965, ApJS, 11, 121
Srinivasan, B., & Anders, E. 1978, Science, 201, 51
Takahashi, K., Witti, J., & Janka, H.-Th. 1994, A&A, 286, 857
Tang, M., & Anders, E. 1988, Geochim. Cosmochim. Acta, 52, 1235
Tuli, J. 1990, Nuclear Wallet Cards (Brookhaven: Brookhaven National Laboratory)

Woosley, S. E., & Hoffman, R. 1992, ApJ, 395,202
Woosley, S. E., & Howard, W. M. 1978, ApJS, 36, 285
Woosley, S. E., Wilson, J. R., Mathews, G. J., Hoffman R. D., & Meyer, B. S. 1994, ApJ, 433, 229

Nucleosynthesis and the Nova Outburst

S. Starrfield

*Department of Physics and Astronomy, ASU, P.O. Box 871504
Tempe, AZ 85287-1504*

J. W. Truran

*Department of Astronomy and Enrico Fermi Institute,
University of Chicago, Chicago, IL 60637*

M. Wiescher

*Department of Physics, University of Notre Dame,
Notre Dame, IN 46556*

W. M. Sparks

*XNH, Nuclear and Hydrodynamic Applications, MS F664,
Los Alamos National Laboratory, Los Alamos, NM 87544*

Abstract.

A nova outburst is the consequence of the accretion of hydrogen rich material onto a white dwarf and it can be considered as the largest hydrogen bomb in the Universe. The fuel is supplied by a secondary star in a close binary system while the strong degeneracy of the massive white dwarf acts to contain the gas during the early stages of the explosion. The containment allows the temperature in the nuclear burning region to exceed 10^8K under all circumstances. As a result a major fraction of the CNO nuclei in the envelope are transformed into β^+-unstable nuclei. We discuss the effects of these nuclei on the evolution. Recent observational studies have shown that there are two compositional classes of novae; one which occurs on carbon-oxygen white dwarfs, and a second class that occurs on oxygen-neon-magnesium white dwarfs. In this review we will concentrate on the latter explosions since they produce the most interesting nucleosynthesis. We report both on the results of new observational determinations of nova abundances and, in addition, new hydrodynamic calculations that examine the consequences of the accretion process on $1.0 M_\odot$, $1.25 M_\odot$, and $1.35 M_\odot$ white dwarfs. Our results show that novae can produce ^{22}Na, ^{26}Al, and other intermediate mass nuclei in interesting amounts. We will present the results of new calculations, done with updated nuclear reaction rates and opacities, which exhibit quantitative differences with respect to published work.

1. Introduction: A Nova Explosion as a Thermonuclear Runaway

The outbursts of classical novae are caused by thermonuclear runaways (hereafter, TNR's) proceeding in the accreted hydrogen-rich envelopes of the white dwarf components of nova binary systems (Truran 1982, 1990; Starrfield 1989, 1993, 1995). For the physical conditions of temperature and density that are expected to obtain in this environment, thermonuclear burning proceeds by means of hydrogen burning from either the proton-proton chain (early in the accretion phase) or the carbon, nitrogen, and oxygen (CNO) bi-cycle (late in the accretion phase and through the peak of the outburst). If there are heavier nuclei present in the nuclear burning shell, they will contribute primarily to the nucleosynthesis and only partially to the energy production.

For solar composition material, energy production and nucleosynthesis from the CNO hydrogen burning reaction sequences impose interesting constraints on the energetics of the runaway: in particular, the rate of nuclear energy generation at high temperatures ($T > 10^8$K) is limited by the timescales of the slower *and temperature insensitive positron decays*, particularly ^{13}N ($\tau = 600$s), ^{14}O ($\tau = 102$s), and ^{15}O ($\tau = 176$s). The behavior of the β^+- unstable nuclei holds important implications for the nature and consequences of classical nova outbursts. For example, significant enhancements of envelope CNO concentrations are required to insure higher levels of energy release on a hydrodynamic timescale (seconds for white dwarfs) and thus produce a violent outburst (Starrfield, Truran, and Sparks 1978; Truran 1982; Starrfield 1995).

The large abundances of the β^+-unstable nuclei, at the peak of the outburst, have important and exciting consequences for the evolution. (1) Since the energy production in the CNO cycle comes from proton captures followed by a β^+-decay, the rate at which energy is produced, at temperatures exceeding 10^8K, depends only on the half-lives of the β^+-unstable nuclei and the numbers of CNONeMg nuclei initially present in the envelope. (2) Since convection operates throughout the entire accreted envelope, it brings unburned CNONeMg nuclei into the shell source, when the temperature is rising very rapidly, and keeps the nuclear reactions operating far from equilibrium. (3) Since the convective turn-over time scale is $\sim 10^2$ sec near the peak of the TNR, a significant fraction of the β^+-unstable nuclei can reach the surface without decaying and the rate of energy generation at the surface can exceed 10^{12} to 10^{13} erg gm^{-1} s^{-1} (Starrfield 1989). (4) The β^+-unstable nuclei decay when the temperatures in the envelope have declined to values that are too low for any further proton captures to occur, yielding isotopic ratios in the ejected material that are distinctly different from those ratios predicted from studies of equilibrium CNONeMg burning. (5) Finally, the decays of the β^+-unstable nuclei provide an intense heat source throughout the envelope that helps in driving the material off the white dwarf.

Theoretical calculations of this mechanism show that sufficient energy is produced, during the evolution described above, to eject material with expansion velocities that agree with observed values and that the predicted light curves produced by the expanding material are in reasonable agreement with the observations (Truran 1982; Starrfield 1989, 1995; Starrfield et al. 1992; Politano et al. 1995). Our recent studies have shown that it is the outbursts that occur on oxygen-neon-magnesium (ONeMg) white dwarfs that produce the most interesting nucleosynthesis and here we concentrate on those outbursts. In order to demonstrate that such outbursts exist, in the next section we briefly describe the abundance determinations for novae and provide a table of abundances determined from nebular techniques. We follow that with a section on theoretical studies of outbursts on ONeMg white dwarfs. We end with a summary.

2. Oxygen-Neon-Magnesium Novae

The production of large amounts of ^{22}Na, ^{26}Al, and other intermediate mass nuclei in a nova outburst requires that the outburst occur on an ONeMg white dwarf. Therefore, it is appropriate at this point to present the evidence that such novae exist. Although Law and Ritter (1983) first proposed that such white dwarfs could exist, it was the outburst of V693 Cr A 1981 that demonstrated that they did exist in nova systems (Williams et al. 1985). Ultraviolet spectroscopic data were obtained for this nova with the International Ultraviolet Explorer satellite (IUE) over a six month period. It was clear from the last set of spectra (which showed strong [Ne IV] 1602Å and 2420Å) that neon was enriched in the ejecta. This was later confirmed by detailed analyses (Williams et al. 1985; Andreä et al. 1994). Their results are given in Table 1 along with those for all well studied novae.

Given these data, Starrfield et al. (1986) performed a theoretical study of accretion onto massive white dwarfs in which only oxygen was enhanced (their nuclear reaction network did not include nuclei above fluorine at that time) and showed that such an outburst would appear different from a nova with just carbon and oxygen enhanced (one that occurred on a CO white dwarf). This is because carbon has a much larger nuclear reaction cross-section than oxygen and so the TNR could occur with less accreted material on the white dwarf. At virtually the same time that they were doing their calculations, Gehrz et al. (1985) reported that the slow nova QU Vul 1984 No. 2 showed enhanced neon. These results, in combination with the IUE spectra that also showed strong neon lines (Starrfield 1988), implied that the ejecta of QU Vul were enriched in neon and magnesium as was again confirmed by detailed analyses (Saizar et al. 1992; Andreä et al. 1994). Truran and Livio (1986) then predicted that about one-third of all observed outbursts should be ONeMg novae.

Two recent outbursts have proved to be very important. Nova V838 Her 1991 was discovered on March 24, 1991 at ~5th Mag. and began a rapid decline in the optical. Although there was strong circumstantial evidence that it was an ONeMg nova (Starrfield et al. 1993), confirmation came from an analysis done by Matheson et al. (1993) who reported that both neon and sulfur were enriched in the ejecta. Their abundance determination was in good agreement with results of TNR's on massive ONeMg white dwarfs, as reported in Starrfield et al. (1992).

V1974 Cyg 1992 was the brightest nova found in outburst since V1500 Cyg 1975 and it was observed from γ-ray wavelengths to cm radio. It was observed with HST, IUE, VOYAGER, ROSAT, KAO, and pointed at, but not detected, by COMPTON GRO. The most important data were obtained by ROSAT, which followed it through its entire X-ray outburst (Krautter et al. 1995), and IUE which not only was able to study it in its fireball stage but provided a wealth of information about the evolution of its nebular spectrum (Hauschildt et al. 1994a;

245

TABLE 1
Heavy Element Abundances (by Mass Fraction) in Novae from Optical Spectroscopy

Object	Year	Ref.	H	He	C	N	O	Ne	Na-Fe	Z	$(Z/Z_\odot)$	$(Ne/Ne_\odot)$	$[Ne/Z]$
T Aur	1891	4	0.47	0.40		0.079	0.051			0.13	6.8		
RR Pic	1925	13	0.53	0.43		0.022	0.0058	0.011		0.043	2.3	6.3	13.5
DQ Her	1934	15	0.34	0.095	0.0039	0.23	0.29			0.57	30.		
DQ Her	1934	7	0.27	0.16	0.045	0.29	0.22			0.57	30.		
HR Del	1967	12	0.45	0.48	0.058	0.027	0.047	0.0030		0.077	4.1	1.7	2.0
V1500 Cyg	1975	3	0.49	0.21	0.070	0.075	0.13	0.023		0.30	16.	13.	4.0
V1500 Cyg	1975	6	0.57	0.27	0.058	0.041	0.050	0.0099		0.16	8.4	5.6	3.3
V1668 Cyg	1978	11	0.45	0.23	0.047	0.14	0.13	0.0068		0.32	17.	3.9	1.1
V1668 Cyg	1978	1	0.45	0.22	0.070	0.14	0.12			0.33	17.		
V693 CrA	1981	14	0.29	0.32	0.046	0.080	0.12	0.17	0.016	0.39	21.	97.	23.
V693 CrA	1981	1	0.16	0.18	0.0078	0.14	0.21	0.26	0.030	0.66	35.	148.	21.
V1370 Aql	1982	10	0.053	0.088	0.035	0.14	0.051	0.52	0.11	0.86	45.	296.	32.
V1370 Aql	1982	1	0.044	0.10	0.050	0.19	0.037	0.56	0.017	0.86	45.	296.	34.
GQ Mus	1983	5	0.27	0.32	0.016	0.19	0.19	0.0034	0.0068	0.41	22.	1.9	0.073
PW Vul	1984	8	0.69	0.25	0.0033	0.049	0.014	0.00066		0.067	3.5	0.38	0.52
PW Vul	1984	5	0.54	0.28	0.032	0.11	0.038			0.18	9.5		
PW Vul	1984	1	0.47	0.23	0.073	0.14	0.083	0.0040	0.0048	0.30	16.	2.3	0.70
QU Vul	1984	9	0.30	0.60	0.0013	0.018	0.039	0.040	0.0049	0.10	5.3	23.	21.
QU Vul	1984	1	0.33	0.26	0.0095	0.074	0.17	0.086	0.063	0.40	21.	49.	11.
QU Vul	1984	2	0.36	0.19		0.071	0.18	0.18	0.0014	0.44	23.	100.	22.
V842 Cen	1986	1	0.41	0.23		0.21	0.030	0.00090	0.0038	0.36	19.	0.51	0.13
V827 Her	1987	1	0.36	0.29		0.24	0.016	0.00066	0.0021	0.35	18.	0.38	0.099
QV Vul	1987	1	0.68	0.27		0.010	0.041	0.00099	0.00096	0.053	2.8	0.56	0.98
V2214 Oph	1988	1	0.34	0.26		0.31	0.060	0.017	0.015	0.40	21.	9.7	2.2
V977 Sco	1989	1	0.51	0.39		0.042	0.030	0.026	0.0027	0.10	5.3	15.	14.
V433 Sct	1989	1	0.49	0.45	0.0056	0.053	0.0070	0.00014	0.0017	0.062	3.3	0.80	0.12
V351 Pup	1991	16	0.37	0.25	0.12	0.076	0.19	0.11		0.38	20.	63.	15.
V1974 Cyg	1992	2	0.17	0.29	0.087	0.073	0.25	0.10	0.066	0.54	29.	59.	10.

References. 1: Andreä et al. 1994; 2: Austin et al. 1996; 3: Ferland and Shields 1978; 4: Gallagher et al. 1980; 5: Hassall et al. 1990; 6: Lance et al. 1988; 7: Petitjean et al. 1990; 8: Saizar et al. 1991; 9: Saizar et al. 1992; 10: Snijders et al. 1987; 11: Stickland et al. 1981; 12: Tylenda 1978; 13: Williams and Gallagher 1979; 14: Williams et al. 1985; 15: Williams et al. 1978; 16: Saizar et al. 1996.

Shore et al. 1993, 1994; Austin et al. 1996). In addition, we were able to analyze it's outburst with two new methods. First, Hauschildt et al. (1992, 1994a, 1994b) studied the early fireball stage with a Non-LTE, spherical, expanding, model atmosphere code that was developed to study novae and supernovae. Second, Austin et al. (1996) developed a method for analyzing nova nebular spectra. It uses Cloudy (Ferland 1994) to predict emission line fluxes which are then compared to observed emission line fluxes. Ten thousand trials are run in order to find the initial conditions (electron density, electron temperature, ionizing flux, elemental abundances, ...) so that CLOUDY's predicted emission line fluxes are in "good" agreement with the observed fluxes. The first set of results have been done for V1974 Cyg and can be found in Table 1.

Even a cursory glance at Table 1 shows that novae ejecta are enriched in the elements which drive extremes of nuclear energy generation. The mean heavy element mass fraction for these *well studied* cases is Z = 0.34 (we averaged multiple determinations for the same nova). Determining the cause of the large discrepancies between some of the abundance determinations *for the same nova outburst* warrants further study, which is in progress (Schwarz et al. 1996, in preparation; Vanlandingham et al. 1996, in preparation). We also note that for at least three recent novae, V1370 Aql 1982, V2214 Oph 1988, and V838 Her 1991 (Matheson et al. 1993; Vanlandingham et al. 1996, in preparation) the ejected material was enriched in sulfur as well as neon and magnesium. We emphasize that the source of these large abundance enrichments must be matter dredged up from the underlying ONeMg white dwarf and processed through hot, hydrogen burning. The production of sulfur requires, in addition, that the white dwarf be massive. Therefore, there must be mass differences between the white dwarfs in novae such as V838 Her 1991 and those in novae such as V1974 Cyg 1992 and V351 Pup 1991 (Starrfield et al. 1992; Politano et al. 1995).

3. Theoretical Studies of Oxygen-Neon-Magnesium Novae

Weiss and Truran (1990) and Nofar, Shaviv, and Starrfield (1990), in separate and independent studies with different nuclear reaction networks but similar nuclear reaction rates, have reported the results of calculations which simulate the synthesis of ^{22}Na and ^{26}Al in ONeMg-rich novae. Their calculations were performed with the use of large nuclear reaction networks that utilized temperature, density, and time profiles which were obtained from earlier hydrodynamic simulations of the outburst.

The results of both their studies can be summarized as follows: (1) they confirm earlier findings of Hillebrandt and Thielemann (1982) and Weischer et al. (1986) that extremely low levels of ^{26}Al and ^{22}Na are expected to be formed in nova envelopes with a solar composition. This result implies that slow CO novae are not expected to contribute significantly to the abundances of either isotope in the galaxy. (2) Enrichment of only the CNO nuclei does not significantly increase the production of ^{22}Na or ^{26}Al, although CO novae may be responsible for production of some rare light nuclei. (3) Greatly increased ^{22}Na and ^{26}Al production does result from envelopes with substantial initial enhancements of ^{16}O, ^{20}Ne, and ^{24}Mg. (4) Novae with ejecta rich in material from an ONeMg white dwarf may represent an important source of ^{26}Al in our Galaxy (Starrfield et al. 1993). (5) The abundances of ^{22}Na predicted for the ejecta of novae involving ONeMg white dwarfs are sufficiently high that we may expect nearby ONeMg novae, such as V838 Her 1991, to produce flux levels of ^{22}Na decay γ-rays detectible by GRO (Starrfield et al. 1992). (6) The calculations also indicate that there should be a strong anti-correlation between ^{22}Na and ^{26}Al production in nova outbursts. (7) The degree of enrichment of ^{22}Na and ^{26}Al is a sensitive function of the temperature history (assuming equal initial concentrations of nuclei) and, therefore, the detection of ^{22}Na would provide useful constraints on the evolution of the TNR.

The results from the one zone nucleosynthesis studies must be verified by hydrodynamic evolutionary studies. In an actual event, convective mixing carries material from the nuclear burning region to the surface on very short time scales. This increases the abundances of nuclei that would have been burned to other nuclei if they had not been carried to higher, cooler layers. In addition, convection continuously brings fresh nuclei into the nuclear burning layers from cooler regions close to the surface.

Therefore, it was necessary to develop a large nuclear reaction network which includes 86 nuclei up to ^{40}Ca, and use this network, in combination with our implicit, one-dimensional hydrodynamic computer code, to study the consequences of accretion of hydrogen rich material onto ONeMg white dwarfs (Politano et al. 1996). A description of the current version of this code and references to earlier version can be found in Politano et al. (1996).

Table 2.   Evolutionary Results

Sequence	1	2	3	4
Mass	1.00M$_\odot$	1.25M$_\odot$	1.25M$_\odot$	1.35M$_\odot$
L (10^{-3}L$_\odot$)	9.4	9.7	2.9	9.6
T$_{eff}$ (K)	20,500	25,700	12,400	30,300
Radius (km)	5379	3496	3495	2488
$\dot{M}$ (10^{-9}M$_\odot$yr^{-1})	1.6	1.6	0.8	1.6
τ_{TNR}(10^4 yr)	7.3	2.0	5.1	.9
M$_{acc}$ (10^{-5}M$_\odot$)	10.5	3.2	4.0	1.5
P$_{TNR}$ (10^{19} dynes cm^{-2})	2.5	5.2	6.5	10.0
PEAK ϵ_{nuc} (10^{17} erg gm^{-1} s^{-1})	0.2	1.0	1.1	1.9
PEAK T$_{ICE}$ (10^8K)	2.24	2.90	2.97	3.56
PEAK L (10^4L$_\odot$)	2.2	4.3	2.5	16.3
PEAK T$_{eff}$ (10^5K)	3.4	6.4	5.1	9.0
M$_{ej}$ (10^{-6}M$_\odot$)	<1.0	1.0	2.0	5.2
V$_{max}$ (km s^{-1})	45	560	1770	2320

In the work reported in Politano et al. (1995), we evolved TNR's in accreted hydrogen rich layers of white dwarfs with masses of 1.0M$_\odot$, 1.25M$_\odot$, and 1.35M$_\odot$. In Table 2 and 3 the evolutionary sequences labeled 1 and 4 are identical to the 1.0M$_\odot$ and 1.35M$_\odot$ sequences reported in Politano et al. (1995). Since those calculations were done, however, we have updated both the nuclear reactions and the opacities but concentrated on evolutionary studies only at 1.25M$_\odot$ since that is the most probable mass for the white dwarf in the V1974 Cyg binary system (Starrfield et al. 1995). However, we redid the calculations for the $1.25 M_\odot$ mass white dwarf reported in Politano et al. with the only change being to increase the convective efficiency from α equal 1.0 to 2.0 (ratio of mixing-length to scale height). This new sequence ejected $\sim 10^{-6} M_\odot$ while the sequence reported in Politano et al. did not eject any material at this mass. Note that for sequences 1, 2, and 4, we assumed that the rate of accretion onto the white dwarf was 10^{17}gm s^{-1} (1.6×10^{-9} $M_\odot$yr^{-1}). For sequence 3, we used a value of half that: 5×10^{16}gm s^{-1} ($8 \times 10^{-10} M_\odot$yr^{-1}). In all four sequences, we used an initial abundance of ONeMg nuclei equal to 50% of the envelope material (by mass). The remaining 50% consisted of a solar mixture of the elements. We assumed that this composition resulted from the mixing of the accreted material with core material.

Table 3. Ejected Abundances (by Mass Fraction)

Sequence	1	2	3	4
Mass	$1.00 M_\odot$	$1.25 M_\odot$	$1.25 M_\odot$	$1.35 M_\odot$
X	.33	.30	.21	.27
Y	.17	.19	.27	.20
^{12}C + ^{13}C (10^{-2})	.94	4.3	1.1	3.4
^{14}N + ^{15}N (10^{-2})	2.0	2.5	7.9	8.8
^{16}O + ^{17}O (10^{-2})	11.8	7.0	5.4	1.1
^{18}F + ^{19}F (10^{-4})	1.7	4.0	7.6	25.9
^{20}Ne + ^{21}Ne + ^{22}Ne	.25	.23	.23	.17
^{22}Na (10^{-3})	.05	1.7	2.7	5.7
^{24}Mg + ^{25}Mg + ^{26}Mg (10^{-2})	5.9	3.8	0.6	4.4
^{26}Al (10^{-3})	19.6	9.4	1.0	7.4
^{27}Al (10^{-3})	14.0	16.0	13.7	19.0
^{28}Si + ^{29}Si + ^{30}Si (10^{-2})	1.6	5.8	9.9	5.3
^{31}P (10^{-4})	0.02	40.2	82.0	202.0
^{32}S (10^{-4})	1.1	29.4	55.0	289.0
^{36}Ar (10^{-5})	1.9	2.1	.5	40.9

There were two other major differences between sequence 3 and the other three sequences. This evolutionary sequence was done using the *carbon-rich* OPAL opacities (Iglesias and Rogers 1993) and updated nuclear reaction rates (Van Wormer et al. 1994; Herndl et al. 1995). Both of these changes had major effects on the evolution and a detailed report will appear elsewhere (Starrfield et al. 1996, in preparation). Here we mention two effects. First, the OPAL opacities are larger than those used in our previous studies. Therefore heat is trapped more effectively within the nuclear burning layers, the temperature rises faster, and it takes less time to reach the TNR. For equal mass accretion rates, the sequence with the OPAL opacities accretes less material. Second, the new reaction rates, as reported in Van Wormer et al. and Herndl et al., are much larger around mass 26 than those used previously. This implies that, for a given white dwarf mass, the amount of ^{26}Al produced during the evolution will be smaller than we found in our previous studies (see Table 3).

Here, we discuss the results for the evolutionary sequence which examined the consequences of accretion onto a 1.0M$_\odot$ white dwarf (Sequence 1 in both Tables) since it is that sequence which produced the largest amount of ^{26}Al. We have not done any new studies at this white dwarf mass but expect that the amount of ^{26}Al will be only slightly reduced from the value reported in Table 3. We terminated the accretion phase when the temperature at the interface between the core and the accreted envelope (hereafter, ICE) had reached 45 million degrees and the rate of energy generation had reached, $\epsilon_{nuc} \sim 1.0 \times 10^9$ erg gm^{-1} s^{-1}. At this time, the abundance of ^{26}Al, which was zero in the initial model, had increased to 5.2×10^{-7} (all abundances in this paper are quoted as mass fractions). The abundance of ^{27}Al, which was initially 1.6×10^{-5}, had not changed at this time. We also found that the convective region, which first appeared when T_{ICE} reached about 30 million degrees, had grown to a region which extended to 200 km above the ICE (75 km below the surface).

By the time T_{ICE} had reached to 2×10^8K, the abundance of ^{26}Al had increased to 1.3×10^{-2} at the ICE. It was being produced by the reaction sequence: ^{24}Mg$(p,\gamma)^{25}$Al$(\beta^+,\nu)^{25}$Mg$(p,\gamma)^{26}$Al. The isomeric state was not important. The peak luminosity is given in Table 2 and the final abundances (mass fraction) for some of the nuclei in our network are given in Table 3. Except for ^{26}Al, ^{27}Al, and a few other nuclei, we summed over the most abundant isotopes in order to keep the table small. Since most of the ^{26}Al comes from the decay of ^{25}Al followed by the proton capture on ^{26}Mg, its abundance continued to increase even after peak temperature had been reached.

The expansion of the nova envelope gradually slowed until, after a few hours, the surface layers were moving at speeds of only a few km s^{-1}. It was clear that the energy production during the TNR was insufficient to *explosively* eject the shells. In addition, the luminosity was only about half the Eddington luminosity (solar mixture, electron scattering opacity) so that it appeared that radiation pressure driven mass loss was unimportant. That this is not true, however, was shown in an important paper by Hauschildt et al. (1994). They found that the very large opacity from the iron group elements ("iron" curtain) in an expanding medium reduces the "effective" Eddington luminosity by factors of as much as 100. This implies that radiation pressure is sufficient to drive off a significant fraction of the envelope when the iron group opacity is largest. This occurs around the time of maximum light in the optical (Shore et al. 1994). The expanding iron group opacity is not in the OPAL tables and was not included in our calculations. We expect, however, when it is included a large fraction of the accreted envelope will be ejected during the early stages of the outburst.

We also neglected common envelope evolution in these calculations which should act on material that extends past about 10^{11}cm. Therefore, although in Table 2 we list the amount of mass ejected as nearly zero, in fact, more than 1.4×10^{-5}M$_\odot$ lie beyond a radius of 10^{11}cm and will ultimately be ejected in a slow nova outburst. The abundances given in Table 3 have been determined assuming that all of the material at radii exceeding 10^{11}cm has been ejected.

The results of our simulations show that, as we go to higher mass white dwarfs, the violence of the outburst increases. In fact, the 1.35M$_\odot$ simulation results in material being ejected during the explosive phase of the outburst and we do not have to depend on either enhanced opacity caused by the "iron" curtain or common envelope evolution to drive mass off the white dwarf. We also note that, for a $1.35 M_\odot$ white dwarf, the peak temperature during the evolution is high enough ($T = 3.56 \times 10^8 K$) for significant nucleosynthesis, involving the intermediate mass nuclei, to occur.

This is confirmed by an examination of the abundances of some of the more massive nuclei shown in Table 3. Note, first, that the abundance of ^{26}Al declines as the white dwarf mass increases while the abundance of ^{22}Na increases as the mass of the white dwarf increases. This suggests that the novae which produce the largest enrichment of ^{26}Al may not be the same novae that produce enhanced ^{22}Na. In addition, as we proceed to higher white dwarf masses, the abundances of ^{31}P, ^{32}S, and ^{36}Ar increase to very large values. All of these nuclei must be produced as the result of "slow" (relative to the rates which dictate energy generation) proton captures on ^{24}Mg over the few minute lifetime of the explosion. We also call attention to the behavior of the light nuclei in our results. ^{12}C increases in abundance with white dwarf mass, but ^{16}O decreases in abundance. In contrast, the total neon abundance remains virtually constant. This could be the explanation of the puzzling feature that all of the observed ONeMg novae always show strong neon lines even when the O, Mg, or Al lines are weak.

4. Summary

In this paper we have examined the consequences of accretion of hydrogen rich material onto ONeMg white dwarfs with masses of $1.0 M_\odot$, $1.25 M_\odot$, and $1.35 M_\odot$. These results, in combination with one zone nucleosynthesis studies have demonstrated that novae produce ^{22}Na, ^{26}Al, and other intermediate mass nuclei in astrophysically interesting amounts. Specifically: 1) Hot hydrogen burning on ONeMg white dwarfs can produce as much as 2% of the ejected material as ^{26}Al and 3% as ^{22}Na. 2) The largest amount of ^{26}Al is produced in the lowest mass white dwarfs, which according to our evolutionary calculations, should eject the largest amount of material. The observations of QU Vul, a slow ONeMg nova, indicate that it has ejected about $10^{-3} M_\odot$. It must be emphasized, however, that such a large amount of ejected mass is inconsistent with the amount predicted to be accreted onto a typical ONeMg white dwarf whose mass is expected to exceed $1.25 M_\odot$. 3) The largest amount of ^{22}Na is produced by the highest mass ONeMg white dwarfs nova systems. These novae are predicted to be among the fastest and most luminous novae. In addition, the abundance of ^{22}Na is sufficiently high for its presence to be included in the energy budget of the ejected material. 4) There cannot be large numbers of QU Vul and V1974 Cyg type novae (slow ONeMg) in the galaxy since, if there were, the observed γ-ray emission from ^{26}Al would be far higher. 5) If QU Vul and V1974 Cyg novae are rare, then we are faced with the disquieting possibility that the ^{26}Al in the solar system, observed through the abundance of its daughter, ^{26}Mg, may have either come from a single event, or not from novae at all.

Acknowledgments. We would like to thank P. Hauschildt, J. Krautter, S. Shore, G. Sonneborn, G. Shaviv, K. Vanlandingham, and R. M. Wagner for their help. This work was supported in part by NSF and NASA grants to the University of Chicago, Notre Dame University, and Arizona State University and by the DOE.

References

Andreä, J, Drechsel, H., and Starrfield, S. 1994. A&A, 291, 869.
Austin, S. J., Wagner, R. M., Starrfield, S., Shore, S. N., Sonneborn, G., and Bertram, R. 1996, AJ, in press.
Ferland, G. 1993, *Hazy - an Introduction to Cloudy*, Department of Physics and Astronomy internal report (Kentucky).
Ferland, G. J., and Shields, G. A. 1978, ApJ, 226, 172.
Gallagher, J. S., et al. 1980, ApJ, 237, 55.
Gehrz, R.D., Grasdalen, G.L., and Hackwell, J.A. 1985, ApJ, 298, L163.
Hassall, B. J. M., et al. 1990, in *The Physics of Classical Novae*, ed. A. Cassatella and R. Viotti, (HD: Springer), p. 202.
Hauschildt, P. H., Wehrse, R., Starrfield, S., and Shaviv, G. 1992, ApJ, 393, 307.
Hauschildt, P. H., Starrfield, S., Austin, S. J., Wagner, R. M., Shore, S. N., and Sonneborn, G. 1994a, ApJ, 422, 831.
Hauschildt, P. H., Starrfield, S., Shore, S. N., Gonzalez-Riestra, R., Sonneborn, G., and Allard, F. 1994b, AJ, 108, 1008.
Hauschildt, P. H., Starrfield, S., Shore, S. N., Allard, F. and Baron, E. 1995, ApJ, 447, 829.
Herndl, H., Görres, J., Wiescher, M., Brown, B. A., Van Wormer, L. 1995, Phys. Rev. C, 52, 1078.
Hillebrandt W., and Thielemann F.-K. 1982, ApJ, 255, 617.
Iglesias, C. A., and Rogers, F. J., 1993, ApJ, 412, 752.
Lance, C. M., McCall, M. L., and Uomoto, A. K. 1988, ApJS, 66, 151,
Law, W. Y., and Ritter, H. 1983, A&A, 63, 265.
Matheson, T., Filippenko, A., and Ho, C. 1993, ApJ, 418, L29.
Nofar, I., Shaviv, G., and Starrfield, S. 1991, ApJ, 369, 440.
Petitjean, P., Boisson, C., and Pequignot, D. 1990, A&A, 240, 433.
Politano, M., Starrfield, S., Truran, J. W., Sparks, W. M., and Weiss, A. 1995, ApJ, 448, 807.
Saizar, P. Starrfield, S., Ferland, G. J., Wagner, R. M., Truran, J. W., Kenyon, S. J., Sparks, W. M., Williams, R. E., and Stryker, L. L. 1992, ApJ, 367, 310.
Saizar, P., Starrfield, S., Ferland, G. J., Wagner, R. M., Truran, J. W., Kenyon, S. J., Sparks, W. M., Williams, R. E., and Stryker, L. L. 1992, ApJ, 398, 651.
Saizar, P., et al. 1996, MNRAS, in press.
Shore, S. N., Sonneborn, G., Starrfield, S., Gonzalez-Riestra, R., and Polidan, R., ApJ, 421, 344.

Snijders, M. A. J., et al. 1987, MNRAS, 228, 329.

Starrfield, S., 1988, in *Multiwavelength Observations in Astrophysics*, ed. F. A. Cordova (Cambridge: University Press), 159.

Starrfield, S., 1989, in *The Classical Novae*, ed. M. Bode and A. Evans, (Wiley: NY), 39.

Starrfield, S. 1993, in *The Realm of Interacting Binary Stars*, ed. J. Sahade, G. E. McCluskey, and Y. Kondo (Dordrecht: Kluwer), 209.

Starrfield, S. 1995, in *Physical Processes in Astrophysics*, ed. I. Roxburgh, and J. L. Masnou, (Springer: Heidelberg), 99.

Starrfield, S., Krautter, J., Shore, S. N., Idan, I., Shaviv, G., Sonneborn, G. 1995, in *Proceedings of IAU 152*, ed. R. Malina (Cambridge: University Press), in press.

Starrfield, S., Shore, S. N., Sparks, Sonneborn, G., W. M., Politano. M., and Truran, J. W. 1992, ApJ, 391, L71.

Starrfield, S., Sparks, W. M., and Truran, J. W. 1986, ApJ, 303, L5.

Starrfield, S., Truran, J. W., Politano, M., Sparks, W. M., Nofar, I., and Shaviv, G. 1993, Physics Reports, 227, 223.

Starrfield, S., Truran, J. W., and Sparks, W. M., 1978, ApJ, 226, 186.

Stickland, D. J., et al. 1981, MNRAS, 197, 107.

Truran, J. W. 1982, in *Essays in Nuclear Physics*, eds. C. A. Barnes, D. D. Clayton and D. N. Schramm (Cambridge: Cambridge U. Press), p. 467.

Truran, J. W. 1990, in *The Physics of Classical Novae*, ed. A. Cassatella and R. Viotti, (HD: Springer), p. 373.

Truran, J. W. and Livio, M. 1986, ApJ, 308, 721.

Tylenda, R. 1978, Acta Astronomica, 28, 333.

Van Wormer, L., Görres, J., Iliadis, C., Wiescher, M., and Thielemann, F. K. 1994, ApJ, 432, 326.

Wiescher M., Görres J. Thielemann F.-K., Ritter H. 1986, A&A, 160, 56.

Weiss, A. and Truran J. W. 1990, A&A, 238, 178.

Williams, R. E., and Gallagher, J. S. 1979, ApJ, 228, 482.

Williams, R. E., et al. 1978, ApJ, 224, 171.

Williams, R.E., Ney, E.P., Sparks, W.M., Starrfield, S., Truran, J.W. 1985, MNRAS, 212, 753.

Cosmic Abundances
ASP Conference Series, Vol. 99, 1996
Stephen S. Holt and George Sonneborn (eds.)

Nucleosynthesis in Supernovae

S. E. Woosley[1]

Astronomy Board of Studies, University of California, Santa Cruz, CA 95064

Abstract.
Our understanding of key nuclear physics and supernova models has advanced to the point that one may realistically attempt a quantitative summary, not only of the processes, but specific sites responsible for producing each of the stable isotopes. Most, though not all, are made in supernovae, but there are various kinds of supernovae, and within these, different kind of events occur - pre-explosive burning, shock processing, neutrino-irradiation, *etc*. Within this context, the origin of the isotopes of elements lighter than zinc is briefly reviewed with emphasis on the uncertainties in our current understanding. A major one is the amount of fall back that occurs during the explosion of more massive stars. Another is nuclear physics that affects the neutrino process.

1. Introduction

We begin with the answer. Fig. I summarizes the nucleosynthesis currently expected from stars more massive than 11 $M_\odot$. This figure uses the ejecta from supernovae of various metallicities as described by Woosley & Weaver (1995; hence WW) as input to a Galactic chemical evolution model described by Timmes, Woosley, & Weaver (1995; hence TWW). The difference between this figure and the one given by Timmes (TWW and this volume) is that only the contribution from massive stars and the Big Bang have been included. Actually that is not exactly correct. Some secondary input from intermediate mass stars and Type Ia supernovae is unavoidable because the massive stars studied by WW used as their starting composition, at each metallicity, abundances that included the products of all previous evolution. So the ^{14}N made in massive stars is influenced, to some extent, by the ^{12}C made in earlier, low mass stars. But within the small uncertainty of such "secondary" contributions, Fig. I is what massive stars make. The results are compared to Anders & Grevesse (1988), which for present purposes is assumed to sample the average interstellar medium at the time the sun was born. To the extent that it does not, as other papers in this volume address, corrections must be applied.

---

[1] Currently at the Max Planck Institut für Astrophysik, Karl Schwarzschild Strasse 1, D-85740 Garching, Germany.

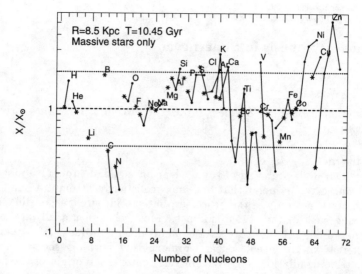

Figure 1. Stable isotopes from hydrogen to zinc synthesized only by massive stars sampled at a time (4.6 Gyr ago) and place where (8.5 kpc Galactocentric radius) the sun was born. The x-axis is the atomic mass number. The y-axis is the logarithmic ratio of the model abundance to the Anders & Grevesse (1988) mass fraction. (Timmes, private communication)

The remainder of the paper will discuss the implications of this figure - the uncertainties in its calculation and possible explanations for a number of gross underproductions.

It may be noted that, for the current value of $^{12}C(\alpha,\gamma)^{16}O$, massive stars are capable of producing one-third to one-half of the Anders & Grevesse carbon abundance. While low mass stars may dominate in producing carbon, the production in massive stars is certainly not negligible. Non-trivial amounts of ^{13}C and ^{15}N are also made in massive stars, and about one-third of the solar ^{14}N. More ^{14}N would be made in stars of higher metallicity (oxygen content) and massive stars could ultimately dominate in ^{14}N production. This figure also does not include the primary nitrogen production in massive stars which happens for certain choices of convective overshoot. The figure indicates that, even without Type Ia supernovae, massive stars could account for most of the iron in the sun (though this is not to say the same would be true at all metallicities), but, as we shall see, the error bar for iron production in massive stars is considerable. They may make as little as one-third of the solar iron. Type Ia is also clearly needed to raise the production of ^{55}Mn and some other species of the iron group.

2. Nuclear physics uncertainties

Exclusive of weak interactions, the nuclear physics uncertainties in supernova nucleosynthesis may be grouped into: 1) $^{12}C(\alpha,\gamma)^{16}O$ and 2) everything else. Alpha capture on carbon plays a key role, not only in setting the ratio of carbon and oxygen that comes out of helium burning, but, because carbon is the fuel for carbon (and ultimately neon) burning, the entire structure of the star is affected, including the final iron core mass and explosion mechanism. This reaction is also unique in that it acts in competition with another, the 3α-reaction, to determine the carbon abundance. Unlike other major burning reactions, its uncertainty cannot be compensated for by a slight recalibration of the burning temperature.

The error bar for this reaction has historically been large, especially 5 years ago. However, recent measurements by Azuma et al. (1994) have helped to pin down the S-factor at 300 keV for the E1 part to a value near 80 keV-barns with a total error bar of about 20 - 25 keV-barns. Much more uncertain is the E2 part of the rate. Barnes (1995) presents arguments that this too should have a value near 70 keV barns and, when combined with other weaker resonances, a total S-factor at 300 keV of 169 ± 55 keV barns is recommended. This is essentially 1.7 times the Caughlan, Fowler, & Zimmerman (1988) value and exactly what Weaver & Woosley (1993) had previously determined as optimal for producing the solar abundances. The latter reference also gives the sensitivity of nucleosynthesis to variations in this rate, the result being that a determination to about 20% would be highly desirable.

Other cross sections typically affect one or a few nuclei, but an exception is those rates that determine the neutron fluence during the s-process. During helium burning and, to a lesser extent, during carbon and neon burning, neutrons are produced by (α,n) reactions on the neutron-rich isotopes of oxygen and especially neon. Most of these neutrons go on neutron poisons such as ^{25}Mg. The remainder drive a limited s-process. This s-process is apparently too strong in our models (see also Käppler et al. 1994). The large abundances in Fig. 1 for $^{60,62,64}Ni$, ^{65}Cu, and ^{68}Zn are merely the leading edge of this, as Fig. II shows. Käppler et al. expressed the hope that advanced burning stages, especially oxygen burning, would destroy the excess s-process component. In this 25 $M_\odot$ star, that did not occur to the necessary extent to decrease the overproduction. A similar rerun of a 15 $M_\odot$ supernova (not shown) also gave an excess of nuclei in the same A = 60 to 86 mass region, but at a lower level, about a factor of two. Although the average over IMF is still to be carried out, these preliminary calculations suggest problems ahead that may require better cross sections for their resolution.

Above magnesium, thousands of reaction rates are employed in a typical stellar nucleosynthesis calculation, most of which come from Hauser-Feshbach theory calibrated (below zinc) against several dozen laboratory measurements. Of these myriads of rates, only a small fraction matter, most of the nuclear flow being governed by well known barrier penetration factors. The chief nuclear uncertainty is the photon transmission function. To gauge its effect Hoffman et al. (1996) have begun comparison calculations in which stellar and supernova models identical to those in WW are tracked using the Thielemann reaction set above A = 24. Preliminary results for a 25 $M_\odot$ supernova show identical yields for all nuclei lighter than A = 75 to the 30% level except for ten. The

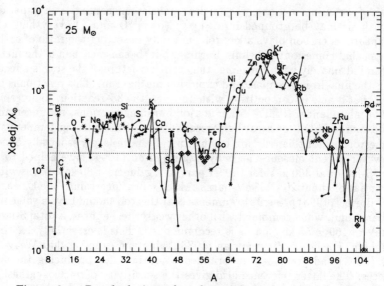

Figure 2. Recalculation of nucleosynthesis in a 25 $M_\odot$ solar metallicity supernova studied by Woosley & Weaver (1995) using a much larger reaction network (476 isotopes vs. 200) that extended above A = 105. Up to A = 64, the nucleosynthesis of WW is reproduced. At higher atomic weight though the new calculation shows evidence for a very strong helium burning s-process. The star was evolved from the main sequence through all burning stages and exploded as described in WW.

largest variation is a factor of two - twice as much ^{44,46}Ca is produced using the Thielemann rates, most of the former as ^{44}Ti, and one-half as much ^{52}Cr. The other seven discrepant nuclei agree to better than a factor of two.

3. Treatment of Convection

So long as nuclear burning proceeds at a leisurely pace compared to the convective turnover time, the theory one uses for mixing the composition is not very important, except, of course, at convective boundaries where overshoot, undershoot, and semi-convection can change both the nucleosynthesis and structure of the star. How these boundary effects are treated is a major source of the variation seen in models calculated by different groups (or by the same group using different physics). So far as nucleosynthesis is concerned, increased semiconvection has two important effects - it postpones the transition to a red supergiant to late in helium burning with particularly important consequences for binary mass transfer, and also gives, for a given helium core, a larger carbon core mass. Some implications of semiconvection for the nucleosynthesis have been discussed in Weaver & Woosley (1993). The treatment of convective overshoot in the Kepler code, is not very physical and needs revision (planned for the next year). One possible consequence is primary nitrogen production in massive metal deficient stars. This has been seen by Weaver & Woosley on various occasions, but quantitative results have not been published because of their unphysical sensitivity to model zoning.

During the advanced burning stages, particularly oxygen and silicon burning, convection and burning occur at comparable rates making the use of mixing length theory, even time-dependent mixing length theory, problematic. Bazan & Arnett (1994ab, 1995) have begun a preliminary exploration of this problem using two dimensional hydrodynamics. They find that mild entropy barriers that might have inhibited the linking of convective shells are less effective when the inertia of convective motion is included, and that convective boundaries are not as precisely defined. Substantial density inhomogeneities are developed that may form the seed for subsequent mixing instabilities after shock passage. A physically complete picture awaits three dimensional models with a realistic depiction of the sub-grid turbulent cascade, but it is expected that trace nuclei, such as ^{26}Al, made in the convective oxygen burning shell, may have their yield appreciably altered. The overall nucleosynthetic picture may not change that much however, as the reasonably good fit in Fig. I indicates.

4. Mass Loss and Rotation?

So long as the presupernova star does not lose its entire hydrogen envelope, the final nucleosynthesis of species other than the few made in the envelope - ^{13}C, ^{14}N, ^{17}O, ^{23}Na - should be unchanged. This is because the structure of the dense helium core and the advanced burning stages are insensitive to removing the envelope. However, for those stars in close binary systems, perhaps as much as 1/3 of all massive stars, and for those single stars of solar metallicity above about 35 M$_\odot$, envelope removal may lead to an epoch of rapid Wolf-Rayet mass loss (Woosley, Langer, & Weaver 1993, 1995; WLW). While presupernova cores having a variety of masses are expected, both observationally and theoretically, one expects a pile up of final masses around 4 $M_\odot$, *i.e.*, the progenitors of the Type Ib and Ic supernovae. The net effect is to remove stars from the upper end of the IMF in the chemical evolution calculation and replace them, nucleosynthetically speaking with extra stars of about 15 $M_\odot$. Because nucleosynthesis in stars above 25 $M_\odot$ is uncertain anyway (see below) and because their number is small, this is probably not a great alternation to the nucleosynthesis given in Fig. I. However a recomputation using the results of WLW is underway by Timmes.

Perhaps the greatest effect of mass loss is on dY/dZ and the synthesis of light elements ejected in the wind. Because the WR star can lose a great deal of mass that might otherwise have ended up in oxygen, say, dY/dZ is reduced. We shall also see in the next section that fall back can have a similar effect.

There has been a recent resurgence of interest in the effects of rotation in massive stars following early work by Endal & Sofia (1978). Fliegner & Langer (1994, 1995) have studied hydrogen and helium burning with variable amounts of rotation and a reasonable prescription for angular momentum transport. Because convection tends to concentrate angular momentum at the shell's outer boundary and thereby provoke considerable mixing, one of the effects of rotation resembles convective overshoot. For a given main sequence mass, the helium cores are a little larger. The hydrogen envelope becomes helium-rich and quite massive stars make an early transition to the WR stage while still burning hydrogen in the core. Future calculations that include the effects of rotational mixing during the advanced burning stages of massive stars are needed.

5. The Explosion and Fall-Back

Probably the biggest uncertainty affecting the nucleosynthesis of iron and the intermediate mass elements is our still primitive understanding of the explosion mechanism. Recent multi-dimensional models of the explosion (Herant, Benz, & Colgate 1992; Herant *et al.* 1994; Burrows & Fryxell 1993; Burrows, Hayes, & Fryxell 1995; Janka & Müller 1995; Miller, Wilson, & Mayle 1993), typically in 2D for a 15 or 20 $M_\odot$ star, have given results which probably qualitatively describe the explosion physics correctly. However, only a very limited set of masses has been explored, the calculations are 2D not 3D, and the neutrino transport is overly simple. All current models are typified by the overly efficient ejection of neutron-rich nuclei. Such calculations do not yet provide an adequate basis for a theory of Galactic chemical evolution.

Thus the stars used as input in Figs. I and II were exploded using pistons as described in WW. Uncertainty in the location of the piston and explosion energy obviously greatly affects nuclei made deep in the star, such as ^{44}Ti and ^{56}Ni. Less obvious, but very important is the interaction of the shock with the overlying mantle and the amount of "fall back" that occurs well after a successful shock has been launched. As the shock encounters extensive regions where the quantity ρr^3 increases outwards, it slows and so too does the matter behind it. If the shock lacks adequate energy, a large mass is decelerated to below the escape velocity by this interaction. Several solar masses can fall back in the more massive stars, even in an explosion that, optically at least, looks like an ordinary Type IIp supernova, but lacks the radioactive tail. This phenomenon is discussed by WW and illustrated in Table 1. The piston in each case was situated at the outer boundary of the iron core.

Table 1. Fall back in solar metallicity supernovae of 15 - 40 $M_\odot$

Mass Model	Piston ($M_\odot$)	Remnant ($M_\odot$)	KE_∞ (10^{51} erg)	$M_{ej}(^{56}Ni)$ ($M_\odot$)
S15A	1.29	1.43	1.22	0.115
S20A	1.74	2.06	1.17	0.088
S25A	1.78	2.07	1.18	0.129
S30A	1.83	4.24	1.13	0
S30B	1.83	1.94	2.01	0.440
S35A	2.03	7.38	1.23	0
S35B	2.03	3.86	1.88	0
S35C	2.03	2.03	2.22	0.568
S40A	1.98	10.34	1.19	0
S40B	1.98	5.45	1.93	0
S40C	1.98	1.98	2.57	0.691

As the table shows, the amount of fall back varies considerably with the mass of the star and the explosion energy. Larger stars have denser, more tightly bound mantles that are difficult to eject. An explosion which yields a kinetic energy at infinity of 1.2×10^{51} erg (net, after the binding energy has been subtracted) can eject most, but not all of the mass external to the iron core of stars under 25 $M_\odot$. Above 25 $M_\odot$, an increasing fraction of the mantle falls back, producing, in many cases, both a bright supernova and a black hole remnant. Since the interaction that leads to the fall back occurs well after the successful shock has already been launched, the engine cannot know ahead of time just how much energy it must provide in order to avoid becoming a black hole. Unless the power of the explosion, for unknown reasons, increases greatly as one goes to the more massive stars, one expects large amounts of nickel, oxygen, and intermediate mass elements to fall into the hole and be lost. Clearly this affects dY/dZ and Galactic chemical evolution in general and adds an element of uncertainty to Fig. I.

6. The Neutrino Process

Considerable discussion at the meeting was given to the synthesis of Li, Be, and B, including a presentation (Duncan et al., this volume) of recent determinations of Be and B abundances in metal deficient stars. These measurements indicate that B is primary, scaling linearly with the metallicity, which surprised some. Even more surprising were indications that Be may track B in metal deficient stars in roughly solar proportions. How this might be achieved in a model that uses proton irradiation in Orion-type clouds, rather than the customary cosmic ray spallation in the interstellar medium (which would give B and Be both secondary), was discussed by Ramaty and by Casse at the meeting.

The production of Li and B by neutrino irradiation in a 25 $M_\odot$ model is given in Table 2 extracted from WW. Nucleosynthesis by neutrino inelastic scattering during the first minute of the supernova is included in all of the models that were used to compute Figs. I and II. The cross sections employed are based upon experimental data, where available, (mostly photo-spallation cross sections), by Wick Haxton and are summarized in Woosley et al. (1990). From these studies, one expects ^{11}B to be primary in origin, since it is made from ^{12}C in the same supernova where the carbon is made, so that boron is observed to be primary comes as no surprise. However, for the cross sections employed, very little ^{9}Be is made by the neutrino process. One would have expected a variable ratio of Be/B as time passed (smaller at earlier time). If the data showing that the ratio is constant is correct, then either there is a separate source in nature of primary ^{9}Be and that source makes only part of ^{11}B, or else a mechanism for making ^{9}Be in supernovae has been overlooked.

Table 2. Neutrino nucleosynthesis in a 25 $M_\odot$ supernova

Isotope	w/o ν	with ν $T_\nu = 8$	with ν $T_\nu = 6$
^{7}Li	2.13E-08	7.31E-07	2.81E-07
^{10}B	2.51E-09	2.75E-09	2.58E-09
^{11}B	1.18E-08	2.35E-06	1.03E-06
^{15}N	3.39E-05	2.33E-04	1.29E-04
^{19}F	3.53E-05	1.48E-04	9.15E-05
^{26}Al	9.38E-05	1.29E-04	1.14E-04

It seems difficult to avoid making substantial ^{11}B by the neutrino process. One could reasonably turn the neutrino temperature down to 6 MeV (Table 2) and thus reduce the synthesis by a factor of two (WW used 8 MeV). This would still make most of the solar abundance (Fig. I). There remains the possibility that the cross sections for $^{12}C(\nu,\nu'n)^{11}C$ and $^{12}C(\nu,\nu'p)^{11}B$ or the spectrum of high energy neutrinos has been seriously overestimated. This seems unlikely, but warrants investigation and might imply that the neutrino process plays no major role in nucleosynthesis, not only of boron, but of anything. Another possibility is that the production of ^{9}Be (and perhaps ^{6}Li and ^{10}B as well) by the neutrino process has been seriously underestimated. Since the meeting Wick Haxton and I have looked into this possibility and found reasons to believe that ^{9}Be synthesis may have been underestimated, e.g., by neutrino spallation of ^{14}N. This is an area where laboratory measurements can help and where more work is needed. It certainly would be too soon to count the neutrino process out.

It is interesting that the neutrino process is capable of making a non-negligible amount of ^{15}N. This is the only nucleus for which, at the present time, one must invoke an origin in classical novae. Perhaps adjustments in the cross sections or neutrino spectra in the other direction might ultimately make all of the ^{15}N in supernovae.

7. Nucleosynthesis in Type Ia

As has been long recognized, Type Ia supernovae are a major source of iron in our Galaxy and in others. For a version of Fig. I that includes iron group nucleosynthesis from a typical Type Ia carbon deflagration (Nomoto, Thielemann, & Yokoi 1984), see TWW. It may also be that Type Ia's are responsible for a neutron-rich component of nucleosynthesis that includes ^{48}Ca, ^{50}Ti, ^{54}Cr, and ^{58}Fe. All these nuclei are notably deficient (except perhaps ^{58}Fe) in Fig. I, and can be produced in the inner 0.01 $M_\odot$ of a Chandrasekhar mass explosion if the central density at ignition exceeds 4×10^9 g cm^{-3} (Woosley & Eastman 1995). Such an ignition density might be reasonable for very low accretion rates. Low values of Y_e are needed, around 0.42. The slow flame and high density in these models promotes electron capture which generates such a value naturally. Regions having the same amount of electron capture can be found near the mass cut in massive stars, but there the higher gravitational potential of the neutron star, plus the dominance of radiation pressure, implies high entropy and a freeze out from nuclear statistical equilibrium that is dominated by α-particles (see next section). ^{48}Ca cannot be made in such an environment.

Other varieties of Type Ia supernovae, the so called "sub-Chandrasekhar mass models", may be required to produce a few other isotopes that are only made well in high temperature explosive helium burning, notably ^{44}Ca (as ^{44}Ti), ^{47}Ti, and ^{51}V. These isotopes are also deficient in Fig. I, but can be made in those supernovae where a detonation wave passes through helium (Woosley & Weaver 1994). It is currently contriversial what fraction of Type Ia supernovae are of this variety.

8. Nucleosynthesis in Neutrino Driven Winds - ^{64}Zn

Elsewhere in this volume, Meyer discusses the neutrino driven wind and its relation to r-process nucleosynthesis. Hoffman *et al* (1996) have explored the kind of nucleosynthesis that might be expected at early times ($\sim$ 1 s as opposed to late - $\sim$10 s - when the r-process is made). For values of Y_e in the range 0.48 to 0.49 and dimensionless entropies per baryon of $\sim$50, a strong α-rich freeze out produces the light p-nuclei ^{64}Zn ^{70}Ge, ^{74}Se, ^{78}Kr and ^{84}Sr, ^{90,91}Zr, and ^{92}Mo. Note that while these are called p-nuclei, they really have an excess of neutrons, e.g., ^{92}Mo has 42 protons and 50 neutrons. This combination of moderate entropy and neutron-excess seems very likely to be achieved in the common supernova after the shock is launched, when the neutrinos drive a radially streaming wind of steadily increasing entropy and declining mass flux. If so, all the nuclei just mentioned would be primary, produced in about the same amounts in a supernova that is extremely metal-deficient as nowadays. ^{64}Zn is not only deficient in Fig. I, which did not include the neutrino wind, but also the most abundant isotope of zinc. One would expect to see evidence for its primary nature in metal deficient environments.

This work has been supported by the NSF(94-17171), and, in Germany, where the paper was prepared, by the Alexander von Humboldt Foundation.

References

Anders, E., & Grevesse, N. 1989, *Geochim. Cosmochin. Acta*, **53**, 197
Azuma, R. E., Buchman, L., Barker, F. C., Barnes, C. A., and 13 others 1994, Phys.Rev.C, **50**, 1194
Barnes, C. A. 1995, *Proc. Symp. on the Physics of Unstable Nuclei, Nucl. Phys. A*, 588, 295c
Bazan, G. & Arnett, W. D. 1994, ApJ, **433**, L41
Bazan, G., & Arnett, W. D. 1994, *BAAS*, **26**, 1444 and in preparation for ApJ
Bazan, G., & Arnett, W. D. 1995, ApJ, submitted.
Burrows, A., & Fryxell, B. A. 1993, ApJ, **418**, L33
Burrows, A., Hayes, J., & Fryxell, B. A. 1995, ApJ, **450**, 830
Endal, A. S., & Sofia, S. 1978, ApJ, **220**, 279
Fliegner, J., & Langer, N. 1994, in *Pulsation, Rotation, and Mass Loss in Early-Type Stars*, IAU Symp. 162, eds. L. Balona, H. Henrichs, and J. Le Contel, (Kluwer Academic: Netherlands), p. 147
Fliegner, J., & Langer, N. 1995, in *Wolf-Rayet Stars: Binaries, Colliding Winds, Evolution*, IAU Symp. 163, eds. K. van der Hucht and P. Willaims, (Kluwer Academic: Netherlands), p. 326
Caughlan, G. R., & Fowler, W. A. 1988, *ADNDT*, **40**, 283
Herant, M., Benz, W., & Colgate, S. A. 1992, ApJ, **395**, 642
Herant, M., Benz, W., Hix, J., Colgate, S. A., & Fryer, C. 1994, ApJ, **435**, 339
Hoffman, R. D., Woosley, S. E., Fuller, G., & Meyer, B. 1996, ApJ, in press.
Janka. T., & Müller, E. 1995, A&A, in press
Käppler, F., Wiescher, M., Giesen, U., Görres, and 7 others, 1994, ApJ, **437**, 396.
Miller, D. S., Wilson, J. R., & Mayle, R. 1993, ApJ, **415**, 278
Nomoto, K., Thielemann, F.-K., & Yokoi, 1984, A&A, **286**, 644
Timmes, F. X., Woosley, S. E., & Weaver, T. A. 1995, ApJS, **98**, 617 (TWW)
Woosley, Hartmann, D. H., Hoffman, R., & Haxton, W., 1990, ApJ, **356**, 272
Weaver, T. A., & Woosley,S. E. 1993, *Physics Reports*, **227**, 65
Woosley, S. E., & Weaver, T. A. 1994, ApJ, **423**, 371
Woosley, S. E., & Weaver, T. A. 1995, ApJS, **101**, 181 (WW)
Woosley, S. E., and Eastman, R. G. 1995, *Proceedings of Menorca School of Astrophysics*, eds. E. Bravo, R. Canal, J. Ibanez, and J. Isern, *Societat Catalana de Fisica*, p. 105
Woosley, S. E., Langer, N., & Weaver, T. A. 1993, ApJ, **411**, 823
Woosley, S. E., Langer, N., & Weaver, T. A. 1995, ApJ, **448**, 315

Cosmic Abundances
ASP Conference Series, Vol. 99, 1996
Stephen S. Holt and George Sonneborn (eds.)

Observational Evidence for Nucleosynthesis by Supernovae

Robert P. Kirshner

Harvard-Smithsonian Center for Astrophysics

Abstract. Supernovae transform the elements and broadcast them into the interstellar medium. Observations give strong reasons for believing Type Ia supernovae (SN Ia) are incinerated white dwarfs that produce several tenths of a solar mass of iron-peak nuclei, but new data indicate that not all SN Ia are identical. Type II supernova (SN II) explosions produce about 0.1 $M_\odot$ of iron-peak elements, and young remnants like CasA and N132D show massive amounts of oxygen and oxygen-burning products. The connection between the chemical yield of supernovae and the supernova remnants they produce has been explored with ASCA X-ray spectra of young supernova remnants in the LMC. The "Balmer-dominated" remnants have hot interiors whose X-ray abundances match the expected products of SN Ia. Older remnants lose their chemical identity, but can be used to establish the integrated effect of chemical enrichment, as in recent studies of M33. Schematically, the pattern of iron enrichment through SN Ia and oxygen enrichment by SN II can be matched to the chemical abundance patterns in galaxies to infer the mix of enrichment sources. Supernova explosions can now be detected at redshifts of $z = 0.5$ or more: a look-back time of about 5 billion years. The differences between the long lifetimes of SN Ia and short lives for SN II progenitors could be reflected in chemical abundance patterns for galaxies (or their precursors) at large redshift.

1. Introduction

Supernovae are widely reputed to play a very important role in the chemical enrichment of the Universe (Trimble 1982,1983). I hear myself saying this every year in no uncertain terms to elementary astronomy classes. It must be true. Because theoretical models provide such presuasive and detailed predictions for the element production from supernova explosions, our belief in this process is even stronger than the empirical evidence. Since the human mind is fallible, so are its products, including supernova models, and it is important to check this elaborate story where we can.

The evidence comes from supernovae and from supernova remnants: indirectly from the time history of the luminosity (the light curve) and more directly from the spectrum. The path from supernova observations to supernova abundances is explored in detail by Dick McCray's contribution to this conference, but a few general points are worth noting here. Supernovae are classified by their spectra near the maximum of the light curve: Type I supernovae have **no** hydrogen in their spectra while Type II supernovae show strong hydrogens lines. The quantitative analysis of supernova spectra is far from simple (Eastman, Schmidt and Kirshner 1996, Höflich 1995), but even the qualitative features go a long way toward indentifying the explosions with possible paths of stellar evolution. Supernova light curves are also powerful clues to nucleosynthesis because much of the energy released as light comes from radioactive elements synthesized in the explosion. The most important of these is ^{56}Ni which decays to ^{56}Co with a 6.1 day half-life and then to stable ^{56}Fe with a 77.7 day half-life.

The classification of supernova spectra now includes reasonably well-defined SN Ib and perhaps SN Ic. These are useful additions to the classifications scheme because they help separate objects which are superficially similar (all the SN I lack hydrogen in their outer layers) but which are probably from very different stars. We associate SN Ia with thermonuclear explosions in white dwarfs and SN Ib/c with core collapse in massive stars (much like SN II) after they lose their hydrogen envelopes.

2. SN Ia: Abundant Iron

Type Ia supernovae are observed in spiral, elliptical, and irregular galaxies. Their presence in ellipticals, where current star formation is rare, is taken as evidence that SN Ia arise from an old, long lived population. White dwarfs in binaries have been the focus of theoretical attention, but identifying the present-day systems which will become SN Ia has not been easy (Kenyon et al. 1993). Despite these observational uncertainties, the working theoretical picture is the detonation of a carbon-oxygen white dwarf at the Chandrasekhar mass through accretion from a binary companion. This well-defined physical event leads to definite predictions: an expectation the star will burn about half a solar mass to nuclear statistical equilibrium and produce a uniform luminosity. Observations show that this picture is too simple: while iron peak elements are copiously produced, the amount is not unifrom, as reflected in variations in the spectra and light curves.

2.1. Early Spectra

Early observations of SN Ia showed that they comprised a reasonably homogneous spectroscopic class (Oke and Searle 1974). Aside from the absence of hydrogen, a SN Ia at maximum shows lines of intemediate-mass elements, most conspicuously the Si II lines observed in the expanding atmosphere near 6150 Å. The presence of intermediate mass elements places a strong constraint on the propagation of the burning wave in the stellar interior. The most successful generic models require a sub-sonic burning wave ("deflagration") to match the observed velocity range of silicon, magnesium, and oxygen (Nomoto, Thielemann, and Yokoi 1984). Techniques for computing synthetic spectra allow the predictions for a wide variety of explosion models, including double detonations and other burning shcemes to be compared with the data (Höflich and Khokhlov 1996).

Observations of SN 1991T and SN 1991bg show that the range of SN Ia is larger than previously appreciated. SN 1991T had no lines of Si, Ca, or S and this is probably due to low abundances of those elements in the outer

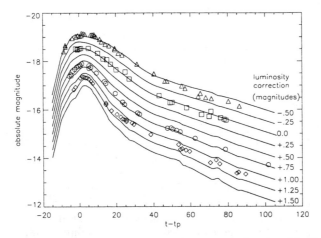

Figure 1. Light curves of SN Ia. The range in luminosities is a factor of 5 and may reflect the mass of ^{56}Ni produced. The variation in light curve shape with luminosity provides a way to estimate the luminosity and distance of SN Ia (Riess, Press, and Kirshner 1995).

layers (Phillips et al. 1992, Jeffery et al. 1992). SN 1991T was also exceptionally luminous. At the other extreme, SN 1991bg (Filippenko et al 1992, Leibundgut et al 1993) showed an unusual spectrum with numerous strong titanium lines at maximum, and this SN Ia was exceptionally dim. The connection between supernova spectra and supernova luminosities has been explored by Nugent et al. (1995). While the details of supernova rates for SN Ia of each variety are not known, models for the chemical enrichment of the Galaxy (Timmes et al. 1995) use a simpler picture of SN Ia abundances than Nature produces.

2.2. Luminosity

Type Ia supernova explosions expand from the compact dimensions of a white dwarf to over 10^{14}cm as they approach maximum light. The ferocious adiabatic losses in this expansion are balanced by the delayed energy input from nickel, then cobalt decay during the two to three week rise to maximum. For this reason, the light curves of SN Ia, though they are affected by the complexities of diffusion through an expanding medium of changing opacity, are not too remote from the process of synthesizing iron-peak elements. While observers were analyzing SN 1991T and SN 1991bg, Branch (1992) used a Chandrasekhar mass model, a computation of the nickel mass and a schematic diffusion model to compute the peak luminosity of SN Ia with the aim of determining the Hubble constant. Aside from important technical improvements in this approach (Höflich and Khokhlov 1996), the empirical evidence is now quite strong that there is a range of lumnosities for SN Ia, as shown in figure 1, so a single theoretical model does not account for SN Ia. The Hubble Constant needs actual observations for calibration (Riess, Press, and Kirshner 1995).

This range of luminosities, especially the intrinsically faint objects like SN 1991bg, suggests that the mass of radioactive ^{56}Ni produced may vary from object to object. It may even be the case that the progenitors are not at the

Chandrasekhar mass (Woosley and Weaver 1994, Ruiz-Lapuente et al. 1995, Livne and Arnett 1995). If alternative models for evolution to an exploding white dwarf are taken seriously, then the lifetimes of their progenitors may be important as we look back over cosmological time (Canal et al. 1996).

3. SN II: A Collapse and a Blast of Fresh Air

For SN II, the spectrum near maximum and the luminosity near maximum are not closely related to the chemical abundances of the ejecta. The spectrum is dominated by the hydrogen-rich outer layers of the star, whose abundances may be slightly shifted by CNO processing and mass loss, but which fundamentally reflect the gas out of which the massive progenitor of a SN II formed just 10^7 years in the past. The energy for SN II comes from core collapse, so the luminosity is not closely coupled to nuclear events. SN II are important for nucleosynthesis, as the site of the r-process for exotic uranium, gold, and lead and as the source of intermediate-mass elements such as oxygen, neon, and silicon which are produced in post-main sequence energy generation. While theory predicts the details of nucleosynthesis, observation tests the outline of this picture, and holds the promise of supplying some important details of stellar evolution that are difficult to predict. The difficulty, as explained in McCray's talk, is making the connection between the observed spectra and the abundances in the exotic astrophysical setting of a supernova.

3.1. Early Spectra

The spectra observed in the first month or so of a SN II are dominated by the diffusion of radiation through a massive stellar envelope. This masks the interesting chemistry of the interior, but it does have one useful side-effect. As recently shown by Eastman, Schmidt, and Kirshner (1996), the continuum emission from a SN II is not much affected by the details of the atmosphere structure and chemistry so individual SN II's can be utilized as "custom yardsticks" for extragalactic astronomy. Every cloud has a silver lining, especially if the silver is radioactive.

While atmospheres for a wide range of red giant progenitors produce similar observables, one important lesson from the observations of SN 1987A in the LMC and of SN 1993J in M81 is that mass loss can be very significant, and will affect the observed spectrum and light curve. In the fortunate case of SN 1987A, the structure (Plait et al. 1995, Burrows et al. 1995) and chemistry (Panagia et al. 1996) of the pre-supernova mass loss is accessible through HST observations, but this is not likely to be the case for many supernovae.

3.2. Late Spectra

Dick McCray's paper in this volume treats the astrophysics of a radioactively heated nebula, such as we see in SN 1987A today. The goal is to understand the astrophysics of the debris well enough to interpret the spectrum to yield abundances of the inside of the star (Wang et al. 1996). In the case of SN 1993J, there is powerful interaction between the expanding debris and a dense circumstellar surrounding. As discussed by Houck and Fransson (1996), ionizing radiation that arises from shocks can excite the inner debris for SN 1993J. The

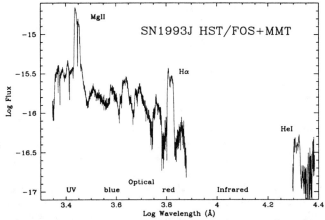

Figure 2. Spectrum of SN 1993J showing UV data from HST, and ground-based optical and IR spectra. The broad emission lines are excited by the interaction with circumstellar matter. Matter from deep within the star should be conspicuous in future emission.

models indicate that emission we see in 1995 should come from the helium-rich zone of the outer core, and that deeper layers should soon be probed.

3.3. Luminosity: the iron peak and the tail

Although SN Ia have a closer link between the radioactive material and the optical luminosity observed at maximum, for SN II there are some clues to the production of iron-peak elements in the hot region just outside the forming neutron star. These hints reside in the luminosity of the "tail" of the SN II light curve, where the energy for the optical light comes from the thermalization of radioactive decay products from ^{56}Co. For SN 1987A, the prediction of 0.07 solar masses of radioactive cobalt was well-matched by the observations of the light curve, and subsequently by the X-rays and γ-rays observed directly from the nuclear decays (see Arnett et al. 1988 for references). There is now a significant handful of SN II for which we can estimate the ^{56}Ni production from the late-time luminosity (Blanton et al. 1995). For SN 1990E and SN 1969L, the ^{56}Ni mass agrees within the errors with the 0.07 solar mass found for SN 1987A. For SN 1991G, the result is about 0.02 ± 0.01 solar mass. Since there are other clues suggesting that SN 1991G had a low explosion energy, there may be a connection between the explosion mechanism and the iron-peak production. Since hydrodynamic models for core collapse are not particularly clear on the location of the "mass cut" between material that goes in to form the neutron star and the material that is ejected (and cooked to nuclear statistical equilibrium), these empirical measures may help.

4. Supernova Remnants: The Core of the Matter

The deep interior of the pre-supernova star becomes the expanding debris of a young supernova remnant. With typical velocities of several thousand kilometers per second, this debris is heated by its encounters with the surrounding material and it produces shocks in the interstellar gas. For young supernova remnants, the debris reflects the properties of the explosion. For old ones, it allows a measurement of abundances in the interstellar medium. Optical and X-ray observations of Tycho's SNR and Cas A in our galaxy and of N132D and a handful of Tycho-like SNRs in the LMC now show how to connect the observations of supernova remnants with the supernovae that produced them.

One interesting question is the boundary between "young" remnants which retain the properties of the source and "old' remnants which are sampling the mixed interstellar products of many supernovae. On the average, an expanding supernova remnant will encounter several solar masses of interstellar matter in just 1000 years. At that point, with a radius of just a few parsecs, we should expect significant dilution of the debris. However, this may be too conservative a view. The effect of a SN II ejecting a solar mass of oxygen, whose interstellar abundance is of order 10^{-3}, will be significant over 1000 solar masses of the ISM. For a SN Ia, ejecting nearly a solar mass of iron, whose ISM abundance is lower, the affected volume could be larger. While one-dimensional computations of the hydrodynamics of supernova debris with the ISM suggest that the debris does not travel far, the observations show that the material from the core is chemically and physically inhomogenious and some dense lumps may travel much farther than the average. At least in Puppis A (Winkler et al. 1988) and in Vela (Aschenbach 1995) there are fast-moving fragments which may be ejecta at distances up to 20 parsecs.

4.1. SN Ia: X-rays to the Rescue

Supernova classifications are spectroscopic categories of the spectra at maximum light. Despite Tycho Brahe's undoubted skill in observing the galactic supernova of 1572, he did not take any spectra. However, the preponderance of evidence favors the view that Tycho's supernova is the result of a SN Ia explosion. The optical emission from Tycho's remnant is underwhelming (Smith et al. 1991), but curiously informative. Despite the fact that SN Ia are expected to be copious sources of iron, the optical spectra of the faint filaments at the rim of the remnant show only Balmer lines of hydrogen. However, closer inspection shows that these "Balmer dominated" spectra have two components to the H-α line. One is narrow, but the other is a 2000 kms^{-1} broad line whose origin lies in charge exchange of fast-moving protons behind the supernova shock with neutral hydrogen atoms which drift through the plasma-mediated shock zone. In the case of Tycho's remnant, the combination of the proper motion of the Balmer-dominated filament and the inferred shock velocity leads to an estimate for the distance and absolute magnitude for the supernova.

X-ray surveys of the LMC led to the identification of a few SNR which showed the same type of "Balmer-dominated" spectrum refs here. Analysis of their line widths (Smith et al. 1991) showed that these SNR's were young. But were they really the result of Type Ia supernova explosions? The answer is "yes" and it comes from ASCA spectra of the emission from the interior of these objects (Hughes et al. 1995). As described by Rob Petre in this volume, the X-ray spectra show that the emission seen is plausibly from the chemical mix expected from a SN Ia explosion.

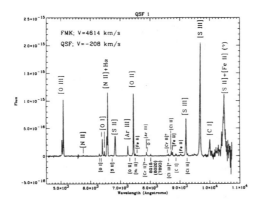

Figure 3. Spectrum of Cas A. A fast moving knot of debris (at 4614 kms^{-1}) shows lines of elements from carbon through iron. The emssion at Hα comes from slow moving gas that may be related to mass loss before the explosion.

4.2. SN II: Shot from Guns

Two young supernova remnants have been investigated in enough detail to provide some information about the process of nucleosynthesis. Both Cas A in our Galaxy and N132D in the LMC show the unmixed debris from massive star explosions. In the case of Cas A, the pattern of abundances has long been known to correspond to the oxygen that ought to be present in the interior of a massive star plus the products of oxygen burning: sulfur, argon, and calcium (Chevalier and Kirshner 1979). These fragments of the star are moving at velocities of 4000-8000 kms^{-1} and most show no trace of hydrogen or helium. Fesen and Becker (1991) suggest that Cas A may have had a thin coat of hydrogen when it exploded– making it more like a SN Ib than a conventional SN II.

In any case, the interior is expected to be mixed by the neutrino-driven convection that couples the core collapse to the explsoion. As recent two-dimensional hyrdodyanmical computations have shown (Herant et al. 1994), chemical inhomogeneities in the debris may have their origin in this macroscopic mixing. For SN 1987A, the early appearance of the X-rays is consistent with mixing of iron-peak elements far out in the star's mantle. Observations of Cas A show that in this remnant, a single spectrum of a single filament shows elements from carbon through iron at the same space velocity. This suggests that for Cas A, too, the mixing was extensive, but not necessarily microscopic.

N 132D is harder to resolve, since it is 25 times as far away, but does not have the intervening absorption that makes Cas A a challenge. Individual filaments can be observed, and the overall kinematics of the remnant are now understood (Morse, Winkler, and Kirshner 1995). N132D has some filaments with pure emission of oxygen and neon, with an overall expansion velocity of 1650 kms^{-1} and a kinematic age of 3150 years. The presence of neon in the oxygen-rich zone is a difference from Cas A that may have its origin in the mass and burning history of the pre-supernova star. HST observations of N132D are underway which will provide a sharper view of this object and will pave the way for ultraviolet spectra of chemically distinct filaments.

4.3. M33: The Last 10^5 Years

Supernova remnants in Local Group galaxies beyond the LMC are worth finding. Although they are generally too large and therefore too old to illuminate the chemical contributions of supernova explosions, SNR's in other galaxies provide a convenient record of past supernova activity, a check on the energy production of supernova, and an alternative way to measure the gas-phase abundances in the ISM of another galaxy. Our work has concentrated on M33, using an extensive interference-filter survey of the galaxy (Long et al. 1990) together with VLA maps of the non-thermal radio sources (Gordon et al. 1996) to identify supernova remnants. Most are on the range from 10 parsecs to 100 parsecs, so unfortunately the sample does not include young remnants whose emission might be dominated by their own ejecta.

The picture based on Gordon's thesis (1995) is that M33's SNR's probably obey the simple Sedov law for adaibatic expansion ($r \sim t^{2/5}$), so that counting remnants as a function of size yields an estimate of the supernova rate. For M33, that is approximately one supernova (remnant) every 300 years. The SNR shocks heat the interstellar gas, and the subsequent hydrogen recombination produces a partially-ionized zone where collisional excitation and collisional ionization prevail for most species. By looking at the strength of lines that are formed in the recombination zone, many of the modelling uncertainties are reduced. For example, the [N II] line strength at 6548 Å and 6584 Å can be compared to the Hα line right between them to provide a good estimate of the nitrogen abundance. For M33, the nitrogen abundance gradient found this way agrees well with the work based on H II regions. As noted by Smith et al. (1993), though the gradient agrees well, the absolute value of the nitrogen abundance found from SNRs appears to be 0.4 dex larger. The origin of this discrepancy remains unresolved.

5. High z Supernovae: an Enriching History

Searches for supernovae now extend up to redshifts of 0.5 and greater (Kirshner et al. 1995, Perlmutter et al. 1995). While investigations such as carried out by Timmes et al. (1995) aim at matching the present-day abundance patterns in galaxies by an artful combination of SN Ia and SN II abundance contributions, these distant searches raise the possiblity of seeing directly the effects of the lifetimes of supernova progenitors on the mix of supernovae detected at high redshift. A recent computation by Canal et al. (1996) illustrates the effects that might be observed. Chemical analyses of Lyman-alpha absorbers that might be ancient disk galaxies (see the contribution by Lu, Sargent, and Barlow in this volume) may also provide a clue to the role of long-lived iron-producing SN Ia and fast-burning oxygen-producing SN II in the early chemical history of galaxies.

Acknowledgments. Many people have contributed to this work, especially Shawn Gordon, Frank Winkler, Jon Morse, Jim Herrnstein, Adam Riess, Peter Challis, Pilar Ruiz-Lapuente, Ron Eastman, Brian Schmidt, Peter Garnavich, and Jason Pun. Studies of supernovae and supernova remnants at Harvard are supported by the NSF through grant AST-9218475 and by NASA through grant GO-2563 from the Space Telescope Science Institute.

References

Aschenbach, B. 1995, in *Proc. of 17th Texas Symposium*, H. Böringer et al. eds., The New York Academy of Sciences, p.196.
Blanton, E.L. et al. , 1995, AJ, 110, 2868.
Branch, D. 1992, ApJ, 392, 35.
Burrows, C., et al. 1995, ApJ, 452, 680.
Canal, R, Ruiz-Lapuente, P., & Burkert, A., 1996, ApJ, 456, L101.
Chevalier, R.A. & Kirshner, R.P. 1979, ApJ233, 154.
Eastman, R., Schmidt, B. and Kirshner, R. 1996, ApJ, in press.
Fesen, R.A. & Becker, R. 1991, ApJ, 371, 621.
Filippenko et al. , 1992, AJ, 104, 1543.
Gordon, S.M. 1995, Ph.D. Thesis, University of New Mexico.
Gordon, S.M. et al. , 1996, ApJ, in press.
Herant,M., et al. , 1994, ApJ, 435, 339.
Höflich 1995, ApJ,443, 533
Höflich, P. & Khokhlov, A., 1996, ApJ, in press.
Houck, J.C. and Fransson, C. 1996, ApJ, 456, 811.
Hughes, J.P. et al. , 1995, ApJ, 444, L81.
Jeffery et al. 1992, ApJ, 397, 304.
Kenyon, S. et al. , 1993, ApJ, 407, l81.
Kirshner, R.P. et al. 1995, IAU Circular 6270
Leibundgut et al 1993, AJ, 105, 301.
Livne, E. & Arnett, D., 1995, ApJ, 452, 62.
Long, K.S., Blair, W.P., Kirshner, R.P., and Winkler, P.F., 1990, ApJS, 72, 61.
Morse, J.A., Winkler, P.F., & Kirshner, R.P., 1995, AJ, 109, 2104.
Nomoto, K., Thielemann, F.-K. & Yokoi, K. 1984, ApJ, 286, 644.
Nugent, P. et al. 1995, 455, L147.
Oke, J.B. and Searle, L. 1974, Ann. Rev. Astron. Astrophys., 12, 315.
Panagia, N., et al. 1996, ApJ, 458, 000.
Perlmutter, S. et al. 1995, IAU Circular 6267
Phillips et al. 1992, AJ, 103, 1632.
Plait, P.C., Lunqvist, P., Chevalier, R.A., & Kirshner, R.P., 1995, ApJ, 439, 730.
Riess, A.G, Press, W.H., & Kirshner, R.P., 1995, ApJ, 438, L17.
Ruiz-Lapuente, P. /etal 1995, ApJ, 439, 60.
Smith, R.C., Kirshner, R.P., Blair, W.P., Long, K.S., & Winkler, P.F., 1991, ApJ, 372, 531.

Smith, R.C., Kirshner, R.P., Blair, W.P., Long, K.S., & Winkler, P.F., 1993, ApJ, 407, 564.

Timmes, F. X., Woosley, S.E., & Weaver, T.A., 1995, ApJS, 98, 617.

Trimble, V. 1982, Reviews of Modern Physics, 54,1183.

Trimble, V. 1983, Reviews of Modern Physics, 55, 511.

Wang, L. et al. , 1996, ApJ, in press.

Winkler, P.F., Tuttle, J.H., Kirshner, R.P., and Irwin, M.J. 1988, in *Supernova Remnants and the Interstellar Medium*, R.S. Roger and T.L. Landecker, eds., p.197.

Woosley, S.E, & Weaver, T.A., 1994, ApJ, 423, 371.

Inferring Abundances from the Spectra of Supernovae

Richard McCray

JILA, University of Colorado, Boulder, CO 80309-0440

Abstract. Traditional methods for inferring abundances from spectra don't work well for supernovae. The newly-synthesized elements may fail to appear in the X-ray spectra of supernova remnants. Standard spectroscopic methods are confounded by the inhomogeneity in composition and temperature and by optical depth in emission lines. Methods that should work are based on the principle that the gamma rays from newly-synthesized radioactive elements illuminate the envelope uniformly after a year or so.

1. Introduction

We got to get this theory out of things.

Willy Fowler enjoyed quoting the above remark by former Alabama Governor George Wallace, and I thought it appropriate to repeat it here, since Willy's spirit pervades this conference. Moreover, the remark is entirely appropriate to the subject at hand. Everyone here is confident that supernovae are the main sources of heavy elements in the universe, and this notion gains impressive support from calculations of the sort described here by Stan Woosley and Frank Timmes. But we shouldn't be so entranced by this work that we forget: most of what we know about supernova nucleosynthesis comes from observations, not of *real* supernovae – the kind that have exploded in interstellar space, but of *virtual* supernovae – the kind that have exploded in supercomputers.

In this paper, I will address the question: how well can we determine the nucleosynthesis yields of real supernovae from their observed spectra? The answer will be: up to now, not well at all; however, we can be optimistic about the potential of future observations.

Before addressing the main subject of the interpretation of supernova spectra, I remind the reader of one of the main conclusions that one may draw from the talk by Timmes: SN Ia's – the type believed to result from the explosions of white dwarfs – may contribute significantly to the observed cosmic abundance of iron, but they produce far too much iron to account for the observed abundances of other elements. Therefore, as regards nucleosynthesis it is more important to learn to interpret the spectra of SN II's and SN Ib's – the types that come from core collapse of massive stars. The more important question regarding SN Ia's is the astronomical one: in a given cosmic system, what are the relative contributions (integrated over time) of SN Ia's to SN II's and SN Ib's? To first order, all we need to know about SN Ia's is that they make mostly iron. That's good, because it's going to be a lot tougher to read nucleosynthesis yields from the spectra of SN Ia's than from the spectra of SN II's, and the latter problem is tough enough for now.

Therefore, I shall concentrate on the problem of interpreting the spectra of SN II's. In doing so, I will emphasize what we have learned from SN 1987A. For decoding the language of supernova spectra, SN 1987A is a Rosetta Stone. It is unique because we have observed its entire electromagnetic spectrum, from radio to gamma rays, and we have been able to observe its (UVOIR) spectrum for almost nine years after explosion. As the reader will see, the most reliable information about nucleosynthesis comes from the spectrum at late times.

Before doing that, however, I would like to make a few comments about the problem of inferring abundances from the spectra of supernova remnants.

2. Supernova Remnants

At this conference Rob Petrie reviewed the tremendous progress that has been achieved in observing the X-ray spectra of supernova remnants. Emission lines from O, Ne, Si, S, Ca, and Fe are clearly evident in the spectra of several bright SNR's. Can we make quantitative inferences about supernova nucleosynthesis yields from such observations?

I am not optimistic. There are two major obstacles. Substantial progress has been made to overcome the first, but the second is even more severe.

In order to discuss these problems, I must emphasize a major paradigm shift that become increasingly apparent during the past two decades: *the circumstellar environment of SN II's and SN Ib's is determined by mass loss from their progenitor stars.* Massive stars will have fast $(V_w \sim 500 - 3000$ km s$^{-1})$ stellar winds during their main sequence or post-main sequence evolution. These winds will displace the ambient interstellar gas into a shell surrounding a low density cavity to form an "interstellar bubble." Moreover, since massive stars are formed in clusters, a given supernova is likely not to be the first in the neighborhood, and thus the interstellar ambient medium is likely to be the low density interior of a "superbubble" formed by the action of all the previous supernovae in the cluster (McCray & Kafatos 1987). But the progenitor is likely to expel several solar masses of its outer envelope in a slow $(V_w \sim 10-30$ km s$^{-1})$ stellar wind during its red giant stage. *The radiation from young supernova remnants almost certainly comes from the shock interaction of the supernova ejecta with the slow-moving gas that has been ejected from the progenitor itself.*

The first major obstacle to interpreting the X-ray spectra of supernova remnants is to model accurately the complicated shock dynamics and radiation from the interaction. This is difficult because the X-rays are emitted both by the circumstellar gas that has been shocked by the forward blast wave and by the supernova ejecta that have been shocked by an inward-propagating reverse shock. Moreover, the X-ray emissivity depends not only on the instantaneous state of the gas, but also on the history of the shock dynamics, since the emitting gas is probably not in ionization equilibrium.

Recent work on modeling the X-ray emission from Kepler (Borkowski et al 1994) shows that these formidable technical challenges can be overcome, and gives hope that one can construct realistic physical models for the X-ray spectra of such remnants.

The second obstacle is more severe, however. We cannot be confident that we are seeing all of the supernova ejecta. Here is the reason: as a rule of thumb, the mass that has been overtaken by the reverse shock will be comparable to the mass overtaken by the forward shock (cf. Chevalier 1982). But the circumstellar mass expelled by the progenitor may be substantially less than that of the ejecta. In that case, the blast wave will overtake all of the circumstellar matter before the reverse shock has overtaken all of the ejecta. A rarefaction wave will then propagate inward and rob the reverse shock of its driving pressure. If so, much of the newly synthesized matter in the inner ejecta may fail to emit X-rays.

3. Three Classes of Supernova Spectra

We can identify three major components that appear in the spectra of supernovae. The first, called the *photospheric spectrum*, usually dominates during the first few months after outburst (until maximum, and during the plateau phase of the light curve). The photospheric spectrum is characterized by a strong blackbody-like continuum with a characteristic temperature that levels off at ~ 5500 K after a week or two. The optical continuum is punctuated by strong P-Cygni lines, mainly hydrogen lines in SN II and metal lines in SN II.

By definition, the photospheric spectrum represents emission by atoms above a photosphere, below which we can discern no information about element abundances. Although it is possible to infer anecdotal information about relative abundances of elements from such spectra (for example, SN Ia spectra are rich in metal lines and lack hydrogen lines), it is difficult to make quantitative inferences, especially since many lines are saturated so that their strengths become independent of the respective atomic densities.

As the supernova envelope expands and thins out, the photosphere recedes to the center and the continuum fades. At this time, the supernova spectrum becomes a *nebular spectrum*, in which most of the luminosity appears as emission lines and the optical continuum is difficult to discern from the many overlapping weak emission lines. The nebular spectrum is characterized by lines from neutral and singly-ionized elements and linewidths with FWHM $\sim 3000-4000$ km s^{-1}. The spectrum of SN 1987A after four months (McCray 1993) is the best-observed example of a nebular spectrum. The nebular spectrum provides the best opportunity to infer nucleosynthesis yields, since it clearly comes from the glowing radioactive interior of the supernova envelope where most of the newly synthesized elements reside.

Many supernova spectra display a third component, which I shall call the *impact spectrum*. It is also characterized by strong emission lines, but it is very different from the nebular spectrum of SN 1987A. An impact spectrum typically displays emission lines of elements ionized twice or more. The line profiles are often asymmetric, and substantially broader (FWHM $\gtrsim 5000$ km s^{-1}) than those characteristic of a nebular spectrum. It is evidently caused, not by the radioactive glow of the inner supernova debris, but by the shocks that result from the impact of the debris with circumstellar matter that has been ejected by the progenitor. An impact spectrum is typically accompanied by X-ray emission.

In fact, the impact spectrum actually represents the early onset of the supernova remnant phase. It is of little use for inferring nucleosynthesis yields, since it comes only from the outermost part of the supernova ejecta. For purposes of understanding nucleosynthesis yields, the impact spectrum is nothing more than a nuisance, since it only confounds the analysis of the underlying nebular spectrum.

Evidence for impact with circumstellar matter is often seen in the optical spectrum and X-ray emission from supernovae almost immediately after explosion. This would be expected if the supernova progenitor was a red giant, as must be the case, for example, with SN 1993J (Wheeler & Filippenko 1995). In this respect, SN 1987A is exceptional because its progenitor was a blue giant star and has no dense circumstellar gas in its immediate vicinity. Evidently, the blue giant wind has pushed aside the gas from its prior red giant phase into a bipolar nebula, which we see as the remarkable triple ring system. The impact spectrum of SN 1987A will appear sometime between AD 1999 – 2005, when the supernova blast wave first reaches the inner ring (Luo, McCray & Slavin 1994; Chevalier & Dwarkadas 1995). In the meantime, the emission lines in the spectrum of SN 1987A will come mainly from the innermost part of the supernova ejecta, where most of the newly-synthesized elements reside.

4. The Failure of Standard Nebular Diagnostics

The terms "nebular spectrum" and "abundances" in the same context call to mind the vast body of work on the interpretation of the spectra of H II regions and planetary nebulae. One immediately turns to the classic text by Osterbrock (1989) for a detailed account of the underlying physics and techniques for interpreting such spectra. But one soon discovers that the techniques described there are of little use for inferring abundances from the nebular spectra of supernovae.

There are three major problems: (1) the supernova envelope is not chemically homogeneous; (2) by the time the nebular spectrum is apparent, temperatures in the supernova envelope are too cool for optical emission lines to provide useful estimates of abundances; and (3) supernova envelopes have such high densities that the emission line strengths become independent of atomic densities. I'll briefly describe each of these problems. For more details, see McCray (1993, 1995).

4.1. Inhomogeneity

The progenitor of a SN II is believed to have a laminar structure, with nested spherical shells composed of Fe, Si + S + Ca, O + Ne, O + C, C + He + O, He, H, respectively (e.g., Woosley *et al* 1995). Instabilities after the explosion certainly cause different element groups to become mixed, but the mixing is macroscopic, not microscopic – *stirred*, not *blended*. At the atomic level, each different element group retains its chemical integrity. Then, when the ejecta become transparent in the continuum, the temperature of each chemically distinct clump is determined by a balance between gamma ray heating and radiative cooling. The heating is more-or-less uniform (see §5) but the cooling, which takes place mainly through atomic and molecular line emission, depends very strongly on chemical composition. As a result, the temperatures of clumps of different composition can differ by a factor of two or more at any given time. For example, in SN 1987A, at 400 days after explosion, the temperature inferred from the luminosity of the [O I]$\lambda\lambda$6300,6364 doublet is $T_O \approx$ 3000 K (Li & McCray 1992); while that inferred from the CO bands is $T_{CO} \approx$ 1500 (Liu & Dalgarno 1995); from the Ca II lines, $T_{Ca} \approx$ 5000 (Li & McCray 1993); and from the Fe, Co, Ni lines, $T_{Fe} \approx$ 3000 (Li et al 1993). These marked differences in temperature are compelling evidence that these lines are emitted by chemically distinct regions.

It follows that any attempt to interpret the supernova emission line spectra using the assumption that different elements have the same temperature will be futile. We can make reliable inferences from the spectrum only if we are sure that all the emission lines in question come from a region of the same chemical composition.

4.2. Low Temperatures

The essence of the problems caused by low temperature can be illustrated with the following simple example. Suppose that we have a total mass, M_Z, of atoms of a given element having atomic mass, m_Z, and this atom has a low-lying excited state with energy E_u that is populated in LTE and has an optically thin forbidden transition, $h\nu_{ul}$, to some lower state, l. Then, the luminosity of the corresponding emission line is given by

$$L(\nu_{ul}) = \frac{M_Z}{m_Z} A_{il} h\nu_{ul} \frac{\exp(-E_u/kT)}{G(T)}, \qquad (1)$$

where A_{ul} is the radiative decay coefficient and $G(T)$ is the partition function.

The LTE assumption will be good provided that the electron density, n_e, exceeds some critical density, $n_{cr} \approx A_u/C_i$, where A_u is the net rate for spontaneous decays and C_u is the net rate coefficient for electron impact de-excitations. This is the case for all the strong forbidden lines seen in the nebular spectrum of SN 1987A. Then, the line luminosity is a function of only two unknowns, M_Z and T.

Now, suppose that the atom has two excited states, E_u and E_v, both of which have such forbidden transitions. Then, the luminosities $L(\nu_{ul})$ and $L(\nu_{vl})$ of the two emission lines would have different temperature dependencies and we could infer M_Z and T independently from the observed luminosities of the two lines.

Here's the problem of low temperatures. Take a typical optical forbidden line, say, [O I]λ6300. The excitation energy of the excited state is 1.968 eV, and so the exponential in equation (??) can be written $\exp(-22,840 K/T)$. But, for a typical temperature $T_O \sim 3,000$ K, a 10% increase in temperature will cause the same increase in $L(6300)$ as a doubling of the oxygen mass. We see that the luminosities of optical lines are far too sensitive to temperature to give a good measure of mass. Instead, they are very good thermometers: even with a factor of two uncertainty in assumed mass, the temperature can be inferred within 10%.

Since optical line luminosities are so sensitive to temperature, the luminosity emitted by a small mass of hotter material can mask that emitted by a much greater mass of cooler material. Therefore, if the temperature is not uniform, one could easily underestimate the emitting mass by a substantial factor.

The luminosities of infrared emission lines from fine structure levels are much less sensitive to temperature than optical lines. For example, [Co II]10.2 μm, which is very prominent in the spectrum of SN 1987A, has an excitation energy $E_i/k = 1410$ K, and so its luminosity is not very sensitive to temperature. (In fact, it is less sensitive than $\exp(-1410/T)$ because $G(T)$ also increases with temperature.) Therefore, one might hope for more success in interpreting supernova spectra in the infrared band than in the optical band, and that is certainly true in the case of SN 1987A, the first supernova for which excellent infrared spectra of the nebular phase have been available. But, as we discuss below, there is yet another problem that confounds the inference of nucleosynthesis yields from emission lines.

4.3. Line Opacity

As I remarked, equation (??) is valid provided that the upper state is populated in LTE and the line is optically thin. Moreover, the LTE assumption is good for most forbidden lines. But how good is the latter assumption? This assumption is generally valid for optical and infrared emission lines in the spectra of familiar emission nebulae, such as H II regions, planetary nebulae, supernova shells, etc. But, at the high atomic densities ($n_Z \sim 10^7 - 10^8$ cm^{-3}) of supernova envelopes, we *cannot* safely assume that forbidden lines are optically thin. The longer the wavelength, the greater the likelihood that emission lines will be optically thick. (Radio astronomers are familiar with the consequences of this fact. For example, the ratio of the emission line strength of 2.6 mm emission line of $^{13}C^{16}O$ to that of $^{12}C^{16}O$ from molecular clouds is always much greater than the actual abundance ratio of $^{13}C/^{12}C$ because the $^{12}C^{16}O$ is optically thick.)

Generally, it's tricky to describe line transfer in nebulae because the line opacities are very sensitive to line profiles and, hence, to the dynamics of emitting gas. But, in the case of supernova envelopes we have a break, because we know the gas dynamics exactly. It is the simplest possible case: free hypersonic expansion. Because of this expansion, the line scattering at any given wavelength is localized to a relatively small region, of thickness $\delta r \sim a_s t$, where a_s is the thermal molecular velocity and t is the time since explosion. Since such a region would have moved a distance $\Delta r \sim V_{exp} t$ since the explosion, the fractional thickness of the scattering region is $\delta r/\Delta r \sim a_s/V_{exp} \lesssim 10^{-3}$ for typical conditions in supernova envelopes.

In such an environment, the line transfer can be described accurately by the Sobolev approximation (for further details and references, see McCray 1993). The optical depth of the line is given by:

$$\tau_s = \frac{\lambda_0^3 t g_u A_{ul} n_l}{8\pi g_l} \left[1 - \frac{g_l n_u}{g_u n_l}\right], \qquad (2)$$

where the $n_{u,l}$'s and the $g_{u,l}$'s are the atomic densities and statistical weights of atoms in the upper and lower states, respectively, in the resonance scattering zone.

Note that $\tau_S \propto t^{-2}$ (because the free homologous expansion implies that $n_l \propto t^{-3}$). Therefore, an optically thick emission line will eventually become optically thin as the envelope expands and thins out. Note also that the $\tau_S \propto \lambda_0^3$ dependence implies that infrared lines tend to have greater optical depths than optical lines.

When the emission lines are optically thick, $\tau_{lu} \gg 1$, we find that equation (??) must be replaced by

$$L(\nu_{ul}) = \frac{4\pi\nu_{ul}V_Z}{ct}B_\nu(T),\qquad(3)$$

where V_Z is the net volume occupied by atoms of element Z and $B_\nu(T)$ is the Planck Function. Since $V_A \propto t^3$ for free expansion, we find that $L(\nu_{ul}) \propto t^2$ for constant T. The fact that the luminosities of several emission lines from SN 1987A actually do increase approximately as t^2 for $t \lesssim 200$ days (cf. McCray 1993) is clear evidence that the lines are optically thick and that the temperature is fairly constant.

Note the main difference between the optically thin limit and the optically thick limit. In the former case (eq. [??]), the luminosity is proportional to the *mass* of the emitting element; in the latter case case (eq. [??]), it is proportional to the *volume occupied* by that element, and independent of the mass. Therefore, until we can be sure that the emission lines are optically thin, we can't use their luminosities to infer masses even if we know the temperature. In the case of SN 1987A, many infrared lines remain optically thick for three years or more.

5. Methods That Should Work

Up to now, everything I have said is bad news for those of us who wish to infer abundance yields from supernova spectra. But, despite all that I have said, I am optimistic about the prospects. Why?

The empowering principle is this: *after about one year, the gamma rays illuminate all elements with approximately equal intensity.* Gamma rays have fairly long mean free paths; for example, the optical depth of the envelope of SN 1987A to the 1.24 MeV gamma rays from the decay of ^{56}Co is $\tau_C \sim 0.3(t/1 \text{ yr})^{-2}$ (McCray 1993). Moreover, the gamma ray energy deposition per unit mass is nearly independent of composition. That is true because the gamma rays deposit their energy by Compton scattering, and the electrons receive nearly the same recoil energy whether they are free or bound. Therefore, if we call f_Z the fraction of the total gamma ray luminosity that is deposited in the emitting region and absorbed by element Z, we find that

$$f_Z \approx \frac{M_Z}{M_H + M_{tot}}, \text{ and} \qquad(4)$$

$$f_H \approx \frac{2M_H}{M_H + M_{tot}},\qquad(5)$$

where M_Z, M_H and M_{tot} are, respectively, the masses of element Z, hydrogen, and all elements in the emitting region. (Note that the emitting region is that region where the gamma rays deposit most of their energy and does not include the outer hydrogen envelope.) The difference between equations (4) and (5) stems from the fact that hydrogen has one electron per nucleon, while all other elements have approximately 0.5 electrons per nucleon.

[Equation (4) is not quite true for the clumps composed of Fe, Co, and Ni because these elements have exceptionally large photoelectric opacities. For details, see Li *et al* (1993).]

5.1. Bolometric Method

The most obvious example of the bolometric method is the measurement of the mass of newly synthesized ^{56}Co from the supernova light curve during its exponential decay phase. For SN 1987A, this mass, $M(^{56}\text{Co}) = 0.069 \pm 0.003 M_\odot$ (Bouchet *et al* 1991) is by far the most accurately known nucleosynthesis datum. It is straightforward to infer the nucleosynthesis yields of ^{56}Co (hence ^{56}Fe) from the light curves of other supernovae.

But we can also use the bolometric method to infer the fractional abundances of other elements. The essence of the method is simple: *as long as the radiative cooling time is short compared to the age of the supernova, the luminosity radiated by clumps of a given composition must be equal to the luminosity deposited by the gamma rays.* Therefore, if we can identify the fraction of total UVOIR luminosity that is radiated in spectral features coming from these clumps, we may equate this fraction with f_Z and use equation (4) to solve for the mass fraction contained in the clumps.

This method should work. Since it depends on energy conservation, the answers will not be sensitive to details of atomic physics and line transfer. For example, if we know that the oxygen-rich clumps cooled primarily by radiation of [O I]$\lambda\lambda$6300,6364 and the CO 4.6 μm and 2.3 μm bands, we can estimate the mass fraction of these clumps by simply equating f_O in equation (4) to the fraction of the UVOIR luminosity radiated in these spectral features. One doesn't need to know the electron fraction, the excitation rate coefficients, or whether the features are optically thick or thin. It suffices to know that they are the main cooling agents.

But beware! It's not always obvious which spectral features should be attributed to which element composition. Without some understanding of the mechanisms for radiative cooling of different element compositions, one can easily blunder. Here's the trap: *the radiative transitions responsible for radiative cooling of a given clump are not necessarily those of the most abundant elements in the clumps.*

This point is illustrated by the strong Ca II emission lines in the spectrum SN 1987A. Li & McCray (1993) showed that these emission lines do not come from the roughly $(3-10) \times 10^{-3} M_\odot$ of newly-synthesized calcium that is expected in the envelope of SN 1987A. Instead, they must come from the $\sim 2 \times 10^{-4} M_\odot$ of primordial calcium that is contained in the hydrogen- and helium-rich clumps in the emitting region. A greater mass of newly-synthesized calcium can be present in the envelope, but there cannot be enough mass of calcium-rich material to account for the observed fractional luminosity seen in the Ca II emission lines, according to equation (4). The hydrogen- and helium-rich clumps emit a substantial fraction of their luminosity as Ca II lines because these lines provide one of the most efficient radiation channels. In these clumps, the hydrogen and helium act as buffers to thermalize the gamma ray luminosity and convert it to the observed Ca II lines.

If we understand how a gas of a given composition will radiate, we can use the bolometric method in some cases to obtain detailed information about the composition. This point is illustrated very nicely by the study by Liu and Dalgarno (1995) of the temperature of the oxygen-rich clumps in SN 1987A. Note that the temperature of these clumps inferred from the [O I]$\lambda\lambda$6300,6364 emission lines substantially exceeds the temperature inferred from the CO bands. Liu & Dalgarno have a lovely explanation for this fact. The oxygen-rich clumps are not uniform in composition (Woosley et al 1995). Some parts are composed of O + C but no He (Si may be present also), while other parts are composed of O + C + He. The latter parts are devoid of CO because He$^+$ produced there by gamma ray illumination will rapidly dissociate CO. When CO is present, its emission bands will dominate the radiative cooling; if not, the gas temperature must rise so that the [O I]$\lambda\lambda$6300,6364 lines are strong enough to provide the necessary cooling. It follows from equation (4) that the ratio of the luminosity of the CO emission bands to the [O I]$\lambda\lambda$6300,6364 lines is a measure of the ratio of the mass of oxygen-rich matter that is devoid of helium to the mass that contains traces of helium.

5.2. Nonthermal Excitation

Another promising method to infer abundances comes from analyzing nonthermal emission lines produced by gamma ray illumination. This method has the advantage over the bolometric method that the nonthermal emission line from a given element provides a direct measure of the abundance of that very element. But it is limited to those emission lines which we know are excited primarily by nonthermal rather than thermal electrons. The method has been used with some success to infer the abundances of hydrogen, helium, and oxygen in SN 1987A.

The basic tool needed for this method is a detailed calculation of the fractions of the energy of a fast primary electron that are deposited in thermal heating, ionization, and excitation of atomic states. That calculation has been done by Kozma & Fransson (1992) for gases having composition mixes characteristic of supernova envelopes.

The first example is the infrared recombination lines of hydrogen in SN 1987A. The luminosities of these lines provide a measure of the net ionization rate of all hydrogen atoms by gamma ray energy deposition, and the ratio of this rate to the total gamma ray energy deposition (the UVOIR bolometric luminosity) gives a measure of the fraction of the emitting gas that is hydrogen. Xu et al (1992) show that this number is about 30%, or about $M_H \approx 3 M_\odot$ assuming a core mass of $10 M_\odot$. (Remember that this mass is only the mass of hydrogen that is co-mingled with newly-synthesized elements in the emitting region and does not include the outer hydrogen envelope.)

In using this argument, one must take care to ensure that the hydrogen is actually being ionized primarily by nonthermal electrons. Xu et al show that this is not always the case in SN 1987A. For $t \lesssim 600$ days, the population of atomic hydrogen in the $n = 2$ state is sufficiently high that the Balmer continuum is optically thick, in which case photoionization of H*($n = 2$) by Balmer continuum photons can exceed ionization by fast electrons by several times. But after 600 days, the Balmer continuum optical depth drops rapidly and the method should work.

A second example is the He I 2.06 μm ($2\,^1P \to 2\,^1S$) emission line in SN 1987A. The $2\,^1P$ state of He I can be populated only by recombination of He$^+$ and direct excitation of singlet states of He I by fast electrons. In ordinary nebulae the He I 2.06 μm line is very weak because the $2\,^1P$ state is depopulated rapidly by the He I λ584 ($2\,^1P \to 1\,^1S$) resonance line. But, at the high densities of supernova envelopes the latter transition can be suppressed by resonant trapping in the supernova envelope, sufficiently that every excitation or cascade to the $2\,^1P$ will be followed by emission of a He I 2.06 μm photon instead of He I λ584. If so, the luminosity of the He I 2.06 μm line provides a measure of the mass of helium in the emitting region. Li & McCray (1995) estimate that $M_{He} \approx 3 M_\odot$ in SN 1987A.

There is an interesting nuance to this argument. The mass of helium measured this way does not include any helium that is mixed microscopically with primordial hydrogen. The reason is that H I atoms will photoabsorb the He I λ584, permitting the $2\,^1P$ to decay by this channel instead of by He I 2.06 μm. In fact, for $t \lesssim 450$ days, C I is sufficiently abundant to photoabsorb He I λ584 and suppress He I 2.06 μm. Therefore, to measure the helium mass by this method we must use observations of He I 2.06 μm for $t \gtrsim 450$ days.

The final example of nonthermal excitation is [O I]$\lambda\lambda$6300,6364 at late times. In SN 1987A, for $t \lesssim 750$ days, this doublet is evidently excited thermally and its light curve decreases rapidly as the temperature of the oxygen clumps decreases. But at ~ 750 days, the decay rate decreases suddenly and the subsequent luminosity of the doublet remains approximately a fixed fraction of the UVOIR bolometric luminosity (McCray 1993). Fransson et al (1995) recognized that this behavior implied that the oxygen clumps had become too cool to radiate [O I]$\lambda\lambda$6300,6364 by thermal excitation, and therefore that one could infer the mass of oxygen from the nonthermal excitation calculations by Kozma & Fransson (1992). They find that the net mass of oxygen atoms must be roughly $M_O \approx 1.5 M_\odot$. Note that M_O inferred by this means includes the entire mass of oxygen in the emitting region, unlike the mass that would be inferred from the [O I]$\lambda\lambda$6300,6364 lines by the bolometric method at earlier times.

There is, however, a problem with the nonthermal excitation method. We know that absorption of emission lines by dust became important in the envelope of SN 1987A for $t \gtrsim 450$ days (McCray 1993). Since the nonthermal method requires observations at times later than this, we must correct the observed emission line luminosities for dust absorption. Up to now, this has not been done carefully and as a result the masses inferred by the nonthermal method have fairly large uncertainties.

6. Summary

Thanks mainly to SN 1987A, we have come a long way toward understanding what the spectra of supernovae can tell us about nucleosynthesis; but we still have a long way to go. Even now, there is still much to be learned from further studies of SN 1987A, both with HST spectrometry and more thorough modeling of existing data.

Large new telescopes such as Keck and Gemini will permit us to obtain late-time nebular spectra of relatively nearby supernovae of sufficient quality to infer nucleosynthesis yields using the methods described in §5. Perhaps even more important are rapid advances in the technology to observe the infrared spectra of supernovae. SN 1987A has shown us how valuable this spectral range can be for inferring supernova abundances. SOFIA may play an important role in enabling us to observe the 4.6 μm fundamental band of CO, which we have seen is particularly important for determining abundances of the oxygen-bearing zones.

References

Borkowski, K. J., Sarazin, C. L., & Blondin, J. M. 1994, ApJ, 429, 710

Bouchet, P., et al 1991, A& A, 245, 490

Chevalier, R. A. 1982, ApJ, 258, 790

Chevalier, R. A., & Dwarkadas, V. V. 1995, ApJ, 452, L45

Fransson, C., Houck, J., & Kozma, C. 1995, in Supernovae and Supernova Remnants, R. McCray & Z-R. Wang, Cambridge: Cambridge U. Press, 211

Kozma, C., & Fransson, C. 1992, ApJ, 390, 602

Li, H.-W., & McCray, R. 1992, ApJ, 387, 309

Li, H.-W., & McCray, R. 1993, ApJ, 405, 730

Li, H.-W., McCray, R., & Sunyeav, R. A. 1993, ApJ, 419, 824

Liu, W. & Dalgarno, A. 1995, ApJ, 454, 472

Luo, D., McCray, R., & Slavin, J. 1994, ApJ, 430, 264

McCray, R. 1993, ARA&A, 31, 175

McCray, R. 1995, in Supernovae and Supernova Remnants, R. McCray & Z-R. Wang, Cambridge: Cambridge U. Press, 223

McCray, R., & Kafatos, M. 1982, ApJ, 317, 190

Osterbrock, D. E. 1989, *Astrophysics of Gaseous Nebulae and Active Galactic Nuclei* (Mill Valley, CA: University Science Books)

Wheeler, J. C. & Filippenko, A. V. 1995, in Supernovae and Supernova Remnants, R. McCray & Z-R. Wang, Cambridge: Cambridge U. Press, 241

Woosley, S. E., Weaver, T. A., & Eastman, R. G. 1995, in Supernovae and Supernova Remnants, R. McCray & Z- R. Wang, Cambridge: Cambridge U. Press, 137

Xu, Y., McCray, R., Oliva, E., & Randich, S. 1992, ApJ, 386, 181

X-Ray Observations of Supernova Remnants (and what they tell us about nucleosynthesis)

R. Petre

Laboratory for High Energy Astrophysics, NASA/Goddard Space Flight Center, Greenbelt, MD 20771 USA

Abstract. The current generation of X-ray spectrometers, most notably those on the *ASCA* Observatory, have facilitated the clearest view of the X-ray spectra of supernova remnants. These spectra are dominated by emission lines from highly ionized nucleosynthesis products. While absolute abundance determinations remain difficult to perform, it has become possible to make reliable relative abundance measurements for a number of remnants both in the Galaxy and in the Magellanic Clouds. For young supernova remnants it is possible to use these these measurements infer the type of explosion that produced them. We review *ASCA* and other observations of supernova remnants, with an emphasis on those observations which have led to the determination of a progenitor type.

1. X-Ray Emission from Supernova Remnants

A supernova explosion imparts $\sim 10^{51}$ ergs of kinetic energy to its ejecta, sending a substantial fraction of a stellar mass hurtling into the surrounding interstellar medium at velocities in excess of 1,000 km s^{-1}. As the ejecta encounter the surrounding medium, strong shock waves are propagated into both the medium and the ejecta, heating them to a temperature of $\sim 10^7$ K and ionizing them. The radiation from gas at this temperature is emitted primarily in the X-ray band. Thus X-ray observations provide the most direct probe of the products of stellar nucleosynthesis that have been returned to the interstellar medium (ISM) via supernovae.

The X-ray emission from shock-heated gas has two components: lines and continuum. The lines are produced by atomic transitions of the metals in the gas. The X-ray band between 0.5 and 10 keV contains the strongest transitions from the He-like and H-like ions of the most abundant nucleosynthesis products from oxygen to iron. The continuum emission arises from electron-ion bremsstrahlung, and provides a direct measure of the electron temperature. The relative line strengths from one element provide information regarding temperature and ionization state. Comparison of line strengths from different elements allow inferences about relative abundances, while the strength of the line emission relative to the continuum gives information about absolute abundances.

1.1. Complexities

Despite the attractiveness of using X-ray spectroscopy to study the composition of supernova remnants, a number of complications have severely limited the amount of detailed study that has actually been performed. The main complications arise from nature. A supernova remnant is in a dynamic state, evolving in a multicomponent ISM. The results of observations must be compared with the spectra predicted from hydrodynamic modeling of supernova remnants that takes into account time-dependent ionization (Hamilton, Sarazin & Chevalier 1983; Hughes & Helfand 1985). These non-equilibrium ionization (NEI) models are characterized by two parameters: the shock temperature T and an ionization parameter nt, the product of the ambient interstellar density n and the time t since the material was shocked. What is worse, early results from *ASCA* (see below) indicate the need for multiple NEI components in each remnant. Additionally, it is necessary to take into account the energy-dependent absorption by the ISM. The column density to some Galactic SNR is high enough to render them invisible below 1 keV (making measurements of the O and Ne lines impossible). The column density can also vary significantly across the surface of a remnant (e.g., Rho *et al.* 1994). Finally, the presence of contaminating sources of emission can compromise measurments. the and contribution by a non-thermal continuum component (as in SN1006).

A second source of complication is the presence of inaccuracy in the values of the strengths of some of the lines in the spectral models as a consequence of uncertainty in the underlying atomic physics. This is especially the case for the complex Fe L band, arising from Be-like to Ne-like iron, whose lines form a virtual forest between 0.8 and 1.3 keV.

Finally, a prime complication has been our inability to make unambiguous spectral measurements. Prior to the launch of *ASCA*, only a few remnants were sufficiently bright to facilitate detailed study with sufficient spectral resolution. Most previous spectrometers were unable to resolve line blends, making difficult the resolution of weak lines from the continuum or nearby strong lines.

1.2. Current Capabilities

In young supernova remnants, the X-ray emission is dominated by the reverse-shocked ejecta. For these remnants especially we hope to be able to infer abundances, and by comparing these with the predictions of nucleosynthesis models learn something about the progenitor star. (If, on the other hand, we knew the progenitor mass, measuring the abundances via the X-ray emission would in principle allow us to test the reliability of the models.) The ejecta from remnants older than $\sim$2,000 y are usually diluted by the ISM. In that case, it is possible to use the X-ray emission as a probe of the aboundances in the ISM. There are older remnants, however, such as Puppis A, in which either the enrichment of the ISM was pronounced, or discrete knots of ejecta can still be identified. In these remnants it is still possible to infer the progenitor mass (Canizares & Winkler 1981).

The limitations mentioned above should make apparent the difficulty associated with making definitive absolute metal abundance measurements of supernova remnants using X-ray spectroscopy. For many remnants, however, it is possible to minimally determine the nature of the progenitor; i.e., whether the

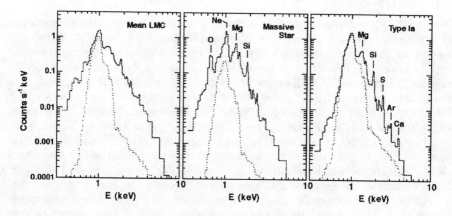

Figure 1. Simulated *ASCA* spectra for plasma with $kT=1.5\times10^7$ K and $nt=10^{11}$ cm^{-3} s, and three assumed sets of abundances. Panels from left to right show mean abundances in the LMC (0.3 solar), those from a core collapse supernova of a 25 M$_\odot$ star, and those for a Type Ia supernova. The dotted curves in each figure represent the emission from Fe. The spectra of the two types of supernovae are qualititatively different. From Hughes *et al.* (1995).

supernova explosion was caused by core collapse of a massive star or deflagration/detonation of a white dwarf. It is possible to do this because the two types of explosions produce very different nucleosynthesis yields, which are reflected in their X-ray spectra. Type Ia supernovae produce relatively large amounts of Mg, Si, S and Fe, and virtually no O and Ne (e.g., Nomoto *et al.* 1986). Core collapse explosions, on the other hand, produce an overabundance of O and Ne compared with Si, S and Fe (e.g. Thielemann *et al.* 1994). The qualitatively different resulting X-ray spectra are shown in Fig. 1.

2. SNR in the Magellanic Clouds

The Magellanic Clouds are among the most convenient locations for studying the composition of supernova remnants, for a number of reasons. The Clouds contain quite a number of high luminosity remnants; they are sufficiently nearby that SNR have fluxes suitable for observation by existing spectrometers; the low column density to the Clouds means that, unlike most Galactic remnants, the spectra of Cloud remnants are not subject to the severe absorption of low energy X-rays; and the metal abundances in the Cloud ISM are considerably less than solar, making differences due to enrichment by supernovae easier to detect in principle.

The most definitive work to date for LMC remnants (or any other remnants) using the X-ray spectral signature to determine the explosion type is that of Hughes et al. (1995). Their comparison of two remnants whose optical emission

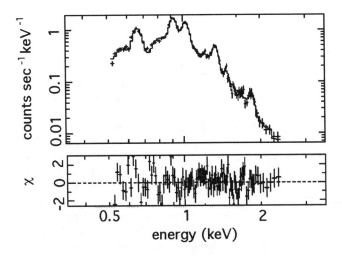

Figure 2. *ASCA* spectrum of E0102-72. The spectrum is dominated by strong O and Ne lines from ejecta, a clear indication that the progenitor was a massive star. From Hayashi *et al.* (1994).

is dominated by Hα from nonradiative shocks, and one with a peculiar X-ray morphology, shows that all are products of type Ia explosions.

In Figure 2 we show the X-ray spectrum of the brightest remnant in the SMC, E0102-72 (Hayashi*et al.* 1994). In contrast to the LMC remnants in Hughes *et al.*, its strong oxygen and neon lines signal an origin from a core collapse event. This remnant offers a perfect case study of why it so difficult to make any inferences beyond this simple one (e.g., estimate the progenitor mass based on the abundances in the X-ray emitting gas). Its nearly symmetrical, limb-brightened X-ray morphology suggests a symmetric explosion into a relatively uniform interstellar medium (Hughes 1994). Its small angular diameter (40 arc seconds) makes possible a comparison between its integrated spectrum and those produced by NEI models (Hamilton, Sarazin & Chevalier, 1983; Hughes & Helfand 1985). However, we find that even in this apparently symmetrical object no single component NEI model provides a satisfactory representation of the spectrum. In order to achieve an acceptable fit, at least three different (T, nt) components are required, one for each prominent set of lines. This unanticipated spectral complexity appears consistently in the supernova remnants observed by *ASCA*, and constrains our ability to make more detailed comparisons to the predictions of nucleosynthesis models. In E0102-72 it indicates that each element is encountering different shock conditions, and suggests that the ejecta are stratified.

3. Galactic SNR

3.1. Tycho

Tycho (SN1572) is the archetype of Type Ia events. Its X-ray spectrum is one of the most closely studied of that of any SNR, and probably the best understood and the most consistent with conventional wisdom. The integrated X-ray spectrum can be understood in the context of a model with a forward shock front propagating into a uniform ISM, and a reverse shock propagating into uniform density ejecta consisting of pure heavy elements, stratified into layers (Hamilton *et al.* 1986). The total ejecta mass of 1.4 $M_\odot$, about $1 M_\odot$ of which is either Si group or Fe group metals, is clearly consistent with that expected from a Type Ia progenitor. Subsequent observations have confirmed the general predictions of this model. A BBXRT observation showed that the ionization conditions experienced by the Fe are different from those of the other metals, suggesting that the Fe filling the interior has lower density than the outer ejecta layers (Petre *et al.* 1993). A comparison between the *Einstein* and *ROSAT* HRI images (Vancura *et al.* 1995) and narrow band imaging using *ASCA* data (Hwang & Gotthelf 1995) suggests the presence of minor temperature and abundance variations within the remnant. While the presence of the abundance variations indicate that at most incomplete mixing has taken place between the ejecta layers, modeling of the radial profiles of the *ASCA* narrow band images requires that some mixing must have occurred.

3.2. SN1006

One of the most exciting *ASCA* results to date involves SN1006. Koyama *et al.* (1995) demonstrate that the dominant, featureless spectral component that has long posed a mystery, is isolated to two segments of the limb, and arise as synchrotron radiation from electrons accelerated within the shock to energies as high as a few hundred TeV. As the physical processes responsible for accelerating relativistic particles to these high energies do not differentiate between electrons and ions, they infer that the non-thermal X-ray spectrum in the rims of SN1006 is the first strong evidence for the acceleration of cosmic rays by supernova shocks. The *ASCA* data showed that everywhere else in the remnant, the X-ray emission has a thermal spectrum, with strong line emission typical of other supernova remnants. The strengths of the lines from the intermediate α-burning elements, magnesium, silicon, and sulfur, are significantly enhanced, requiring abundances well in excess of solar (Fig 3). This is the first X-ray evidence that SN1006 was a Type Ia supernova, as inferred from its historical light curve (Stephenson *et al.* 1977).

3.3. Kepler

The nature of the progenitor of Kepler's supernova remnant (SN1604) has been a source of controversy. While the historical light curve and its location well out of the Galactic plane argue for a Type Ia supernova, the large amount of circumstellar matter required to produce the high X-ray luminosity (White & Long 1983) suggests the progenitor was a massive star that underwent appreciable pre-supernova mass loss (Borkowski *et al.* 1994). In Fig. 4 we show the inferred abundances for silicon through nickel from the recent *ASCA* observa-

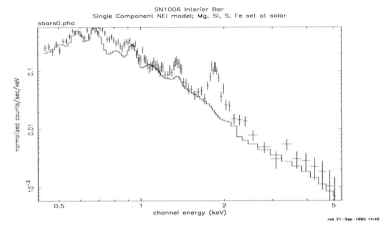

Figure 3. *ASCA* spectrum of the interior of SN1006. The model (solid curve) represents a plasma in which the abundances of the intermediate α burning elements (Mg, Si, & S) have been fixed equal to that of oxygen (which in this case has solar abundance), as would be expected if the emission arose exclusively from shocked interstellar material. The strong excess of these elements suggests a Type Ia origin.

tion, Based on the inferred abundances, the current X-ray data favor a Type Ia progenitor (Tsunemi et al. 1995).

3.4. Cas A

In contrast to the three Galactic remnants discussed above, Cas A is thought to have originated from a massive progenitor on the basis of heavy element enrichment observed in the optical emission lines and the substantial X-ray emitting mass (Raymond 1984; Fabian et al. 1980). While no definitive X-ray abundance measurements incorporating the full power of the NEI models have been made, Borkowski et al. (1995) have shown that the ionization conditions responsible for the Fe K emission are different from those encountered by lighter elements. Spatially-resolved spectral mapping using *ASCA* has revealed a strong differential Doppler shift from northwest to southeast of maximum amplitude $\sim$2,000 km s^{-1} (Holt et al. 1994). This is direct confirmation of the earlier inference based on *Einstein* FPCS data (Markert et al. 1983) that the bulk of the X-ray emission is confined to an expanding, inclined ring. Such a geometry could arise from the explosion of a rapidly-rotating, massive star. Narrow band maps centered on strong line features reveal only slight structural differences, suggesting mixing of ejecta. In contrast, a map of the high energy continuum emission appears more similar to the radio continuum map than the X-ray line maps. This could be interpreted as due to the dominance of forward-shocked gas over reverse-shocked material in the radio-bright regions. Alternatively, the correlation with the radio emission might indicate the presence of a non-thermal X-ray emitting component, similar to that seen in SN1006 (and therefore due to synchrotron radiation from shock-accelerated electrons).

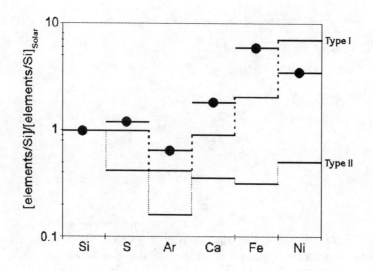

Figure 4. Inferred metal abundances for Kepler from analysis of *ASCA* spectrum, compared with abundances expected for Type Ia and core collapse supernovae. Abundances are more consistent with a Type Ia origin. From Tsunemi *et al.* (1995).

3.5. W49B

W49B has one of the more spectacular X-ray spectra among supernova remnants, with intense lines from silicon through iron. Narrow band images constructed using *ASCA* data show that while the Si and S line emission arise in a shell, the Fe emission is centrally concentrated (Fujimoto *et al.* 1995). This is clear evidence for metal stratification, similar to that predicted by the Hamilton *et al.* (1986) model of Tycho, and thus indirect evidence for a Type Ia progenitor. The X-ray emission suffers significant line-of-sight absorption below ∼1.3 keV due to high column density, however, rendering us unable to observe the strength of the O and Ne lines, and the corresponding high optical extinction makes impossible a search for optical line emission from O-rich ejecta. Thus the nature of the W49B progenitor remains undetermined.

3.6. G292.1+1.8

G292.1+1.8 is a young (<1,000 y) SNR with an unusual X-ray morphology; it appears as an ellipsoidal disk of nearly uniform surface brightness, bisected by a bar of high surface brightness. Optical spectra of the bar show it to be oxygen-rich and expanding at high velocity (Goss *et al.* 1979), which has led to the suggestion that it is a ring of ejecta, similar to that observed in the X-ray in Cas A, but viewed at low inclination (Clark & Tuohy 1982). Using the spectra obtained by the *Einstein* SSS and IPC and the *EXOSAT* ME, Hughes & Singh (1994) showed that the inferred abundances are most consistent with those expected from a 25 $M_\odot$ progenitor. Unlike Cas A, there is no evidence for a difference in the ionization conditions of the various metals.

3.7. Puppis A

Puppis A is the most prominent example of a "middle-aged" ($t \sim 4{,}000$ y) remnant for which information about the progenitor can be inferred. Despite its age, the strengths of the oxygen and neon lines relative to those of iron from *Einstein* FPCS measurements require a progenitor mass of at least 25 $M_\odot$ (Canizares & Winkler 1981). Confirmation that the Puppis A progenitor was a massive star comes in the form of the discovery of a central stellar remnant, most likely a neutron star (Petre *et al.* 1995).

4. X-Ray Observations of Supernovae

The study of the X-ray emission from supernovae themselves is a recent development, due in part to the availability of sensitive X-ray instrumentation and in part to good fortune (e.g., the proximity of SN1987a in the LMC and SN1993j in M81). While these observations have revealed a number of surprises, one thing they have not revealed is the composition of the ejecta. Some models of the early evolution of core-collapse supernovae (e.g., Chevalier & Fransson 1994; Chugai 1993) predict strong X-ray line emission from reverse-shocked ejecta or a shocked, clumpy wind. While this may be the mechanism responsible for the production of X-ray luminosities on the order of 10^{40} erg s^{-1} in the peculiar supernovae SN1978k (Ryder *et al.* 1993) and SN1986j (Bregman & Pildis 1993), the low X-ray fluxes of these distant objects makes virtually impossible the accumulation of a spectrum of high enough quality to infer abundances. Iron K lines were detected during the early stages SN1987a and SN1993j. The *ASCA* spectrum of SN1986j, the highest quality spectrum of a supernova obtained, shows clear evidence for a strong, velocity-broadened Fe K line, but the inferred metal abundances are consistent with the solar values (Houck & Bregman 1995). In SN1987a the Fe line was produced not by thermal emission, but by In SN 1993j, the Fe emission might arise from reverse-shocked ejecta, but no inference regarding Fe abundance has yet been made (Kohmura *et al.* 1994).

5. Summary

The high spectral sensitivity of *ASCA*, coupled with its capability for spatially-resolved spectroscopy, has provided a major advance toward our goal of making reliable metal abundance measurements in SNR. It is now possible in many cases to unambiguously "type" the progenitors of SNR, based on their X-ray spectra. For some remnants of core-collapse supernovae (e.g., Puppis A and G292.1+1.8), it has become possible to constrain the progenitor mass based on the X-ray spectra. The next generation of X-ray obseratories (AXAF, XMM, Astro-E) will offer far more powerful spectral capabilities, that will facilitate more reliable abundance measurements.

Acknowledgments. We gratefully acknowledge the contributions of U. Hwang and E.V. Gotthelf to the material presented here, and to the many members of the *ASCA* team. We especially thank J. Hughes for helpful discussions and for supplying Fig. 1, H. Tsunemi for supplying the material presented in Fig. 4, and J. Houck and J. Bregman for sharing their results on SN 1986j in advance of publication. We also thank P. Tyler for her assistance in preparing some of the figures.

References

Becker, R.H., Szymkowiak, A.E., Boldt, E.A., Holt, S.S., & Serlemitsos, P.J., 1980, ApJ240, L33
Borkowski, K.J., Sarazin, C.L. & Blondin, J.M. 1994, ApJ, 429, 710
Borkowski, K.J., Szymkowiak, A.E., Blondin, J.M. & Sarazin, C.L.. 1995, ApJ, submitted
Bregman, J.N., & Pildis, R. 1992, ApJ, 398, L107
Chevalier, R.A., & Fransson, C. 1994, ApJ, 420,268
Chugai, N.N. 1993, ApJ, 414, L101
Clark, D.H., & Touhy, I.R. 1982, in "Supernova Remnants and Their X-Ray Emission," eds. J. Danziger & P. Gorenstein (Dordrecht: Reidel) p. 153
Fabian, A.C., Willingale, R., Pye, J.P., Murray, S.S., & Fabbiano, G. 1980, MNRAS, 193, 175
Fujimoto, R., et al., 1995, PASJ, 47, L31
Hughes, J.P., & Helfand, D.J., 1985, ApJ291, 544
Becker, W., 1995, these proceedings
Goss, W.M., Skellern, D.J., Watkinson, A., & Shaver, P.A. 1979, A&A, 151, 52
Gotthelf, E.V., et al. 1995, ApJ, submitted
Hamilton, A.J.S., Sarazin, C.L., & Chevalier, R.A., 1983, ApJS51, 115
Hamilton, A.J.S., Sarazin, C.L., & Szymkowiak, A.E., 1986, ApJ300, 713
Hayashi, I., Koyama, K., Ozaki, M., Miyata, E., Tsunemi, H., Hughes, J.P., & Petre, R., 1994, PASJ46, L121
Holt, S.S., Gotthelf, E.V., Tsunemi, H., & Negoro, H., 1994, PASJ46, L151
Hughes, J.P., & Helfand, D.J., 1985, ApJ291, 544
Hughes, J.P., & Singh, K.P., 1994, ApJ422, 126
Hughes, J.P., 1994, in "The Soft X-Ray Cosmos", eds. E.M. Schlegel & R. Petre, (New York: AIP), p. 144
Hughes, J.P., et al., 1995, ApJ444, L81
Hwang, U., & Gotthelf, E.V. 1995, ApJ, submitted
Kohmura Y., *et al.* 1994, PASJ, 46, L157
Koyama, K., Petre, R., Gotthelf, E.V., Hwang, U., Matsuura, M., Ozaki, M., & Holt, S.S., 1995, Nature, 378, 255
Markert, T.H., Canizares, C.R., Clark, G.W., & Winkler, P.F. 1983 ApJ, 268, 134
Nomoto, K., Thielemann, F.-K., & Yokoi, K. 1884, ApJ, 286, 644

Petre R. et al. 1993, in UV and X-Ray Spectroscopy of Laboratory and Astrophysical Plasmas, eds. E. Silver & S. Kahn (Cambridge: Cambridge University Press) p. 424.

Petre, R., Becker, C.M., & Winkler, P.F. 1995 ApJ, submitted

Raymond, J.C. 1984 ARA&A, 22, 75

Ryder, S.L., et al. 1993, ApJ, 416, 167

Stephenson, F.G., Clark, D., & Crawford, D. 1977, MNRAS, 180, 567

Thielemann, F.-K., Nomoto, K., & Hashimoto, M. 1994, in Supernovae, Les Houches, Session LIV, eds. S. Bludman, R. Mochkovitch, & J. Zinn-Justin (North Holland: Elsevier) p. 629

Tsunemi, H., et al. 1995, in preparation

Vancura, O., Gorenstein, P., & Hughes, J.P. 1995, ApJ, 441, 680

White, R.L., & Long, K.S. 1983, ApJ, 264, 196

The Cassiopeia A Supernova Remnant: Dynamics and Chemical Abundances

K. J. Borkowski, J. M. Blondin

Department of Physics, North Carolina State University, Raleigh, NC 27695

A. E. Szymkowiak

NASA/Goddard Space Flight Center, Code 666, MD 20771

C. L. Sarazin

Department of Astronomy, University of Virginia, P.O. Box 3818, Charlottesville, VA 22903

Abstract. We model the Cas A supernova remnant in the framework of the circumstellar medium (CSM) interaction picture. The comparison of our X-ray emission calculations with the ASCA spectrum suggests that the X-ray continuum and the Fe Kα line are dominated by the shocked CSM, but prominent Kα complexes of Mg, Si and S must be produced by supernova ejecta with strongly enhanced abundances of these elements. An explosion of a stellar He core is consistent with these findings.

1. The Circumstellar Shell

Cassiopeia A is the youngest known supernova remnant (SNR) in our Galaxy. Its progenitor is believed to be a massive star which lost most of its H-rich envelope shortly ($\sim 10^5$ yr) before the explosion. Optical spectroscopy has revealed two classes of emission knots: slowly-moving (less than few hundred km s^{-1}) "quasi-stationary flocculi" (QSF), and fast (few thousand km s^{-1}) knots of supernova (SN) ejecta. The QSFs were found to be nitrogen- and helium-rich (Kirshner & Chevalier 1977; Chevalier & Kirshner 1978), which secured their interpretation in terms of shock-excited circumstellar matter (CSM) ejected by the SN progenitor prior to its explosion (Peimbert 1971; Peimbert & van den Bergh 1971).

A substantial presupernova mass loss is supported by an apparent low SN luminosity (Ashworth 1980), which is the expected outcome if a progenitor star lost most of its outer H-rich envelope and became a blue supergiant before the explosion (Chevalier 1976). The high He and N overabundance observed in the QSFs suggests that most of the H-rich envelope was expelled before the explosion, leaving behind a compact blue supergiant, possibly a Wolf-Rayet star of WN8 subtype (Fesen, Becker, & Blair 1987; Fesen, Becker, & Goodrich 1988).

Chevalier & Liang (1989) pointed out the consequences of a high presupernova mass loss on the observable characteristics of SNRs. For such remnants, exemplified by Cas A, the fast wind of the blue supergiant progenitor sweeps the mass lost during the preceding red supergiant stage into a dense CSM shell. Upon the passage of the SN blast wave through the shell and during subsequent interaction stages this shell will be accelerated outward. The amount of acceleration depends on the shell density and its mass – a relatively dense and massive shell will be accelerated slowly, leading to a slowly expanding SNR shell. Chevalier & Liang (1989) noted that the slow expansion age measured for Cas A radio-emitting material ($\sim$ 900 yr – Anderson & Rudnick 1995) is consistent with the presence of the dense CSM shell. The interaction of SN ejecta with this shell creates one of the most luminous SNRs in our Galaxy.

We have undertaken a comprehensive effort to model the Cas A SNR in the framework of the CSM interaction picture. Here, we present results from our one-dimensional calculations performed with the Virginia Hydrodynamics PPM (VH-1) code. We use recent X-ray observations acquired by the Advanced Satellite for Cosmology and Astrophysics (ASCA) to constrain our models.

2. Interaction of SN Ejecta with the CSM shell

We assume the following CSM density distribution: (1) a low-density wind-blown bubble with radius r_c and density n_b, (2) a shell of swept red supergiant wind material with density n_s, extending from r_c to r_s, followed by (3) a wind with velocity $v_w = 15$ km s^{-1} and the mass-loss rate $\dot{M}$. We found good agreement with observations for a model with $n_b = 0.02$ cm^{-3}, $n_s = 15$ cm^{-3}, $r_c = 1.25$ pc, $r_s = 1.6$ pc, the shell mass $M_s = 4.6 M_\odot$, and $\dot{M} = 7.8 \times 10^{-5} M_\odot$ yr^{-1}.

We use a simple model for the SN ejecta density distribution valid for a massive SN (Chevalier & Liang 1989). SN ejecta expand homologously, with velocity $v = r/t$. They consist of the two distinct components: the outer component with $\rho \propto r^{-6}$, and the constant density inner component. The ejecta mass M_{ej} must not exceed several solar masses, because most of the H-rich envelope was ejected before the explosion. In the following discussion, we take $M_{ej} = 3 M_\odot$ and the standard value of 10^{51} ergs for the ejecta kinetic energy.

The blast wave hits the shell at 26 yr with velocity 23,500 km s^{-1}. It then splits into the transmitted shock and the reflected shock upon encountering the shell boundary. The transmitted shock velocity is equal to 2000 km s^{-1}, which means that the shell is impulsively accelerated to 1500 km s^{-1} at this time. The 2000 km s^{-1} transmitted shock completes its transit of the shell 210 yr after the explosion, and at this time it starts to propagate into the undisturbed red supergiant wind. The model has two distinct shells: the hotter, more tenuous ejecta (reverse shock) and the denser and cooler CSM shell, surrounded on the outside by the less dense shocked wind gas. At present, the mass of X-ray emitting gas is equal to $7.9 M_\odot$ ($4.77 M_\odot$ in the shell, $1.72 M_\odot$ in the shocked ejecta and $1.42 M_\odot$ in the shocked wind). The current shell radius is equal to 1.78 pc, and its velocity is 2200 km s^{-1}, in agreement with observations.

3. X-Ray Spectrum and Chemical Abundances

The recent observations of Cas A with the ASCA satellite have provided a high quality X-ray spectrum (Holt et al. 1994; Figure 1). The spectrum is dominated

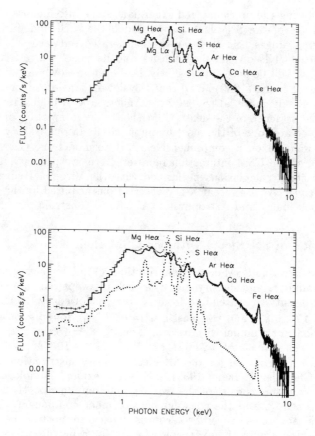

Figure 1. The spatially-integrated spectrum of Cas A obtained with the ASCA Solid-state Imaging Spectrometers. The calculated X-ray spectrum is shown by the *solid curve* in the top panel, while the bottom panel shows the contributions from the CSM (*solid curve*) and from the SN ejecta (*dotted curve*) separately.

by Heα complexes of cosmically abundant elements, with particularly strong Si and S lines, but fainter Lyα lines of Mg, Si, and S are also clearly visible.

We calculated X-ray spectra for our hydrodynamical models as described in our earlier work on Kepler's SNR (Borkowski, Sarazin, & Blondin 1994). The CSM is chemically homogeneous, but for the SN ejecta we use a two-zone, chemically stratified model, with the inner zone enriched in heavy elements. The results are presented in Figure 1, for our hydrodynamical model discussed above. For these spectra, there is no collisionless electron heating, with Coulomb collisions between ions and electrons being the sole heating source for electrons.

X-ray emission from the shocked CSM alone accounts for the observed continuum emission, as well as for the Fe Kα line. We obtain the following abundances (with respect to the solar values) in the CSM: 0.40 for Fe and Ni, and 0.84 − 0.89 for other elements heavier than Ne. These abundances are consistent with the expected approximately solar abundances of heavy elements in the CSM. The ionization state of the CSM shell, as implied by the position of the Fe Kα line, is high enough to produce both Lα and Heα lines of Mg, Si and S with approximately the same strength (Fig. 1, solid curve in the bottom panel). But these high Lα/Heα line ratios do not fit the ASCA spectrum, which is dominated by prominent Heα lines of Mg, Si, and S. The very strong Heα lines must originate in a different region within the remnant – in our models they are produced in the recently shocked ejecta. (This might also be the case for Ar but Ar Heα line was not included in our calculations – this line was simply fitted by a Gaussian.) As can be seen in Figure 1 (dotted curve in the bottom panel), the Lα/Heα line ratios in the ejecta are much lower than in the CSM, indicating a low ionization state of the ejecta. A low ionization state is indeed expected for the recently shocked ejecta with densities lower than the CSM shell density.

The determination of the chemical abundances in the shocked ejecta of Cas A is highly model dependent. At this time we can merely state that Si and S, and to a lesser degree Mg are strongly enhanced in the SN ejecta. For the model shown in Figure 1, we need about $0.1 M_\odot$ each for both Si and S, and an order of magnitude less for Mg. The relative abundances of these elements are somewhat more reliable. The S/Si ratio is 1.4 solar, while the Mg/Si ratio is only 0.10 solar. There is little (if any) Fe in the shocked ejecta, and the same is true of Ne. The O abundance is uncertain because of the high interstellar absorption toward Cas A, but the O/Si ratio might also be subsolar. Hydrodynamical simulations with more realistic, chemically stratified SN models are necessary for improving upon these preliminary numerical results.

References

Anderson, M. C, & Rudnick, L. 1995, ApJ, 441, 307

Ashworth, W. B. 1980, J. Hist. Astr., 11, 1

Borkowski, K. J., Sarazin, C. L., & Blondin, J. M. 1994, ApJ, 429, 710

Chevalier, R. A. 1976, ApJ, 208, 826

Chevalier, R. A., & Kirshner, R. P. 1977, ApJ, 218, 142

Chevalier, R. A., & Liang, E. P. 1989, ApJ, 344, 332

Fesen, R. A., Becker, R. H., & Blair, W. P. 1987, ApJ, 313, 378

Fesen, R. A., Becker, R. H., & Goodrich, R. 1988, ApJ, 329, L89

Holt, S. S., Gotthelf, E. V., Tsunemi, H., & Negoro, H. 1994, PASJ, 46, L151

Kirshner, R. P., & Chevalier, R. A. 1977, ApJ, 218, 142

Peimbert, M. 1971, ApJ, 170, 261

Peimbert, M., & van den Bergh, S. 1971, ApJ, 167, 223

Chemical Evolution of Galaxies

F. X. Timmes

Laboratory for Astrophysics and Space Research, Enrico Fermi Institute, University of Chicago, 5640 South Ellis, Chicago, IL 60637 and Department of Physics and Astronomy, Clemson University, Clemson, SC 29634

Abstract. Stars with M $\gtrsim$ 10 M$\odot$ and their associated supernovae are the dominant contributors to most of the heavy element abundances in our Galaxy and other galaxies. Refined numerical modeling, both of the presupernova phases of stellar evolution and of the explosions in the ensuing supernova events, are now providing increasingly realistic predictions of the nucleosynthetic yields. Coupled with an increase in our knowledge of the general physical properties of the Milky Way, in addition to more refined dynamical and chemical evolution scenarios, several long standing questions are receiving fresh attention. For example, comparison of the calculated isotopic solar composition to the observed values from these modern stellar-chemical evolution models is discussed and constitutes one of the main points of this paper. Direct confirmation of massive star nucleosynthesis are (and can be) provided by observations of SN 1987A in the Large Magellanic Cloud, SN 1993J in M81, and the most metal deficient stars in the Galactic halo. The short lifetimes of massive stars, $\tau < 10^8$ years, allows their ejecta to enrich the Galactic halo gas at the very earliest epochs, permitting inferences to be drawn concerning the timescale and star formation history of our Galactic halo. Abundance trends identified in the early enrichment history of the Milky Way may hold implications for trends in the chemical evolution of the QSO absorption line systems as a function of redshift. Simple transformations allow the abundance trends observed in our Galaxy to be converted into redshift space, where they may be compared with metal enrichment found in the damped Lyman-α systems. Discussion of these transformations and trends are the second main point of this paper.

1. Discussion

Massive stars and core collapse supernovae are thought to be significant sources of nuclei ranging from ^{7}Li through uranium. They are the primary source of the elements from oxygen to calcium, and produce approximately one third to one half of the iron peak nuclei. In the heavy element region beyond the iron peak, these stars are believed to be the formation site of the s-process neutron capture products from approximately copper through strontium and of the r-process nuclei through ^{238}U. Several detailed nucleosynthesis studies are now available for both the progenitor and explosive phases of massive star evolution. Woosley & Weaver (1995; henceforth WW95), Thielemann, Nomoto & Hashimoto (1995), and Arnett (1995) have each published detailed isotopic yields from more or less realistic stellar explosions. Each project has been independent, with each somewhat different in its method, scope and goals. While there are some differences in the predicted yields, it is encouraging that the quantitative agreement between these three projects is so good.

Burbidge et al. (1957) and Cameron (1957) painted a broad and compelling paradigm of how the elements are synthesized in stars. They coined the names of the various process that operate in the stellar interior (e.g., the s-, r-, and p-processes) and identified the chief nucleosynthetic products from most of the hydrostatic burning stages. Many of the details have changed, however, especially in the light of new physics that was unknown in the late 1950's. For example, scattering by the Z_0 intermediate vector boson gives rise to neutral current reactions, which add a significant source of neutrino cooling. This cooling affects the core structure of a massive star, which in turn determines to a large extent the nucleosynthesis. Burbidge et al. and Cameron, however, posed a very important question: can the nucleosynthesis that takes place in stars and is forcefully ejected, eventually, after many rounds of star formation, produce the solar composition? By the early 1980's various groups had run large nuclear reaction networks on specific stages of stellar evolution – core silicon burning, shell oxygen burning, and s-processing during dredge-up to name just a few. These studies hinted that a sizable portion of the solar composition could be synthesized by stars. Supernova 1987A suggested many observational tests of massive star evolution and nucleosynthesis, along with providing several unexpected features. In the early 1990's the index n in Moore's Law (computer speed doubles and price halves every 18 months) had become significant enough to allow the routine use of large nuclear reaction networks in very finely gridded stellar models. Coupled with an increase in our knowledge of the general physical properties of our Galaxy, plus more refined dynamical and chemical evolution scenarios, the question posed by Burbidge et al. and Cameron is receiving fresh attention.

Figure 1 shows the result of a calculation by Timmes, Woosley & Weaver (1995, henceforth TWW95) for all isotopes lighter than gallium. The most abundant isotope of a given element is marked by an asterisk and isotopes of the same element are connected by solid lines. If the all the isotopes had an ordinate of unity, the solar composition would have been exactly replicated. This perfect case is denoted by a dashed horizontal line. The horizontal dotted lines denote factors of two variation around the ideal case. Although nucleosynthesis from intermediate/low mass stars and standard Type Ia supernovae were included in the calculation, massive star nucleosynthesis dominants the functional form of the results. The isotopes below calcium show less scatter than those isotopes above since these isotopes are chiefly produced by hydrostatic burning processes before the explosion, while heavier isotopes are sensitive to the uncertain modeling of the explosion mechanism and hence have a larger scatter. There is probably no systematic pattern to the scatter. Up to calcium, it generally makes little difference whether the presupernova or the exploded yields are used. Massive stars can only account for about 1/4 to 1/3 of the solar abundances of the carbon isotopes and ^{14}N (see Woosley, this volume). This is consistent with the bulk of

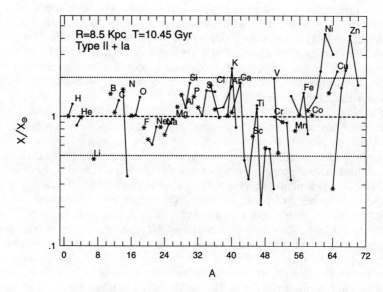

Figure 1. Stable isotopes from hydrogen to zinc present in the interstellar medium at a time when (4.6 Gyr ago) and place where (8.5 Kpc Galactocentric radius) the Sun was born. The x-axis is the atomic mass number. The y-axis is the logarithmic ratio of the model abundance to the Anders & Grevesse (1988) mass fraction. Figure from TWW95.

these isotopes being produced during CNO processing and dredged-up material from helium shell flashes in intermediate and low mass stars.

Almost all of the isotopes below calcium, and nearly all of the most abundant isotopes of an element show good agreement with solar values. In terms of absolute solar abundances, the stable isotopes from hydrogen to zinc range over some 10 orders of magnitude. There remain many uncertainties (e.g., the treatment of convection, details of the explosion mechanism, residual disagreement on key nuclear reaction rates, function form of the star formation rate, the measured abundances themselves) that effect the spread and pattern in Fig 1. However, that the detailed nucleosynthesis used in the stellar–chemical evolutions compress the ratios of the calculated abundances to the solar abundances to within a factor of two is very encouraging. In stronger terms, the isotopic solar composition from hydrogen to zinc is replicated.

Additional clues to the nature of massive star nucleosynthesis are provided by our knowledge of the abundances in the oldest and most metal deficient stars in our Galaxy: the field halo and globular cluster stars. Reviews of interesting abundance trends as a function of metallicity have most recently been provided by Wheeler, Sneden & Truran (1989) and Lambert (1989). Discussions by Edvardsson et al. (1993) of abundance trends in disk stars, and by McWilliam et al. (1995) of abundances at metallicities down to [Fe/H] $\simeq$ -4 dex, provide important observational information beyond that contained in the reviews cited above. TWW95 gathered the available abundance determinations in stars from lithium to zinc and compared them to modern stellar–chemical evolution calculations. There are significant abundance trends to be inferred from metal deficient stars that are relevant to both charged–particle and neutron–capture nucleosynthesis processes.

Stellar abundance determinations generally show that oxygen through titanium are enriched in metal-deficient halo dwarfs ([Fe/H] $\lesssim$ -1.0 dex), by a roughly a factor of three. As has been recognized for some time, these α-chain abundance patterns are consistent with such theoretical predictions for the ejecta of core collapse supernovae whose stellar progenitors possess a short lifetime. In contrast to the α-chain elements, some even proton number nuclei in the iron group (Cr, Fe, Ni, Zn) seem to be produced in approximately solar relative proportions, at least down to metallicities [Fe/H] $\simeq$ -3. Similarly, trends in the abundances of the odd-Z iron peak elements relative to iron are found to be generally consistent with the expected metallicity dependences (even-odd effect) of explosive nucleosynthesis processes.

The recent investigation of abundances in extremely metal deficient stars by McWilliam et al. (1995), which extends down to metallicities [Fe/H] $\simeq$ -4 dex, has some identified remarkable trends in the α-chain elements (Mg, Si,Ca), iron peak elements (Sc, Ti, Cr, Mn, Co, Ni) and heavy elements (Y, Sr, Ba, Eu). On average, these stars are relatively rich in cobalt (relative to iron), but low in chromium and nickel. For most of the stars, the s-process element abundances are weak, consistent with the relative lack of available neutron flux. There are fascinating exceptions, such as CS 22898-027 with [Fe/H] = -2.35 dex, where barium and other s-process elements are overabundant by $\simeq$ 250. This s-process production factor excess is larger than one might expect for a star with this metallicity. Thorburn & Beers (1992) have interpreted this star as a CH subgiant, for which mass exchange with a binary companion may account for a pre-enrichment of ^{22}Ne, the likely neutron source. Large dispersions are found in the McWilliam et al. (1995) survey. One may interpret these variations and trends as a signature of an epoch of Galactic evolution before significant mixing of stellar ejecta with the ambient interstellar medium had occured. Audouze & Silk (1995) have used this constraint to show that there may be a natural metallicity threshold of Z/Z$\odot$ $\simeq$ 10^{-4} below which one would expect to find few, if any, halo stars.

Metal-poor halo stars exhibit signatures of massive star nucleosynthesis. Indeed, it seems quite reasonable that the oldest stars (which presumably are the most metal poor) in our Galaxy should reflect contamination by the earliest products of supernova activity. Besides ordinary massive star nucleosynthesis in a Salpeter-like initial mass function, nothing else is required to produce the abundances of the elements seen in the oldest stars. One may be curious to investigate if the nucleosynthesis patterns found in the halo dwarfs are similiar to (and can be transformed into) the patterns observed in the damped Lyman-α systems.

Gas enriched by nucleosynthetic processes is detected out to redshifts z $\simeq$ 4 by the metal absorption lines they produce in the spectra of the background quasi-stellar objects (QSOs). Analyses of the absorption spectra produced by the gas which is suspected to reside in galaxies can uniquely measure a variety of physical parameters associated with the gas: temperature, column density, chemical abundances, grain content, and average supernova rates. Redshifts measured in the absorption systems presumably sample the gas at different points in their roughly common evolutionary path, and the abundance patterns in these systems offer an understanding of their nucleosynthetic evolution. The heavy element abundances found in the QSO absorbers span the metallicity range 10^{-3} Z$\odot \leq$ Z $\leq$ 0.3 Z$\odot$, which overlaps the range found in Galactic disk, thick disk, and halo dwarf stars (Lauroesch et al. 1995; Timmes, Lauroesch & Truran 1995, henceforth TLT95). If most galaxies form in a fashion similar to the Milky Way, the distinctive abundance patterns exhibited by Galactic stars should be discernible in the absorption line spectra produced by the gas which (accidentally) intercepts the line of sight of the background QSO.

The general expression for the lookback time – redshift relation is

$$H_o t = \int \frac{dz}{(1+z)\, E(z)}$$

where H_o is the Hubble constant. The function $E(z)$ is given by (e.g., Peebles 1993)

$$E(z) = \left[\Omega(1+z)^3 + \Omega_R(1+z)^2 + \Omega_\Lambda\right]^{1/2}$$

subject to the constraint that

$$\Omega + \Omega_R + \Omega_\Lambda = 1$$

Attention is focused here on cases where the cosmological constant is zero, so that the only free parameter is the density of matter Ω. The above expressions are easy enough to integrate numerically; the results are shown in Fig. 2 for two choices of the density parameter (Ω = 1.0 and 0.2) and three choices of the Hubble constant (H_o = 50, 75 and 100 km sec^{-1} Mpc^{-1}). For these parameters, the age of the universe (z=∞) ranges from approximately 6 to 16 Gyr.

A first–order model, and perhaps the simplest model, for the QSO absorption line systems assumes that the absorbers follow an abundance history and dynamical evolution which is diffeomorphic to the Galaxy (i.e., the Galaxy is a standard candle). This model is readily testable (and falsifiable!) by comparing the abundances observed in QSO absorption line systems with the abundance behavior, suitably transformed into redshift space, found in the Galaxy. Knowledge of a Galactic age–metallicity relationship can be mapped into a redshift–metallicity relation, for a chosen cosmology, by using Fig. 2. Hence, any abundance–metallicity trend can be transformed into an unambiguous abundance–redshift trend. Other than the choice of cosmology, the procedure described so far has no free parameters. We now add a single parameter, called the delay time. The delay time τ_{delay} is the total time between the start of the Big Bang and formation of a thin disk. The choice $\tau_{delay} = 0$ reduces to the no free parameter case. It is important to recognize that we will be transforming observed abundance histories, not chemical evolution model results, into redshift space.

The small plot in the upper right hand corner of Fig. 3 shows the age–metallicity (time versus [Fe/H]) relationship of the Galaxy inferred from spectral observations of dwarf stars in the solar vicinity. This small plot constitutes the

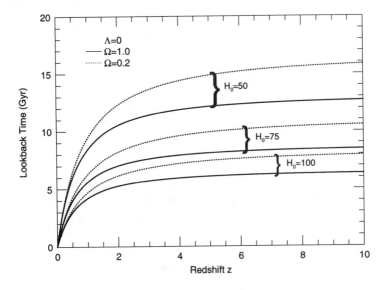

Figure 2. Lookback time as a function of redshift, from TLT95.

input. Note that it is scaled between 0 and 1 by dividing by the maximum age of the universe, whatever that value may be for a chosen cosmology. Transposing the time coordinate into a redshift coordinate (see Fig. 2) yields the main plot shown in Fig. 3. Curves are shown for $\Omega=1.0$ and $\Omega=0.2$ cosmologies and three values of the delay, time $\tau_{delay} = 0$, 1 and 3 Gyr. In this simple picture, no abundance evolution occurs between the start of the Big Bang and τ_{delay}. At $\tau_{delay} + \delta$, the evolution shown in the upper right hand corner commences. The time required to reach a metallicity of [Fe/H] $\simeq$ -1.0 dex, the canonical halo-disk transition period, is a few times 10^8 years, which is consistent with dynamical collapse timescales for the halo. Since it is required that the entire age–metallicity relationship conform to the age of the universe (minus any delay time), the redshift–metallicity evolutions are independent of the Hubble constant and depend only on Ω.

Spectroscopic abundance determinations of the zinc to hydrogen ratio in damped Lyman-α absorption line systems are shown in Fig. 3 as the solid circles, and represent a compilation of recent measurements. If one is to compare the [Zn/H] observations with the [Fe/H] calculations, two conditions must be met. First, zinc and iron must track each other, i.e., [Zn/Fe] = 0, over the entire metallicity range. Several high resolution, low noise surveys that employ digital spectra of the Zn I lines at 4722 Å and 4810 Å have shown that the zinc to iron ratio in Galactic halo and disk stars is indeed solar, with a very small star-to-star scatter over the metallicity range. This result is robust with respect to variations in the adopted effective temperatures and surface gravities. Theoretical stellar-chemical evolution calculations also suggest that zinc and iron

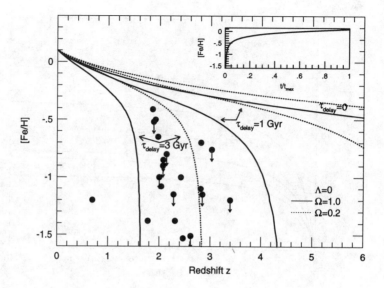

Figure 3. Iron to hydrogen ratio evolution as a function of redshift, from TLT95.

are produced in tandem over the entire metallicity range (WW95, TWW95). The second condition to be met is that zinc should not be significantly depleted onto dust grains. Extensive surveys of the interstellar medium have shown that zinc, relative to hydrogen, is generally present in the gas phase in nearly solar proportions, and that it is depleted onto dust by at most a factor of 2–3 in the densest clouds surveyed. Thus, zinc is not readily incorporated onto dust grains. Hence, it appears justifiable to compare the [Zn/H] observations with the [Fe/H] calculations of Fig. 3.

Evolution of the α–chain abundances (oxygen, neon, magnesium, silicon, sulfur, argon, calcium and titanium) are shown in Figure 7. The small plot in the upper right hand corner forms the input and represents a schematic evolution of the α–chain elements with the metallicity [Fe/H] as inferred from surveys of solar neighborhood stars. The precise form of this evolution is, of course, slightly different for each of the different elements. Still, α–chain elements generally follow the pattern indicated by the small plot. Shifting the metallicity dependence of the small plot into redshift space yields the main plot of Fig. 4, where the α–chain abundances, relative to hydrogen, are shown as a function of redshift for Ω=1.0 and Ω=0.2 cosmologies and three values of the delay time $\tau_{delay} = 0$, 1 and 3Gyr. Spectroscopic abundance determinations of the silicon to hydrogen ratio [Si/H] in the damped Lyman–α systems are shown in Fig. 4 as the solid circles. The observations are a compilation of recent silicon abundance determinations. All of the silicon abundance determinations utilized the observed Si II column densities, and the any ionization corrections were removed to put all

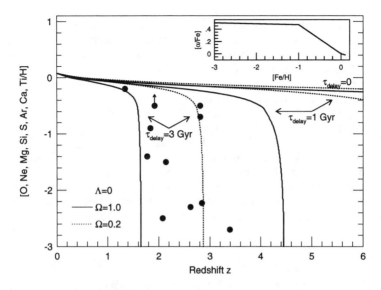

Figure 4. Abundance evolution of oxygen, neon, magnesium, silicon, sulfur, argon, calcium and titanium with redshift. Figure from TLT95.

the measurements on an equal footing. Silicon probably has similar or perhaps larger grain depletion factors than zinc, but these depletion corrections only shift the observations in the vertical direction.

The main point of Figs. 3 and 4 is that for all reasonable choices of Ω in a $\Lambda=0$ cosmology, the observations of the zinc and silicon abundances in damped Lyman-α systems are consistent, for more than two orders of magnitude in abundance, with a $\simeq 3$ Gyr delay time between the start of the Big Bang and the formation of a thin disk. However, caution is advisable. The scatter in the abundances and the narrow redshift window over which they have been observed could conspire to give the illusion that chemical enrichment began at these same redshifts. There may be selection effects such as dust obscuration or significant line of sight variations that could substantially broaden the present redshift window. Nor is it clear that observational trends of stars in the solar vicinity are the best ones to use in the transformations – perhaps an average over radial or vertical distance would be more appropriate. If this turns out to be the case, then our conclusion about a few Gyr hiatus before abundance evolution commences in the disks of galaxys may have to be abandoned, or at least modified.

The redshift–abundance patterns given above used a delay time which was interpreted to mean that no abundance evolution occurred between the start of the Big Bang and τ_{delay}. This is the simplest and most unambiguous case. Any other interpretation or modification, while yielding the next order correction to the model, opens a Pandora's box of complications and uncertain parameters.

As an example, consider the case where the total time delay τ_{delay} is the sum of a no abundance evolution phase and a halo phase. The halo phase is characterized by how long τ_{halo}, a classical halo dwarf abundance pattern, persists. For times larger than τ_{delay}, the abundance evolution is taken to follow the paths they did in the previous figures. The difference between τ_{delay} and τ_{halo} is the time period where no abundance evolution takes place. Unlike the previous abundance evolutions, the abundance evolutions will now have the metallicity pleasantly decreasing with redshift. The expense of this improvement mild improvement in the simple first-order model is the opening of a Pandora's box: the time alloted to the no evolution phase and the halo phase must be partitioned, a disk-halo transition point must be selected, and the evolutions now depend on the value of the Hubble constant. To turn the situation around, as the number of high quality abundance determinations in QSO absorption line systems increase, perhaps they will be able to distinguish between the uncertainties and assist in selecting the correct parameters for extensions to the simple paradigm.

References

Anders, E., & Grevesse, N., 1989, Geochim. Cosmochin. Acta, 53, 197

Arnett, W. D., 1995, ARAA, 33, 115

Audouze, J., & Silk, J., 1995, ApJ, 451, L49

Burbidge, E. M., Burbidge, G. R., Fowler, W. A., & Hoyle, F., 1957, Rev. Mod. Phys., 29, 547

Cameron, A. G. W., 1957, Chalk River Report, CRL-41

Edvardsson B., Andersen J., Gustafsson B., Lambert D. L., Nissen P. E., & Tomkin J., 1993, A&A, 275, 101

Lambert, D. L., 1989, Cosmic Abundances of Matter; AIP Conference Proceedings 183, C. J. Waddington, New York, Amer. Inst. Phys., 168

Lauroesch, J. T., Truran, J. W., Welty, D. E., & York, D. G., 1995, PASP, in press

McWilliam, A., Preston, G., Sneden, C., & Searle, L., 1995, AJ, 109, 2757

Peebles, P. J. E., 1993, Principles of Physical Cosmology, Princeton, Princeton Univ. Press

Thielemann, F. -K., Nomoto, K., & Hashimoto, M., 1995, ApJ, in press

Thorburn, J., & Beers, T., 1992, BAAS, 24, 1278

Timmes, F. X., Lauroesch, J., & Truran, J. W., 1995, ApJ, 451, 468

Timmes, F. X., Woosley, S. E., & Weaver, T. A., 1995, ApJS, 98, 617

Wheeler, J. C., Sneden, C., & Truran, J. W., 1989, ARAA, 27, 279

Woosley, S. E., & T. A. Weaver, 1995, ApJS, 101, 181

Abundances and Globular Cluster Ages

B.E.J. Pagel

NORDITA, Blegdamsvej 17, Dk-2100 Copenhagen Ø, Denmark

Abstract. Turn-off ages of globular clusters have raised questions as to how far they constrain cosmology and whether there is an age-metallicity relation, but these are subject to various uncertainties including the way to handle non-solar element mixtures, precise location of the turnoff and especially the distances. A new approach, based on HB and RGB morphology, gives ages of about 13 Gyr for clusters ranging in metallicity from 47 Tuc to M 92.

1. Introduction

Globular cluster ages have recently been reviewed by Chaboyer (1995), who gives a total uncertainty range from 11 to 21 Gyr when all possible random and systematic errors are considered, and by Bolte & Hogan (1995), who give an age of $15.8 \pm 2.1(1\sigma)$ Gyr for M 92, which they point out is hard to reconcile with Einstein–de Sitter cosmology. This raises an issue that is worth discussing. Another issue is the existence or otherwise of an age-metallicity relation among globular clusters, which is defended by some authors (*e.g.* Chaboyer, Sarajedini & Demarque 1992; Chaboyer, Demarque & Sarajedini 1995), but disputed (Sandage 1993) or left as an open question (Salaris, Chieffi & Straniero 1993) by others. In this paper I review some results and problems of the main-sequence turn-off method which has been most commonly used and describe some new work that is based on horizontal-branch (HB) and red-giant branch (RGB) morphology.

2. Ages from main-sequence turnoff

The age of a cluster is usually based on the luminosity at the turnoff from the main sequence, where hydrogen becomes exhausted at the centre and the star reaches its maximum effective temperature, using a relation that can be schematically represented as

$$<L>t = 0.007 mc^2 X q, \qquad (1)$$

where L is the luminosity, t the age, m the mass, X the initial hydrogen mass fraction and q the effective mass fraction of the core. $<L>/m$ depends on the chemical composition through opacity and energy generation by the CNO cycle. There is also the effect of helium on the molecular weight, but there are good reasons to believe the helium abundance to be fixed at $Y = 0.23$ or 0.24. From

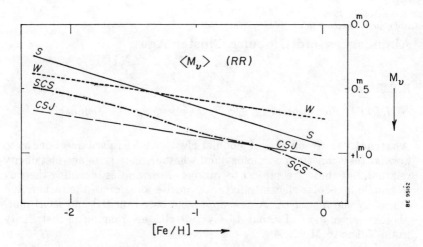

Figure 1. Calibrations of RR Lyrae luminosities as a function of metallicity. S: Sandage (1993); W: Walker (1992); SCS: Salaris, Chieffi & Straniero (1993); CSJ: Carney, Storm & Jones (1992).

evolutionary models, assuming a solar mix of heavy elements, one can deduce an error budget

$$-\frac{\Delta t}{t} \simeq \frac{\Delta L}{L} + 0.2\frac{\Delta Z}{Z} \simeq 4\frac{\Delta m}{m} - 0.1\frac{\Delta Z}{Z}, \qquad (2)$$

where Z is the mass fraction of heavy elements. The second part of eq (2) represents an alternative approach to ages which will be described later. While in some cases the turnoff luminosity L_{TO} is derived by fitting nearby subdwarfs with parallaxes (see Bolte & Hogan 1995), the more usual method (and one that has the advantage of being unaffected by interstellar absorption and reddening) is to measure directly from the HR diagram the visual magnitude interval ΔV_{TO}^{HB} from the turnoff up to the horizontal branch, which in turn is calibrated in M_v either by the use of theoretical HB models or from empirical or semi-empirical calibrations of the luminosities of RR Lyrae variables as a function of metallicity [Fe/H]. These in turn are based on a variety of criteria, notably

1. Baade-Wesselink pulsation analysis (*e.g.* Carney, Storm & Jones 1992).

2. A relation (theoretical or semi-empirical) that fits (supposed) luminosities of extragalactic cepheids (Walker 1992; Lee, Freedman & Madore 1993).

3. The Oosterhoff period-shift effect (Sandage 1993).

As can be seen from Fig 1, there are severe discrepancies amounting to a range of $0^m.4$, which corresponds to a 40 per cent difference in age. Added to this is the difficulty in actually defining the turnoff to better than $\pm 0^m.1$ or so, since the isochrones are almost vertical there. Thus Carney, Storm & Jones (1992) argue that, although ΔV from Buonanno, Corsi & Fusi-Pecci (1989) is

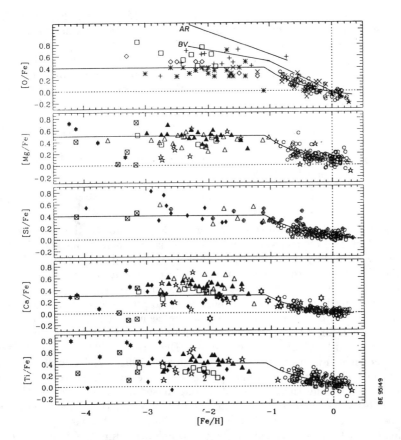

Figure 2. Oxygen and α-element to iron ratios as a function of [Fe/H], adapted from Pagel & Tautvaišienė (1995). BV denotes the Fe, O relation adopted by Bergbusch & VandenBerg (1992) and AB denotes that of Abia & Rebolo (1989), used by Carney, Storm and Jones (1992).

smaller for M 68 than for M 92, corresponding to an age difference of 4 or 5 Gyrs, the HR diagrams of the two clusters are indistinguishable and therefore M 68 should be treated as being as old as M 92; I think they could with equal logic have argued that M 92 should be as young as M 68.

In recent years, it has become apparent that the heavy-element abundance in low-metallicity objects cannot simply be derived by scaling down from $Z_\odot$ according to spectroscopic measurements of [Fe/H], because oxygen and α-particle elements (Ne, Mg, Si, S, Ca) are overabundant relative to iron. However, there is disagreement both as to how overabundant they are, especially in the case of oxygen (see Fig 2), and as to how the overabundances are to be handled. Bergbusch & VandenBerg (1992) argue that only the oxygen enhancement should be taken into account, because at low metallicity none of these elements contributes significant internal opacity, but oxygen contributes to energy generation through the CNO cycle. Assuming [O/Fe] = 0.75 for M 92 (with [Fe/H] $\simeq$ −2), they derive an age of 14 Gyr, low compared to many other estimates, but an equally significant reason for this low age is the adoption of an apparent distance modulus $(m - M)_v = 14^m.7, 0^m.2$ higher than typical values in the literature (*cf.* Table 1), which in itself leads to a reduction of about 20 per cent in the age. Salaris, Chieffi & Straniero (1993) argue, on the other hand, that since (within uncertainties) the equality

$$[X_C + X_N + X_O + X_{Ne}] = [X_{Mg} + X_{Si} + X_S + X_{Ca} + X_{Fe}] \quad (3)$$

holds, the isochrones can be very well mimicked by standard solar-mix isochrones using a heavy-element abundance

$$Z = Z_0(0.638 f_\alpha + 0.362) \quad (4)$$

where Z_0 is the solar Z scaled according to [Fe/H] and f_α the average enhancement factor of oxygen and α-elements. This argument is backed up with evolutionary models by them and by Chaboyer, Sarajedini & Demarque (1992), where the effects on internal opacity are taken into account in some detail and it is argued further that the low-temperature envelope opacities (which cannot be split into contributions from different elements) do not have a significant effect on the isochrones up to the turnoff. As can be seen from Fig 2, my favourite value of f_α is 2.5 for [Fe/H] < −1, corresponding to $Z/Z_0 = 2$, and this value has also been used by Chaboyer, Sarajedini & Demarque (1992), whose adopted ages are shown in Fig 5. Carney, Storm & Jones (1992) used two alternative prescriptions. Their less radical one, [α/Fe] = [O/Fe] = 0.3, led to ages of up to 21 Gyr for M 92 and their more radical one, based on the findings of Abia & Rebolo (1989) shown in Fig 2, led to an age of 17 Gyr for the same cluster and rather little trend of age with diminishing metallicity. Both investigations find a significant scatter in age at a given [Fe/H], which according to Chaboyer *et al.* is related to galactocentric distance, but Salaris, Chieffi & Straniero cast some doubt on this because it is not confirmed by $\Delta(B - V)$, the colour difference between the turnoff and the base of the RGB, which should be a good criterion of relative age at a fixed metallicity.

There are thus numerous uncertainties relating to turnoff ages, among which the distance is probably the major source. Additional uncertainties relate to the mixing length and to the effects of helium diffusion and the Debye-Hückel equation of state, each of which can reduce all age estimates by 1 Gyr or so (Chaboyer 1995). Thus an age of as little as 12 Gyrs is not ruled out, even for M 92, which is generally found to be among the oldest clusters, and even if one ignores completely some more exotic hypotheses such as that of significant mass loss along the main sequence (Willson, Bowen & Struck-Marcell 1987).

3. A new approach: HB and RGB morphology

The spread of stars along the HB is mainly due to previous mass loss which varies stochastically from one star to another (Rood 1973). The range of colours where zero-age HB stars are found is a function of metallicity (the "first parameter") and of the range of ZAHB masses. More precisely, the ZAHB colour at given

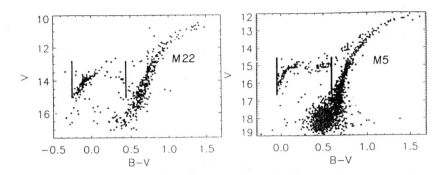

Figure 3. HR diagrams for M 5 and M 72, showing the adopted extent of the horizontal branch. Adapted from Jimenez et al. (1995).

metallicity depends on both the star's total mass and the ratio of core mass to total mass, but the core mass is essentially fixed by the physics of the helium flash and is quite insensitive to the mass and metallicity. For a given average mass loss, the average final mass is thus a decreasing function of age, which is therefore a popular candidate for the "second parameter" (Searle & Zinn 1978), although other candidates such as CNO abundance have also been suggested. A strong case for age as the chief (though perhaps not necessarily the only) second parameter has been made by Lee, Demarque & Zinn (1994), who find a tendency for the clusters to be younger in the outer Galactic halo. Jørgensen & Thejll (1993), using analytical fits to a variety of RGB models and following evolution along the RGB with mass loss treated by Reimers's (1975) formula, showed that, for clusters with narrow RGB's (the majority), star-to-star variations in initial mass, metallicity or mixing-length parameter can be ruled out as a source of the spread along the HB, leaving as likely alternatives only either variations in the Reimers efficiency parameter η (or some equivalent) or a delayed helium flash caused by differential internal rotation. The latter alternative would lead to a fuzzy distribution of stars at the RGB tip.

These considerations suggest the possibility of finding the turnoff mass, and hence the age, from RGB and HB morphology. The use of HB morphology is not completely new: Crocker, Rood & O'Connell (1988) have used theoretical HB models to determine the mass distribution along the HB of M 92, and Dorman, VandenBerg & Laskarides (1989) and Dorman, Lee & VandenBerg (1991) have used details of the evolution from the ZAHB to determine the helium abundance in 47 Tuc and M 15 respectively, deriving an age of 13.5 Gyr for 47 Tuc which is in good agreement with the turnoff age by Chaboyer, Sarajedini & Demarque (1992) and Sandage (1993). However, for most clusters the statistical data from the literature are too sparse to test details of RGB and HB morphology, and there have not been enough consistent evolutionary models giving an adequate fit to the RGB, including effects of mass loss. These matters are addressed in a new investigation by Jimenez et al. (1995), of which a brief description is given below.

CCD photometry was carried out on 5 clusters (M 5, M 22, M 68, M 72 and M 107) using the Danish 1.5m telescope on La Silla (see Fig 3), and analysed

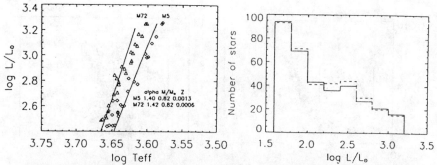

Figure 4. *Left panel*: Fits to the RGB's of two globular clusters. *Right panel*: Combined luminosity function along the RGB for three clusters, M 5, M 68 and M 72, where the data make it possible to distinguish RGB from AGB stars; predicted and observed numbers are shown by the full-line and broken-line histograms respectively. After Jimenez et al. (1995).

together with data from the literature for M 92, M 3 and 47 Tuc using Kurucz (1993) model atmospheres to determine reddenings, effective temperatures and luminosities from UBV data and distance moduli taken from the literature (see Table 1). A new set of RGB evolutionary models was calculated (Jimenez & MacDonald 1995) and represented by analytical formulae similar to those of Jørgensen & Thejll (1993) so that the effects of mass loss could be accurately followed by switching at each stage from one model series to another with slightly lower mass. Fig 4 shows the fits to some of the RGB's, illustrating the existence of a well-defined RGB tip (which rules out internal rotation as a source of spread along the HB), and a slight bending towards lower temperatures near the tip which may be due to the mass loss. The mixing-length parameter is found to be close to 1.4 in all cases and the RGB tips defined by the histograms probably provide the most robust indicator of distances that is available at the present time.

The distribution of ZAHB masses comes from fitting the (B-V) colours along the HB to ZAHB models (Sweigart 1987; Bencivenni et al. 1989; Castellani, Chieffi & Pulone 1991; Dorman 1992ab; Dorman, Rood & O'Connell 1993), using the method of Crocker, Rood & O'Connell (1988). The distribution of masses is quite symmetrical (*cf.* Crocker, Rood & O'Connell), with stars that have lost virtually no mass appearing at 2σ above the mean. Representing the mass loss by Reimers's (1975) formula, there is good agreement with a cluster-independent average value of his efficiency factor $<\eta>= 0.4$ which is in good accord with direct observations of mass loss from red giants, and the average mass plus 2σ gives the mass along the RGB which is the same as the turnoff mass. This is the most delicate part of the age determination, however, since there are uncertainties in the physics of HB models and much depends on the chemical composition details like oxygen and α-element enhancement. Taking the models at face-value, Jimenez et al. assume [O/Fe]=[α/Fe]=0.3 for [Fe/H] < -1, corresponding to $Z/Z_0 = 1.5$, and note that the α-element enhanced models of Bencivenni et al. confirm the contention of Salaris, Chieffi & Straniero that oxygen and α-element enhancement can be mocked up by the

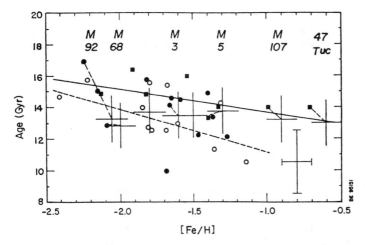

Figure 5. Ages determined as a function of metallicity by Chaboyer, Sarajedini & Demarque (1992), compared to preliminary ages from Table 1 (crosses).

use of solar-scaled models (Sweigart 1987; Castellani, Chieffi & Pulone 1991) with Z appropriately enhanced. The various models are in good agreement for a given chemical composition, but equation (2) indicates the sensitivity of the age to any errors in the mass estimate. Provisional results by Jimenez *et al.* are shown in Table 1.

Table 1. Cluster distances and ages from RGB and HB morphology

Cluster	[Fe/H]	Z	$(m-M)_v$ lit.	$(m-M)_v$ RG tip	$<m>_{HB}$ $(m_\odot)$	m_{TO} $(m_\odot)$	Age Gyr
M 92	-2.1	.0002	14.5	14.45	0.71	0.81	13.2
M 68	-2.0	.0003	15.2	15.25	0.73	0.82	12.7
M 22	-1.8	.0007	13.6	13.50	0.78	0.81	13.7
M 3	-1.6	.0008	15.0	15.01	0.72	0.82	13.5
M 72	-1.5	.0009	16.5	16.50	0.71	0.82	13.5
M 5	-1.3	.0021	14.5	14.53	0.70	0.82	13.8
M 107	-0.9	.0046	15.0	14.97	0.77	0.89	13.2
47 Tuc	-0.6	.008	13.46	13.46	0.78	0.91	13.0

Fig 5 shows preliminary ages from Table 1 with the turnoff ages found by Chaboyer, Sarajedini & Demarque (1992). The agreement is excellent for all objects in common except M 92, and at present it is not clear where the error lies. Jimenez et al.'s results are in agreement with the contention of Sandage (1993) that there is no age-metallicity relation for the globular clusters, but their sample is only a small one. Chaboyer, Demarque & Sarajedini (1995) have recently discussed a much larger sample of 43 clusters, and show rather convincingly that either there is an age-metallicity relation, or at least there is significant dispersion in the ages for metallicities [Fe/H] between -1.3 and -2.2, and rather little correlation with galactocentric distance. Their absolute ages are larger than in their 1992 paper, but subject to the various difficulties mentioned in Section 2 above. The results of Jimenez et al., which may also be subject to some corrections for helium diffusion and the Debye-Hückel equation of state, confirm the conclusion of Chaboyer (1995) that an age of 11 Gyr for the oldest clusters—and hence an Einstein–de Sitter universe—is still not ruled out, although it does require most of the uncertainties to be pushed in one direction.

References

Abia, C. & Rebolo, R. 1989, ApJ, 347, 186
Bencivenni, D., Castellani, V., Tornambé, A. & Weiss, A. 1989, ApJS, 71, 109
Bergbusch, P.A. & VandenBerg, D. 1992, ApJS, 81, 163
Bolte, M. & Hogan, C.J. 1995, Nature, 376, 399
Buonanno, Corsi, C.E. & Fusi Pecci, F. 1989, A&A, 216, 80
Carney, B.W., Storm, J. & Jones, R.V. 1992, ApJ, 386, 663
Castellani, V., Chieffi, A. & Pulone, L. 1991, ApJS, 76, 911
Chaboyer, B. 1995, ApJ, 444, L9
Chaboyer, B., Demarque, P. & Sarajedini, A. 1995, preprint
Chaboyer, B., Sarajedini, A. & Demarque, P. 1992, ApJ, 394, 515
Crocker, D.A., Rood, R.T. & O'Connell, R.W. 1988, ApJ, 332, 236
Dorman, B. 1992a, ApJS, 80, 701
Dorman, B. 1992b, ApJS, 81, 221
Dorman, B., Lee, Y.-W. & VandenBerg, D.A. 1991, ApJ, 366, 115
Dorman, B., Rood, R.T. & O'Connell, R.W. 1993, ApJ, 419, 596
Dorman, B., VandenBerg, D.A. and Laskarides, P.G. 1989, ApJ, 343, 750
Jimenez, R. & MacDonald, J. 1995, in prep.
Jimenez, R., Thejll, P., Jørgensen, U.G., MacDonald, J. & Pagel, B.E.J. 1995, MNRAS, in press.
Jørgensen, U.G. & Thejll, P. 1993, A&A, 272, 255
Kurucz, R.L. 1993, CDROM 13
Lee, M.G., Freedman, W.L. & Madore, B.F. 1993, ApJ, 417, 553
Lee, Y., Demarque, P. & Zinn, R. 1994, ApJ, 423, 248
Pagel, B.E.J. & Tautvaišienė, G. 1995, MNRAS, 276, 505
Reimers, D. 1975, Mem Soc Roy Sci Liège, 8, 369
Rood, R.T. 1973, ApJ, 184, 815
Salaris, M., Chieffi, A. & Straniero, O. 1993, ApJ, 414, 580
Sandage, A.R. 1993, AJ, 106, 719
Searle, L. & Zinn, R. 1978, ApJ, 225, 357
Sweigart, A.V. 1987, ApJS, 65, 95
Walker, A.R. 1992, ApJ, 390, L81
Willson, L.A., Bowen, G.H. & Struck-Marcell, C. 1987, Comm. Ap., 12, 17

Abundances in the Galactic Halo Gas

Blair D. Savage

*Department of Astronomy, University of Wisconsin-Madison,
475 N. Charter Street., Madison, WI 53706*

Kenneth R. Sembach

*Center for Space Research, Massachusetts Institute of Technology,
77 Massachusetts Avenue, Cambridge, MA 02139*

Abstract. Abundance measurements of Mg, Si, Fe, S, Ni, Cr, Mn, Ti, and Zn in the Galactic halo gas are reviewed. The observations refer mostly to the warm neutral gas of the low halo at $|z| < 2$ kpc, which is probed through optical and ultraviolet absorption line observations of distant halo stars. There is a progression toward increasing gas-phase abundances of the refractory elements from the Galactic disk to the halo. The small observed variation in halo cloud abundances of the depleted refractory elements supports the idea that resilient dust grain cores exist in the halo clouds. The inferred composition of the grain cores implies the existence of silicates, oxides, and possibly iron grains. The processes moving matter from the disk into the low halo evidently are not violent enough to completely destroy the interstellar grain cores. Abundances for several elements have been obtained for gas in Galactic high velocity clouds (HVCs). Observations of the HVCs along the sight lines to NGC 3783 and Fairall 9 suggest that these particular clouds are likely the result of tidally stripped matter from the Magellanic Clouds or some other extragalactic object.

1. Introduction

The general interstellar medium of the Milky Way contains cool, warm, and hot gas. These different gas phases interact with each other and probably exist in rough pressure equilibrium. The stirring of matter in the Galactic disk by supernova explosions, cloud–cloud collisions, stellar winds, and other processes causes the matter in the various phases to extend different distances into the Galactic halo. The stratification of this "Galactic atmosphere" reaches to ≈ 0.1 kpc for the cool clouds, ≈ 0.5 kpc for the warm neutral gas, ≈ 1 kpc for the warm ionized gas, and ≈ 5 kpc for the hot ionized gas. For reviews of the theoretical and observational aspects of gas in the Galactic halo see Spitzer (1990), McKee (1993), and Savage (1995).

In the low halo ($|z| < 2$ kpc) warm neutral clouds are typically moving at velocities less than 100 km s^{-1} and the flow is mostly directed toward the Galactic plane. The motions may be the result of the circulation processes associated with a Galactic fountain (Shapiro & Field 1976) in which hot over-pressurized gas is created in regions of OB associations by the collective effects of supernova explosions and stellar winds (Bruhweiler et al. 1980). This hot ionized gas moves into the Galactic halo where it cools, recombines, and rains down onto the Galactic plane. A low temperature Galactic fountain (Houck & Bregman 1990) seems capable of explaining the motions of the intermediate velocity gas of the Galaxy while the behavior of some of the high velocity clouds with $|v| > 100$ km s^{-1} requires a more energetic Galactic fountain (Bregman 1980) or alternate explanations (see Wakker 1991). The relative importance of fountain-type processes in establishing the observed distribution of gas away from the plane of the Milky Way is not known, and there are a number of alternate explanations for the support and ionization of gas in the halo (e.g., see Slavin & Cox 1993; Hartquist 1994; Raymond 1992).

It has been clear for many years that accurate measures of elemental abundances in the gas of the Galactic halo are crucial for determining the history of the gas and the processes responsible for the existence of gas at large distances from the Galactic plane. In this brief review of the abundances of the gas in the Galactic halo, we first discuss the ground based observations of absorption by Ca II and Ti II and then turn to the interstellar abundance measures of Mg, Si, Fe, S, Ni, Cr, Mn, and Zn in the ultraviolet obtained by the *Goddard High Resolution Spectrograph* (GHRS) on the *Hubble Space Telescope* (HST). Sections 2, 3, and 4 refer to gas in the low halo with $|z| < 2$ kpc. In Section 5 we review abundance results for HVCs at large distances from the Galactic plane.

2. Abundances in Gas Clouds Situated in the Low Halo ($|z| < 2$ kpc)

2.1. Optical Absorption Line Abundance Studies

The earliest work on the existence of heavy elements in the gas of the Galactic halo involved studies of absorption in the optical H and K lines of Ca II at 3933 and 3968 Å (Adams 1949; Munch & Zirin 1961). Spitzer (1956) proposed the possible existence of a high temperature Galactic corona based on the presence of Ca II absorbing clouds at large distances from the Galactic plane which were believed to be in pressure equilibrium with a hotter medium filling the interstellar space between the clouds. Unfortunately, Ca II is a trace ionization state of Ca in the neutral interstellar gas and is not a good diagnostic of elemental abundances since the ionization correction for Ca II is large and uncertain. A further complication in determining a gaseous Ca abundance arises because Ca is readily incorporated into interstellar dust.

The only species that can be observed in absorption from the ground that is in its dominant ionization state in H I regions is Ti II. With an ionization potential of 13.58 eV, Ti II is closely coupled to H I (IP = 13.60 eV). Therefore, optical studies of Ti II have yielded important information about the abundance of this highly refractory element in the gas of the Galactic disk and halo (Stokes 1978; Albert 1983; Edgar & Savage 1989; Albert et al. 1993; Lipman & Pettini 1995).

One technique for determining the properties of gas in the halo from integrated measurements of column densities along various sight lines is to study the behavior of N sin$|b|$ versus $|z|$ for a large number of stars, where b is the Galactic latitude of the star, z is the distance of the star from the Galactic plane, and N is the total atomic column density to the star. The $|z|$–distance at which N sin$|b|$ stops increasing is a measure of the scale height of the particular species being studied. By combining the results for a large number of sight lines from several investigations, Lipman & Pettini (1995) have established that the Galactic scale height of gaseous Ti II is 1.5 kpc. For sight lines to stars in the halo at $|z| > 0.5$ kpc, N(Ti II)/N(H I) is typically 0.1 times the solar abundance. In contrast, for sight lines to stars in the disk N(Ti II)/N(H I) is found to span a very large range from ~ 0.05 solar to ~ 0.001 solar. For disk directions, sight lines with small values of N(Ti II)/N(H I) contain diffuse clouds with substantial amounts of molecular hydrogen (Edgar & Savage 1989). These diffuse clouds have substantial gas densities, which favors the removal of Ti from the gas phase due to accretion of Ti onto dust grains (Jenkins 1987). For halo sight lines, various grain destruction processes liberate up to $\approx 10\%$ of the Ti from the grains. The very large scale height of gaseous Ti is due to the extreme deficiency of Ti in the gas of the disk.

2.2. Ultraviolet Absorption Line Abundance Studies

The bulk of the information on abundances in halo gas has come from ultraviolet spectrographs orbiting above the Earth's absorbing atmosphere. While a few results have come from the *Copernicus* and the *International Ultraviolet Explorer* (IUE) satellites, most of the abundance information has been provided in the last few years by the GHRS on the HST.

The *Copernicus* satellite obtained valuable information about abundances in the interstellar matter of the Galactic disk (Bohlin et al. 1983; Jenkins, Savage & Spitzer 1986; Jenkins 1987). However, *Copernicus* was unable to obtain spectra of fainter stars in the halo and obtained only a limited amount of information on the abundances within halo gas. Savage & Bohlin (1979) noted that the abundance of Fe in the gas toward several halo stars exceeded that for the gas toward stars in the Galactic disk. Similar results were suggested based on data for halo stars from the IUE satellite for Fe and several other elements (Jenkins 1983; van Steenberg & Shull 1988) and in the direction of the LMC for intermediate and high velocity gas (Savage & de Boer 1981). However, because of the relatively low spectral resolution of the IUE (25 km s^{-1}) there were substantial uncertainties associated with line blending and line saturation effects in these studies. It was clear that an accurate study of element abundances in halo gas required spectra with higher resolution and the ability to probe absorption toward faint halo stars.

The launch of the GHRS on the HST in 1990 has greatly extended the possibility of obtaining accurate abundance data for gas situated in the Galactic disk and the low halo. For reviews of GHRS interstellar abundance results see Cardelli (1994) and Savage & Sembach (1996a). The characteristics of the GHRS and its inflight performance are described by Brandt et al. (1994) and Heap et al. (1995). The GHRS contains echelle and first order gratings capable of obtaining high signal–to–noise ultraviolet interstellar absorption line spectra of faint halo stars at resolutions of 3.5 or 10 to 20 km s^{-1}. The GHRS has been used to probe absorption through the entire halo toward bright extragalactic

objects such as quasars and AGNs. Table 1 summarizes the different sight lines
that have been studied to provide information on gas in the low halo.

The interstellar species suitable for halo gas abundance studies accessible by
the GHRS over the wavelength region from 1150 to 3200 Å are Mg II, Si II, Fe II,
S II, Ni II, Cr II, Mn II, and Zn II. These ions are the dominant ionization stages
in neutral hydrogen regions since these elements have first ionization potentials
< 13.6 eV and second ionization potentials >13.6 eV. We assume the ionization
corrections are small and that $[X/H] = \log \{N(X\ II)/N(H\ I)\} - \log (X/H)_c$,
where subscript c denotes the cosmic abundance. The values of N(H I) in this
equation are from the Lyα absorption line, 21–cm emission measures, or estimates based on N(Zn II) assuming $[Zn/H] \approx [Zn/H]_c$. The errors associated
with these ionization assumptions are discussed by Sembach & Savage (1996)
and appear to be approximately 0.05 to 0.1 dex.

Although lines of O I, C II, and N I are also observable by the GHRS, the
absorption lines are either very strong and saturated or very weak and not easily
detected in halo gas clouds, which typically have $N(H\ I) < 3\times10^{20}$ atoms cm^{-2}.

Representative spectra for the HD 116852 sight line are shown in Figure 1
in the form of continuum normalized intensity versus LSR velocity. The spectra
shown are from the ground for Na I and Ca II and from the GHRS for Fe II
and Mn II. For this interesting sight line ($l = 304.9°$, $b = -16.1°$, $d = 4.8$
kpc, $z = -1.3$ kpc), the effects of Galactic rotation produce the extension of
absorption to negative LSR velocities approaching -70 km s^{-1}. This allows a
study of halo gas conditions and abundances at ≈ 0.5 kpc and ≈ 1.0 kpc below
the Sagittarius and Norma spiral arms, respectively (see Sembach & Savage
1996).

Table 1.   The GHRS Halo Gas Absorption Line Data Base

Object	l (°)	b (°)	d (kpc)	z (kpc)	log N(H I)	Reference
HD 18100	217.9	−62.7	3.1	−2.8	20.14	Savage & Sembach (1996b)
HD 22586	264.2	−50.4	2.0	+1.5	20.35	Jenkins & Wallerstein (1996)
HD 38666	237.3	−27.1	1.1	−0.50	19.85	Sofia, Savage, & Cardelli (1993)
HD 93521	183.1	+62.2	1.7	+1.5	20.10	Spitzer & Fitzpatrick (1993)
HD 116852	304.9	−16.1	4.8	+1.3	20.96	Sembach & Savage (1996)
HD 120086	329.6	+57.5	1.0	+0.84	20.41	Jenkins & Wallerstein (1996)
HD 149881	31.4	+36.2	2.1	+1.2	20.57	Spitzer & Fitzpatrick (1995)
HD 167756	351.5	−12.3	4.0	+0.85	20.81	Cardelli, Sembach, & Savage (1995)
3C 273	290.0	+64.4	...	...	20.10	Savage et al. (1993)

An overall summary of elemental abundance results for clouds in the Galactic disk and halo is provided in Figure 2 and Table 2. Figure 2 from Sembach
& Savage (1996) shows [X/Zn] for different interstellar paths as indicated by
the legend on the figure. The paths include: (1) the cool disk clouds in the
direction of ζ Oph and ξ Per; (2) the warm disk clouds toward HD 93521,
μ Col, and ζ Oph; (3) the warm disk and warm halo clouds toward HD 18100
and HD 167756; and (4) the warm halo clouds toward HD 116852, HD 149881,
HD 93521, μ Col, and the QSO 3C 273. The sources of the various measurements are found in Table 1 and Sembach & Savage (1996). Since Zn is generally
undepleted in the warm neutral gas, it is expected that $[X/Zn] \approx [X/H]$.

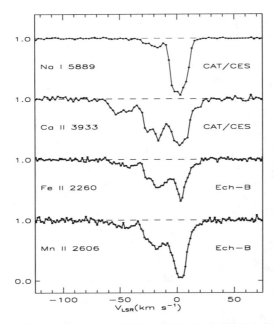

Figure 1. Halo gas absorption toward HD 116852 ($l = 304.9°$, $b = -16.1°$, $d = 4.8$ kpc, $z = -1.3$ kpc). The optical data are from the ESO Coudé Echelle Spectrograph and have a resolution of 4.5 km s^{-1}. The Fe II and Mn II profiles are GHRS echelle mode data and have a resolution of 3.5 km s^{-1}. Absorption from ≈ -10 to -35 km s^{-1} provides information about halo gas ≈ 0.5 kpc below the Sagittarius spiral arm while absorption from ≈ -35 to -70 km s^{-1} traces gas ≈ 1 kpc below the Norma spiral arm. (From Sembach & Savage 1996).

Table 2. Diffuse Cloud Gas-Phase Abundance Summary[a]

[X/H]	Type of Cloud			
	Halo	Disk+Halo	Warm Disk	Cool Disk
Mg	($< -0.28, -0.56$)	($-0.59, -0.62$)	($-0.73, -0.90$)	($-1.24, -1.56$)
Si	($-0.09, -0.47$)	($-0.23, -0.28$)	($-0.35, -0.51$)	(-1.31)
S	($-0.23, +0.16$)	($+0.03$)	($-0.03, +0.14$)	(~ 0.00)
Mn	($-0.47, -0.72$)	(-0.66)	($-0.85, -0.99$)	($-1.32, -1.45$)
Cr	($-0.38, -0.63$)	($-0.72, -0.88$)	($-1.04, -1.15$)	($-2.08, -2.28$)
Fe	($-0.58, -0.69$)	($-0.80, -1.04$)	($-1.19, -1.24$)	($-2.09, -2.27$)
Ni	($-0.77, -0.91$)	(-1.15)	($-1.44, -1.48$)	($-2.46, -2.74$)

[a]The values listed represent the range of [X/H] found by Sembach & Savage (1996). For some sight lines we assumed [X/H] $\approx$ [X/Zn] $\approx$ [X/S] since no H I or H$_2$ estimates were available for the individual clouds studied. Zn and S are nearly undepleted in such environments. For cases where only one value is listed, the element was measured for only one sight line.

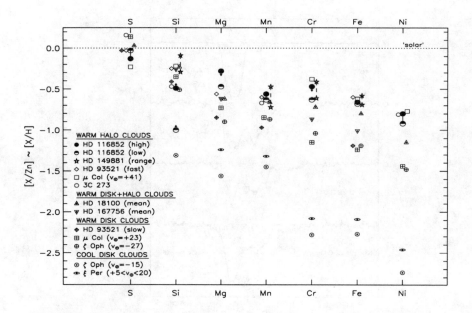

Figure 2. A comparison of gas–phase abundances in clouds found in the Galactic disk and halo. $[X/Zn] = \log(X/Zn) - \log(X/Zn)_\odot$ is plotted for S, Si, Mg, Mn, Cr, Fe, and Ni. The solar reference abundances are from meteoritic data (Anders & Grevesse 1989). Since Zn is normally only slightly depleted in the warm ISM, $[X/Zn]$ should be $\approx [X/H]$. Table 1 gives the source of the observations for the halo gas data. Note the trend of increasing gas–phase abundances from cool disk clouds, to warm disk clouds, to warm halo clouds. The variations for the warm halo clouds are remarkably small given that the sight lines sample halo gas in the solar vicinity as well as gas in the halo under the Sagittarius and Norma spiral arms. The abundance results shown here are summarized in Table 2. (From Sembach & Savage 1996)

The abundance trends illustrated in Figure 2 and summarized in Table 2 are very interesting. There is a general progression toward increasing gas–phase abundances of the depleted elements from the cool disk clouds, to the warm neutral gas of the disk, to the warm neutral clouds of the halo. The variations in the gas–phase abundances among the different halo clouds are small. For example, the seven data points for [Fe/H] in the warm halo cloud gas range from −0.58 to −0.69 dex (see Table 2). This small variation is quite surprising since the various clouds studied span halo gas in the solar neighborhood and gas under the Sagittarius and Norma spiral arms. There evidently is no systematic dependence of these gas–phase halo abundances on Galactocentric distance from $R_g \approx 7$ to 10 kpc.

3. Implications for the Composition of Grain Cores

The upper envelope of abundances for the halo clouds shown in Figure 2 implies that the refractory elements are depleted (deficient) from the gas. The lack of abundance variation coupled with the observed level of depletion supports the idea that the grains in halo clouds have been eroded down to grain cores that are difficult to destroy. The processes that move gas from the disk to the halo apparently are not violent enough to completely destroy the dust (Sembach & Savage 1996).

We explore the composition of these grain cores in Table 3 We list values of $10^6(X/H)_c$, $10^6(X/H)_g$, and $10^6(X/H)_d$ where c, g, and d denote cosmic, gas, and dust, respectively. The dust-phase abundance of a particular element is obtained by assuming $(X/H)_d = (X/H)_c - (X/H)_g$. The assumption of cosmic (current epoch) disk abundances is probably valid if the gas of the low halo is supplied from the disk through processes that circulate the matter on a relatively short time scale ($\leq 10^9$ years). For cosmic abundances we use the solar system meteoric abundances from Anders & Grevesse (1989). To explore the possibility the Sun is overabundant by ≈ 0.2 dex compared to current epoch Population I abundances (see Mathis, this volume), we list in the footnote to the table the effect of reducing the solar system reference abundance by 0.2 dex for each element listed.

Table 3.   Gas- and Dust-Phase Abundances for Halo Clouds[a,b]

	Mg	Si	Fe	S	Ni	Cr	Mn
$10^6(X/H)_c$	38	36	32	19	1.8	0.48	0.34
$10^6(X/H)_g$	11	20	7.4	17	0.26	0.15	0.083
$10^6(X/H)_d$	27	16	25	2.0	1.5	0.33	0.26

[a]We assume $(X/H)_d = (X/H)_c - (X/H)_g$. For the cosmic reference abundance, we assume meteoritic abundances from Anders & Grevesse (1989). If the true current epoch reference abundances are 0.2 dex smaller for all the elements, the seven values of $10^6(X/H)_d$ are from left to right: 12, 2.9, 13, 0.0, 0.87, 0.15, and 0.13.

Assuming solar abundances are valid references, the numbers in Table 3 imply that $\approx$ 69%, 44%, and 78% of the Mg, Si, and Fe in the halo clouds is locked in dust. Since these are the most abundant heavy elements (other than C, N and O for which we have no information for halo cloud gas or dust), it is of interest to investigate the implication of this particular mixture of elements for the composition of the dust grain cores. The existence of silicate grains is well-established in the ISM of the Galactic disk from the 9.7 and 18 μm SiO stretch and bend features observed along high extinction sight lines (Roche & Aitken 1985). The features have strengths that imply most of the Si must exist in grains (Draine & Lee 1984). It is believed these silicate grains may be in the form of (Mg, Fe)SiO$_3$ (pyroxene) and/or (Mg, Fe)$_2$SiO$_4$ (olivine). For pure pyroxenes the expected value of (Mg+Fe)/Si is 1.0. For pure olivines the expected value is 2.0. From Table 3 we see the observed value of (Mg+Fe)/Si is $(27+25)/16 = 3.3$ for solar reference abundances and $(12+13)/2.9 = 8.6$ for reference abundances 0.2 dex smaller than solar (see the notes to Table 3). In either case the value of the ratio is substantially larger than the expected range from 1.0 to 2.0 for pure silicate grains. In addition to being present in silicates, the Mg and Fe in these halo clouds must exist in some other type of dust grain core. Likely possibilities include pure Fe grains and various oxides, such as MgO, Fe_2O_3, and Fe_3O_4 (Nuth & Hecht 1990; Fadeyev 1988). However, Fe grains are so easy to destroy (see section 4) the presence of Mg and Fe in oxides seems more likely.

The results discussed here for Mg rely on the adoption of the factor of 4.7 revision to the f–values for the Mg II doublet at 1239 and 1240 Å recommended by Sofia, Cardelli, & Savage (1994). If the f–values from Hibbert et al. (1993) are used instead, the values of (Mg+Fe)/Si in the dust cores are $(0+25)/16 = 1.6$ and $(0+13)/2.9 = 4.5$ for solar abundances and abundances lower than solar by 0.2 dex, respectively. The case for oxides or pure iron bearing grains is not strong in this case, which is favored by Spitzer & Fitzpatrick (1993, 1995). Additional laboratory or theoretical determinations of the Mg II f–values are required to settle this issue.

4. Implications for the Processes Supplying Gas to the Halo

The abundance pattern shown in Figure 2 is consistent with a more severe processing of dust in the halo clouds than in the disk clouds. This processing may result from either more frequent or more severe shocking of the halo clouds compared to the disk clouds. The GHRS results confirm, in a quantitative way, previous optical and IUE studies that showed that the vertical Galactic scale heights of refractory elements (Ca, Ti, Fe) are larger than those of H I (Edgar & Savage 1989; Lipman & Pettini 1995).

The halo cloud grain core composition indicates that localized enrichment of the gas behind a shock is probably most pronounced when the grain *cores* are destroyed. The resiliency of the refractory grain cores in the halo gas is somewhat surprising since transfer of gas into the halo is most likely accomplished through supernova explosions in the disk. The thermalization of supernova kinetic energy and betatron acceleration of grains in the post-shocked regions behind supernova blast waves lead to grain–grain collisions and sputtering (both thermal and non–thermal). These processes are most efficient in the warm, neutral interstellar medium (Draine & Salpeter 1979) where the gas densities are moderate ($n_H \sim 1$ cm^{-3}) and temperatures are high (T $\sim 10^4$ K), as is found for some of the halo clouds studied (see Spitzer & Fitzpatrick 1993, 1995).

It is interesting that grain destruction models predict that pure Fe grains are far more easily destroyed by fast shocks than silicate grains due to both their higher velocities in the post-shocked gas and their higher mass densities (Jones et al. 1994; McKee et al. 1987). When combined with our conclusion that there must be another Fe–bearing population or component to the grain constituency besides silicate grains to account for the large observed values of (Mg+Fe)/Si, this prediction favors oxides as the likely additional carrier of the Fe in the halo cloud dust.

GHRS searches for the enrichment of Fe–peak elements (Fe, Ni) relative to α–process elements (Mg, Si) in halo gas by Type Ia supernovae at large distances from the Galactic plane suggest that the gas transport processes cycling material between the disk and halo operate effectively on time scales short enough (t $\sim 10^{7-8}$ yr) to mask any vertical abundance gradients from SN ejecta (Jenkins & Wallerstein 1996). Circulation processes include, but are probably not limited to, general turbulence in the disk (Lockman & Gehman 1991), photo–levitation of diffuse clouds (Ferrara et al. 1989), and the movement of material by the flow of a Galactic fountain (Bregman 1980). The small spread in the Fe abundance of the halo gas (see Table 2) supports the idea of an efficient mixing of the interstellar material. The <10% change in [Fe/H] in halo clouds between R_g = 7 to 10 kpc may therefore be more reflective of the general tendency for mass to be exchanged between the disk and halo than it is of *in situ* processes.

It is possible that some of the sub-solar gas-phase abundances observed in halo clouds could be due to a lower average metallicity in the halo compared to the disk rather than due to incorporation of material into dust grains. The infall of unprocessed gas or gas with low metal abundances would dilute the halo material and reduce the gaseous abundances below expected values. Such a scenario has been proposed to explain the apparent deficiency ($\approx$ a factor of 2) of interstellar gas-phase oxygen toward stars in the solar neighborhood (Meyer et al. 1994). Observations of metal abundances in HVCs at large distances from the Galactic plane (see Section 5) suggest that some of these clouds have metal abundances of approximately 0.1 to 0.3 solar. While it is possible that the HVCs could contribute to lowering the halo metallicity, it remains to be determined if there is a large enough population of such clouds (or their lower velocity counterparts) to sufficiently counteract the opposing tendency to drive the metallicity toward solar when the disk and halo gases mix.

5. Abundances in Galactic High Velocity Clouds

The high velocity clouds (HVCs) of the Milky Way are clouds detected in H I 21-cm emission with $|v_{LSR}| > 100$ km s^{-1} whose radial velocity is inconsistent with Galactic rotation. The origin of the HVCs is poorly understood, but it is clear that both distance and abundance measurements are crucial for understanding the HVCs and the role they play in Galactic phenomena. The distance to only one HVC has been determined; Complex M lies between $z \approx$ +1.7 and +4.8 kpc (Danly, Albert & Kuntz 1993; Keenan et al. 1995). However, large lower limits have been determined for others (de Boer et al. 1994). The properties of HVCs are reviewed by Wakker (1991) and van Woerden (1993). We discuss the HVCs as a topic separate from that of gas in the low halo since it appears that many of the HVCs could be quite distant. For example, high velocity clouds in the Magellanic Stream, which is a ~180° band of H I emission extending from the Magellanic Clouds to the south Galactic pole and beyond, are almost certainly associated with the Clouds and therefore provide information about processes occurring in the outermost regions of the Milky Way.

While a number of the HVCs have been detected in the absorption lines of Ca II toward extragalactic objects, the measurements are not suitable for abundance studies because of the large and uncertain ionization corrections and the need to allow for the possible presence of Ca in dust. The GHRS and *Faint Object Spectrograph* (FOS) on the HST are beginning to produce abundance measurements for gas in the HVCs. Nine HVCs have been detected in the strong Mg II absorption doublet at 2800 Å (see Table 8 in Savage & Sembach 1996a). Many of these observations yield only lower limits to [Mg/H] because of the effects of absorption line saturation and/or line blending. However, the measurements demonstrate that processed elements are regularly detected in the HVCs. Observations of the -147 km s^{-1} Mg II absorption toward the Seyfert galaxy Mrk 205 yield [Mg/H] = $-0.59^{+1.09}_{-0.38}$, implying that substantial abundances of processed elements exist in Complex C, an extensive HVC in the Northern sky (Bowen & Blades 1993; Bowen, Blades, & Pettini 1995).

The highest quality data for abundances in HVCs are for the paths to NGC 3783 (l = 287.5°, b = 23.0°) and Fairall 9 (l = 295.1°, b = -57.8°). For NGC 3783, Lu et al (1994a) detected the HVC at 240 km s^{-1} in the absorption lines of S II and Si II. Sulfur is the most valuable element for deriving an abundance for this HVC since the observed lines of S II are not saturated, and sulfur is not readily depleted onto interstellar dust. Furthermore, with an ionization potential of 23 eV, S II is expected to be a dominant stage of ionization in the HVC, which has N(H I) = 1.21×10^{20} cm^{-2}. If the ionization corrections are not large, the observed measure of N(S II)/N(H I) = 2.8×10^{-6} implies [S/H] = $-0.82^{+0.12}_{-0.18}$. For the same cloud, the Si II line is saturated and yields only a lower limit [Si/H] > -2.22. These abundances indicate that the HVC in the direction of NGC 3783 is most likely associated with gas stripped from the Galaxy by an extragalactic object(s) such as the Magellanic Clouds (Lu et al. 1994a).

Fairall 9 lies in the direction of the Magellanic Stream. The high velocity H I emission in this direction has two components, one with N(H I) $\approx 2 \times 10^{19}$ cm^{-2} at $v_{LSR} \approx 160$ km s^{-1} and the other with N(H I) $\approx 6 \times 10^{19}$ cm^{-2} at $v_{LSR} \approx 200$ km s^{-1} (Morras 1993). Lu et al. (1994b) used the GHRS to study absorption by S II and Si II in these two clouds. S II was not detected and Si II produced saturated absorption. Assuming the ionization corrections are not large, the observations yield [Si/H] > -0.7 and > -1.15 and [S/H] < -0.05 and < -0.5 in the two clouds, respectively. These limits rule out the Magellanic Stream as primordial (metal deficient) gas and are consistent with an origin closely tied to gas in the Magellanic Clouds.

A large uncertainty in studies of HVC abundances results from sampling differences between absorption line data, which measure absorption over an infinitesimal solid angle, and reference H I 21-cm emission line data, which usually sample a relatively large beam (FWHM = 34' for the NGC 3783 and Fairall 9 H I data discussed above). It is important to map at higher angular resolution those HVCs for which UV data are now becoming available.

Acknowledgments. Much of the new information on elemental abundances in Galactic halo gas is being provided by the GHRS on the HST. We thank the many individuals who contributed to the construction and operation of this instrument. KRS acknowledges support from grant GO-05883.01-94A through the Space Telescope Science Institute, which is operated by AURA under NASA contract NAS5-26555. BDS appreciates support from NASA under grant NAG5-1852.

References

Adams, W.S. 1949, ApJ, 109, 354

Albert, C.E. 1983, ApJ, 272, 509

Albert, C.E., Blades, J.C., Morton, D.C., Lockman, F.J., Proulx, M., & Ferrarese, L. 1993, ApJS, 88, 81

Anders, E. & Grevesse, N. 1989, Geochim. Cosmochim. Acta, 53, 197
Bohlin, R.C., Hill, J.K., et al. 1983, ApJS, 51, 277
Bowen, D. & Blades, J.C. 1993, ApJ, 403, L55
Bowen, D., Blades, J.C. & Pettini, M. 1995, ApJ, 448, 662
Brandt, J.C., Heap, SR., et al. 1994, PASP, 106, 890
Bregman, J.N. 1980, ApJ, 236, 577
Bruhweiler, F.C., Gull, T.R., Kafatos, M., & Sofia, S. 1980, ApJ, 238, L27
Cardelli, J.A. 1994, Science, 265, 209
Cardelli, J.A., Sembach, K.R., & Savage, B.D. 1995, ApJ, 440, 241
Danly, L., Albert, C.E., & Kuntz, K.D. 1993, ApJ, 416, L29
de Boer, K.S., Altan, A.Z. et al. 1994, A&A, 286, 925
Draine, B.T. & Lee, H.M. 1984, ApJ, 285, 89
Draine, B.T. & Salpeter, E.E. 1979, ApJ, 231, 77
Edgar, R.J. & Savage, B.D. 1989, ApJ, 340, 762
Fadeyev, Y. 1988, in *Atmospheric Diagnostics of Stellar Evolution*, ed. K. Nomoto (Berlin:Springer-Verlag), 174
Ferrara, A., Franco, J., Ferrini, F., & Barsella, B. 1989, in *Structure and Dynamics of the Interstellar Medium*, IAU Colloquium 120, eds. G. Tenorio-Tagle, M. Moles, & J. Melnick (Berlin: Springer Verlag), 54
Hartquist, T.W. 1994, Ap&SS, 216, 185
Heap, S.R., Brandt, J.C., et al. 1995, PASP, 107, 871
Hibbert, A., Dufton, P.L., Murray, M.J., & York, D.G. 1983, MNRAS, 205, 535
Houck, J.C. & Bregman, J.N. 1990, ApJ, 352, 506
Jenkins, E.B. 1983, in *Kinematics, Dynamics and Structure of the Milky Way*, ed. W.L.H. Shuter (Dordrecht: Reidel), 21
Jenkins, E.B. 1987, in *Interstellar Processes*, eds. D.J. Hollenbach & H.A. Thronson (Dordrecht: Reidel), 533
Jenkins, E.B., Savage, B.D., & Spitzer, L. 1986, ApJ, 301, 355
Jenkins, E.B. & Wallerstein, G. 1996, ApJ, in press
Jones A.P., Tielens, A.G.G.M., McKee, C.F., & Hollenbach, D.J. 1994, ApJ, 433, 797
Keenan, F.P., Shaw, et al. 1995, MNRAS, 272, 599
Lipman, K. & Pettini, M. 1995, ApJ, 442, 628
Lockman, F.J. & Gehman, C.S. 1991, ApJ, 382, 182
Lu, L., Savage, B.D., & Sembach, K.R. 1994a, ApJ, 426, 563
Lu, L., Savage, B.D., & Sembach, K.R. 1994b. ApJ, 437, L119
McKee, C.F. 1993, in *Back to the Galaxy*, eds. S. G.Holt & F.Verter (New York: AIP), 499
McKee, C.F., Hollenbach, D.J., Seab, C.G., & Tielens, A.G.G.M. 1987, ApJ, 318, 674
Meyer, D.M., Jura, M.J., Hawkins, I., & Cardelli, J.A. 1994. ApJ, 437, L59
Morras, R. 1983, AJ, 88, 62

Munch, G. & Zirin, H. 1961, ApJ, 133, 11
Nuth, J.A. & Hecht, J.H. 1990, Ap&SS, 163, 79
Raymond, J.C. 1992, ApJ, 384, 502
Roche, P.F. & Aitken, D.K. 1985, MNRAS, 215, 525
Savage, B.D. 1995, in *The Physics of the Interstellar and Intergalactic Medium*, eds. A. Ferrara, C.F. Mckee, C. Heiles & P.R. Shapiro, (San Francisco: ASP Conf. Series), 233.
Savage, B.D. & Bohlin, R.C. 1979, ApJ, 229, 136
Savage, B.D. & de Boer, K.S. 1981, ApJ, 243, 460
Savage, B.D., Lu, L., Weymann, R., Morris, S.L., & Gilliland, R.L. 1993, ApJ, 404, 134
Savage, B.D. & Sembach, K.R. 1996a, ARA&A, in press
Savage, B.D. & Sembach, K.R.. 1996b, ApJ, submitted
Sembach, K.R. & Savage, B.D. 1996, ApJ, in press
Shapiro, P.R. & Field, G.B. 1976, ApJ, 205, 762
Slavin, J.D. & Cox, D.P. 1993, ApJ, 417, 187
Sofia, U.J., Cardelli, J.A. & Savage, B.D. 1994, ApJ, 430, 650
Sofia, U.J., Savage, B.D., & Cardelli, J.A. 1993, ApJ, 413, 251
Spitzer, L. 1956, ApJ, 124, 20
Spitzer, L. 1990, ARA&A, 28, 71
Spitzer, L. & Fitzpatrick, E.L. 1993, ApJ, 409, 299
Spitzer, L. & Fitzpatrick, E.L. 1995, ApJ, 445, 196
Stokes, G.M. 1978, ApJS, 36,115
van Steenberg, M. & Shull, J.M. 1988, ApJ, 330, 942
van Woerden, H. 1993, in *Luminous Blue Stars at High Latitude*, ed. D.D. Sasselov, (San Francisco: ASP Conf. Series), 11
Wakker, B.P. 1991, in *The Interstellar Disk-Halo Connection in Galaxies*, IAU Symposium 144, ed. H. Bloemen (Dordrecht: Kluwer), 27

The Composition of Interstellar Dust

John S. Mathis
University of Wisconsin-Madison, 475 N. Charter St., Madison WI 53706; e-mail mathis@madraf.astro.wisc.edu

Abstract. The implications of spectral features in interstellar dust are discussed. Several reasons are given for believing that the "reference abundance" of the interstellar medium, or abundance of elements relative to H in the gas plus dust, is significantly lower than solar. The change in the reference abundance, along with recent measurements of gas-phase abundances, place very significant restraints on dust models. A recent model is shown.

1. Introduction

There have been valuable recent reviews of abundances in the interstellar medium (ISM) by Tielens (1996), Dorschner & Henning (1996), and Savage & Sembach (1996). Wilson & Rood (1994) have reviewed the isotopic composition of the gas-phase ISM.

The observations needed to determine the grain size distribution over most of its range are in the ultraviolet (UV) part of the spectrum, either in the diffuse ISM or in the outer parts of molecular clouds. Except for one line of sight (HD 29647; Whittet et al. 1988), no ices are observed along the sight lines used determine UV extinctions. We will confine ourselves to only this ice-free dust, some of which is in rather dense regions containing H_2 besides in the diffuse ISM.

2. Spectral Features in Dust Emission and Extinction

Tielens (1996) discussed the spectroscopic signatures of various spectral features, including his estimate of the volume fractions of various components. Spectral features are also summarized in Mathis (1990, 1993) and Dorschner & Henning(1996).

2.1. The 217.5 nm Feature

The chief problems in explaining the 217.5 nm "bump" are (a) its great strength, which severely limits the possible carriers; (b) the relative invariance of its central wavelength, λ_0, for which the observations of Fitzpatrick & Massa (1988) show a standard deviation of only 9 Å; and (c) its FWHM, γ_0, that is distributed between 0.8 and 1.2 μm^{-1} (Draine 1989).

Only about 1/8 of the solar carbon in the form of graphite is required for the bump. About 30% of the atoms of the next most abundant refractory species (Mg, Si, or Fe) are required (Draine 1989) even if f (the oscillator strength of the transition) has the large value of unity. These elements are needed for the silicates or oxides seen in the 9.7 and 18 μm absorption bands. Steel & Duley (1987) suggested that the bump might arise from an UV absorption of OH$^-$ in silicates in a low-coordination (surface) sites, but $f = 0.13$ for the transition (Fritz, Lüty, & Anger 1963), requiring 2 OH$^-$ ions for each Si atom, all at the surface of small silicate grains. Observations show < 0.007 O–H bonds per Si atom in the star VI Cyg No. 12, with apparently normal dust (see below).

Laboratory measurements of amorphous carbon (Bussoletti et al. 1987) show essentially no 217 nm feature until extensive annealing has occurred (Menella et al. 1995a,b), and the bump is rather weak over the smooth background absorption. The optical constants would require about C/H = 250 $\times$ 10^{-6}, or 250 ppM (I will hereafter quote all abundances relative to 10^6 H, or parts per million H). Furthermore, the peak wavelength of the feature in amorphous C shifts markedly with the degree of annealing.

The observations of the bump seem explainable (Mathis 1994) by the hypotheses that (a) the carriers are all small, in the sense $a \leq 0.006$ μm, so that the variation of λ_0 is small, and (b) coatings broaden the bump without shifting it. Restrictions, possibly met by several materials, exist on the optical constants of the coatings – a less satisfactory situation than if one and only one material had the needed properties. The coatings hypothesis is supported by two more bump properties: (a) The narrowest bumps are found in H II regions, where the large radiation field might be expected to erode coatings; and (b) the three most peculiar bumps are along lines of sight through dark, quiescent ISM, just where coatings would be expected to be largest. Draine & Malhotra (1994) concluded that icy coatings could not produce the needed characteristics, but ices do not have the proper behavior of the optical constants.

It has been suggested (Joblin et al. 1992) that polycyclic aromatic hydrocarbons (PAHs) or small carbonaceous non-graphitic grains can produce the bump by a bulk absorption. This hypothesis seems dubious to me because of (a) the lack of an absorption at about 300 nm as shown in the spectra of the PAHs, and (b) the constancy in λ_0, if the bump is caused by a mixture of widely varying materials, seems very surprising.

Because of uncertainties in the optical constants of astronomical graphite, the shape of the bump carriers is hardly constrained. The largest problem with the small-graphite hypothesis, in my opinion, is that both the optical constants and grain shapes (not sizes) must be closely the same along the many lines of sight , since λ_0 depends upon them (Gilra 1972).

2.2. Carbonaceous Emission Bands

It is well established that the infrared emission features at 3.28, 6.2, 7.7, and 11.3 μm are caused by aromatic carbons in molecules with some H atoms attached to the edges. The molecules must be small enough to be significantly heated after absorption of a photon. The strengths of the various transitions are known (Tielens & Allamandola 1991), and Tielens (1996) estimates that only 1% of the cosmic C (i.e., 4 ppM) is needed to produce the transitions. The somewhat larger carbonaceous grains that are required to produce the IRAS 12, 25, and much of the 60 μm fluxes require about 30 ppM (Désert et al. 1990). These very large molecules/small grains might include the carriers of the Extended Red Emission (Witt 1989; Witt & Boroson 1990; Furton & Witt 1990; Schmidt & Witt 1991) that probably results from fluorescence from hydrogenated amorphous carbon.

3. Interstellar Depletions

The "depletion" of an element in interstellar gas toward a certain star is the observed fraction of the element that is *not* seen in the gas phase by the interstellar absorption lines in the spectrum of that star. The strengths of absorption lines of the *dominant* stage of ionization of various elements (the main stage in the neutral ISM or, separately, in the diffuse ionized gas) are measured. Various interstellar velocity components ("clouds") can usually be identified. Then the column density of the various ions are compared to that of H (from Lyman-α if available), or to a very lightly depleted element such as Zn if H is not available.

3.1. The carbon gas-phase abundance

Carbon is one of the most problematic elements for depletion studies because the column density of its dominant stage, C^+, must be determined from a single very weak line, C II] $\lambda 2325$. Before the Goddard High Resolution Spectrograph (GHRS) on the Hubble Space Telescope (HST), there was only one line of sight (towards δ Sco) for which C/H had been measured; now there are several (Cardelli et al. 1996). Among four GHRS measurements, $(C/H)_{gas} = (140 \pm 20)$ ppM. The least secure measurement, towards ξ Per, is $(C/H)_{gas} = 260 \pm 70$ ppM. The solar abundance is ≈ 400 ppM.

3.2. The Reference Abundance.

The amount of the element in dust is obtained by subtracting the gas-phase abundance from a "reference abundance", or the abundance of the element in both gas and dust. Until recently, this reference abundance has been assumed to be solar.

There are at least four reasons for believing that the reference abundance is appreciably less than solar, discussed separately below. In addition to these arguments, another suggestion comes from the noble gas abundances (Kr and Ar), for which the ISM gas-phase abundances are about 60% of solar. It is unlikely that they would be incorporated into grains, but the uncertainties in the solar values might be sufficient to account for the apparent discrepancies.

The Interstellar Oxygen Abundance. The solar abundance is 850±70 (Anders & Grevesse 1989) or 740 ± 90 (Grevesse & Noels 1993). Cardelli et al. (1996) have discussed the gas-phase O/H obtained from the GHRS spectra of six stars. The derived values are quite uniform, yielding $(O/H)_{gas-phase} = 310 \pm 20$ ppM.

Of course, the silicates in grains contain oxygen, basically in the form of oxides: MgO, Fe_2O_3, and SiO_2. All other elements have abundances small in comparison to those three elements. With solar abundances, 150 ± 30 ppM of O are in the silicates/oxides, accounting for O/H = 460 ppM in gas plus dust. *Where is the remaining 350 - 400 ppM?*

Water has a strong O–H stretch band at 3.07 μm that is not seen in the diffuse ISM and in the outer parts of molecular clouds. The band is not visible in the spectrum of VI Cyg No. 12 (Sandford et al. 1991); with the cross section for the band (Tielens & Allamandola 1991) and the standard value of N(H)/A(V), I estimate that the O/H in ice for this star is < 0.2 ppM – utterly negligible. Towards the Galactic center source Sgr A ($N(H) \approx 10^{23}$ cm^{-2}), Tielens et al. (1996) find O/H $\approx$ 20 ppM in water, apparently as water of hydration in silicates.

Interstellar O is not bound in the molecules CO or O_2 in the diffuse ISM. Towards the Galactic center, Tielens et al. (1996) find $CO/H_2O \approx 0.1$. For ζ Oph, with a relatively large fraction of its H in H_2, $CO/C^+ \approx 0.01$ (Federman et al. 1993; Lambert et al. 1994). CO is probably less abundant when there is a smaller fraction of H_2. Molecular O is even less abundant than CO (Fuente et al. 1993), as expected from theory (van Dishoeck et al. 1993).

The problem of the " missing" O disappears if we assume a reference abundance of about 500 ppM, or $\approx$ 60% of solar, since then we can account for all of the O.

"Organic refractory" mantles on silicate cores (Greenberg 1979), if they exist, must be almost entirely carbonaceous (possibly partially hydrogenated), very similar to the material injected directly into the ISM by carbon stars. Whether an appreciable amount of this carbon results from the organic residue left behind after photolysis of ices within the clouds is an open question.

The Interstellar Nitrogen Abundance. The N/H abundance ratio is discussed in Cardelli et al. (1996). There is a strong N–H stretch at 2.96 μm, on the wing of the O–H stretch, that is weak or missing in the diffuse ISM. Ferlet (1981) and York et al. (1983) find gas-phase N/H $\approx$ 60 ppM; solar is 93 $\pm$ 16 (Grevesse & Noels 1993). The N/H in dust towards VI Cyg No. 12, as judged from the 2.96 μm N–H stretch band, is < 1 ppM. Thus, *the gas-phase nitrogen, about 0.6 of solar, is the total for the ISM, since there is little N in dust.*

Abundances of Young Stars. Sofia et al. (1994), Mathis (1995), Cardelli et al. (1996), and Savage & Sembach (1996) have discussed the measured O/H ratios for B stars (Gies & Lambert 1992; Cunha & Lambert 1994; Kilian 1992; Kilian, Montenbruck, & Nissen 1992). Mean values of O/H for non- supergiant stars range from 295 ppM (Kilian 1992, field stars) to 505 ppM (Cunha & Lambert 1994, Orion Association), with a scatter that probably reflects real variations. An abundance substantially sub-solar (740 ppM) is indicated by all authors, whether using non-LTE atmospheres or not.

Averages of N/H in B stars are tabulated in Venn (1995). The averages for non-supergiants are N/H = 67 $\pm$ 20 ppM (NLTE stellar models, Gies & Lambert 1992), or 62 $\pm$ 25 (standard deviation) for the average of all 70 non-supergiants. The spread is apparently partially real, and the standard deviation of the mean is probably not meaningful. While the two values differ by less than one standard deviation of a single measurement, the large number of similar values of (N/H) strongly indicates a real deficiency of (N/H) in stars as compared to solar. The mean stellar abundance of N agrees with the gas-phase ISM value, consistent with there being very little N in dust and the reference being 2/3 solar.

A similar statement holds for (C/H) in stars. Snow & Witt (1995) estimate C/H = 225 $\pm$ 50 ppM from an extensive survey of various stellar abundance determinations. Venn (1995) lists 170 ± 60 for non-supergiants. The solar is 360 – 440 ppM, so the stellar C/H is 0.5 – 0.4 solar.

H II Regions. Abundances in H II regions are conventionally determined from the [O III]4363/5007 ratio providing "the" electron temperature of the nebula. From this temperature the emissivities of both the ionic collisional lines and of recombination lines such as Hβ follow. The lines of [O II] and [O I] provide the abundances of O^+ and O^0, so for oxygen there is no correction for unseen stages of ionization. This procedure is widely followed, straightforward, and possibly wrong. We will discuss the evidence against it shortly.

The results of the standard procedure in the Orion Nebula are O/H = 384 ppM, with a total spread (not standard deviation) of five determinations (Baldwin et al. 1991, Rubin et al. 1991, Osterbrock et al. 1992, Dufour et al. 1992, and Peimbert 1987) of 0.16 dex. Peimbert et al. (1993) find O/H in M 8 = 310 ± 45 ppM.

If we take the standard analysis at face value, it is clear that even if the amount of O in dust is the same as it is in the diffuse ISM, about 150 ppM, the total O/H is definitely subsolar: 530 ppM in Orion and 460 ppM in M 8.

The possible problem with the standard analysis is the presence of large fluctuations in nebular electron temperature (T) that are much larger than those predicted by photoionization models. The fluctuations are suggested by the observed strengths of faint recombination lines of O^+ and C^+, arising from interactions of O^{+2} and C^{+2}, respectively, with electrons. These lines have emissivities varying as $T^{-0.8}$. The collisionally excited [O III] lines arise from the same reactants, but with emissivities that increase about as T^4, so the ratio of the lines is a strong function of T. The recombination lines in H II regions and planetary nebulae (reviewed in Peimbert 1995), while weak, seem much stronger than predicted by the standard analysis. If the present observations are correct, apparently the recombination lines arise from regions that are considerably cooler than those from which the [O III] lines are produced, and the true nebular abundances are considerably larger than those from the standard analysis. However a quantitative analysis of the errors in the standard analysis is impossible without a detailed theory of the distribution of T among various nebular densities. A commonly used algorithm (Peimbert (1967) is rather arbitrary, but a more accurate one requires physical constraints on the fluctuations.

There are good reasons for being cautious about the accuracy of the measurements of the recombination lines (Mathis 1996; Maciejewski et al. 1996), including radio measurements in the Orion Nebula that support the T determined from [O III] lines (Wilson & Jäger 1987) and also the Balmer continuum/Hβ ratio (Liu et al. 1995).

In summary, the standard analysis of H II regions suggests abundances of order 50% of solar, but the possible presence of recombination lines of O^+ and C^+ is worrisome. These lines might suggest that there are large fluctuations in T in ionized nebulae, and that their true compositions are possibly solar. We tentatively assume that the heavy-element recombination line strengths have somehow been overestimated and proceed on the assumption that the ISM reference abundance is about 2/3 solar.

4. The Nature of Grains in the Diffuse ISM

With a lower reference abundance and a rather high gas-phase abundance of C, we are faced with a major challenge to explain the extinction per H atom observed in the ISM (Bohlin, Savage, & Drake 1978).

To me, there seems to be good evidence that large grains are probably composite in nature, with the various materials intermixed, along with voids or vacuum. Small grains that have their own structural integrity can be chemically homogeneous, as shown by the isotopically anomalous inclusions within meteorites. Fluffy large grains are observationally indicated by the X-ray scattering halos surrounding point sources, such as X-ray binaries (Woo et al. 1994) and V1974 Cyg (Mathis et al. 1995). Another reason for suggesting that grains are fluffy comes from the systematic decrease (Cardelli, Clayton, & Mathis 1989) of the "far UV extinction", $A(1200 \text{ Å})/A(V)$, from the diffuse ISM to the outer parts of clouds. This decrease is caused by a dearth of small grains that cannot be explained by the accretion of the gas-phase atoms, but must be coagulation (Jura 1980). I feel that grains will become composite because maintaining their chemical integrity during coagulation seems difficult. The composite grains will contain at least a modest degree of vacuum because the coagulating particles will not fit together compactly.

Sofia et al. (1994), Spitzer & Fitzpatrick (1993), and Sembach & Savage (1996) have studied lines of sight towards objects at high Galactic latitude with the GHRS, thereby probing the composition of dust in the Galactic halo (see Savage, this volume). They use Zn as a reference abundance because it is lightly depleted into grains, and the H column densities for individual clouds are not possible to determine from the very strong Lyman-α. They all find that there is a surprising regularity of the gas-phase abundances of elements in the clouds well separated in velocity from the gas in the plane; the gas contains about 30% of the Mg, 55% of the Si, and 28% of the Fe, if we assume a solar reference abundance. Thus, *the Fe, Mg, and Si are not contained in equal proportions in the grains, as they would be if all of them were in silicates.*

The contrast is, of course, more impressive if the reference abundance for the modest-velocity clouds is sub-solar. If the reference abundance is 70% solar, almost none of the halo Mg and Fe atoms are in the gas, while 1/3 of the Si is in some form (SiO ?) that goes into the gas phase more readily than the Fe and Mg. Silicates, long a mainstay of almost all theories of interstellar dust, are not the only form of silicon in the ISM.

5. Grain Models with Reduced Carbon and Silicates

The reduction of the abundances of the refractory elements in grains provides a severe challenge to grain models and might lead to falsifying all except those close to being correct. At present, there are a fair number of solar-abundance models that can fit the observations. The reduction of carbon is, of course especially severe. I have tried to fit the extinction per H atom with 70% of solar abundances of all elements, plus 130 ppM of carbon in the gas. Thus, I go from solar (400 ppM) to $0.7 \times 400 - 130 = 150$ ppM. With 50 ppM for the bump, we are reduced from C/H = 350 ppM (solar) to 100 ppM in the non-bump grains!

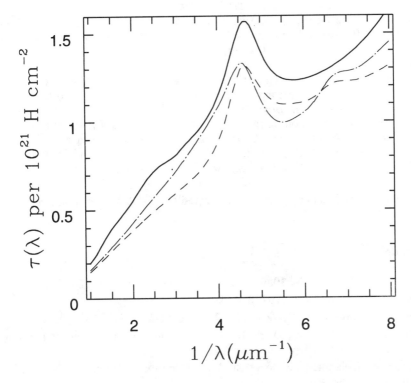

Figure 1. Optical depths per 10^{21} H atoms plotted against wavenumber. *Solid line*: observations (uncertainties: 9%); *dot-dashed*: model using graphite within composite grains; *dashed*: amorphous carbon in composite grains.

The observations of $E(B-V)$ per H^0 atom are discussed in Diplas & Savage (1994). They find a fractional error in the mean $N(\text{H I})/E(B-V)$ of 5.7%, which I take as the fractional error, including the correction for H$^+$ and H$_2$. I assume the Bohlin et al. (1978) values of $N(\text{H})/E(B-V)$, which takes H$_2$ into account in a statistical way. This 5.7 % error estimate is, then, quite conservative. The conversion from $E(B-V)$ to $A(V)$ (i.e., $R = 3.0 - 3.2$) involves about a 3.2% error, so the "standard" $A(V)/N(\text{H})$ contains at least a 9% error and probably substantially more.

Figure 1 shows various optical depth per 10^{21} H nuclei, plotted against wavenumber. The solid line is the observation (Bohlin et al. 1978), with a 9% error indicated. Also shown are two models, each with (a) fluffy grains, that improve the extinction per H in the visual; (b) silicates, using all of the available Mg, Si, and Mg; (c) 50 ppM of carbon in small graphite grains, to produce the bump; and (d) the rest of the carbon in one of two forms. The solid curve has the remaining carbon in graphite in the composite grains. The dashed curve has the remaining carbon in amorphous carbon with the most absorbing optical constants (i.e., best fit) of the several forms I tested: the "Be" form tabulated by Rouleau & Martin (1991). The cross sections of the composite grains were calculated using effective optical constants derived by the Maxwell-Garnett rule (Bohren & Huffman 1983). The size distribution was not assumed to be a power-law, but rather was adjusted to provide the best logarithmic fit to the extinction.

We see that the computed curves fall somewhat short of the observed. In view of the uncertainties in the optical constants, it is not clear whether or not the discrepancy is serious. However the use of the Maxwell-Garnett rule does not maximize the extinction per gram (Stognienko et al. 1995), and future work will enable a better fit to the observations. The feature shown at $1/\lambda = 6.5$ μm^{-1} is an artifact of the optical constants used for the silicate (Kim & Martin 1995). Future models with a better fit will be forthcoming.

This research has been partially supported by NASA and by assistance from the organizers.

References

Anders, E., & Grevesse, N. 1989, Geochim. Cosmochem. Acta, 53, 197
Baldwin, J. A., et al. 1991, ApJ, 374, 580
Bohlin, R. C., Savage, B. D., & Drake, J. F. 1978, ApJ, 224, 132
Bohren, C. F., & Huffman, D. R. 1983 Absorption and Scattering of Light by Small Particles (New York: Wiley)
Busolletti, E., Colangeli, L., Borghesi, A., & Orofino, V. 1989, A&AS, 70, 257
Cardelli, J. A. 1994, Science, 265, 209
Cardelli, J. A., Clayton, G. C., & Mathis, J. S. 1989, ApJ, 245, 345
Cunha, K., & Lambert, D. L. 1994, ApJ, 426, 170
Désert, F. X., Boulanger, F., and Puget, J. L. 1990, A&A, 237, 215
Diplas, A. , & Savage, B. D. 1994, ApJ, 427, 274
Dorschner, J., & Henning, Th., 1996, A&A Rev., in press
Draine, B. T. 1985, ApJS, 57, 587
Draine, B. T. 1989, in IAU Symp. 135, *Interstellar Dust*, ed. L. J. Allamandola & A. G. G. M. Tielens (Dordrecht: Reidel), 313
Draine, B. T., & Lee, 1984, ApJ, 285, 89 (DL)
Draine, B. T., & Malhotra, S. 1993, ApJ, 414, 632
Dufour, R. J., Shields, G. A., & Talent, R. J., Jr. 1982, ApJ, 252, 461
Federman, S. R., Sheffer, Y., Lambert, D. L., & Gilliland, R. L. 1993, ApJ, 413, L51
Ferlet, R. 1981, A&A, 98, L1
Fitzpatrick, E. L., & Massa, D. 1988, ApJ, 328, 734
Fritz, B., Lüty, F., & Anger, J. 1963, Zs. Phy., 174, 240
Fuente, A., Cernicharo, J., Garcia-Burillo, S., & Tejero, J. 1993, A&A, 275, 558
Furton, D. G., & Witt, A. N. 1990, ApJ, 364, L45
Gies, D. R., & Lambert, D. L. 1992, ApJ, 387, 673

Gilra, D. P. 1972, in *The Scientific Results from the Orbiting Astronomical Observatory OAO-2*, ed. A. D. Code, NASA SP-310, 295
Greenberg, J. M. 1979, in Stars and Starsystems, ed. B. E. Westerlund (Dordrecht: Reidel), 173
Grevesse, N., & Noels, A. 1993, in Origin and Evolution of the Elements, ed. N. Prantzos, E. Vangioni-Flam, & M. Cassé (Cambridge: Cambridge Univ. Press), 15
Joblin, C., Léger, A., & Martin, P. 1992, ApJ, 393, L79
Jura. M. 1980, ApJ, 235, 63
Kilian, J. 1992, A&A, 262, 171
Kilian, J., Montenbruck, O., & Nissen, P. E. 1994, A&A, 284, 437
Kim, S.-H., & Martin,. P. G. 1995, ApJ, 442, 172
Lambert, D. L., Sheffer, Y., Gilliland, R. L., & Federman, S. R. 1994, ApJ, 420, 756
Liu, X.-W., Barlow, M. J., Danziger, I. J., & Storey, P. J. 1995, ApJ, 450, L59
Maciejewski, W., Mathis, J. S., & Edgar, R. J. 1996, ApJ, in press
Mathis, J. S. 1990, ARAA, 28, 37
Mathis, J. S. 1993, Rep. Prog. Phys., 56, 605
Mathis, J. S. 1994, ApJ, 422, 176
Mathis, J. S., Cohen, D., Finley, J. P., & Krautter, J. 1995, ApJ, 449, 320
Mathis, J. S. 1996, Rev. Mex. Astr. Astrof., in press
Mathis, J. S., Rumpl, W., & Nordsieck, K. H. 1977, ApJ, 217, 425 (MRN)
Menella, V., Colangeli, L., Blanco, A., Bussoletti, E., Fonti, S., Palumbo, P. 1995a, ApJ, 444, 288
Menella, V., Colangeli, L., Bussoletti, E., Monaco, G., Palumbo, P., & Rotundi, A. 1995b, ApJS, 100, 149
Osterbrock, D. E., Tran, H. D., & Veilleux, S. 1992, ApJ, 389, 305
Peimbert, M. 1967, ApJ, 120, 22
Peimbert, M. 1987, in Star Forming Regions, ed. M. Peimbert & J. Jugaku (Dordrecht: Reidel), 111
Peimbert, M. 1995, in The Analysis of Emission Lines, ed. R. E. Williams (Cambridge: Cambridge Univ. Press), in press
Peimbert, M., Torres-Peimbert, S., & Dufour, R. J. 1993, ApJ, 418, 760
Rubin, R. H., Simpson, J. P., Haas, M. R., & Erickson, E. F. 1991, ApJ, 374, 564
Rouleau, F., & Martin, P. G. 1991, ApJ 377,526
Sandford, S. A., Allamandola, L. J., Tielens, A. G. G. M., Tapia, M., & Pendleton, Y. 1991, ApJ, 371, 607
Savage, B. D., & Sembach, K. R. 1996, ARAA, in press
Sembach, K. R., & Savage, B. D. 1996, ApJ, in press
Schmidt, G. D., & Witt, A. N. 1991, ApJ, 383, 698
Snow, T. P., & Witt, A. N. 1995, Science, in press
Sofia, U. J., Cardelli, J. A., & Savage, B. D. 1994, ApJ, 430, 650

Spitzer, L., Jr., & Fitzpatrick, E. L. 1993, ApJ, 409, 299
Steel, T. M., & Duley, W. W. 1987, ApJ, 315, 337
Stognienko, R., Henning, Th., & Ossenkopf, V. 1995, A&A, 296, 797
Tielens, A. G. G. M. 1996, in *the Role of Dust in Star-Forming Regions*, ed. H. U. Käufl & R. Siebenmorgen, (Berlin: Springer-Verlag), in press
Tielens, A. G. G. M., & Allamandola, L. J. 1991, in Solid State Astrophysics, School E. Fermi, 111 (ed. E. Bussoletti & G. Strazzula (Amsterdam: North Holland), 29
Tielens, A. G. G. M., Wooden, D. H., Allamandola, L. J., Bregman, J., & Witteborn, F. C. 1996, ApJ, in press
van Dishoeck, E. F., Blake, G. A., Draine, B. T., & Lunine, J. I. 1993, in Protostars and Planets III, ed. E. H. Levy & J. I. Lunine (Tucson: Univ. of Arizona Press), 163
Venn, K. A. 1995, ApJ, 449, 839
Whittet, D. C. B., Adamson, A. J., McFadzean, A. D., Bode, M. F., & Longmore, A. J. 1988, MNRAS, 233, 321
Wilson, T. L., & Jäger, B. 1987, A&A, 184, 291
Wilson, T. L., & Rood, R. T. 1994, ARAA, 32, 191
Witt, A. N. 1989, in Interstellar Dust, ed. L. J. Allamandola & A. G. G. M. Tielens (Dordrecht: Kluwer), 87
Witt, A. N., ,& Boroson, T. A. 1990, ApJ, 355, 182
Woo, et al. 1994, ApJ, 436, L5
York, D., Spitzer, L., Jr., Bohlin, R. C., Hill, J., Jenkins, E. B., Savage, B. D., & Snow, T. P. 1983, ApJ, 266, L55

Cosmic Abundances
ASP Conference Series, Vol. 99, 1996
Stephen S. Holt and George Sonneborn (eds.)

Abundances in Gaseous Nebulae

Harriet L. Dinerstein

Department of Astronomy, University of Texas, Austin, TX 78712

Abstract. I review progress in the study of elemental abundances in gaseous nebulae in our own and other galaxies, emphasizing developments in the last five years. Advances in instrumentation have led to significant new results from optical, ultraviolet, and infrared spectroscopy. Recent re-examinations of the assumptions and calibrations underlying familiar techniques have made it clear that our present knowledge of nebular abundances is limited primarily by our incomplete understanding of the physical conditions in nebulae, particularly inhomogeneities and the mechanisms that produce them.

1. Introduction

As is true for many of the speakers at this conference, I am faced with the daunting task of reviewing an extremely broad topic within the confines of limited time and space. I have been asked to cover abundances in gaseous nebulae both within the Milky Way and in external galaxies. In order to make this task managable, I will treat this talk as an update to two comprehensive reviews published in 1990, by Dinerstein (1990) and Shields (1990). The former focused on abundances in extragalactic H II regions; the latter reviewed the properties of extragalactic H II regions in general. Either can serve as the starting point for the present review.

The last five years have seen a dramatic expansion of the observational data base on H II regions, especially extragalactic regions. As is so often true in astronomy, this expansion has been primarily driven by instrumental developments. Optical spectroscopy of nebular emission lines can now be performed on large numbers of galaxies, or alternatively on large numbers of H II regions within particular galaxies, using efficient CCD spectrometers. The ability to study faint objects with ultraviolet spectroscopy has been greatly advanced by the availability of the *Hubble Space Telescope* (HST), a trend that should continue at least as dramatically when the next generation UV spectrometer (the *Space Telescope Imaging Spectrograph*, or STIS) is installed on HST. The main progress in infrared spectroscopy has been in studies of the Milky Way, and has come mostly from the *Infrared Astronomy Satellite* (IRAS) and observations made from the *Kuiper Airborne Observatory* (KAO). Other than a few of the brightest sources, infrared emission-line spectroscopy of extragalactic H II regions awaits the input of the *Infrared Space Observatory* (ISO) and the *Stratospheric Observatory for Infrared Astronomy* (SOFIA).

The interpretation of nebular emission line strengths is now on a firmer footing, due to the availability of new atomic data (see the review of Pradhan & Peng 1995 and references therein). However, many issues remain unresolved. What may be called the "internal" reliability of calculated models of nebulae (ionization structure, emergent line intensities) has been assessed through such exercises as the recent Lexington, Kentucky workshop on nebular models and simulations (Ferland et al. 1995). External or systematic effects are much harder to evaluate. These include the issue of inhomogeneities in temperature and density, which has cropped up repeatedly in the history of this field (see Peimbert 1995 and references therein). Nevertheless, the last few years have seen many intriguing and substantial new results. Some of the most important work has been on systematic effects, such as how abundances depend on a nebula's position within a galaxy, on the Hubble type of the host galaxy, and on the galaxy's environment. All of these questions will be essential to interpreting observations of galaxies at large redshifts, as well as providing constraints for theories of galaxy formation and evolution.

2. Observational Progress since 1990

2.1. Abundances in Milky Way Nebulae

First let us set the stage by reviewing the status of work on the abundance gradient of the Milky Way, circa 1990. In a benchmark paper, Shaver et al. (1983) combined optical and radio information to deduce abundances for H II regions across the Galactic disk, from the obscured inner Galaxy to slightly beyond the solar circle. These authors concluded, essentially in agreement with earlier optical studies of H II regions within a few kpc of the Sun, that the Galactic abundance gradient in oxygen is $d\log(O/H)/dR = -.07$ dex kpc^{-1}, while the nitrogen gradient is slightly steeper, $d\log(N/H)/dR = -.09$ dex kpc^{-1}. Their results on ratios of other species relative to oxygen were not as definitive, but they found Ne/O and Ar/O to be constant, and S/O to perhaps increase slightly with increasing R_G. In the 1980's, several groups conducted surveys of Galactic H II regions in the infrared, from the KAO and ground-based telescopes. Mid-infrared (5 - 20 μm) studies examined emission lines of Ne, Ar, and S, which were found to be elevated in the inner galaxy (e.g. Lester et al. 1981; Herter, Helfer, & Pipher 1983). Far-infrared spectroscopic studies sampled doubly-ionized N and O, and found a higher N/O ratio in the inner Galaxy (Lester et al. 1987). Both effects appeared to be about a factor of two or three, but with significant uncertainty due to the lack of accurate collision strengths at that time.

Fich & Silkey (1991) attempted to extend the work of Shaver et al. to 18 kpc in the outer Galaxy. Unfortunately, their results show large scatter and large error bars, due mainly to the fact that they were unable to directly determine values for the electron temperature (T_e) in many of their objects. Their major claimed result was a suggestion that the N/H gradient flattens at large R_G. Most of the recent progress on Galactic H II regions in the last few years has been in the area of infrared spectroscopy. Using low-resolution spectra obtained by IRAS, Simpson & Rubin (1990) derived Galactic gradients in Ne and S from a large sample of H II regions, finding values quite similar to those in O and N: $d\log(Ne/H)/dR = -.08$ dex kpc^{-1} and $d\log(S/H)/dR = -.05$

dex kpc^{-1}. The most recent major compendium of abundances derived from airborne observations of infrared lines is that of Simpson et al. (1995), who find formal gradients of $-.10$, $-.08$, and $-.07$ dex kpc^{-1} for $\log(N/H)$, $\log(Ne/H)$, and $\log(S/H)$ respectively, but comment that the observed pattern is more accurately described as a step function, such that all abundances inside $R_G = 6$ kpc are about a factor of two higher than at large R_G, as suggested by Lester et al. (1987) for the N/O ratio. Simpson et al. do find, as seen in external spiral galaxies, that the nebular ionization (and hence presumably the temperatures of the exciting stars) increases with R_G. Their study included regions out to 10 kpc. Infrared observations of three H II regions in the outer Galaxy by Dinerstein et al. (1993; 1996) find that N/O continues to decrease beyond the solar circle. Further observations of H II regions in the outer galaxy are presented by Rudolph et al. (this conference). The use of new collision strengths for many infrared lines (Blum & Pradhan 1992; Pradhan & Peng 1995) make the recent analyses more reliable. It also, for example, brings determinations of N/O from the infrared lines into better agreement with the most recent determinations in the optical (Baldwin et al. 1991).

Elemental abundances in Galactic planetary nebulae were reviewed by Clegg (1989) and Peimbert (1990). Planetary nebulae cannot be treated in the same way as H II regions, since they arise from stellar populations with a range of ages. Furthermore, much of the interest in studying their composition lies in probing nucleosynthesis and mixing processes within their progenitor stars, as well as the evolution of elemental abundances with time in the Galaxy (Perinotto 1991; Kingsburgh & Barlow 1994). Nevertheless, a Galactic gradient can be estimated for samples of planetary nebulae selected to belong to a homogeneous population; when this is done, the inferred gradient, for example for O/H, is similar to that found for H II regions (e.g. Amnuel 1993; Maciel & Köppen 1994).

2.2. Gradients in External Spiral Galaxies

The standard, or "direct," method of determining elemental abundances in nebulae from emission line observations has been described by many authors; a concise "flow chart" is shown in Dinerstein (1990, p. 260). In the simplest form of this method, a value for T_e is derived from an observed line intensity ratio, usually [O III] 4363/5007Å, and the nebula is assumed to be isothermal. Temperature-dependent line emissivities are calculated and used to convert the observed intensity ratio of lines from two different elements (e.g. O and H) into an ionic ratio. Correction for unobserved ions is made by multiplying by an "ionization correction factor" or *icf*, which is based either on calculations of nebular ionization structure or some empirically-determined formula. This method is practical only for relatively bright and low-metallicity H II regions.

For faint or metal-rich nebulae in which the 4363Å line is too weak to measure, other methods must be invoked. The main alternate method relies on the fact that, to a large extent, extragalactic H II regions define a tight, well-defined sequence in a variety of line-ratio diagrams formed from bright lines such as [O III] 5007Å and Hβ (e.g. McCall et al. 1985). Pagel et al. (1979) showed that, if calibrated against abundance, one can use easily observed quantities such as $R_{23} = [F([O\ III]\ 4959+5007Å) + F([O\ II]\ 3726+3729Å)]/F(H\beta)$ as abundance indicators. These so-called "bright-line" or "empirical" methods have several shortcomings. First, the relation given above is not monotonic, but rather, double-valued; as log O/H decreases from ≥ 9.0, T_e and therefore R_{23} increase due to the lower efficiency of cooling by the O^{++} ion. However, at O/H of a few tenths of solar, the trend reverses and the line ratio begins decreasing again; now the declining number of ions able to emit in the lines begins to dominate the behavior. For a given H II region, it may be hard to determine on which "branch" of the relation the object lies. Second, there are several different calibrations of the "bright-line" ratios in the literature (e.g. Edmunds & Pagel 1984; McCall et al.; Dopita & Evans 1986; Skillman 1989). At the model-dependent high-abundance end, they differ by up to 0.6 dex (see Edmunds 1989 for a comparison and discussion). Finally, in certain regimes the relation may not be unique, but may depend on other parameters such as the filling factor or so-called "ionization parameter" U, which is essentially the ratio of the number density of photons to the gas density (see Shields 1990).

Before 1990, only a few large external galaxies had been observed extensively enough to determine not only *an* abundance, but also trends in abundances across their disks. The last few years have finally seen the publication of several studies based on substantially larger samples. Vila-Costas & Edmunds (1992) summarize data on 30 galaxies from the literature. They confirm the well-known "mass-metallicity" relation (see Dinerstein 1990); in particular they find that the central abundances, as extrapolated from the observed radial gradients, correlate well with galaxy mass. They also find relationships between abundances and galaxy type: barred galaxies have shallower slopes (as was known previously, e.g. from M83); and non-barred spirals of later Hubble types (Sc, Sd) show steeper gradients than do earlier types. However, the number of very early type galaxies (Sa and Sb) was small, a deficiency addressed by Oey & Kennicutt (1993), who observed 15 early-type spirals. These authors found systematically higher metallicities in the H II regions of the early type galaxies, but an apparent break-down (or flattening) of the mass-metallicity relation at the high-mass end. (This may, however, be an artifact of systematic effects, since the abundances were necessarily derived from the "bright-line" method, which is particularly model-dependent at high metallicities.) In the most ambitious study to date, Zaritsky et al. (1994) combined new with pre-existing data to examine abundances in 39 galaxies with at least five H II regions each. They compare the gas-phase abundances as measured at a fiducial radius (not the extrapolated central abundance) to several global parameters of the galaxies, and find that the abundances correlate best with circular velocity (a measure of total mass), but also with luminosity and t-type (which are themselves at least somewhat correlated with mass). They suggest that the steepest gradients are found in intermediate-type galaxies rather than very early or late ones, but point out that it is difficult to compare gradients in a self-consistent way, since different authors choose different radial normalizations.

Complementary to these large studies of many galaxies are in-depth studies of individual galaxies, in which abundances are determined for a very large number of H II regions. A good example is the study of NGC 628 and NGC 6946 by Belley & Roy (1992). These authors examine the very best-studied galaxies, in which at least 50 H II regions have been observed, and claim that they all have very similar radial gradients, $\mathrm{dlog(O/H)/dR} = -.08 \pm .01$ dex kpc^{-1}. Perhaps the best-studied external spiral of all is M101. Earlier references are given by Dinerstein (1990) and Shields (1990). Scowen et al. (1992) used emission-line imaging to derive line ratios and therefore abundances at many positions within M101. They call attention to a "threshold effect"; the line ratios are systematically different at positions of higher surface brightness, and therefore so are the abundances derived from the bright-line method. This is probably an ionization effect, because they use only [O III] lines, rather than lines of both [O III] and [O II]. Nevertheless, routine application of one particular bright-line calibration has the curious effect of showing an apparent "break" (change of slope) in the radial O/H gradient at around 10 kpc for the high-surface brightness regions, but not for the fainter ones! This has been countered by both Garnett & Kennicutt (1994) and Henry & Howard (1995), who argue that there is no such break, but rather that the slope of the radial gradient is constant over the entire disk of the galaxy. In the first of a series of papers, Kennicutt & Garnett (1996) re-examine the H II regions of M101, looking at such issues as whether there is intrinsic dispersion at a given radius. They derive O/H abundances via the bright-line method using three different calibrations and show that the presence or absence of a break depends on the choice of calibration; they also suggest that there may be a global asymmetry in the gradient, associated with the interaction of M101 with neighboring galaxies.

2.3. Abundances in Dwarf Irregular Galaxies

Next we come to studies of small, generally metal-poor, galaxies dominated by a single giant H II region. These objects have suffered from a surfeit of different names, including "isolated extragalactic H II regions," "H II galaxies," and "blue compact galaxies." Essentially, they are small, gas-rich galaxies in which vigorous star formation is occurring at present, but relatively little star formation and chemical enrichment has occurred in the past. They offer the appealing opportunity of watching the early stages of the progressive enrichment of a galaxy in metals, as well as finding material whose helium abundance is relatively unchanged since the Big Bang. (The subject of primordial or cosmological helium abundances will not be addressed here.) These were the objects which first inspired the idea of "bursting" or episodic star formation (Searle & Sargent 1972). A general review of their properties is given by Kunth (1989). Unlike many of the H II regions in large spirals, those in the dwarf galaxies are generally high-excitation, low-metallicity objects in which it is usually possible to determine T_e directly, so that observers need not rely on the bright-line method.

A major quest of observers has been to find the lowest-metallicity dwarf galaxy, perhaps one even more metal-poor than I Zw 18, with O/H = 1.5×10^{-5}. Skillman et al. (1988; see their Figure 3) showed that the "mass-metallicity" relation extends down to low values of both, and suggested that an effective strategy for finding metal-poor objects was to look at low-luminosity galaxies. This was borne out by Skillman et al. (1989), who looked at low-luminosity members of the Local Group. Continuing efforts to find additional objects of this kind have produced larger samples, but have not qualitatively changed earlier conclusions. The frequency distribution of the O/H abundances in dwarf galaxies peaks at $\log(O/H)+12 = 8.2$, with a spread from 7.5 to 8.5 (Peña et al. 1991; Terlevich et al. 1991). Galaxies with O/H as low as that of I Zw 18 are found occasionally (e.g. Izotov et al. 1991) but are rare, and objects selected for their low luminosities have systematically lower abundances (Moles et al. 1990). One interesting fact is that that N/O shows large variations at a given O/H (Marconi et al. 1994), which may be related to different "clocks" driving the enrichment of N and O (which are produced in different astrophysical sites), and to episodic star formation, which is clearly a major characteristic of the star formation activity in these galaxies.

3. Fundamental Physics and Calibrations

3.1. Uncertainties in the Bright-Line Method

In §2.2 above we discussed some of the uncertainties associated with the bright-line method which is the basis for nearly all measurements of faint and metal-rich nebulae. The relationship between O/H abundance and R_{23} is intrinsically double-valued; this leads to an uncertainty of at least ±0.3 dex, a factor of two, near a value of (O/H)≈ 0.2 solar (see Skillman 1989 for a discussion). Unfortunately, this is a very "interesting" regime, relevant to both H II regions in dwarf galaxies and those in the outer regions of spiral disks (e.g. Kennicutt & Garnett 1996). In certain regimes, the bright-line ratio is also sensitive to properties such as the ionization parameter U, stellar temperature T_{eff}, and possibly even depletion of gas-phase species into dust (see §3.3 below). These effects are largest for high-metallicity, low-excitation regions, which is why the derived abundances for such objects are said to be model-dependent. The calibrations of R_{23} diverge most strongly at this end (a spread of up to 0.6 dex, a factor of 4), which is precisely where one most needs the bright-line method, because the 4363Å line cannot be observed. This calibration uncertainty has been discussed by Edmunds (1989; see his Figure 1), Oey & Kennicutt (1993), and Kennicutt & Garnett (1996), among others. A recent step towards resolving this disagreement was taken by Kinkel & Rosa (1994), who measured temperature-sensitive line ratios of N^+, O^+, and S^{++} in the "benchmark" metal-rich H II region Searle 5 in M101, and derived log(O/H)+12 = 8.9, on the low end of the various calibrations.

3.2. The Temperature Structure of Real Nebulae and the Saga of t^2

Abundances derived from collisionally excited optical and ultraviolet lines are strongly dependent on the gas temperature, because of the exponential dependence of the excitation cross-section. This is not a problem for infrared lines (Dinerstein 1986; 1995), which are sensitive to density instead (Rubin 1989). This temperature sensitivity places strong demands on the observer to measure the value of T_e as accurately as possible. However, even this does not take care of the whole problem, since there is no guarantee that a unique value of T_e gives a good description of conditions throughout the whole nebula. The actual temperature structure *within* a nebula is important for the abundance determination. "Structure" can be broken down into two categories: macroscopic (e.g. systematic gradients through the ionization region) and microscopic (over physically small length scales). Nebular ionization models predict, for example, that the outer regions of metal-rich nebulae are hotter, due to hardening of the radiation field because of radiative transfer effects and the diminishing abundance of the efficient coolant O^{++} (Stasińska 1980). These effects, however, do not fully explain the observed differentials between T_e values measured from different ions (Garnett 1992).

The thorniest problem in the temperature structure of real nebulae is whether there are variations or imhomogeneities in temperature (and density) on small scales, which would have the effect of introducing a broader spectrum of T_e values within the nebula. A two-parameter characterization was introduced by Peimbert (1967), who defined a density-weighted (by $n_e n_i$) mean temperature T_0, and a similarly weighted rms "temperature fluctuation" parameter t^2, which effectively measures the width of the temperature distribution. The latter can be related to observed line intensity ratios, and if non-zero, has a significant impact on the inferred abundances. Values of order $t^2 = 0.03 - .04$ will elevate the abundances of species such as O, N, Ne, and S by about 0.2–0.4 dex. Interestingly, this is enough to bring the nebular abundances into agreement with the solar values. This analysis was initially applied to well-studied H II regions like Orion (e.g. Peimbert & Costero 1969), but was not adopted by most investigators for extragalactic H II regions, in part because of the lack of sufficient spectral information to derive *in situ* values for more than a single temperature parameter. Therefore, essentially all extragalactic abundance determinations have assumed either isothermal nebulae, or nebulae with only macroscopic temperature structure (as described by photoionization models).

Recent work by Peimbert and his collaborators has raised the issue of temperature fluctuations in nebulae anew. The primary impetus has come from studies of intrinsically faint, but only weakly temperature-dependent, recombination lines of O and N. New calculations of recombination rates, and new observations in the literature, have prompted a renewed argument that t^2 has a value of 0.04–0.045 in H II regions and planetary nebulae (Peimbert 1993; Peimbert et al. 1993; Peimbert 1995; see also Dinerstein et al. 1985; 1995). Ironically, this comes at a time when some researchers have concluded that the Sun has higher elemental abundances than the local interstellar medium, and should therefore not be used to define the "reference" abundances (see §4.1. below, and Mathis 1996, this conference). Another problem is that the origin of such temperature inhomogeneities is unclear. Peimbert (1993) suggests several possibilities, including local variations in the dust-to-gas ratio, the presence of small amounts of shocked gas, and inhomogeneities in the abundances of important coolant species. Of these, shocks may be the most plausible, particularly in nebulae that host stars with strong winds, which is often the case both for the massive O stars that ionize giant H II regions and for the central stars of planetary nebulae.

3.3. Other Effects

There are a number of other physical properties that can affect derived abundances in similar ways to the temperature inhomogeneities discussed above. For example, density inhomogeneities will affect the emergent line intensity ratios, and can mimic temperature inhomogeneities (Rubin 1989). However, the density contrast required to have a significant effect on optical lines is fairly extreme, such that inclusions of order $n = 10^6$ cm^{-3} may be required (Viegas & Clegg 1994). Condensations with such high densities are in fact seen in the Orion

Nebula (Bautista et al. 1994), although it is not clear whether they contribute significantly to the [O III] emission.

Another idea that has been explored recently is that the degree of depletion into dust grains of refractory elements, such as Si, Mg, and Fe, may affect the cooling rate and thereby alter the gas temperature (Henry 1993). The singly-ionized forms of these elements are efficient coolants because they possess easily-populated low-lying energy levels which give rise to infrared emission lines. A nebula in which some of the grains have been destroyed, liberating these species back into the gas phase, will cool more efficiently than one in which nearly all of these atoms are tied up in the dust; this can mimic a higher elemental abundance. However, Shields & Kennicutt (1995) have reconsidered the role of dust, and find that it produces several different effects, some of which cancel; they conclude that the uncertainty in the gas-phase depletions introduces a somewhat smaller uncertainty in abundances determined from the usual optical lines than proposed by Henry.

Another aspect of giant extragalactic H II regions is that they are ionized by large ensembles of stars, which probably include associations with a range of ages. As pointed out by Masegosa et al. (1991), this should produce spatial variations in the ionization parameter U and therefore in the line ratios, an effect which has not yet been significantly addressed or modelled.

4. Some Outstanding Questions

4.1. What are the Standard or "Reference" Abundances?

We have already alluded to a long-term issue regarding the absolute abundance scale (metal-to-hydrogen ratios) in nebulae: the fact that standard procedures (assuming isothermality) yield values that are lower by about 0.2-0.3 dex than the solar value. How well do the nebular values agree with the abundances in young stars, which should be good indicators of present-day nebular abundances (without suffering from dust depletion effects)? A number of investigators have measured abundances in B stars, and find log(O/H) values significantly lower than the "canonical" solar value of 8.9; these studies yield values of about 8.65 (Fitzsimmons et al. 1990; Cunha & Lambert 1992), although the adoption of a significant microturbulence parameter for the model atmospheres can elevate the derived abundances to nearly the solar value (Fitzsimmons et al. 1992). Additionally, these authors and others (e.g. Kaufer et al. 1994) find no evidence for a radial gradient in the metallicity of B stars of the magnitude suggested for Galactic H II regions, although there seems to be a substantial dispersion in metallicity at large galactocentric distances (Rolleston et al. 1994). Stellar and interstellar abundances can also be derived and compared in the Magellanic Clouds, where they appear to be in reasonable agreement, with the possible exception of carbon (Spite & Spite 1990; Pagel 1993).

We are thus presented with a dilemma: should we be worried about the discrepancy between the nebular and stellar abundance scales in the solar neighborhood? If we adopt a moderately large value for t^2, then the nebular abundances rise to nearly the solar value. Secondly, we may choose a different set of comparison abundances (designated "reference abundances" by Mathis 1996, this conference), depending on whether we use the Sun or the B stars. Furthermore, close examination of the studies of even such a well-observed H II region as Orion reveals that different authors find differences of about a factor of two for several elements, even when taken relative to oxygen. This situation is summarized in Table 1 below, which gives a sobering reminder that the uncertainties in the nebular values are probably large enough to accomodate modest apparent discrepancies between the nebular and stellar abundance scales.

Table 1. Comparison of Abundances Derived for Orion

Ratio	Baldwin et al.	Rubin et al.	Osterbrock et al.	Peimbert
O/H	3.8×10^{-4}	4×10^{-4}	...	5.7×10^{-4}
C/O	0.56	0.85	...	0.59
N/O	0.23	0.17	0.21	0.13
Ne/O	...	0.20	0.18	0.20
S/O	0.035	0.021	...	0.029
Ar/O	0.0055	0.011	0.0088	0.0060

Key to References: Baldwin et al. (1991); Rubin et al. (1991); Osterbrock, Tran, & Veilleux (1992); Peimbert (1993) for $t^2 = .04$.

4.2. Which factor dominates the systematics?

Which factors control or determine the abundances of elements in nebulae? Is it the nature of the host galaxy, and if so, is the total galaxy mass or the morphological type more important? From the discussion in §2.2, it appears that in general the total galaxy mass, or some surrogate of it (such as the luminosity or circular velocity) is the main determining factor. Barred spirals have shallower gradients (or none), which is expected on dynamical grounds; there may be a trend for later-type spirals to have steeper gradients than early types, but the scatter is quite large and such a trend is not well-determined. Even the correlation of abundance with galaxy mass has exceptions. Several recent studies of the relatively inactive (in the sense of star formation), although physically large and gas-rich, spirals of the class known as "low surface brightness" (LSB) galaxies have shown them to have anomalously low gas-phase metallicities for their masses (McGaugh 1994; Rönnback & Bergvall 1995).

The question of whether gas-phase abundances depend on the environment of the host galaxy has been an active area recently, and has been mainly addressed by studying galaxies in the Virgo cluster. The first salvo was fired by Shields et al. (1991), who found systematically higher O/H abundances in five spirals in the Virgo cluster, as compared with field galaxies. This conclusion was questioned by Henry et al. (1992), who found a "normal" O/H value for another galaxy in Virgo, NGC 2403. The contradiction was more apparent than real, however. These two groups used different calibrations for the bright-line method, and chose a different set of field galaxies to provide the comparison. Also, NGC 2403 lies on the outskirts of the Virgo cluster, and so may have suffered less from environmental impacts than objects closer to the cluster center. Henry et al. (1994) found a high O/H value for NGC 4254, in better agreement with Shields et al. Skillman et al. (1996) examined H II regions in nine Virgo cluster spirals. They use the H I deficiency as an indicator of the severity of the environmental influence, and find the highest abundances for galaxies in the most H I-deficient group, which presumably have experienced the strongest stripping of their interstellar gas by outside influences.

4.3. Ratios of other elements to oxygen

Finally, we turn to the abundances of other elements relative to oxygen. The abundance ratio N/O has been studied extensively, particularly since it can be measured from the ground. It has been established that there must be multiple nucleosynthetic sources and processes contributing to the nitrogen abundance, which behaves differently in the low- and high-abundance regimes. In the low-metallicity regime there appears to be a "floor" of $\log(N/O) = -1.5$, presumably due to a primary process that can produce N from a star without seed metals (Garnett 1990). On the other hand, in objects with metallicities comparable to that of the Sun, N/O varies approximately with O/H, behavior that would be exhibited by a species that is produced by a secondary mechanism. Between these two limits, a large scatter is seen; Vila-Costas & Edmunds (1993) propose that there may be a "delayed-primary" mechanism as well as a secondary mechanism in operation, and that in different galaxies these mechanisms contribute to different extents.

With regard to sulfur, the situation is unclear. A number of authors studying both the Milky Way and external galaxies have suggested trends in S/O, or alternatively, a shallower radial gradient for S than for O. However, in many cases these S abundances are derived from limited observational material ([S II] lines only, or temperature-sensitive [S III] lines). The present uncertainties in most derived S abundances are sufficiently large that it is premature to give strong weight to any claimed trend (Díaz et al. 1991).

The study of C and Si in extragalactic H II regions has taken a great step forward with the application of ultraviolet spectroscopy from HST. Garnett et al. (1995a) present newly determined C abundances for several extragalactic H II regions. They find that C/O increases with O/H, but C/N shows non-monotonic behavior, suggesting that enrichment in these two elements is decoupled. Garnett et al. (1995b) measured Si/O for a number of extragalactic H II regions, and find a uniform, high value of $\log(Si/O) = -1.6$, which suggests that the gas-phase abundance of silicon is not severely depleted in these H II regions; perhaps the grains are destroyed by stellar winds or supernova shocks (see also Garnett 1996).

5. Looking Ahead

What developments can we anticipate in the field of nebular abundances, in the next few years? Clearly, the flood of new data will continue, if not increase. Optical studies will further clarify the roles and effects of the global properties of galaxies and their environments on the abundances in the H II regions that they host. Ultraviolet spectroscopy with new instruments on HST should yield much more information on the abundances of C, Si, and N from their doubly-ionized states, providing accurate abundances relative to O and clarifying the operation of chemical evolutionary processes. In the infrared, spectroscopic observations of S, Ne, and Ar with ISO should provide much more reliable measurements of these elements, perhaps to be followed up later on with SOFIA and SIRTF. Better atomic parameters and more sophisticated modelling, particularly addressing the issue of inhomogeneities in temperature and density, are to be hoped for, and may resolve some of the current unsettled issues of calibration and consistency. I expect (and certainly hope!) that any review of the subject of abundances in nebulae that is written 5 or 10 years from now will present a substantial number of new conclusions!

Acknowledgments. I thank the organizers of this meeting for giving me the opportunity to draw together and summarize the current state of knowledge on nebular abundances; I also thank G.A. Shields for helpful discussions while preparing this review.

References

Amnuel, P.R. 1993, MNRAS, 261, 263

Baldwin, J.A., Ferland, G.J., Martin, P.G., Corbin, M.R., Cota, S.A., Peterson, B.M., & Slettebak, A. 1991, ApJ, 374, 580

Bautista, M.A., Pradhan, A.K., & Osterbrock, D.E. 1994, ApJ, 432, L135

Belley, J. & Roy, J.-R. 1992, ApJS, 78, 61

Blum, R.D., & Pradhan, A.K. 1992, ApJS, 80, 425

Clegg, R.E.S. 1989, in *Planetary Nebulae, IAU Symp.* 131, S. Torres-Peimbert, Dordrecht: Kluwer, 139

Cunha, K. & Lambert, D.L. 1992, ApJ, 399, 586

Díaz, A.I., Terlevich, E., Vilchez, J.M., Pagel, B.E.J., & Edmunds, M.G. 1991, MNRAS, 253, 245

Dinerstein, H.L. 1986, PASP, 86, 979

——————. in *The Interstellar Medium of Galaxies*, H.A. Thronson, Jr. & J. M. Shull, Dordrecht: Kluwer, 257

——————. 1995, in *The Analysis of Emission Lines*, R.E. Williams & M. Livio, Cambridge: Cambridge U. Press, 134

Dinerstein, H.L., Haas, M.R., Erickson, E.F., & Werner, M.W. 1993, BAAS, 25, 850

——————. 1995, in *Airborne Astronomy Symposium on the Galactic Ecosystem: From Gas to Stars to Dust*, M.R. Haas, Davidson, J.A., & Erickson, E.F., San Francisco: ASP, 365

——————. 1996, in preparation

Dinerstein, H.L., Lester, D.F., & Werner, M.W. 1985, ApJ, 291, 561

Dopita, M.A. & Evans, I.N. 1986, ApJ, 307, 431

Edmunds, M.G. 1989, in *Evolutionary Phenomena in Galaxies*, J.E. Beckman & B.E.J. Pagel, Cambridge: Cambridge U Press, 356

Edmunds, M.G. & Pagel, B.E.J. 1984, MNRAS, 211, 507

Ferland, G., et al. 1995, in *The Analysis of Emission Lines*, R.E. Williams & M. Livio, Cambridge: Cambridge U. Press, 83

Fich, M. & Silkey, M. 1991, ApJ, 366, 107

Fitzsimmons, A., Brown, P.J.F., Dufton, P.L., & Lennon, D.J. 1990, A&A, 232, 437

Fitzsimmons, A., Dufton, P.L., & Rolleston, W.R.J. 1992, MNRAS, 259, 489

Garnett, D.R. 1990, ApJ, 363, 142

———————. 1992, AJ, 103, 1330

———————. 1996, RevMexA&A, in press

Garnett, D.R. & Kennicutt, R.C., Jr. 1994, ApJ, 426, 123

Garnett, D.R., et al. 1995a, ApJ, 443, 64

———————. 1995b, ApJ, 449, L77

Henry, R.B.C. 1993, MNRAS, 261, 306

Henry, R.B.C. & Howard, J.W. 1995, ApJ, 438, 170

Henry, R.B.C., Pagel, B.E.J., & Chincarini, G.L. 1994, MNRAS, 266, 421

Henry, R.B.C., Pagel, B.E.J., Lasseter, D.F., & Chincarini, G.L. 1992, M N R A S, 258, 321

Herter, T., Helfer, H.L., & Pipher, J.L. 1983, A&AS, 51, 195

Izotov, Yu.I., Lipovetsky, V.A., Guseva, N.G., Kniazev, A.Yu., & Stepanian, J.A. 1991, A&A, 247, 303

Kaufer, A., Szeifert, Th., Krenzin, R., Baschek, B., & Wolf, B. 1994, A&A, 289, 740

Kennicutt, R.C., Jr. & Garnett, D.R. 1996, ApJ, in press

Kingsburgh, R.L. & Barlow, M.J. 1994, MNRAS, 271, 257

Kinkel, U. & Rosa, M.R. 1994, A&A, 282, L37

Kunth, D. 1989, in *Evolutionary Phenomena in Galaxies*, J.E. Beckman & B.E.J. Pagel, Cambridge: Cambridge U. Press, 22

Lester, D.F., Bregman, J.D., Witteborn, F.C., Rank, D.M., & Dinerstein, H.L. 1981, ApJ, 248, 524

Lester, D. F., Dinerstein, H. L., Werner, M. W., Watson, D. M., Genzel, R., & Storey, J. W. V. 1987, ApJ, 320, 573

Maciel, W.J. & Köppen, J. 1994, A&A, 282, 436

Marconi, G., Matteucci, F., & Tosi, M. 1994, MNRAS, 270, 35

Masegosa, J., Moles, M., & del Olmo, A. 1991, A&A, 249, 505

Mathis, J. 1996, this volume

McCall, M.L., Rybski, P.M., & Shields, G.A. 1985, ApJS, 57, 1

McGaugh, S.S. 1994, ApJ, 426, 135

Moles, M., Aparicio, A., & Masegosa, J. 1990, A&A, 228, 310

Oey, M.S. & Kennicutt, R.C., Jr. 1993, ApJ, 411, 137

Osterbrock, D.E., Tran, H.D., & Veilleux, S. 1992, ApJ, 389, 305

Pagel, B.E.J. 1993, in *New Aspects of Magellanic Cloud Research*, G. Klare, Berlin: Springer-Verlag

Pagel, B.E.J., Edmunds, M.G., Blackwell, D.E., Chun, M.A., & Smith, G. 1979, MNRAS, 189, 95

Peimbert, M. 1967, ApJ, 150, 825

———————. 1990, Rep. Prog. Phys. 53, 1559
———————. 1993, Rev. Mex. A&A, 27, 9
———————. 1995, in *The Analysis of Emission Lines*, R.E. Williams & M. Livio, Cambridge: Cambridge U. Press, 165
Peimbert, M. & Costero, R. 1969, Bol. Obs. Tonantzintla y Tacubaya, 5, 3
Peimbert, M., Storey, P.J., & Torres-Peimbert, S. 1993, ApJ, 414, 626
Peña, M., Ruiz, M.T., & Maza, J. 1991, A&A, 251, 417
Perinotto, M. 1991, ApJS, 76, 687
Pradhan, A.K. & Peng, J. 1995, in *The Analysis of Emission Lines*, R.E. Williams & M. Livio, Cambridge: Cambridge U. Press, 8
Rönnback, J. & Bergvall, N. 1995, A&A, in press
Rubin, R. H. 1989, ApJS, 69, 897
Rubin, R.H., Simpson, J.P., Haas, M.R., & Erickson, E.F. 1991, ApJ, 374, 564
Rolleston, W.R.J., Dufton, P.L., & Fitzsimmons, A. 1994, A&A, 284, 72
Rudolph, A.L., Simpson, J.P., Erickson, E.F., Haas, M.R., & Fich, M. 1996, this volume
Scowen, P.A., Dufour, R.J., & Hester, J.J. 1992, AJ, 104, 92
Searle, L. & Sargent, W.L.W. 1972, ApJ, 173, 25
Shaver, P.A., McGee, R.X., Newton, L.M., Danks, A.C., & Pottasch, S.R. 1983, MNRAS, 204, 53
Shields, G.A. 1990, ARAA, 28, 525
Shields, G.A., Skillman, E.D., & Kennicutt, R.C., Jr. 1991, ApJ, 371, 82
Shields, J.C. & Kennicutt, R.C., Jr. 1995, ApJ, 454, 807
Simpson, J.P., Colgan, S.W.J., Rubin, R.H., Erickson, E.F., & Haas, M.R. 1995, ApJ, 444, 721
Simpson, J.P. & Rubin, R.H. 1990, ApJ, 354, 165
Skillman, E.D. 1989, ApJ, 347, 883
Skillman, E.D., Kennicutt, R.C., Jr., Shields, G.A., & Zaritsky, D. 1996, ApJ, in press
Skillman, E.D., Melnick, J., Terlevich, R., & Moles, M. 1988, A&A, 196, 31
Skillman, E.D., Terlevich, R., & Melnick, J. 1989, MNRAS, 240, 563 Spite, M. & Spite, F. 1990, A&A, 234, 67
Stasińska, G. 1980, A&AS, 85, 359
Terlevich, R., Melnick, J., Masegosa, J., Moles, M., & Copetti, M.V.F. 1991, A&AS, 91, 285
Viegas, S.M. & Clegg, R.E.S. 1994, MNRAS, 271, 993
Vila-Costas, M.B. & Edmunds, M.G. 1992, MNRAS, 259, 121
———————. 1993, MNRAS, 265, 199
Zaritsky, D., Kennicutt, R.C., Jr., & Huchra, J.P. 1994, ApJ, 420, 87

Temperature Fluctuations in the Planetary Nebula NGC 6543

Robin L. Kingsburgh, J.A. López

IAUNAM, P.O. Box 439027, San Diego, CA, 92143-9027 U.S.A.

and M. Peimbert

IAUNAM, Apdo Postal 70-264, México 04510 D.F, México

Abstract. We present abundances derived from echelle spectroscopy of the planetary nebula NGC 6543 from the forbidden, collisionally excited lines, and from the permitted, recombination lines, and find a discrepancy of a factor of 4 between the O^{2+}/H^+ ratio derived by the recombination lines and the the forbidden lines. This discrepancy suggests the presence of temperature fluctuations in NGC 6543, where the collisionally excited lines are exponentially weighted to regions of higher temperature. A value for the root-mean-square temperature fluctuation parameter $t^2(^{OII}_{abun})= 0.057$ is derived. We have also estimated t^2 by comparing the derived Balmer Jump temperature $T_e(BaJ)= 7100^{+1200}_{-900}$K with the temperature derived from the [O III] 5007/4363Å ratio $T_e[O\ III]= 7950\pm100$K. This comparison would suggest $t^2(BaJ)= 0.026$, a factor of 2 lower than $t^2(^{OII}_{abun})$, however the two temperatures do agree within the formal errors.

1. Introduction

Perhaps the most important parameter needed to derive accurate abundances in nebulae is the electron temperature. The classical method of deriving abundances uses the temperature derived from the ratio of collisionally excited lines (e.g. T_e[O III] derived from the 5007/4363Å ratio). These lines, however, have an exponential dependence on temperature and would tend to arise from areas of higher temperature, compared to recombination radiation, which has a more slowly varying temperature dependence. Abundance ratios derived from recombination lines are nearly insensitive to temperature, and provide a more accurate means for deriving abundances compared with the collisionally excited lines. Discrepancies between the abundances derived from the recombination lines and from the collisionally excited lines have been found for other objects, and point to the existence of temperature fluctuations in nebulae (Peimbert et al. 1993 Liu et al. 1995, this work).

The presence of temperature fluctuations in nebulae has long been debated in the literature. From Peimbert (1967) who defined the RMS temperature fluctuation parameter t^2 and found discrepancies between electron temperatures (T_e) derived from collisional lines and recombination radiation, to Barker (1979) who found no difference between T_e[O III] and T_e derived from the Balmer Jump, through to Dinerstein et al. (1988) who found T_e derived from IR [O III] lines to be lower than T_e[O III] derived from the optical lines. Recent additional evidence for the presence of temperature fluctuations comes from the present work on recombination lines, and from Peimbert et al. (1995a,b) based on T_e derived from helium and carbon lines. The presence of temperature fluctuations in nebulae has implications for all previous studies of nebular abundances using the traditional methods, and ultimately for ISM abundances adopted for chemical evolution models.

We have initiated a survey of echelle spectroscopy of PN at the Observatorio Astronómico Nacional in San Pedro Mártir, Baja California, to observe the faint recombination lines. Here we present results for the first object, NGC 6543.

2. Observations and Reductions

The echelle spectra for NGC 6543 were obtained on June 11-13, 1994, with the REOSC spectrograph at the 2.1-m (f/7.5) telescope with a Tektronix CCD as detector. The spectral coverage was $\sim$3300 Å to 6450 Å. A 13″ slit was used, which avoided order overlaping at the shortest wavelengths. The slit width was 2″, and the position angle was 90°.

The data were reduced using the IRAF echelle reduction package. The spectra were bias subtracted, flat-fielded, corrected for background scattered light, wavelength calibrated with a Th-Ar arc, atmospheric extinction corrected, and flux calibrated with the standard star HD192281. Each order was then extracted, and merged into a continuous spectrum. Line fluxes were then measured using IRAF's gaussian fitting routines.

3. Analysis

The reddening was derived by comparing the Balmer decrement to their case B predicted values; c(Hβ)=0.16. The line fluxes were dereddened using the reddening law of Seaton (1979). Members of the higher order Balmer decrement were found to be within 5% of their predicted values. The electron temperature T_e was derived via the [O III] 4959/4363 Å and [N II] 6583/5755Å ratios, where the 6583/ Hα ratio was taken from a low resolution spectrum of NGC 6543 obtained at the same position (Kingsburgh, in preparation). The electron density n_e was derived using the [O II] 3726/3729 Å, [Ar IV] 4711/4740 Å and [Cl III]5517/5537 Å ratios. T_e(BaJ), the electron temperature given by the height of the Balmer Jump was derived following Barker (1979). The diagnostics are as follows: T_e[O III]= 7950$\pm$100K, T_e[N II]= 9600$\pm$400K, T_e(BaJ)= 7100^{1200}_{-900}K, n_e[O II]= 6300$\pm$200 cm^{-3}, n_e[Ar IV]= 3980$\pm$400 cm^{-3} and n_e[Cl III]= 3850$\pm$600 cm^{-3}. Generally in PN, T_e[O III]>T_e[N II] (e.g. see Fig. 3 Kingsburgh & Barlow 1994), however here we have T_e[O III]/T_e[N II] = 0.83. This would imply that the electron temperature is increasing outwards from the central star for NGC 6543, as the ionization potential to get N$^+$ (14.5 eV) is lower than that to get O^{2+} (35 eV), however our T_e[N II] may be more uncertain than given by the formal errors, as the 6583 and 5755Å fluxes were taken from different spectra (but at the same position).

Abundances for the collisionally excited lines were derived by solving the equations of statistical equilibrium, adopting T_e= 7950 K and n_e= 6300 cm^{-3}. Helium abundances were derived following Clegg (1987). The total overall abundances were derived using the ionization correction factor (ICF) scheme of Kingsburgh & Barlow (1994). Table 1 presents the ionic abundances relative to H$^+$ and total abundances relative to H by number, derived by the collisionally excited lines and a 'traditional' analysis.

Recombination line abundances for O^{++} were derived using the recombination coefficients of Storey (1994) and branching ratios of Liu et al. (1995), which take into account the effects of intermediate coupling. Liu et al. (1995) find that the recombination coefficients predicted by LS-coupling do not produce predicted line intensities in agreement with observations, and have re-calculated the recombination coefficients for O II assuming intermediate coupling for the 4f-3d transitions. Much better agreement with their observations and better self-consistency is found with their new calculations. Table 2 presents the ionic abundance ratios together with the multiplet numbers used to derive the abundance. The N^{2+}, N^{3+}, C^{2+} and C^{3+} recombination line abundances were derived using the recombination coefficients of Péquinot et al.(1991). As fewer lines are used to derive the abundance ratios, they are more uncertain than the O^{2+}/H$^+$ ratios, but are presented for interest in Table 2.

4. Discussion

Inspection of Tables 1 and 2 reveals a discrepancy of a factor of 4 between the O^{2+}/H$^+$ abundance derived via the collisionally excited lines and via the recombination lines. Similar discrepancies have been found for the PN NGC 6572 by Peimbert et al. (1993) and for NGC 7009 by Liu et al. (1995). These discrepancies can be explained by the presence of spatial temperature fluctuations over the observed volume of the nebulae; for NGC 6543, we derive a value for the root-mean square temperature fluctuation parameter t^2, of 0.057. Table 3 presents estimates for t^2 based on the O^{2+} recombination line abundances, as well as the Balmer Jump temperature, and the C III temperature. We find the t^2 derived from T_e(BaJ) is a factor of 2 lower than $t^2(^{OII}_{abun})$, however T_e[O III] and T_e(BaJ) do agree within the formal errors. The t^2(C III) was derived using T_e(C III) of Peimbert et al. 1995b and is equal to t^2(BaJ). It is conceivable that T_e(BaJ) and T_e(C III) could be equal to T_e[O III], however, the discrepancy between the O^{2+}/H$^+$ abundance derived from the collisional and recombination lines is definitely problematic. At this point, the mechanism responsible for temperature fluctuations and hence this discrepancy is unknown (see Peimbert 1995), but it is an important problem for nebular astrophysics that needs further investigation.

Table 1. Abundances Derived from Collisionally Excited Lines

element	ion	n_{ion}/n_{H+}	ICF	n_X/n_H
He	He$^+$	0.123	–	0.123
O	O$^+$	3.57×10^{-5}		
	O^{2+}	5.69×10^{-4}	1.0	6.04×10^{-4}
N	N$^+$	3.47×10^{-6}	16.9	5.87×10^{-5}
Ne	Ne^{2+}	1.51×10^{-4}	1.06	1.60×10^{-4}
Ar	Ar^{3+}	4.34×10^{-7}	1.06	4.80×10^{-7}
S	S$^+$	2.96×10^{-7}		
	S^{2+}	5.86×10^{-6}	1.82	1.12×10^{-5}
Cl	Cl^{2+}	1.50×10^{-7}	–	–

Table 2. Recombination Line Abundances

ion	Mult. No.	n_X/n_H
O^{2+}	10,19,20,48,49	2.30×10^{-3}
	53,54,68,93	
	93.01 (19 lines)	
N^{2+}	28,48 (2 lines)	1.83×10^{-4}
N^{3+}	18 (1 line)	5.15×10^{-5}
C^{2+}	6 (1 line)	6.07×10^{-4}
C^{3+}	1 (2 lines)	$>4.15\times10^{-5}$

Table 3. Estimates for t^2

$T_e[OIII] = 7950\pm100$
$T_e[NII] = 9600\pm400$

$T_e(^{OII}_{abun}) = 6000 \rightarrow t^2(^{OII}_{abun})=0.057$
$T_e(\text{BaJ}) = 7100^{+1200}_{-900} \rightarrow t^2(\text{BaJ})=0.024$
$T_e(\text{CIII}) = 7300 \rightarrow t^2(\text{CIII})=0.020$

References

Barker, T. 1979, ApJ, 219, 914
Clegg, R.E.S. 1987, MNRAS, 221, 31p
Dinerstein, H.L. et al., 1988, ApJ, 291, 561
Kingsburgh, R.L. & Barlow, M.J. 1994, MNRAS, 271, 257
Liu, X-W. et al., 1995, MNRAS, 272, 369
Peimbert, M. 1967, ApJ, 150, 825
Peimbert, M. 1995, in: "The Analysis of Emission Lines", ed. RE Williams, Cambridge University Press, in press
Peimbert, M. Storey, P.J. & Torres-Peimbert, S. 1993, ApJ, 414, 626
Peimbert, M. et al., 1995a, RevMexA&A, 31, 131
Peimbert, M. et al., 1995b, RevMexA&A, 31, 147
Péquinot, D. et al., 1991, A&A, 251, 680
Seaton, M.J. 1979, MNRAS, 187, 73p
Storey, P.J. 1994, A&A, 282, 999

Abundances and Stellar Populations in Giant HII Regions

William H. Waller[1], Joel Wm. Parker[1], and Eliot M. Malumuth[2]

NASA Goddard Space Flight Center, Laboratory for Astronomy and Solar Physics, Code 681, Greenbelt, MD 20771

Abstract. We address the question of whether or not metal abundance affects the relative production of low and high-mass stars in giant HII regions. Photometry and spectroscopy of ionizing clusters in the Galaxy, LMC, and SMC reveal no significant relation between metallicity and the slope (Γ) of the stellar initial mass function (IMF). Recent HST/WFPC2 photometry of 3 giant HII regions in M33 also does not show any trend involving Γ and nebular abundance. These results indicate that the observed anti-correlation between metal abundance and nebular "excitation" derives from metallicity-dependent stellar and nebular opacities rather than variations in the stellar IMFs.

1. Introduction

The role of metal abundance in governing the star formation process remains controversial. Several theoretical studies of star formation incorporate metallicity as an important factor affecting cloud fragmentation and protostellar accretion processes which, in turn, may govern the high-mass end of cluster Initial Mass Functions (IMFs) (Kahn 1974; Shields & Tinsley 1976; Silk 1986). Alternatively, the ambient metallicity could play a merely incidental role, while dynamical factors may actually govern the upper masses of clouds, the clusters that form within the clouds, and the stars that populate the clusters (Waller 1990; Waller & Hodge 1991).

Evidence for environmentally-sensitive IMFs is usually based on COMPOSITE spectral indices of stellar population. These composite diagnostics include the broadband visible and UV colors, absorption-line strengths and ratios, hydrogen-line luminosities and equivalent widths, emission-line ratios, and far-infrared "excesses" relative to the radio Bremsstrahlung emission. Claims of metallicity-dependent stellar populations are mostly based on observations of [OIII]/Hβ, [OIII]/[OII] and other emission-line ratios which are sensitive to the nebular "excitation" and, by inference, to the effective temperatures of the ionizing clusters. The sense of the trend is for the nebular excitation and inferred effective temperature to increase with decreasing O/H abundance (Campbell 1987; Vilchez & Pagel 1988; Vilchez et al. 1988; Shields 1990).

---

[1] Hughes STX Corporation

[2] Computer Sciences Corporation

In the absence of strong age gradients within the samples, the effective temperatures of the clusters are thought to trace either the upper-mass limits or the slopes of the cluster IMFs. The anticorrelation of effective temperature with metallicity would then suggest that the IMFs are biased towards hotter, higher mass stars in regions of lower metal abundance. Other studies involving the Hα equivalent width as a tracer of the high-mass IMF indicate no metallicity dependence, but rather a sensitivity to dynamical factors such as shearing and tidal disruption in the disks of galaxies (Waller 1990; Waller & Hodge 1991).

2. Challenges and Opportunities

Such claims of environmentally-sensitive IMFs depend on the assumption that one has correctly interpreted the various composite spectral indices. To evaluate the composite indices in terms of the actual stellar populations, it is first necessary to RESOLVE and photometrically characterize the prominent members of the population. Spectroscopic follow-up of the hottest and brightest members is then necessary to constrain the temperatures, ionizing luminosities and masses of the stars most responsible for powering the composite spectral indices. Only after these steps are taken, can one relate the young stellar populations to the composite spectral indices and to potential influences such as nebular abundances and ambient dynamics.

Giant HII regions and OB associations in the Galaxy, LMC and SMC are sufficiently nearby for their ionizing stellar populations to be resolved via ground-based imaging. Obscuration by dust in the Galaxy and the small size of the Magellanic Clouds restrict the total sample of giant HII regions to a small number. Metallicities of near-solar (Galaxy), $\sim$0.4 solar (LMC), and $\sim$0.1 solar (SMC) can be sampled (Massey et al. 1995).

At a distance of 0.84 Mpc, the Local Group galaxy M33 contains numerous giant HII regions which can be imaged by the HST/WFPC2 down to a linear resolution of 0.4 pc. Metallicities ranging from 1.4 solar to 0.1 solar can be sampled in this medium-size SA(s)cd galaxy. Recent HST programs have yielded fully-resolved multi-band images of 6 giant HII regions in M33 (Hunter et al. 1995; Malumuth, Waller, & Parker 1995a, 1995b; Parker et al. 1995; Waller et al. 1996). Beyond the Local Group is a handful of other late-type galaxies which are sufficiently nearby for their giant HII regions to be partially resolved by HST into individual stars (Drissen 1995).

3. Recent Findings and Future Prospects

Photometry and spectroscopy of ionizing clusters in the Galaxy, LMC, and SMC reveal no significant relation between metallicity and the slope (Γ) of the power-law IMF. The strongest variation in Γ is evident when comparing clustered and field populations — the field stars having significantly steeper IMF slopes (Massey et al. 1995).

HST/WFPC2 photometry of 3 giant HII regions in M33 also does not show any consistent trend involving metal abundance and IMF slope (see **Figure 1**). The upper mass limit appears to be constrained more by cluster age than by anything else. These results are consistent with predictions based on the Hα

equivalent width but differ from those based on analysis of diagnostic emission-line ratios (Malumuth, Waller & Parker 1995a, 1995b; Parker et al. 1995; Waller et al. 1996).

To reconcile the resolved and composite measures of stellar population, it is necessary to re-interpret the effects of metallicity on the nebular emission. If metallicity does not affect the stellar IMF, it can either alter the EUV opacities of the hot stellar atmospheres or of the nebulae themselves. Recent models of ionizing clusters indicate that most of the observed variations in nebular excitation can be attributed to metallicity-dependent EUV opacities in the stellar atmospheres. The deeper photoionization absorption edges in the EUV spectra of higher-metallicity stars results in lower effective temperatures for the same stellar mass (McGaugh 1991; Garcia-Vargas et al. 1995).

The response of the nebulae to the same ionizing radiation field also varies with metallicity. The abundance of metal "coolants" is strongly linked to the electron temperature which, in turn, determines recombination rates, collisional excitation rates, and resulting emission-line ratios (Vilchez & Pagel 1988; Shields 1990; Dinerstein, these proceedings). Besides the well-known cooling effect of metals, absorption of EUV photons by dust grains can significantly "soften" the emergent radiation field, resulting in lower ionization states elsewhere in the HII region (Sarazin 1976; Shields 1990; McGaugh 1991; see however counter-arguments by Shields & Kennicutt 1995). Unfortunately, direct measurements of the EUV emission from O-type stars or of the EUV absorption by cosmic dust have yet to be made.

Future prospects for constraining the metallicity-IMF relation include more and better spectrophotometry of the ionizing stellar populations that underlie and power giant HII regions. In M33, photometric analysis of all 6 giant HII regions that were imaged by HST/WFPC2 will improve the sampling in metallicity. HST imaging of another 3 giant HII regions has been proposed. Groundbased spectroscopy of the hottest and brightest resolved stars is planned. A similar HST study of giant HII regions in NGC 2403 is underway (Drissen 1995).

Acknowledgments. This work was supported by NASA through grant number GO-5384 from the Space Telescope Science Institute.

References

Campbell, A. 1987, in Star Formation in Galaxies, NASA Conf. Pub. 2466, C. J. C. Persson (Wash. DC: NASA), p. 479

Drissen, L. 1995, private communication

Garcia-Vargas, M. L., Bressan, A., & Diaz, A. I. 1995, A&AS, 112, 13

Kahn, F. D. 1974, A&A, 37, 149

Hunter, D. A., Baum, W. A., O'Neill, E. J., & Lynds, R. 1995, in press

Massey, P. Lang, C. C., DeGioia-Eastwood, K., & Garmany, D. 1995, ApJ, 438, 188

Malumuth, E. M., Waller, W. H., & Parker, J. W. 1995a, AJ, in press

Malumuth, E. M., Waller, W. H., & Parker, J. W. 1995b, in The Interplay between Massive Star Formation, the ISM, and Galaxy Evolution, eds. D. Kunth et al., in press
McGaugh, S. S. 1991, ApJ, 380, 140
Parker, J. Wm. & Garmany, C. D. 1993, AJ, 106, 1471
Parker, J. Wm. et al. 1995, in The Interplay between Massive Star Formation, the ISM, and Galaxy Evolution, eds. D. Kunth et al., in press
Sarazin, C. L. 1976, ApJ, 208, 323
Shields, G. A. 1990, ARA&A, 28, 525
Shields, J. C. & Kennicutt, R. C. 1995, ApJ, 454, 807
Silk, J. 1986, in Luminous Stars and Associations in Galaxies, IAU Symp. 116, eds. C. W. H. DeLoore et al. (Dordrecht: D. Reidel), p. 301
Vilchez, J. M. & Pagel, B. E. J. 1988, MNRAS, 231, 257
Vilchez, J. M. et al. 1988, MNRAS, 235, 633
Waller, W. H. 1990, Ph.D Dissertation, University of Massachusetts
Waller, W. H. & Hodge, P. W. 1991, in Dynamics of Galaxies and their Molecular Cloud Distributions, IAU Symp. 146, eds. F. Combes & F. Casoli (Dordrecht: Kluwer Academic Publishers), p. 187
Waller, W. H. et al. 1996, in preparation

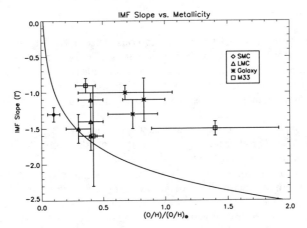

Figure 1. Relation between metal abundance (O/H) and IMF slope (Γ) for ionizing stellar populations in the Galaxy, LMC, SMC, and M33. The O/H abundances are based on measurements of [OII] and [OIII] emission-line strengths relative to those at Hβ (cf. Vilchez et al. 1988). IMF slopes for giant HII regions and OB associations in the Galaxy, LMC and SMC are from the spectrophotometry of Massey et al. (1995) and Parker & Garmany (1993). IMF slopes for the giant HII regions in M33 are from the photometry of Hunter et al. (1995) and Malumuth, Waller, & Parker (1995a & 1995b). The curve represents the relation between metallicity and IMF slope derived from diagnostic emission line ratios (Campbell 1987). For reference, the standard Salpeter IMF has a slope of $\Gamma = -1.35$.

Cosmic Abundances
ASP Conference Series, Vol. 99, 1996
Stephen S. Holt and George Sonneborn (eds.)

Abundance Measurements in the Outer Galaxy

Alexander L. Rudolph[1], Janet P. Simpson[2,3], Michael R. Haas[3], Edwin F. Erickson[3], and Michel Fich[4]

Abstract.
Five H II regions at large distances from the center of the Galaxy (R = 13–17 kpc) have been observed in far-IR emission lines of [O III] (52μm, 88μm) [N III] (57μm), and [S III] (18μm) using the Kuiper Airborne Observatory. Observations of these ions have been combined with Very Large Array radio continuum observations of these sources to determine abundances of these ions. A simple ionization correction scheme has been used to determine the abundances of N, O, and S relative to H, as well as the relative abundance N/O, and these results have been combined with similar results from other studies to determine the abundance gradient of these elements in the Milky Way.

1. Introduction

The relative abundances of the elements constitute a vital clue towards understanding the formation and evolution of the Milky Way and of other galaxies. These abundances vary from place to place and also both within and between classes of objects. Determinations of both absolute and relative abundances as a function of galactocentric radius are a central input to chemical evolution models of galaxies (Matteuci 1991). Numerous recent studies of H II regions have shown that there are chemical abundance gradients with galactocentric radius in the Milky Way (Shaver et al. 1983, Lester et al. 1987, Rubin et al. 1988, Simpson & Rubin 1990, Simpson et al. 1995).

Studies of abundances in H II regions have largely been carried out by observing the fine-structure lines of heavy elements such as N, O, and S, which occur primarily at optical and far-infrared (FIR) wavelengths. In many cases, particularly when studying our own Galaxy, there are many advantages to using the far-infrared (FIR) rather than the optical lines to determine abundances.

The outer Galaxy (the region outside the Solar circle, $R_\odot$ = 8.5 kpc) is an ideal place to study the radial variation of abundance in our Galaxy, both because it is at an extreme position in the Galaxy, and because of the differences

---

[1] Department of Physics, Harvey Mudd College, Claremont, Calif. 91711

[2] SETI Institute, Mountain View, Calif. 94043

[3] NASA-Ames Research Center, Mail Stop 245-6, Moffett Field, Calif. 94035

[4] Guelph – Waterloo Program for Graduate Work in Physics, Physics Department, University of Waterloo, Waterloo, Ontario, Canada N2L 3G1

in physical conditions compared to the Solar neighborhood. This paper reports on observations of the FIR lines of N, O, and S in five H II regions in the outer Galaxy using the Kuiper Airborne Observatory (KAO).

2. FIR v. Optical Abundance Determinations

There are a number of advantages and disadvantages to using optical and FIR lines to determine abundances in H II regions:

Optical Lines

Advantages

- Lines are easy to observe in many H II regions.

Disadvantages

- Corrections must be made for differential extinction between lines at various wavelengths.
- Regions of the Galaxy with very high extinction are inaccessible to optical observations.
- The determinations of N_e, the electron density, from line ratios is very sensitive to the assumed electron temperature T_e, which is often poorly known.
- Observations of high-excitation H II regions (which are often the brightest) require large corrections for the unseen doubly ionized states (Fich & Silkey 1991).

FIR Lines

Advantages

- Almost no extinction corrections must be made, either to individual lines, or between lines at various wavelengths.
- The low extinction in the FIR allows observations to be made throughout the Galaxy, in particular, to the far-inner Galaxy (Rubin et al. 1988, Simpson et al. 1995) and the outer Galaxy (this paper).
- Determinations of N_e are very insensitive to the assumed electron temperature T_e.
- High-excitation H II regions require only small corrections for the unseen, singly-ionized states.

Disadvantages

- Observations are relatively difficult to make, somewhat limiting the number of sources that can be observed.

3. Observations and Results

Observations were made in November 1993 aboard the Kuiper Airborne Observatory. We observed five H II in the outer Galaxy at Galactocentric radii ranging from $R = 13 - 17$ kpc. We observed the far-IR emission lines of [O III] ($52\mu m$, $88\mu m$) [N III] ($57\mu m$), and [S III] ($18\mu m$) in all five sources. We detected the [O III] lines in all five sources, the [S III] line in four of the five sources, and the [N III] line in just two of the sources. However, in all cases of non-detection, our upper limits on the relevant elemental abundance is significant.

The data were corrected for instrumental response and atmospheric transmission, and were flux calibrated using observations of the KL nebula in Orion. The final FIR fluxes were then combined with VLA fluxes (Rudolph et al. 1996a, 1996b, Fich 1986, 1993) for the same sources measured in the same beam to determine ionic abundances. Finally, model corrections were made for unseen ionization states using the models of Rubin (1985) to obtain the final elemental abundances. These abundances are listed in Table 1.

Table 1.   Elemental Abundances

Source	R (kpc)	$12 + \log[O/H]$	$12 + \log[N/H]$	$12 + \log[S/H]$	N/O
S127 A	15.0	$8.10^{+0.05}_{-0.06}$	$7.24^{+0.09}_{-0.12}$	$6.43^{+0.12}_{-0.16}$	0.14 ± 0.04
S128	12.9	$8.22^{+0.06}_{-0.07}$	$7.46^{+0.07}_{-0.09}$	$6.57^{+0.06}_{-0.07}$	0.17 ± 0.03
WB 380	17.0	$8.02^{+0.22}_{-0.49}$	<7.44	<6.35	<0.26
WB 870	15.0	$7.98^{+0.10}_{-0.13}$	<7.10	$6.33^{+0.07}_{-0.08}$	<0.14
S288	14.8	$8.03^{+0.09}_{-0.12}$	<7.08	$6.36^{+0.04}_{-0.05}$	<0.11

Figure 1 shows plots of $\log(S/H)$ and $\log(N/H)$ v. R. The data at R < 10 kpc are from Simpson et al. (1995). We wish to emphasize that all the data on this plot were taken with the same instrument and were analyzed in the same fashion, leading to a homogeneous data set. Also plotted on the figures are the best fit to the gradient of each abundance.

Previous studies in both the optical and the FIR have found a gradient of N/H in the inner Galaxy (Shaver et al. 1983, Simpson et al. 1995). One study, by Fich & Silkey (1991), extended optical observations to the outer Galaxy and found evidence that the N abundance flattens out beyond the Solar circle. Our results contradict this finding, and rather suggest that the abundances in all three elements, N, O, and S, continue to fall in the outer Galaxy. For example, we find the the abundance of N is 2-3 times lower in the outer Galaxy than in the Solar neighborhood, and is 10 times lower than in the inner Galaxy. Our result for S/H confirms the trend found by previous FIR observers (Simpson & Rubin 1990; Simpson et al. 1995) but is in sharp contrast with Shaver et al. (1983), who found no gradient in S/H.

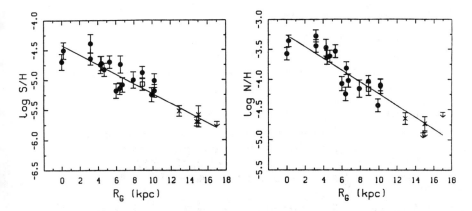

Figure 1. log(S/H) and log(N/H) v. R.

References

Fich, M., 1986, AJ, 92, 787
Fich, M., 1993, ApJS, 86, 475
Fich, M., & Silkey, M., 1991, ApJ, 366, 107
Lester, D.F., Dinerstein, H.L., Werner, M.W., Watson, D.M., Genzel, R., & Storey, J.W.V., 1987, ApJ, 320, 573
Matteucci, F., 1991, Frontiers of Stellar Evolution (ASP Conference Series), David L. Lambert, San Francisco, ASP, 539
Rubin, R.H., 1985, ApJS, 57, 349
Rubin, R.H., Simpson, J.P., Erickson, E.F., & Haas, M.R., 1988, ApJ, 327, 377
Rudolph, A.L., de Geus, E.J., Brand, J., & Wouterloot, J.G.A., 1996a, ApJ, in press
Rudolph, A.L., de Geus, E.J., Brand, J., & Wouterloot, J.G.A., 1996b, in prep.
Shaver, P.A., McGee, R.X., Newton, L.M., Danks, A.C., & Pottasch, S.R., 1983, MNRAS, 204, 53
Simpson, J.P., & Rubin, R.H., 1990, ApJ, 354, 165
Simpson, J.P., Colgan, S.W.J., Rubin, R.H., Erickson, E.F., & Haas, M.R., 1995, ApJ, 444, 721

Cosmic Abundances
ASP Conference Series, Vol. 99, 1996
Stephen S. Holt and George Sonneborn (eds.)

Measuring ISM Molecular Abundances in the Direction of Cassiopeia A by Comparing X-ray and Radio Absorption Studies

Jonathan W. Keohane

The Department of Astronomy, The University of Minnesota and The Laboratory for High Energy Astrophysics, Code 662, Goddard Space Flight Center, Greenbelt, MD 20771; E-mail: jonathan@cassiopeia.gsfc.nasa.gov

Abstract. ASCA derived column densities are compared with $\lambda 21$ cm H I and $\lambda 18$ cm OH optical depths in the direction of Cas A. The H I spin temperature and the ratio N_{OH}/N_{H_2} are measured. Revised abundance ratios are calculated for the molecules: ^{13}CO, ^{12}CO, H_2CO and NH_3.

1. Introduction

Using Cas A as a background source, the interstellar medium has been well-studied in radio absorption. A high resolution (7″) VLA $\lambda 21$ cm absorption study of the Perseus arm velocity feature was carried out by Bieging, Goss and Wilcots (1991). 30″ resolution H I absorption data over a wider velocity range were obtained by Goss and collaborators with the Westerbork array (Schwarz et al. 1995, hereafter SGK). In addition there have been numerous molecular absorption studies — in H_2CO (Goss et al. 1984), in CO (Wilson et al. 1993, hereafter WMMPO), NH_3 (Batrla et al. 1984 and Gaume et al. 1994) and OH (Bieging & Crutcher 1986, hereafter BC). The absorption patterns in the various *molecules* are similar to each other, but very different than that of the H I.

Rasmussen (1995, hereafter R95) studied the spatial dependence of X-ray model parameters using the ASCA satellite, resulting in a total column density (N_H) map. In this paper, we compare the radio absorption data of BC and SGK to the N_H map of R95. From this we measure the scaling relation between column density and equivalent line widths from the radio absorption measurements of SGK and BC. This allows measurements of the average H I spin temperature and the N_{OH}/N_{H_2} abundance of the ISM to be calculated.

2. Analysis

In order to measure the column density (N_H) toward Cas A, we found the equivalent widths ($EW \equiv \int \tau dV$) of the H I and OH absorption measurements of SGK and BC respectively. The total column density can then be given by the relation:

$$N_{H_{Radio}} = D\,(EW_{H\,I}) + E\,(EW_{OH}) + F$$

where the column density is measured in units of $10^{22} cm^{-2}$ and the equivalent widths are measured in km/s. The goal of this analysis is to measure the parameters D, E and F.

In order to match the resolution of SGK's data, we smoothed BC's line and continuum maps to a resolution of 30″, before calculating the optical depths and equivalent widths. This resulted in our obtaining 30″ images of EW_{HI} and EW_{OH}, as well as their respective error maps calculated using standard methods.

R95, performed spatially resolved spectral fits to the 0.5–12 KeV ASCA data, resulting in independently derived column density maps from those discussed above. Spectral fitting works very well to find the total column density towards an X-ray source, because the effective X-ray cross-section is a strong and unique function of photon energy (Morrison & McCammon 1983). The images are shown in figure 1.

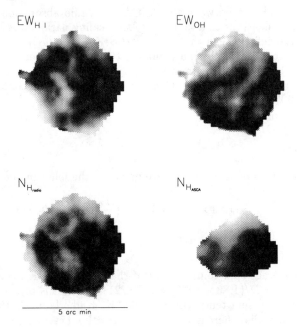

Figure 1. 30″ resolution images of ISM absorption toward Cas A.

The position of the ASCA field was aligned with our radio images by maximizing the cross-correlation of R95's $N_{H_{ASCA}}$ image with the radio absorption using an initial estimate of parameters D and E.

Because an error image was not provided by R95, we were unable to calculate true χ^2 differences. We, therefore, defined our reduced χ^2 such that:

$$\chi^2_{reduced}(D, E, F) \equiv \frac{\sum (N_{H_{ASCA}} - N_{H_{Radio}})^2}{\min \left(\sum (N_{H_{ASCA}} - N_{H_{Radio}})^2 \right)}$$

The best fit values for parameters D, E and F are: $(7.1 \pm 3.8) \times 10^{-3}$, 0.213 ± 0.141, and 0.69 ± 0.19 respectively (70% confidence); the χ^2 dependence on D, E and F is shown in Figure 2. There were 25 independent beams (22 degrees of freedom).

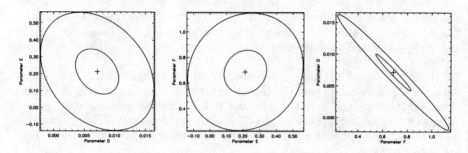

Figure 2. The χ^2 between the ASCA and radio absorption line derived column densities as a function of the scaling parameters D, E and F, which are defined such that $N_H = D(EW_{HI}) + E(EW_{OH}) + F$. All χ^2 values were obtained by individually fitting the parameter not shown in the particular plot. Our reported χ^2 values were scaled such that the minimum reduced χ^2 is unity. Contour levels represent a confidence of 70% and 99%.

3. Discussion

The column densities of H I and OH are given by the following equations:

$$N_{HI} = (1.83 \times 10^{-4}\ EW_{HI}) \left(\frac{T_{spin}}{°K}\right) \times 10^{22} cm^{-2}$$

$$N_{OH} = (2.2 \times 10^{-8}\ EW_{OH}) \left(\frac{T_{ex}}{°K}\right) \times 10^{22} cm^{-2}$$

Therefore the average spin temperature of the H I that SGK measured should be given by $T_{spin} = D/(1.83 \times 10^{-4}) = (39 \pm 21)°K$. This is consistent with the previously estimated spin temperature of 60 °K (Kalberla et al. 1985).

Figure 2 shows that there is very little correlation between parameter E and either parameter D or F. This implies that our best fit parameter E is robust. If we assume a molecular excitation temperature of $T_{ex} = 20°K$ (WMMPO), we find the abundance of OH to be:

$$\frac{N_{OH}}{N_{H_2}} = (4.1 \pm 2.7) \times 10^{-6} \left(\frac{T_{ex}}{20°K}\right).$$

In WMMPO's CO study, they compare their data with the existing literature to find relative abundances of each molecular species studied up to that time. Because there have been no direct measurements of N_{H_2}, their abundances relative to H_2 are uncertain. We calculate new molecular abundances of the ISM using WMMPO's inter-species abundance ratios and the $^{12}C/^{13}C$ ratio of Penzias and Langer (1993) (see Table 1).

Species	N_X/N_{H_2}
OH	$4.1 \times 10^{-6} \pm 2.7 \times 10^{-6}$
^{13}CO	$3.2 \times 10^{-5} \pm 2.5 \times 10^{-5}$
^{12}CO	$2.0 \times 10^{-3} \pm 1.6 \times 10^{-3}$
H_2CO	$2.1 \times 10^{-7} \pm 2.1 \times 10^{-7}$
NH_3	$6.3 \times 10^{-8} \pm 6.2 \times 10^{-8}$

Table 1. The relative abundances of various molecular species assuming WMMPO's molecular excitation temperature ($T_{ex} = 20°K$) and their inter-species abundance ratios ($^{13}CO/H_2CO = 150 \pm 90$, $^{13}CO/OH = 7.7 \pm 3.3$, $^{13}CO/NH_3 = 500 \pm 300$ and $^{12}CO/^{13}CO = 62 \pm 4$). All values will scale directly with the OH excitation temperature.

4. Conclusion

By comparing X-ray and radio spectroscopic absorption measurements, the H I spin temperature and molecular abundances were measured. Future studies of Cas A and other radio and X-ray bright extended objects can significantly enhance our understanding of the ISM, by comparing spatially resolved column densities from either the ROSAT PSPC or ASCA with radio atomic and molecular line absorption data.

Acknowledgments. This work was supported, in part, by the NSF through grants AST-8720285 and AST-9100486. Funding was also provided by NASA/GSFC's Laboratory for High Energy Astrophysics through the Graduate Student Researchers Program. OH absorption data were kindly provided by John Bieging. H I absorption Westerbork data were kindly provided by W. Miller Goss. ASCA data were kindly provided by Andrew P. Rasmussen. I would also like to thank Lawrence Rudnick who contributed a good deal of advice regarding this work.

References

Batrla, W., Walmsley, C.M. & Wilson, T.L. 1984, A&A, 136, 127
Bieging, J.H., & Crutcher, R.M. 1986, ApJ, 310, 853 (BC)
Bieging, J.H., Goss, W.M. & Wilcots, E.M. 1991, ApJS, 75, 999
Gaume, R.A., Wilson, T.L. & Johnston, K.J. 1994, ApJ, 425, 127
Goss, W.M., Kalberla, P.M. and Dickel, H.R. 1984, A&A, 139, 317
Kalberla, P.M.W., Schwarz, U.J. & Goss, W.M. 1985 A&A, 144, 27
Morrison, R. & McCammon, D. 1983, ApJ, 270, 119
Penzias, A. & Langer, W. 1993, ApJ, 408, 539
Rasmussen, A.P. 1995, *in preparation*, (R95)
Schwarz, U.J., Goss, W.M. & Kalberla, P.M.W. 1995 *in preparation* (SGK)
Wilson, T.L., et al. 1993, A&A, 280, 221 (WMMPO)

Energetic Particles and LiBeB

Elisabeth Vangioni-Flam

Institut d'Astrophysique de Paris, CNRS, 98bis boulevard Arago, 75014 Paris, France

Michel Cassé

CE-Saclay, DSM/DAPNIA/Service d'Astrophysique, 91191 Gif sur Yvette cedex, France

Abstract. Galactic Cosmic Rays do not appear anymore as the main agents of beryllium and boron nucleosynthesis in the galaxy. Two other processes have been proposed that weaken their role, both linked to supernovae explosions. They imply, on the one hand, neutrino spallation and, on the other hand, the acceleration of freshly synthesized carbon and oxygen to moderate energies. However, the first process, though fertile in ^{11}B, fails to explain the beryllium evolution, whereas the second one, without exaggerated energy requirements, accounts consistently for i) the linearity of the Be-Fe and B-Fe correlations, particularly at low metallicities (while preserving the constant Li/H ratio at low metallicities) ii) the B/Be ratio in stars and iii) the ^{11}B/^{10}B ratio in meteorites.

1. Introduction

The rare and fragile light nuclei lithium, beryllium and boron are not generated in the normal course of stellar nucleosynthesis, and are, in fact destroyed in stellar interiors. This condition is reflected in the low abundances of these species (Li/H = $1-2\times10^{-9}$, Be/H = $1-3\times10^{-11}$, B/H = $2-8\times10^{-10}$ in the local galactic environment). Fowler, Burbidge and Burbidge (1955) first suggested that light elements bypassed by stellar fusion might be formed on stellar surfaces in collisions of electromagnetically accelerated protons with abundant CNO nuclei. Fowler, Greenstein and Hoyle (1962) and Bernas et al. (1967) put forward models in which they were synthesized by the spallation reactions of stellar protons during the T-Tauri phase of the stellar evolution. But the most pereneous hypothesis was offered by Bradt and Peters (1950) who suggested that cosmic rays might interact with CNO to form LiBeB. Subsequently Reeves, Fowler and Hoyle (1970), Meneguzzi, Audouze and Reeves (1971), Mitler (1972), put this idea on firm quantitative fundations, showing that the solar abundance of ^{6}Li, ^{9}Be, ^{10}B and part of ^{7}Li and ^{11}B could be accounted for if they were produced by spallation reactions of galactic cosmic ray (GCR) nuclei (essentially protons and alpha particles) with medium heavy elements accumulated in the interstellar medium, over the lifetime of the galaxy (see Reeves 1994 for a review).

Up to the nineties, the spallative LiBeB origin was one of the more assured of all astrophysical hypotheses. Notwithstanding ^{7}Li, whose main origin was assumed to be primordial and possibly stellar, the only problem lay in the meteoritic ^{11}B/^{10}B, predicted to be $\simeq 2.5$ in GCR spallation, and observed to be $\simeq 4$ (Shima and Honda 1962, Chaussidon and Robert 1995).

A high flux of low energy cosmic rays was taylored to cure this weakness (Meneguzzi and Reeves 1975, Walker, Mathews and Viola 1985). Its composition was similar to that of usual GCR, i.e. proton and alpha rich, as opposed to the "Orion" component, which is enhanced in C and O (see section 3.). This undetectable flux, prevented to reach the vicinity of the Earth by the solar wind modulation, has remained up to now purely speculative. Allowing for this extra process, the theory of the origin of light elements seemed complete.

However, new facts prompted astrophysicists to reassess the question. In the late eighties, observations of the beryllium abundance in halo stars were achieved down to [Fe/H] ≈ -1.5 (Rebolo et al. 1988, Ryan et al. 1990, Gilmore, Edvardsson and Nissen 1991). Taking into account the progressive enrichment of the ISM in CNO, and assuming that the CR flux is proportional to the supernova rate, i.e. higher in the past, a good fit of the Be evolution was obtained within the limited range explored (Vangioni-Flam et al. 1990). At the same time, inhomogeneous primordial nucleosynthesis models were developed to circumvent the very low production of Be in the standard model (Boyd and Kajino 1989). Data were obtained at even lower metallicities for Be (Ryan et al. 1992, Gilmore et al. 1992, Boesgaard and King 1993), and a few boron abundance measurements were made over the whole metallicity range (Duncan et al. 1992, Edvardsson et al. 1994). These observations indicated a flatter correlation of Be and B vs Fe than expected on the basis of GCR production, increasing the general perplexity (Pagel 1991). Various way-outs were proposed (Duncan et al 1992, Feltzing and Gustafsson 1994) pointing toward a primary origin of light elements. Nonetheless, on the sole basis of the B/Be ratio observed in a few halo stars, irrespectively to the slope of the Be and B vs Fe correlation, the orthodox view in which Be and B were synthesized by spallation reactions between GCR and interstellar nuclei was maintained (Steigman and Walker 1992, Steigman et al. 1993, Fields, Schramm and Truran 1993, Walker et al. 1993, Fields et al 1994, 1995). In a last effort to save a GCR origin, a model was proposed invoking a better cosmic ray confinement in the early galaxy to enhance the light elements production at early times (Prantzos et al. 1993). Other models were proposed along similar lines (Abia, Isern and Canal 1995, Beckman and Casuso 1995).

However very recent observations showing a stricking proportionality between both the beryllium and boron abundances and that of iron from very low metallicities up to solar (Duncan et al. 1995 and Rebull et al. 1995, this meeting) do not encourage to follow this overconfinement model as it produces an essentially quadratic relationship at very low [Fe/H]. More importantly, in the early galaxy, the confinement time should be expected to be smaller than the present one, due to possible early galactic winds and/or the weaker (proto)-galactic magnetic field, depending on its generation timescale (Fields, Schramm and Truran 1993).

2. New developments

Recently, two new production processes, related to supernovae, have been proposed to produce primary light isotopes. Both mechanisms rely on spectacular and rare events: they involve low energy nuclei and supernova neutrinos, respectively. This latter (Dogomatskii and Nadyozhin 1977, Woosley et al 1990) is supported by the fact that SN are generous neutrino suppliers, as observed in the SN 1987A case (Arnett et al. 1989), while the first one is observationally linked to an intense gamma-ray emission in the Orion region (Bloemen et al. 1994), which, admittedly, needs to be confirmed. Indeed, intense gamma-ray lines from C and O deexcitation (4.4 and 6.1 MeV) were observed without counterpart at lower and higher energies, implying the interaction of low energy C and O nuclei with the ambient medium (Bloemen et al. 1994, Vangioni-Flam et al. 1995a,b, Cassé et al. 1995a,b, Ramaty et al. 1995a,b, 1996).

We are now facing a difficult challenge. With three spallative LiBeB production mechanisms in hand, instead of one, how are we to disentangle their respective contribution? To answer this question one must rely on galactic evolutionary models (see section 4.). In this context three drastic constraints should be obeyed :

1. to avoid lithium overproduction at [Fe/H] < −1 for preserving the lithium plateau (e.g. Spite and Spite 1993, Thorburn 1994),

2. to reproduce a B/Be ratio between 10 to 30 (allowing a maximum NLTE correction on the B abundance, see section 4.),

3. to obtain $^{11}B/^{10}B > 4$, globally in the past, so that the ratio after mixing with the GCR spallation process ($^{11}B/^{10}B|_{GCR} \simeq 2.5$) come down to 4.

Again, for the discussion to come, a clear distinction should be made between a primary production mechanism and a secondary one, in the sense of galactic evolution. Indeed, in the plausible hypothesis that the GCR flux Φ, is proportional to the supernova rate (Vangioni-Flam et al. 1990 and subsequent work), the rate of increase of L/H, L standing for Be or B, is:

$$\frac{d(L/H)}{dt} = z(t) \langle \sigma \Phi(t) \rangle, \qquad (1)$$

where z(t) is the evolving CNO fraction by number, and $\langle \sigma \Phi(t) \rangle$ the energy average of the production cross section times the flux. Since $z \propto N_{SN}$, the integrated number of supernovae up to time t, and $\Phi \propto dN_{SN}/dt$, we have, assuming a constant spectrum shape :

$$\frac{d(L/H)}{dt} \propto N_{SN} \frac{dN_{SN}}{dt} \qquad \text{or} \qquad L/H \propto z^2 \qquad (2)$$

On the contrary, freshly synthesized C and O nuclei injected/accelerated in supernovae ejecta would lead to a primary production of LiBeB, independent of the ISM metallicity, by fragmentation of these species on the surrounding H and He nuclei. We emphasize the strong difference between this Orion component and the low-energy "carrot" of Meneguzzi and Reeves (1975), Walker et al.

(1985): the former is p and α poor, while the latter is p and α rich, a composition that is ruled out from gamma-ray line observations and energetic arguments (Ramaty et al. 1995a,b).

We now discuss the relative merits of the different LiBeB production processes in the light of the crucial observations presented in this conference by Duncan and his collaborators. Taken at face value, the clear Be-Fe and B-Fe correlation with slopes of 1 seem to imply a common origin for Be and B, which cannot be readily related to galactic cosmic rays (slope 2). Note that NLTE corrections that are higher at low metallicities (Kiselman 1994) would even aggravate the discrepancy, since the slope of B-Fe correlation would become less than 1.

Virtually no beryllium is generated by the supernova neutrino process (Woosley & Weaver 1995, but see Malaney 1992), whereas the Orion mechanism produces both in adequate proportions (Sec.3); hence a clear advantage is conferred to this latter. However, non-LTE corrections at low [Fe/H] affecting significantly B abundances (Kiselman 1994, Kiselman and Carlsson 1995) but not Be ones (Garcia Lopez, Severino and Gomez 1995) should be carried out to definitively discriminate between the Orion and the neutrino processes (see section 4.). As a further refinement, a correction for a small Be depletion in stars, as indicated by rotational models (Pinsonneault et al. 1992), could be made.

3. The Orion mechanism

C and O nuclei injected/accelerated by massive stars within the Orion cloud, and by extension, within all active star formation regions, lead to a production of primary LiBeB. That the products behave as primaries is easily seen in Eq.(1) for a process that does not depend on the ambient metallicity. More precisely, WC winds and supernovae resulting from the explosion of WC stars could serve as preferential injectors of C and O nuclei in giant molecular clouds, as required by the gamma-ray Orion observation (Ramaty et al. 1996, Fields et al. 1995, Clayton and Jin 1995). We assume that the nuclei propelled to a few MeV/n by winds and explosions are further accelerated up to tens of MeV/n by ambient shocks and/or turbulence in the cloud maintained by the high concentration of active objects, and let them propagate in the cloudy medium, where they suffer dramatic energy losses, and occasionally fertile nuclear collisions with H and He nuclei, producing LiBeB and gamma-ray lines. Less massive stars exploding outside clouds, due to their longer lifetime, should inject nuclei that would be mostly thermalized before encountering propicious shock waves. The formalism of the CR propagation is well known (Meneguzzi, Audouze and Reeves 1971) and need not to be repeated here. It is sufficient to say that at the low energies considered here, ionization losses dominate other losses (escape of the system and nuclear destruction). In this respect the Orion close box model substitutes itself to the famous leaky box model of GCR propagation. As expected, the LiBeB yields depend on the source composition of the fast particles and on the adopted source spectrum. Detailed parametric studies have revealed that the composition of the wind and ejecta of a very massive star, typically of 60 $M_\odot$ on the main sequence (Woosley, Langer and Weaver 1993) associated to a flat spectrum up to 30 MeV/n and decreasing abruptly above could fulfil the constraints set above, this within a reasonable energy budget (Ramaty et al. 1995a,b,1996). Under such favorable circumstances, the observed gamma-ray line luminosity of Orion requires an injection of around $3\ 10^{38}$ erg s^{-1} in the form of fast nuclei. A single supernova is sufficient to sustain the process during $0.5-1\times10^5$ years. Indeed, X ray observations of the Orion–Eridanus superbubble (Burrows et al. 1993) indicate a high SN activity in the past 10^5 yrs in the Orion complex. Also, the maximum mass of stars in Orion being about 40 $M_\odot$, it is not unreasonable to assume that the more massive ones have exploded quite recently (Brown et al. 1994). Going a step further in the generalisation, we assume that the most massive stars explode within or very close to their parental clouds, injecting their C/O rich material in the surrounding medium where it is moderately accelerated in situ by shocks and turbulence generated by stellar winds and supernova explosions.

Under these circumstances light isotopes should be copiously produced all along the galactic lifetime by very massive stars embedded in clouds. As shown in table 1, the Orion mechanism yields $^{11}B/^{10}B = 4.4$ at solar metallicity (4.8 at zero metallicity), to be compared to a GCR ratio of 2.2 (slightly different from the 2.5 commonly admitted). This excess is due to the pecularities of the p + $^{12}C \rightarrow\ ^{11}B$ cross section at low energy (Read and Viola 1984, their graph I). Note that the B/Be and $^7Li/^6Li$ ratios are similar in the GCR and Orion cases.

Table 1. Spallative production rates (in atoms s^{-1} per H atom) and ratios (Lemoine, Vangioni–Flam and Cassé 1995). The target composition is that of the solar system. Three different GCR source spectra have been tested: that of Meneguzzi, Audouze and Reeves (GCR MAR), a lower bound spectrum quite similar to this latter, and an upper bound close to the theoretical ($\propto p^{-2}$) spectrum generated in shock acceleration. The Orion results correspond to the parameters selected by Ramaty et al. 1996. For comparison the original results of MAR (exposed in Audouze and Reeves 1982) are shown in column MAR 1971. For comparison, at zero metallicity, representative of the very early Galaxy, the Orion component yields $^{11}B/^{10}B = 4.8$ and B/Be = 26.3.

	MAR 1971	GCR MAR	GCR lower	GCR upper	Orion
^{6}Li	2.7 10^{-28}	3.1 10^{-28}	1.4 10^{-28}	1.6 10^{-27}	1.3 10^{-24}
^{7}Li	4.0 10^{-28}	4.4 10^{-28}	2.1 10^{-28}	2.2 10^{-27}	2.0 10^{-24}
^{9}Be	6.7 10^{-29}	8.5 10^{-29}	4.0 10^{-29}	2.5 10^{-28}	2.1 10^{-25}
^{10}B	2.9 10^{-28}	3.9 10^{-28}	1.8 10^{-28}	1.3 10^{-27}	9.0 10^{-25}
^{11}B	6.7 10^{-28}	8.9 10^{-28}	4.1 10^{-28}	3.4 10^{-27}	3.9 10^{-24}
Li/Be	10.0	8.8	7.5	14.7	15.5
B/Be	14.4	14.9	12.1	18.4	22.6
^{7}Li/^{6}Li	1.5	1.4	1.5	1.4	1.5
^{11}B/^{10}B	2.3	2.3	2.2	2.6	4.4

The calculated production rate of Li in Orion is $3.3\ 10^{38}$ s^{-1} i.e. 10^{51} atoms in 10^5 yrs as compared to the 10^{56} Li atoms originally present. Thus the recent

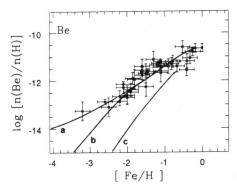

Figure 1. Evolution of beryllium in three cases : a) Orion + GCR, b) overconfined GCR (Prantzos et al. 1993), c) GCR only.

production of lithium has not significantly increased its general abundance in the cloud, in agreement with the solar value found in F and G stars (Cunha et al. 1995).

4. Galactic evolution of LiBeB

On the basis of the local conditions derived from the Orion region, we enlarge the perspective to the whole galaxy. Assuming that the most massive stars (> 60 $M_\odot$ on the main sequence) explode in their parental giant molecular cloud and thereby induce the Orion process, we integrate this process in a standard evolutionary model of galactic evolution (Vangioni–Flam et al. 1995a,b). As usual, the star formation rate is taken proportional to the gas mass fraction. The initial mass function (IMF) is of the Salpeter type (x = 1.7). The time dependent GCR mechanism and the stellar destruction of light isotopes are treated as in Vangioni–Flam et al. (1990). The typical irradiation time, adjusted to fit the Be/H ratio at low [Fe/H] amounts to 5×10^4 yrs, in agreement with the value derived for Orion itself on the basis of purely energetic arguments (see section 3.).

Results are shown in figures 1 to 5. The Orion mechanism seems so efficient and that it would leave little room for any other process. In this framework, GCR should contribute less than 50% to the Be and B abundance in the Solar System. The high boron abundances, observed at [Fe/H] < −1, seem to favor a high B abundance at solar age, i.e. meteoritic (Anders and Grevesse 1989) rather than photospheric (Kohl et al 1977) or stellar, as shown by our calculations (Fig.3). The Pop I stellar value, favored up to now (Reeves and Meyer 1978, Arnould and Forestini 1989) seems too low.

As expected, ^{7}Li is not overproduced below [Fe/H] = -1 by the Orion and GCR processes (fig. 5). Stellar sources of Li, presumably AGB stars (Abia, Isern and Canal 1993) and/or, perhaps neutrinos spallation (Timmes et al. 1995) would still be required to enhance the ^{7}Li abundance to the Pop I level (Li/H ≈ $2 10^{-9}$). Note that both Orion and GCR mechanisms give rise to a slope 1 in the Li-Fe correlation at low Z (fig. 5) due to the operation of the $\alpha + \alpha$ fusion.

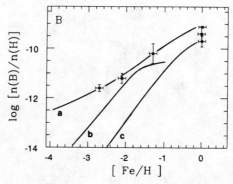

Figure 2. Evolution of boron : a,b,c same as figure 1.

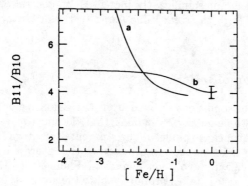

Figure 3. Evolution of $^{11}B/^{10}B$: a) neutrinos + overconfined GCR (Olive et al. 1994), b) Orion + GCR (Vangioni-Flam et al. 1995b).

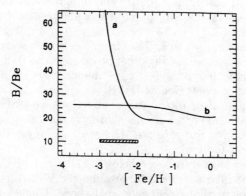

Figure 4. Evolution of B/Be : a,b same as figure 3. Horizontal bar : data from Duncan et al. (this conference) uncorrected for NLTE effect.

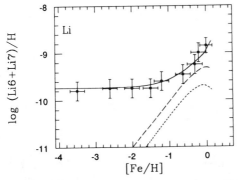

Figure 5. Lithium evolution. Data from Abia, Isern and Canal 1995, Spite and Spite 1993 and Rebolo et al. 1988 (upper envelope above [Fe/H]= -1). dashed line : Orion + GCR, dotted line: GCR only. Full line : primordial + Orion + GCR + stars (Vangioni–Flam et al. 1995b).

The combination of the Orion and GCR mechanisms would lead to small variations of the $^{11}B/^{10}B$ ratio in the galactic evolution, at variance with a neutrino-GCR combination (fig. 3). At first sight, the metal poor observations of Be and B seem to exclude the neutrino scenario (fig. 4). This steems from simple arguments: there is no other possible source of Be than p,α-CNO spallation; hence, if ones produces Be via a spallation mechanism to explain the Be-Fe correlation, one would also produce B. It was shown in table 1 that the B/Be ratio in spallation processes does not strongly depend on the spectrum shape; because this spallation predicted ratio is in agrneement with observations, one could not afford other sources of boron. Note that, in this respect, a more careful analysis of NLTE effects is mandatory to derive an accurate B/Be ratio from the observations; it is premature to draw any definitive conclusion. A more detailed study is underway (Vangioni-Flam et al 1995c). Needless to say, if the neutrino process was shown to produce ^6Li, Be and ^{10}B as well as ^{11}B, the whole situation would have to be reconsidered. For the time being, it seems that the neutrino yields should to be substantially reduced.

Finally, to avoid overproduction of LiBeB would imply, in our present framework, a flattening at low energy of the GCR source spectrum with respect to that derived from diffusive shock acceleration (Lemoine et al. 1995). The suppression of fast nuclei below, say ~ 200 MeV/n, might result from the ionisation energy losses that are not taken into account in shock acceleration simulations. In principle, however, gas infall of primordial composition could be invoked to dilute the light elements that were synthesized in the disk by all the mechanisms operating in conjunction. However, the high rate of infall required to keep the light isotope abundances close to their solar value would introduce a large amount of primordial D in the Galaxy, in contradiction with the need of a large deuterium destruction in the lifetime of the Galaxy (Vangioni–Flam and Cassé 1994) as indicated by the recent confirmation of a high primordial D abundance (Rudgers and Hogan, this conference).

5. Conclusion

The main conclusions of recent studies of the origin and evolution of the LiBeB isotopes are :

First, low energy carbon and oxygen nuclei injected by the wind and the explosion of the most massive stars in giant molecular clouds could well be the major sources of LiBeB in the galaxy, more especially at low metallicites. This would provide a consistent explanation to all observational constraints (B and Be vs Fe correlations, B/Be in stars and ^{11}B/^{10}B in meteorites). Gamma–ray line observations of Orion and other active star forming regions are mandatory to settle this issue. This will be a major target for the INTEGRAL European mission.

Second, little room seems to be left for neutrino spallation, in this case, so that, in all likelihood, the published yields should be significantly reduced. This may set possibly interesting constraints on the detailed supernova explosion mechanism and the related neutrino driven r-process (Vangioni-Flam et al 1995c).

Third, the low energy galactic cosmic ray source spectrum may be flatter than that calculated on the basis of shock acceleration. A revision could be necessary in modelling of particle acceleration at low energies, in particular it would be worhtwhile to include radiative energy losses.

Fourth, a Boron abundance of $4 - 8 \times 10^{-10}$ relative to hydrogen is predicted at solar metallicity.

Further spectroscopic observations are eagerly awaited, concerning the boron abundance in the Sun or Pop I stars, the ^{11}B/^{10}B at low and solar metallicities, and the ^{6}Li/^{7}Li ratio at low metallicities. Refinements and tests of the non-LTE corrections affecting the B abundance are also required to confirm the exclusion or not of neutrinos as main agents of Be and B evolution. Finally, for the sake of clarity, Be-O and B-O correlations, if they could be obtained, would be more appropriate than Be-Fe and B-Fe correlations, as they would remove any effect of type I supernovae.

Acknowledgments. We are endebted to R. Lehoucq for his numerical support.

We thank M. Lemoine, D. Duncan and R. Cayrel for illuminating discussions on galactic cosmic rays, stellar data and future projects, Y. Oberto for constant help. The work of EVF was supported by PICS 114, "Origin and evolution of the light elements", CNRS.

References

Abia, C., Isern, J. and Canal, R. 1993, A&A, 275, 96
Abia, C., Isern, J. and Canal, R. 1995, A&A, 298, 465
Anders, E. and Grevesse, N. 1989, Geochimica et Cosmochimica Acta 53, 197
Arnett, W.D., Bahcall, J.N., Kirshner, R.P. and Woosley, S.E. 1989, ARA&A, 27, 269

Arnould, M. and Forestini, M. 1989, Nuclear astrophysics: Proceedings of the third international summer school, Larabida, Ed.: M. Lozano, M.I. Gallardo and J.M. Arias, p. 48

Audouze,J. and Reeves, H. 1982, in Essay in Nuclear Astrophysics, edts: C.A. Barnes, D.D. Clayton and D.N. Schramm, Cambridge University Press, p. 355

Beckman, J.E. and Casuso, E. 1995, in 'The light element abundances', ESO/EPIC Workshop, ed: P.Crane, Springer, p.105

Bernas,R., Gradsztajn, E., Reeves, H., and Schatzman, E. 1967, Ann. Phys. (N.Y.), 44, 426

Bloemen, H. et al. 1994, A&A, 281, L5

Boesgaard, A. and King, J.R. 1993, AJ, 106, 2309

Boyd, R. and Kajino, T. 1989, ApJ, 336, L55

Bradt, H.L. and Peters,B. 1950, Phys.Rev.A, 77, 54

Burrows, D.N. et al. 1993, ApJ, 406, 97

Brown , A.G.A., de Geeus, E.J. and de Zeeuw, P.T. 1994, A&A, 289, 101

Cassé, M., Vangioni-Flam, E., Lehoucq, R. and Oberto, Y. 1995a, in "Nuclei in the Cosmos III", eds: M. Busso, R. Gallino and C. Raiteri , p.539

Cassé, M., Lehoucq R. and Vangioni-Flam, E. 1995b, Nature, 373, 318

Chaussidon, M. and Robert, F. 1995, Nature, 374, 337

Clayton, D.D. and Jin, L. 1995, ApJ, 451, 681

Cunha, K., Smith, V.V. and lambert, L. 1995, preprint

Dogomatskii, G.V. and Nadyozhin, D.K. 1977, M.N.R.A.S., 178,33P

Duncan D., Lambert D. and Lemke M. 1992, ApJ, 401, 584

Edvardsson B. et al. 1994, A&A, 290, 176

Feltzing, S. and Gustaffson, B. 1994, ApJ,423, 68

Fields, B.D., Schramm, D.N. and Truran, J.W. 1993, ApJ, 405, 559

Fields, B.D., Olive, K.A. and Schramm, D.N. et al. 1994 ApJ, 435, 185

Fields, B.D.,Olive, K. and Schramm, D.N. 1995, ApJ, 439,854

Fields, B., Cassé, M. Vangioni–Flam E. and Nomoto K. 1995, ApJ, accepted

Fowler, W.A., Burbidge, G.R. and Burbidge, E.M. 1955, ApJS, 17,167

Fowler, W.A., Greenstein, J.L. and Hoyle, F. 1962, J. Geophys. Res., 6, 148

Garcia Lopez, R.J., Severino, G. and Gomez, M.T. 1995, A&A, 297, 787

Gilmore, G., Edvardsson, B. and Nissen, P.E. 1991, ApJ, 378, 17

Gilmore, G., Gustafsson B., Edvardsson, B. and Nissen, P.E. 1992, Nature 357, 379

Kiselman, D. 1994, A&A, 286, 169

Kiselman, D. and Carlsson,M. 1995 in 'The light element abundances" ESO/EIPC Workshop, ed: P. Crane, Springer, p. 372

Kohl, J.L., Parkinson W.H. and Withbroe.N. 1977, ApJ, 212, L101

Lemoine, M. Vangioni-Flam, E. and Cassé, 1995, in preparation

Malaney, R.A. 1992, ApJ, 398, L45

Meneguzzi, M. Audouze, J. and Reeves, H. 1971, A&A, 15, 337
Meneguzzi, M. and Reeves, H. 1975, A&A, 40, 99
Mitler, H.E. 1972, Ap&SS, 17, 186
Pagel, B.E. 1991, Nature, 354, 267
Pinsonneault, M.H., Deliyannis, C.P. and Demarque, P. 1992, ApJS, 78, 181
Prantzos, N., Cassé, M. and Vangioni-Flam E. 1993, ApJ, 403, 630
Olive, K.A., Prantzos, N., Scully S. and Vangioni-Flam E. 1994, ApJ, 424, 666
Ramaty, R. , Kozlovsky,. and Lingenfelter R.E. 1995a, ApJ, 438, L21
Ramaty, R. , Kozlovsky, B. and Lingenfelter, R.E. 1995b, Proceedings of the 17th Texas Symposium, New York Acad. Sci., in press
Ramaty, R., Kozlovsky ,B. and Lingenfelter, R. E., 1996, ApJ, in press
Read S.M. and Viola V.E. 1984, Atomic Data and Nuclear Data Tables, 31, 359
Rebolo R., Molaro P., Abia C., Beckman J., 1988, A&A, 193,93
Reeves, H. and Meyer, J.P. 1978, ApJ, 226, 613
Reeves, H., Fowler, W.A. and Hoyle, F. 1970, Nature, 226, 727
Reeves, H. 1994, Rev. Mod. Phys. 66, 193
Ryan, S., Bessel, M., Sutherland, R., Norris, J., 1990, ApJ, 348, 157
Ryan, S., Norris, I., Bessel, M. and Deliyannis, C. 1992, ApJ, 388, 184
Shima, M and Honda M, 1962, J. Geophys. Res., 68, 2849
Spite, F. and Spite, M. 1993, in "Origin and Evolution of the Elements", ed. N. Prantzos, E. Vangioni-Flam and M. Cassé, Cambridge University Press, p. 201
Steigman, G. and Walker T.P. 1992, ApJ, 385, L13
Steigman, G.et al, T.P. 1993, ApJ, 415, L35
Thorburn, J.A. 1994, ApJ, 421, 318
Timmes, F.X., Woosley, S.E. and Weaver, T.A. 1995, ApJ, in press
Vangioni-Flam, E., Cassé, M., Audouze, J. and Oberto, Y. 1990, ApJ, 364, 586
Vangioni-Flam, E. and Cassé,M, 1994, ApJ, 441, 471
Vangioni-Flam, E., Lehoucq, R. and Cassé, M. 1995a, in "The light element abundances" ESO/EIPC Workshop Isola d'Elba, Italy, ed: P. Crane, p. 389
Vangioni-Flam, E., Cassé, M and Ramaty,R. 1995b,ApJ, submitted
Vangioni-Fam, E., Cassé, M., Fields, B. and Olive K. 1995c, in preparation
Walker, T.P., , Mathews, G.J. and Viola, V.E. 1985, ApJ, 229, 745
Walker, T.P. et al 1993, ApJ, 413, 562
Woosley, S.E., Hartmann, D., Hoffman, R., Haxton, W. 1990, ApJ, 356, 272
Woosley, S. E., Langer, N. A., Weaver T. A. 1993, ApJ, 411, 823
Woosley, S. E., and Weaver, T. A. 1995, ApJS, to appear

Abundance Determinations from Gamma Ray Spectroscopy

Reuven Ramaty

Laboratory for High Energy Astrophysics, Goddard Space Flight Center, Greenbelt, MD 20771

Abstract. Gamma ray emission lines resulting from accelerated particle bombardment of ambient gas can serve as an important spectroscopic tool for abundance determinations. The method is illustrated by considering the gamma ray line emission observed from solar flares. The observation of similar gamma ray lines from Orion suggests the existence of large fluxes of low energy Galactic cosmic rays. The role of these cosmic rays in the nucleosynthesis of the light isotopes is discussed.

1. Introduction

The interactions of accelerated particles with ambient matter produce a variety of gamma ray lines following deexcitations in both the nuclei of the ambient medium and the accelerated particles. Astrophysical deexcitation line emission produced by accelerated particle interactions has so far been observed from solar flares (e.g. Chupp 1990; Share & Murphy 1995) and the Orion molecular cloud complex (Bloemen et al. 1994). For a general review of astrophysical gamma ray line emission see Ramaty & Lingenfelter (1995). The solar gamma ray line observations have many applications, including the determination of solar atmospheric abundances (Murphy et al. 1991; Ramaty et al. 1995a; Ramaty, Mandzhavidze, & Kozlovsky 1996a). The Orion observations, even though much less detailed, have nonetheless revealed the existence of large fluxes of low energy cosmic rays in this nearest region of recent star formation (e.g. Ramaty 1996). If such cosmic rays are also present at other sites in the Galaxy, then low energy Galactic cosmic rays may play a very important role in the nucleosynthesis of the light isotopes ^{6}Li, ^{9}Be, ^{10}B and ^{11}B (Cassé, Lehoucq, & Vangioni-Flam 1995; Ramaty, Kozlovsky, & Lingenfelter 1996b).

In the present paper we briefly discuss these topics, referring the reader for more details to the papers mentioned above.

2. Solar Gamma Ray Spectroscopy

The solar flare gamma ray data is now sufficiently detailed to allow the conduct of a meaningful gamma ray spectroscopic analysis of the ambient solar atmosphere. The key is provided by the narrow line emission produced by accelerated protons and α particles interacting with ambient C and heavier nuclei. Owing to their narrower widths, these lines can be distinguished from the broader lines produced by accelerated C and heavier nuclei interacting with ambient H and He. The intensities of the narrow lines depend on the heavy element abundances and thus can be used to determine these abundances. Strong narrow line emission at 4.44, 6.13, 1.63, 1.37, 1.78, and 0.85 MeV, resulting from deexcitations in ^{12}C, ^{16}O, ^{20}Ne, ^{24}Mg, ^{28}Si and ^{56}Fe, respectively, has been observed from many flares. The most recent results, observed with the Solar Maximum Mission (SMM) from 19 flares (Share & Murphy 1995), allow the determination of the abundance ratios C/O, Mg/O and Mg/Ne for all 19 flares, Si/O for 14 flares and Fe/O for 12 flares (Ramaty et al. 1995a; 1996a).

Unlike atomic spectroscopy, nuclear spectroscopy does not require the temperature and ionic state of the ambient gas, neither of which are always well known. On the other hand, abundance determinations by nuclear spectroscopy do require information on the spectrum of the accelerated particles. For the SMM flare analysis (Ramaty et al. 1995a; 1996a) the accelerated particle spectra were constrained by using the 1.63 MeV ^{20}Ne-to-6.13 MeV ^{16}O and the 2.22 MeV neutron capture-to-4.44 MeV ^{12}C line fluence ratios, both of which are strong functions of the particle spectrum.

As in other solar atmospheric abundance studies (e.g. Meyer 1992), it is useful to distinguish two groups of elements depending on their first ionization potential (FIP): low FIP (<10 eV) elements (Mg, Si and Fe) and high FIP (>11 eV) elements (C, O and Ne). The enhancement of low FIP-to-high FIP element abundance ratios in the corona relative to the photosphere is well established from atomic spectroscopy and solar energetic particle observations (e.g. Meyer 1992). Analysis of the gamma ray data has led to the following conclusions (Ramaty et al. 1995a; 1996a):

(i) For the high FIP elements C and O, the derived abundance ratio (by number) is $0.35 \lesssim C/O \lesssim 0.44$. This range is more consistent with the C/O = 0.43±0.05 given by Anders & Grevesse (1989) than with the C/O=0.48±0.1 of Grevesse & Noels (1992). But taking into account the large uncertainty of the latter, there is no real discrepancy. Furthermore, a single value of C/O is consistent with the data for all 19 flares, implying that C/O could have the same value throughout the gamma ray production region. This is in fact not surprising given that C/O is essentially the same in the photosphere and corona.

(ii) For another pair of high FIP elements, O and Ne, the gamma ray data is in better agreement with Ne/O=0.25 than with the commonly adopted photospheric and coronal value of 0.15. Such a low Ne/O could only be accommodated by a very steep accelerated particle spectrum which would take advantage of the very low threshold for the excitation of the 1.63 MeV level of ^{20}Ne. The implied particle spectra, however, are too steep to produce sufficient neutrons to account for observations of the 2.22 MeV neutron capture line. In addition, the energy contained in ions with such steep spectra would be inconsistent with the overall flare energetics. Some EUV and X-ray observations also support a Ne/O higher than 0.15 (Saba & Strong 1993; Schmelz 1993; Widing & Feldman 1995).

(iii) To avoid values of Ne/O larger than 0.3 the accelerated particle energy spectra should be at least as steep as an unbroken power law down to about 1 MeV/nucl. For such power laws, the energy contained in the ions for the 19 analyzed flares ranges from about 10^{30} to well over 10^{32} ergs, and is thus comparable or even exceeds the energy contained in the nonrelativistic electrons that produce the hard X-rays in solar flares. Prior to these recent gamma ray analyses, it was widely believed that a large fraction of the released flare energy is contained in nonrelativistic electrons.

(iv) Considering the abundance ratios between elements of different FIP groups, both Mg/O and Mg/Ne show evidence for variability from flare to flare at about the 3σ level. For Mg/O this variation is confined to a range around the coronal value of 0.2 and does not go down to the photospheric value of 0.045. For Si/O and Fe/O the variations are also confined to a range around their respective coronal values. The fact that the low FIP-to-high FIP abundance ratios derived from gamma ray spectroscopy are enhanced relative to their respective photospheric values shows that the gamma ray production region lies above the photosphere.

3. On the Origin of the Light Isotopes

The discovery of gamma ray line emission from Orion (Bloemen et al. 1994) and the implied existence of large fluxes of low energy Galactic cosmic rays, has led to renewed discussions on the origin of the light elements. It has been known for over two decades that the relativistic Galactic cosmic rays (GCR) may have produced the observed solar system abundances of ^{6}Li, ^{9}Be and ^{10}B (Meneguzzi, Audouze & Reeves 1971; Mitler 1972). These cosmic rays, however, cannot account for the abundances of ^{7}Li and ^{11}B. It is believed that most of the Galactic ^{7}Li is produced in stars (e.g. Reeves 1994). Recent measurements of the boron isotopic ratio in meteorites yielded ^{11}B/^{10}B values in the range 3.84 – 4.25 (Chaussidon & Robert 1995) which exceed the calculated GCR value by a factor of about 1.5. The implications of the Orion gamma ray observations on the origin of the light isotopes have been considered by Cassé, Lehoucq & Vangioni-Flam (1995) and Ramaty et al. (1996b). The advantages of producing the light isotopes with 'Orion-like' low energy cosmic rays are the following:

(i) Low energy cosmic rays can produce B such that ^{11}B/^{10}B $\gtrsim 4$. Energetic arguments favor light isotope production by low energy cosmic rays at typical particle energies around 20 MeV/nucl rather than at lower energies (Ramaty et al. 1996b). At these energies the excess ^{11}B results mostly from ^{12}C via the reactions ^{12}C(p,pn)^{11}C and ^{12}C(p,2p)^{11}B which have lower thresholds than the reaction ^{12}C(p,2pn)^{10}B. At higher energies, and in reactions with ^{16}O, the B isotopic ratio is significantly lower.

(ii) The low energy cosmic rays must be depleted in protons and α particles. The α particle depletion is necessary in order not to overproduce ^{6}Li; the proton depletion ensures a linear dependence of the Be and B abundances on the Fe abundance in stars of various ages. If the low energy cosmic rays are poor in protons and α particles they will produce Be and B only from the breakup of accelerated C and O in interactions with ambient H and He; in this case both the target and projectile abundances could remain constant, leading to a linear growth of the Be and B abundances. On the other hand, the GCR would produce much of the isotopes from the breakup of C and O in the ambient medium whose abundances increase with time, leading to a quadratic growth.

(iii) Arguments of energetics have independently led to the suggestion that the low energy cosmic rays in Orion consist mostly of C and heavier nuclei with the protons and α particles strongly suppressed (Ramaty et al. 1995b; 1996b). This suppression could be the consequence of the particle injection process prior to the acceleration itself. The proposed injection sources are the winds of Wolf Rayet stars (Ramaty et al. 1995b), the ejecta of supernovae from massive star progenitors (Cassé et al. 1995; Ramaty et al. 1996b), and the pick up ions resulting from the breakup of interstellar grains (Ramaty et al. 1996b).

References

Anders, E. & Grevesse, N. 1989, Geochim. et Cosmochim. Acta, 53, 197
Bloemen, H. et al. 1994, A&A, 281, L5
Cassé, M., Lehoucq, R., & Vangioni-Flam, E., 1995, Nature, 373, 318
Chaussidon, M. & Robert, F. 1995, Nature, 374, 337
Chupp, E. L. 1990, Physica Scripta, T18, 15
Mitler, H. E. 1972, Ap&SS, 17, 186
Grevesse, N. & Noels, A. 1992, in Origin and Evolution of the Elements, eds. N. Prantzos et al. (Cambridge: Cambridge Univ. Press), 14
Meneguzzi, M., Audouze, J., and Reeves, H. 1971, A&A, 15, 337
Meyer, J-P. 1992, in Origin and Evolution of the Elements, eds. N. Prantzos et al. (Cambridge: Cambridge Univ. Press), 26
Murphy, R. J., Ramaty, R., Kozlovsky, B., & Reames, D. V. 1991, ApJ, 371, 793
Ramaty, R. 1996, A&A, in press
Ramaty, R. & Lingenfelter, R. E. 1995, in The Analysis of Emission Lines, eds. R. E. Williams and M. Livio, (Cambridge: Cambridge Univ. Press), 180
Ramaty, R., Kozlovsky, B., & Lingenfelter, R. E. 1995b, ApJ, 438, L21
Ramaty, R., Kozlovsky, B., & Lingenfelter, R. E. 1996b, ApJ, in press (Jan 10)
Ramaty, R., Mandzhavidze, N., Kozlovsky, B., & Murphy, R. J. 1995a, ApJ, 455, L193
Ramaty, R., Mandzhavidze, N., & Kozlovsky, B., 1996a, in High Energy Solar Physics, R. Ramaty, N. Mandzhavidze, X.-M. Hua, eds. (AIP: New-York), in press
Reeves, H. 1994, Revs. Modern Physics, 66, 193
Saba, J. L. R. & Strong, K. T. 1993, Adv. Sp. Res., 13 (9)391
Schmelz, J. T. 1993, ApJ, 408, 373
Share, G. H. & Murphy, R. J. 1995, ApJ, 452, 933
Widing, K. G. & Feldman U. 1995, ApJ, 442, 446

Anomalous Cosmic Rays: A Sample of Interstellar Matter

R. A. Mewaldt, R. A. Leske, and J. R. Cummings

California Institute of Technology, Pasadena, CA 91125

Abstract. Anomalous cosmic rays are a sample of the neutral interstellar medium that has been accelerated to energies of ~ 1 to 50 MeV/nuc. A comparison of ^{22}Ne/^{20}Ne measurements from various sources implies that galactic cosmic rays with energies > 100 MeV/nuc are not simply an accelerated sample of the local interstellar medium.

1. Introduction

Anomalous cosmic rays (ACRs) originate from neutral interstellar atoms that have been swept into the heliosphere, ionized by solar UV or charge exchange with the solar wind, convected into the outer heliosphere, and then accelerated to energies of ~ 1 to 50 MeV/nuc (Fisk, Koslovsky & Ramaty 1974). They are mainly singly-charged, and include H, He, C, N, O, Ne, and Ar (see review by Klecker 1994). It has recently been shown that ACRs impinging on the upper atmosphere can be stripped of their remaining electrons and trapped in the Earth's magnetosphere (by the mechanism of Blake & Friesen 1977), where they form a radiation belt composed of interstellar material (see Figure 1). Since its launch in 1992, the Solar, Anomalous, and Magnetospheric Particle Explorer (SAMPEX) has been measuring the composition and energy spectra of ACRs in interplanetary space and in the magnetosphere. These measurements provide a new source of information on interstellar matter.

Cummings & Stone (1995) have used ACR measurements to determine elemental abundances in the neutral interstellar medium (ISM); SAMPEX is extending this study to isotopic abundances (Leske et al. 1995). ACR Isotope measurements are important for studying the evolution of the local ISM since the formation of the solar system, and they are relevant to galactic cosmic ray (GCR) isotope measurements (Mewaldt, Spalding & Stone 1984).

This paper presents measurements of Ne isotopes from three ACR samples in the near-Earth environment, and compares these with the composition of other solar system and galactic material.

2. Anomalous Cosmic Ray Isotopic Composition

Figure 2 shows the composition of Ne isotopes from three regions of SAMPEX's polar orbit. Over the geomagnetic poles (latitudes $\Lambda > 60°$ in Figure 1), where there is a mixture of ACRs and GCRs, the ^{22}Ne/^{20}Ne ratio varies with energy. At energies > 50 MeV/nuc, where GCRs dominate, the ratio is ~ 0.6, compared

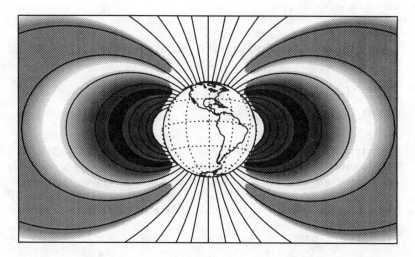

Figure 1. Illustration of the narrow radiation belt composed of interstellar material (*dark band*), embedded in the inner Van Allen belt. The new belt consists mainly of energetic O, N, and Ne with intensities > 100 times that of ACRs in interplanetary space, as well as smaller amounts of C and Ar (Selesnick et al. 1995).

to typical solar system material where ^{22}Ne/^{20}Ne~ 0.1. Even after correction for "secondary" ^{22}Ne produced by cosmic ray spallation, the resulting cosmic ray source ratio of ^{22}Ne/^{20}Ne $\simeq$ 0.45 (e.g., Lukasiak et al. 1994) shows that the nucleosynthesis of cosmic ray and solar system material has differed.

Below 50 MeV/nuc, where ACRs dominate, the ^{22}Ne/^{20}Ne ratio suddenly drops to $\sim$ 0.2. Similar variations in isotopic composition are observed for He, N and O (Mewaldt, Spalding & Stone 1984, Leske et al. 1995), although in these cases cosmic ray spallation makes a greater contribution to the rare isotopes ^{3}He, ^{15}N and ^{18}O. While it is possible to subtract the GCR contributions of ^{20}Ne and ^{22}Ne to obtain a corrected ACR ratio, this analysis is not yet complete. Instruments to be flown on the Advanced Composition Explorer (ACE) in 1997 will extend isotope measurements to lower energy (e.g., 5 to 15 MeV/nuc), where the ACR flux is greater and GCR contamination is minimized, with a collecting power > 30 times that on SAMPEX.

At mid–latitudes (e.g., $50° < \Lambda < 60°$) fully–stripped GCRs are not allowed, but singly–charged ACRs have access because of their greater magnetic rigidity. Here the Earth's field filters out a "pure" sample of ACRs, uncontaminated by GCRs or solar particles. In this region ^{22}Ne/^{20}Ne $\simeq$ 0.1, with a sizable statistical uncertainty.

Finally, at lower latitudes ($\Lambda \sim 45°$) trapped ACRs can be measured. While there are mass–dependent processes (e.g., trapping efficiency and lifetime) to be considered, the ^{22}Ne/^{20}Ne ratio is again $\sim$ 0.1.

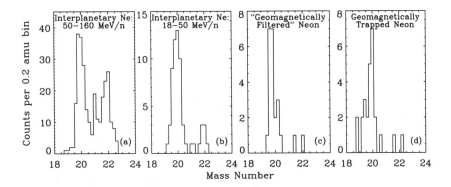

Figure 2. Ne isotope distributions from the MAST instrument on SAMPEX, including: a) 50 to 160 MeV/nuc GCRs from latitudes $\Lambda > 60°$; b) 18 to 50 MeV/nuc cosmic rays (mainly ACRs) from $\Lambda > 60°$; c) "filtered" ACRs from $50° < \Lambda < 60°$; and d) ACRs trapped in the magnetosphere ($\Lambda \sim 45°$).

3. Discussion

To place these measurements in context Figure 3 compares ^{22}Ne/^{20}Ne measurements from several sources. The isotopic composition of solar Ne has long been controversial. Cameron (1983) used the meteoritic component "Neon-A" (with ^{22}Ne/^{20}Ne = 0.122) for his table of solar system abundances, while Anders & Grevesse (1989) chose the solar wind value of ^{22}Ne/^{20}Ne = 0.076 (Geiss et al. 1972). The solar wind value is close to the lunar/meteoritic component "Neon-B", presumed to be implanted solar wind. Recent solar energetic particle (SEP) measurements from SAMPEX (Selesnick et al. 1993) give a ^{22}Ne/^{20}Ne ratio very close to the solar wind value. The GCR source ratio of ~ 0.45 greatly exceeds any of these solar system components. The SAMPEX and Voyager (Cummings, Stone & Webber 1993) values for the local ISM are both ~ 0.1, not sufficiently accurate to differentiate Neon-A and Neon-B, but clearly much less than the GCR source ratio. The values obtained from low energy ACRs over the poles (Figure 2b) and from trapped ACRs (Figure 2d) require additional analysis to estimate possible systematic corrections and uncertainties, but they also clearly favor a ratio of ~ 0.1 over a value as high as ~ 0.4.

These results provide the best evidence to date that cosmic rays are not simply a sample of local ISM that has been accelerated to high energies, as assumed in some models (e.g., Olive & Schramm 1982). Rather, the enhanced ^{22}Ne/^{20}Ne ratio in cosmic rays is evidence for contributions from sources especially rich in ^{22}Ne, such as Wolf-Rayet stars (Prantzos et al. 1986). This work demonstrates the potential of ACRs to provide unique information on the composition of the local ISM. In coming years we can expect improved statistical accuracy from SAMPEX as solar minimum approaches, and improved capability from ACE.

Acknowledgments. We appreciate contributions to this work by A. C. Cummings, R. S. Selesnick, E. C. Stone, and T. T. von Rosenvinge. This work was supported by NASA under contract NAS5-30704 and grant NAGW-1919.

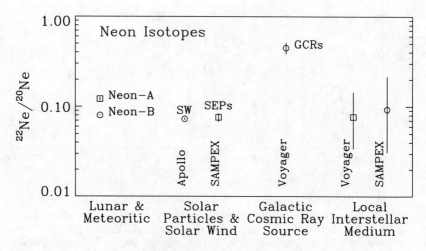

Figure 3. Comparison of ^{22}Ne/^{20}Ne ratios (references in text).

References

Anders, E., & Grevesse, N. 1989, Geochim. Cosmochim. Acta, 53, 197

Blake, J. B., & Friesen, L. M. 1977, Proc. 15th Internat. Cosmic Ray Conf. (Plovdiv) 2, 341

Cameron, A. G. W. 1982, in Essays in Nuclear Astrophysics, C. A. Barnes, D. D. Clayton, and D. N. Schramm, Cambridge Univ. Press, p. 23

Cummings, A. C., & Stone, E. C. 1995, Proc. 24th Internat. Cosmic Ray Conf. (Rome), 4, 497

Cummings, A. C., Stone, E. C., & Webber, W. R. 1991, Proc. 22nd Internat. Cosmic Ray Conf. (Dublin), 3, 362

Fisk, L. A., Kozlovsky, B., & Ramaty, R. 1974, ApJ, 190, L35

Geiss, J., Buehler, F., Cerruti, H., Eberhardt, P., & Filleux, Ch. 1972, Apollo 16 Preliminary Science Report, NASA SP-315, p. 14-1

Klecker, B. 1995, Space Science Reviews, 72, 419

Leske, R. A., Cummings, A. C., Cummings, J. R., Mewaldt, R. A., Stone, E. C., & von Rosenvinge, T. T. 1995, Proc. 24th Internat. Cosmic Ray Conf. (Rome), 2, 606

Lukasiak, A., Ferrando, P., McDonald, F. B., & Webber, W. R. 1994, ApJ, 426, 366

Mewaldt, R. A., Spalding, J. D., & Stone, E. C. 1984, ApJ, 283, 450

Olive, K. A., & Schramm, D. N. 1982, ApJ, 257, 276

Prantzos, N., Doom, C., Arnould, M., & deLoore, C. 1986, ApJ, 304, 695

Selesnick, R. S., Cummings, A. C., Cummings, J. R., Leske, R. A., Mewaldt, R. A., Stone, E. C., & von Rosenvinge, T. T. 1993, ApJ, 418, L45

Selesnick, R. S., Cummings, A. C., Cummings, J. R., Mewaldt, R. A., Stone, E. C., & von Rosenvinge, T. T. 1995, J. Geophys. Res. 100, 9503

Galactic Cosmic Ray Source Elemental Composition

M. A. DuVernois

Enrico Fermi Institute and Department of Physics, The University of Chicago, Chicago, IL 60637 USA

Abstract. The nucleosynthetic origin of the galactic cosmic ray sources (GCRS) is a matter of considerable interest recently (e.g., Leske 1993; Connell & Simpson 1995; DuVernois et al. 1995a). An understanding of the source requires an accurate determination of th e source elemental abundances. These abundances are determined by comparing the University of Chicago Ulysses High Energy Telescope (HET) observations in the heliosphere with results of propagation calculations. By comparing the data, combined with HEAO–3–C2 results, with various propagated source abundances, we derive the best-fit source elemental abundances. The GCRS composition is found to be similar to solar composition with a first ionization potential (FIP) bias.

1. Introduction

The elemental composition of the cosmic rays in the heliosphere has been measured a number of times at various points in the solar cycle (Engelmann et al. 1990). This data has been used to infer source elemental abundances (Webber 1982; Simpson 1983). In this paper, we examine this issue using the high resolution data available from the Ulysses HET and the published data of HEAO–3–C2 (Engelmann et al. 1990; Simpson et al. 1992). The Ulysses HET was designed to resolve iron-group isotopes (Connell & Simpson 1995). Therefore, its elemental identification is quite reliable. In the elemental range from Be to Ni, charge resolution varies from 0.04–0.09 charge units, which is more than adequate to prevent misidentifications.

We use a leaky box model of galactic propagation solved with the weighted slab technique (Ginzburg et al. 1980; García-Muñoz et al. 1987). We augment this model with recent nuclear partial-production cross-section measurements and energy loss enh ancements due to ionized hydrogen (Thayer 1995a, b). The pathlength distribution (PLD) used is a single exponential which fits the Ulysses secondary to primary elemental ratios (DuVernois et al. 1995b, c) for both long (light elements) and short (heavy el ements) pathlengths.

2. Observations

Aboard Ulysses is the University of Chicago High Energy Telescope, a stack of solid-state particle detectors that is an excellent tool for looking at the galactic cosmic rays in the ~40–450 MeV/N range from H–Fe group (Simpson et al.

1992). By usin g quiet-time data from launch through mid-1995, excluding the high-flux Jovian encounter, we have excellent statistics on the heavy ions. Statistical errors are under 10% for each of the elements below iron.

The observed abundances were standardized to an energy of 185 MeV/N. This was done by integrating over the CHIME model spectra (Chenette et al. 1994) for each element at the solar modulation level of $\Phi = 750$ MV, and scaling by the ratio of the spectr al intensities at 185 MeV/N. These normalizations remove the effects of measuring over different energy intervals and allow for direct comparison of abundances. A generous allowance for errors in the spectral shape is included in the reported observed abu ndances. The abundances were then scaled into the Si = 100 format (see the table in DuVernois & Thayer 1995).

3. Propagation

Starting with the leaky box model of galactic cosmic ray propagation, our propagation algorithm handles the energy loss, nuclear spallation, and radioactive decay of the cosmic rays. We solve the differential equations for the energy-dependent propagation of each isotopic species by way of the weighted-slab technique. This separates the creation, loss, and energy-changing terms from the pathlength distribution (PLD). The weighted slab equations

$$\frac{dN_i}{dx} = \frac{\partial}{\partial E}[(\frac{dE}{dx})_i N_i] - \frac{N_0}{\overline{A}}\sigma_i N_i + \sum_{j \ne i}\frac{N_0}{\overline{A}}\sigma_{ij} N_j - \frac{N_i}{\gamma \beta cn \overline{A} T_i} + \sum_{j \ne i}\frac{N_j}{\gamma \beta cn \overline{A} T_j} \quad (1)$$

are solved numerically. $N_i(E, x)$ is the energy and pathlength dependent abundance of the 'i'th species; σ_i, the total nuclear spallation cross-section; σ_{ij}, the partial cross-section between species i and j; n, the number density o f the interstellar material; N_0, Avogadro's number; and T_i, the half-life of the radioactive species. More details of the propagation calculation can be found elsewhere (García-Muñoz et al. 1987; DuVernois et al. 1995b; Thayer 1995b). T he resultant number density is integrated over the PLD to give the local interstellar energy spectrum, $J_i(E)$. The PLD is determined separately by looking at the energy-dependent secondary/primary elemental ratios (Shapiro & Silberberg 1970). The PLD used here is a single exponential, using the energy-dependent mean of García-Muñoz et al. (1987). This spectrum is then modulated in the heliosphere. We model the heliospheric modulation as a spherically symmetric ! field with a time-varying potenti

The results of propagating a trial set of GCRS abundances through our model are compared with the Ulysses data at low energy and the HEAO-3 data at high ($\sim$1–40 GeV/nucleon) energy. The trial set is then modified in an effort to more closely match th e observed abundances and iterated through the propagation until agreement between model and observations are reached.

By starting with solar system abundances, shown to be close to those of the GCRS (Webber 1982; DuVernois et al. 1993; Connell & Simpson 1995; DuVernois et al. 1995a), a small number of iterations were necessary to obtain a source composition that, when p ropagated, fairly reproduces the observed abundances. Agreement between the propagated energy-dependent element ratios and the experimental data was required for all the elements from Be through Ni.

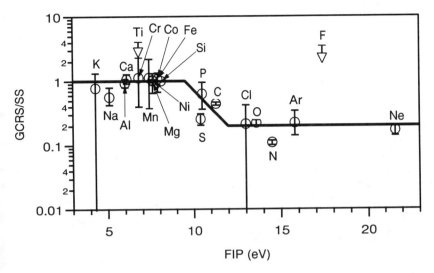

Figure 1. GCRS/Solar Abundance versus FIP

4. Discussion

The source elemental abundances which reproduce the observed elemental abundances in the heliosphere after propagation can be found in DuVernois & Thayer 1995. Errors on the source composition include estimates on the uncertainties of the propagation (PL D, cross-sections, and modulation). These estimates are made in a conservative manner and represent the maximal reasonable errors. It is interesting to compare this composition to the solar system elemental abundances (Anders & Grevesse 1989). Some of th e most pronounced differences are highlighted by the observation of an elemental biasing in the GCRS by FIP (Meyer 1985). A plot of the ratio of GCRS abundances to solar system abundances as a function of FIP is Figure 1. The break in the ratio at 10 eV (from ~ 1 to ~ 0.2) is interpreted as indicative of a 7000 K environment from which ions are picked up and accelerated to become the galactic cosmic rays. This leads to models of the cosmic ray source as a stellar win! d picked up by a shock.

Alternately, there have been suggestions by Meyer (1981, 1995) that the departures from the FIP curve, most notably K and Na, can be explained by their volatility. The volatility bias would arise from a grain origin of the cosmic rays. This possibility sh ould not be ignored.

It's interesting and surprising to note how similar, after the FIP bias, the GCRS and the solar system compositions are. The cosmic ray material is of recent nucleosynthetic origin—a few Myr (Simpson & García-Muñoz 1988; Leske 1993)—while th e solar system matter is ~ 5 Gyr. This appears to rule out a large change in the galaxy's chemical composition over this period of time.

5. Conclusions

The sources of the galactic cosmic rays have an elemental composition which can be calculated from elemental measurements in the heliosphere. The calculation uses the leaky box model of interstellar cosmic ray propagation and is solved using the weighted- slab technique. The elemental source composition is found to be roughly that of solar with the known bias in first ionization potential.

These are the first calculations of the source composition using the new generation of cosmic ray telescopes. Improved charge and mass resolution, large geometrical factors, long collection time, and improved knowledge of propagation parameters allow for an accurate and statistically significant calculation of the cosmic ray source composition. A complete understanding of the nuclear processes creating this composition remains elusive.

Acknowledgments. This research is supported in part by NASA/JPL Grant 955432. I wish to thank J. A. Simpson for his generous support, M. R. Thayer, J. J. Connell, and E. Murphy.

References

Anders, E. & Grevesse, N. 1989, Geochim. Cosmochim. Acta, 53, 197
Chenette, D. L., et al. 1994, IEEE Trans. Nuc. Sci., 41, 2332
Connell, J. J. & Simpson, J. A. 1995, 24th ICRC, OG 5.2.2
DuVernois, M. A., et al. 1993, 23rd ICRC, OG 5.2.7
DuVernois, M. A., et al. 1995a, ApJ, In submission
DuVernois, M. A., Simpson, J. A., & Thayer, M. R. 1995b, ApJ, In submission
DuVernois, M. A., Simpson, J. A., & Thayer, M. R. 1995c, 24th ICRC, OG 5.1.7
DuVernois, M. A. & Thayer, M. R. 1995, ApJ, In submission
Engelmann, J. J., et al. 1990, A&A, 233, 96
Evenson, P., et al. 1983, ApJ, 275, L15
García-Muñoz, M., et al. 1987, ApJS, 64, 269
Ginzburg, V. L., et al. 1980, Astro. & Sp. Sci., 68, 295
Leske, R. A. 1993, ApJ, 405, 567
Meyer, J. P. 1981, 17th ICRC, OG 4-7
Meyer, J. P. 1985, ApJS, 57, 173
Meyer, J. P. 1995, Private communications
Shapiro, M. M. & Silberberg, R. 1970, Ann. Rev. Nucl. Sci., 20, 323
Simpson, J. A. 1983, Ann. Rev. Nucl. Part. Sci., 33, 323
Simpson, J. A. & García-Muñoz, M. 1988, Sp. Sci. Rev., 46, 205
Simpson, J. A., et al. 1992, A&AS, 92, 365
Simpson, J. A., et al. 1995, Science, 268, 1019
Thayer, M. R. 1995a, 24th ICRC, OG 8.1.7
Thayer, M. R. 1995b, In preparation
Webber, W. R. 1982, ApJ, 252, 386

The O/H Abundance Distribution in the Large Spiral Galaxy NGC 4258

Yvan Dutil

Observatoire du mont Mégantic and Université Laval, Département de Physique, Québec, Qc G1K 7P4, Canada dutil@phy.ulaval.ca

Abstract. Employing imaging spectrophotometry with interference filters, we have measured the global abundance gradient of oxygen in the disk of the spiral galaxy NGC 4258 (SABbc). Based on our measurements of diagnostic nebular line ratios in 73 H II regions, the global abundance gradient in NGC 4258 is found to be −0.022 dex/kpc and the central abundance is 12 + log O/H = 8.85. The shallowness of the gradient is consistent with the galaxy being strongly barred, but its mean abundance level appears too low by 0.2 dex, when compared with galaxies of the same mass or Hubble type.

1. Introduction

The study of chemical abundances is a powerful tool to understand the mechanisms wich control the evolution of galaxies. Abundance determinations have been done of the interstellar medium in many disk galaxies (Zaritsky et al. 1994); these studies have revealed the existence of radial abundance gradients. Several processes have been invoked to explain the existence of such gradient: radial variation of the star formation rate and of the gas fraction (Phillips & Edmunds 1991) and flows of gas (Edmunds & Greenhow 1995). However, the present observations do not permit to choose a particular model linking chemical evolution and the morphological types of galaxies. Bars also affect the abundance gradients by inducing large scale flows (Martin & Roy 1994). Most studies are based on a small number of galaxies and in many cases on a handful of H II regions per galaxy. The need for more extensive observations is most crying for the early-type galaxies (Zaritsky et al. 1994).

The lack of measurements for early-type galaxies reflects the intrinsic difficulties of the observations; H II regions are fainter and smaller than in later type galaxies, and the nebular diagnostic lines become very weak at the high levels of abundance observed in these galaxies. Recently, there have been attempts to overcome these difficulties (Oey & Kennicutt 1993; Zaritsky et al. 1994). But the abundance distribution is still only sparsely known for this type of galaxies.

Imaging spectrophotometry can help to overcome the inherent difficulties of these observations. Using this method, we have been able to derive the abundance of oxygen of 73 H II regions in the SABbc galaxy NGC 4258.

2. Observations

The observations were done using a focal reducer (f/8 → f/3.5) on the Mont Mégantic Observatory 1.6m telescope. We had four observing runs: two in February and April 1994, and two in February and April 1995. Images were taken with narrow band interference filters tuned for Hα, Hβ, [OIII]λ5007 and [NII]λ6584 and also in the red and green continua. Only the observations of the northern part of the galaxy are reported here.

Spectroscopic observations of two bright H II regions have been obtained at CFHT in order to flux calibrate the spectra. The line intensities of Hα, Hβ, [OIII] and [NII] were derived interactively using the SPLOT function of the LONGSLIT package of IRAF[1]. Line ratios were corrected for the interstellar reddening by comparing the Hα/Hβ ratio with the theoretical Balmer decrement (case B) as given by Osterbrock (1989). The reddening correction was applied employing the reddening law of Savage & Mattis (1979). Before the reddening correction was applied, the Hβ flux was corrected for the underlying Balmer absorption by adding 2 Å of equivalent width (Belley & Roy 1992).

The oxygen abundance was derived by using the semi-empirical relation between O/H and ([NII]λ6584/([OIII] λ5007) (Alloin *et al.* 1979) and [OIII]$\lambda\lambda$4959, 5007)/Hβ (Edmunds & Pagel 1984). The ([NII]λ6584/ ([OIII]λ5007) line ratio provides a tighter relation with abundances than does ([OIII]$\lambda\lambda$4959, 5007)/Hβ. This comes from the greater intrinsic sensitivity of [NII]/[OIII] to variations of electronic temperature, because the two lines of this ratio are collisionaly excited, and are affected by temperature variations in opposite direction. This indicator is generaly well correlated with the [OIII]/Hβ indicator.

3. Discusion

Using the [NII]/[OIII] ratio, we derived the O/H abundances of 73 H II regions in NGC 4258. The central abundance of oxygen as extrapolated from the gradient is 12 + Log = 8.85, and the global gradient is –0.022 dex/kpc (D = 5.5 Mpc). These values are consistent with those derived by Oey & Kennicutt (1993) but is lower by 0.25 dex from that of Zaritsky *et al.* (1994). This difference can be explained in part by the small sample of only 8 H II regions used by Zaritsky *et al.* and by a different calibration relation.

---

[1]IRAF is distributed by the National Optical Astronomy Observatory wich is operated by the Association of Universities for Research in Astronomy Inc. under contract to National Science Foundation

If we compare the central abundance of NGC 4258 with that of NGC 628 (Belley & Roy 1992), which is a galaxy of similar mass, we note that the central abundance in NGC 4258 is lower by 0.35 dex. This difference apppears to be too large to be explained only by factors stated above. Since NGC 628 and NGC 4258 have a similar M_B (Zaritsky et al. 1994), this implies that it is underabundant in oxygen by a factor of two. On the other hand, if we compare NGC 4258 with NGC 7331, which has a similar mass and morphological type (Oey & Kennicutt 1993; Zaritsky et al. 1994), the same abundance discrepancy appears.

Abundance gradients are not clearly related to the mass nor to the morphological type of galaxies (Edmunds & Roy 1993; Oey & Kennicutt 1993; Zaritsky et al. 1994). A more direct relation exists between bar ellipticity and the slope of the gradient: the more elongated the bar is, the shallower is the gradient (Martin & Roy 1994).

In late type galaxies, the existence of a shallow gradient can be explained by the presence of a bar (Martin & Roy 1994). If the mechanisms controlling the gas flow in early types are the same as in later types, the shallow gradient could betray the presence of a strong bar in NGC 4258. Unfortunatly, measuring directly the strength of the bar in NGC 4258 is difficult because of the high inclination angle of the galaxy ($i = 69°$, Huchtmeier & Richter 1989). However, using the relation found by Martin & Roy (1994) between the slope of the abundance gradient and the strength of the bar, one can infer an ellipticity $\epsilon_B \approx 6.3$; this corresponds indeed to a very strong bar.

4. Conclusion

We have inferred the presence of a very strong bar in NGC 4258 from its shallow abundance gradient. Models suggest that bars cannot form spontaneously by intrinsic dynamical instabilities in early type spiral galaxies; formation of a bar in these galaxies requires a tidal interaction (Noguchi 1995). Moreover, the relatively low mean abundance may be explained by a large inflow of low metalicity gas or by a recent merger. NGC 4258 has a well-known central jet (Cecil, Wilson & Tully 1992; Dutil, Beauchamp & Roy 1995) The large scale gas flow could also feed the observed central jet (Noguchi & Bekki 1994).

This research was funded by a NSERC grant to J.-R. Roy and by the Fonds FCAR of the Government of Québec.

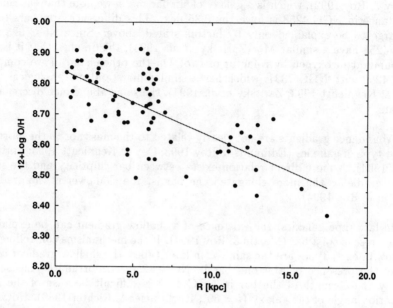

Figure 1. The radial O/H distribution in NGC 4258 derived from the [NII]/[OIII] line ratio.

References

Alloin, D., Collin-Souffrin, S., Joly, M., & Vigroux, L. 1979, A&A, 78, 200
Belley, J., Roy, J.-R. 1992, ApJS, 78, 61
Cecil, G., Wilson, A. S., & Tully, R. B. 1992, ApJ, 290, 265
Dutil, Y., Beauchamp, D., & Roy, J.-R. 1995, ApJ, 444, L85
Edmunds, M. G., Greenhow, R. M. 1995, MNRAS, 272, 241
Edmunds, M. G., Pagel, B.E.J. 1984, MNRAS, 211, 507
Edmunds, M. G., Roy, J-.R. 1993, MNRAS, 261, L17
Huchtmeier, W., Richter, O. 1989, A&A, 210, 1
Martin, P. , Roy, J.-R. 1994, ApJ, 424, 599
Noguchi, M., Bekki, M. 1994, A&A, 290, 7
Noguchi, M. 1996, in IAU coll. 157, BARRED GALAXIES, eds. R. J. Buta et al, ASP Conference Series, in press
Oey, M. S., Kennicutt, R. C. 1993, ApJ, 411, 137
Osterbrock, D. E. 1989, Astrophysics of Gaseous Nebulae and Active Galactic Nuclei (Mill Valley, CA: University Science Books)
Phillips, S., Edmunds, M. G. 1991, MNRAS, 251, 84
Savage, B. D., Mattis, J. S. 1979, ARA&A, 17, 73
Zaritsky, D., Kennicutt, R. C., Huchra, J. P. 1994, ApJ, 420, 87

Abundances in Elliptical Galaxy Hot Interstellar Media

M. Loewenstein[1]

NASA/GSFC Laboratory for High Energy Astrophysics, Code 662, Greenbelt, MD, 20771

Abstract. I review the abundances measured in the hot interstellar gas in elliptical galaxies using *ASCA* x-ray spectroscopy. The metallicities range from 0.1-0.8 solar and generally have Si-to-Fe ratios near solar, in systematic discordance with optical abundance determinations. This discrepancy and other puzzles and implications raised by the x-ray measurements are discussed.

1. Introduction

Since most of the star formation, and therefore most of the heavy element production, in elliptical galaxies occured during the very earliest stages of their development, observations of abundances in these systems provide a window into the galaxy formation process. In this contribution I will review x-ray (and, briefly, optical) determinations of elemental abundances in elliptical galaxies and discuss some of the issues that emerge from these results.

2. Optical Abundances

A review of the recent literature on measurement and analysis of optical spectral line indices can be summarized as follows. (1) The average metallicity of the stars derived from the Mg2 index is 2-3 times solar in the cores of bright elliptical galaxies, and about solar in faint galaxies; however, the Mg-to-Fe ratio is also 2-3 for the bright galaxies. (2) Abundance gradients show great variation, with slopes ranging from -0.1 to -0.7, tend to be steeper for galaxies with large central values of the Mg2 index, and have not been observed to flatten inside of two effective radii (r_e). (3) The global luminosity-averaged metallicity is 0.8 ± 0.4 times solar, where the uncertainties reflect measurement errors as well as a range of assumed behaviour outside the observed region.

3. X-ray Abundances from *ASCA*

Elliptical galaxies are generally filled with hot (~ 0.7 keV) interstellar gas. The extent of the hot gas is sometimes coincident with the optical light and sometimes

---

[1]also with Universities Space Research Association

much more extended. Abundances are derived from $ASCA$ observations by fitting spectra with simple models of optically thin, single-tempetature, thermal plasmas where the abundances are free to vary. The abundances of elements with prominent features in the $ASCA$ bandpass can be fixed at their solar proportions, or varied independently if the signal-to-noise ratio is high enough.

3.1. Advantages and Disadvantages of $ASCA$-derived Abundances

The major advantages x-ray abundance determinations in elliptical galaxies have over their optical counterparts are the relative straightforwardness and simplicity of the abundance derivation process and the ease with which non-solar ratios can be introduced – determination of optical abundances must include assumptions about the initial-mass-function and star formation history in the galaxy, and requires accurate and detailed models of post-main sequence stellar evolution and stellar atmospheres. On the other hand, the optical spectral indices are direct measures of stellar metallicities, while the hot gas observed in x-rays may be contaminated by type Ia supernova (SNIa) explosions or accreted intergalactic gas. While the optical measurements are generally restricted to $\sim r_e$ the x-ray measurements extend to many r_e; however, the spatial resolution of $ASCA$ is greater than $2r_e$ and often only one average spectrum is obtained.

3.2. $ASCA$ Results

Excepting two galaxies observed with BBXRT (Serlemitsos et al. 1993), $ASCA$ has provided the first accurate and precise measurements of elemental abundances in elliptical galaxy interstellar media. The abundances listed in Table 1 are compiled from analysis of performance verification phase data by the $ASCA$ team (primarily, R. Mushotzky and M. Loewenstein at the NASA/Goddard Space Flight Center, K. Makishima and K. Matsushita at the University of Tokyo, and H. Awaki and H. Matsumoto at the University of Kyoto), supplemented by analysis of public and my own guest-observer observations.

Table 1. $ASCA$ Abundances in Elliptical Galaxies

Galaxy	Metallicity	Limits[a]	Source/Remarks
NGC 1404	0.14	0.11-0.17	Loewenstein et al. 1994
NGC 4365	< 0.1	...	very preliminary[b]
NGC 4374	0.14	0.10-0.22	Loewenstein et al. 1994
NGC 4406	0.45	0.37-0.58	Awaki et al. 1994
NGC 4472	0.63	0.52-0.79	Awaki et al. 1994
NGC 4552	0.10	0.07-0.18	preliminary
NGC 4636	0.38	0.32-0.45	Awaki et al. 1994
NGC 4649	0.28	0.24-0.34	preliminary
NGC 4697	< 0.1	...	very preliminary[b]

[a]90% confidence limits
[b]analysis complicated by strong hard component

4. Puzzles and Implications

The low hot gas abundances reported in the previous section raise the following issues.

4.1. Optical/X-ray Discrepancy

In every case where there is accurate x-ray and optical abundances, the global (luminosity-weighted) averages are discrepant in the sense that the hot gas is underabundant with respect to the stars, typically by factors of 2-5. A number of possible explanations for this disagreement have been advanced.

Is it possible that the x-ray-derived abundances are simply incorrect? It has been suggested that uncertainties in the atomic physics of the Fe L-shell line complex could cause large errors in the x-ray abundances. However, the latest evidence from consistency checks using stars and clusters that have both Fe L and Fe K features, as well as incorporation of updated atomic parameters, indicate that the present codes produce abundances accurate to $\sim 30\%$. Another possibility is that an additional soft x-ray continuum associated with the stars in the galaxy is present and causing an underestimate of the hot gas abundances by diluting spectral features. I have examined this possibility in detail by simultaneously fitting *ROSAT* PSPC – with its superior low energy response – and *ASCA* spectra of NGC 4552 and NGC 4697. Based on this analysis, a soft component is an unlikely explanation for the low hot gas abundances, although more work is needed to rule it out.

Can the interstellar abundances be diluted by accretion of unenriched intergalactic gas? As Bertin and Toniazzo (1995) have pointed out, the accreted gas results in enhanced x-ray emission. In fact, the oppposite trend to what they predict is observed: the metallicity tends to be higher in galaxies with larger ratios of x-ray-to-optical luminosity. Accretion is probably occuring in some systems, but cannot explain the low abundances in general.

4.2. Where Have All the Supernovae Gone?

The Fe abundance in the hot gas will be enriched to 3-6 times solar for the "standard" early-type galaxy SNIa rate. The *ASCA* spectra strongly rule out any such enhancement. Either the standard rate is an overestimate by greater than a factor of ten or the SNIa ejecta do not efficiently mix into the hot interstellar medium.

4.3. Relation to Enrichment of Intracluster Gas

Recent detailed chemical evolutionary models appearing in preprints by Elbaz et al. and Matteucci and Gibson tuned to produce the abundances seen in the intracluster medium (ICM) overproduce the interstellar Fe by a large factor. It remains to be seen whether these models can be revised in a reasonable way that will simultaneously reproduce both interstellar and ICM abundances.

4.4. The Silicon-to-Iron Ratio

The average Si-to-Fe ratio for six galaxies with spectra that have sufficient statistics for these elements to be measured separately is 1.3 in solar units (the 90%

uncertainties in the Fe and Si abundances are typically 20% and 40%, respectively). By contrast, this ratio is typically about 2 in the ICM (Mushotzky et al. 1996), as is the Mg-to-Fe ratio in the stars. This mild Fe overabundance could be produced by SNIa exploding at 0.025 times the standard rate.

4.5. Galaxy Formation Clues

There is a tentative correlation between the hot gas metallicity and the product of the x-ray extent and ISM temperature that is a crude measure of the total dark matter content. This correlation is the x-ray equivalent of the color-magnitude relation, and indicates that more massive galaxies are better able to retain their metal enriched gas against ejection by galactic wind and thereby initiate additional cycles of star formation.

5. Final Remarks and Speculations

I would like to emphasize the following three aspects of this work.

1. These results are preliminary – a considerably larger and more homogeneous sample of $ASCA$-derived elliptical galaxy abundances will be available in the near future.

2. The x-ray data offers a global perspective on the sorts of correlations (e.g., total mass vs. metallicity) that optical observers have been investigating inside the effective radius and, mostly, inside the core.

3. The low abundances are puzzling but are probably correct to 30-50%.

There are two distinct morphologies of x-ray emitting gas halos in elliptical galaxies: halos that are roughly co-spatial with the stars and halos with a much larger scale-height. It could very well be that the low abundances in these two classes have different origins. The very extended x-ray halos in luminous galaxies seem to be repositories for material that has not been as fully enriched as the bulk of the stars. The even lower abundances in low-luminosity galaxies may only be explainable if these galaxies lose their interstellar gas in a supernovae-driven galactic wind and then re-accrete more poorly enriched intergalactic gas.

Acknowledgments. I would like to thank Richard Mushotzky and Jesus Gonzalez for sharing their expertise in x-ray and optical spectroscopy of elliptical galaxies, respectively.

References

Awaki, H., et al. 1994, PASJ, 46, L65
Bertin, G, & Toniazzo, T. 1995, ApJ, 451, 111
Loewenstein, M., et al. 1994, ApJ, 436, L75
Mushotzky, R. F., et al. 1996, ApJ, in press
Serlemitsos, P. J., et al. 1993, ApJ, 413, 518

Galaxies in Clusters: Implications for Abundances

Joseph Silk

Departments of Astronomy and Physics, and Center for Particle Astrophysics, University of California, Berkeley, CA 94720, USA

Abstract. Enrichment associated with early star formation can affect both the stellar metallicities of giant ellipticals and the intracluster gas phase abundances. The initial mass function in a major merger that triggers a starburst must be top-heavy in order to explain the amount of intracluster iron. Implications include the inevitability of wind-driven mass loss, the explanation of the systematic increase in mass-to-light ratio with luminosity inferred from the fundamental plane in terms of compact remnants, Type II supernova yields both in the stellar populations of luminous ellipticals and in the intracluster gas, and a potentially detectable diffuse cosmic far infrared background signal.

1. Introduction

The intracluster medium in rich clusters provides a reservoir for all ejecta from cluster galaxies over the past Hubble time. The abundances of heavy elements in the intracluster gas consequently provides a time integral, in conjunction with stellar metallicities, of the chemical history of cluster galaxies. This talk develops the common theme of how enrichment associated with early star formation can affect both the stellar metallicities of giant ellipticals and the intracluster gas phase abundances.

I begin by reviewing the relevant properties of cluster galaxies. Section 3 discusses intracluster and intragroup gas, and in Section 4 I describe a model for early-type galaxy formation. A final section provides various predictions.

2. The Galaxies

Rich clusters are dominated by ellipticals and S0's. Ellipticals formed stars rapidly (relative to a Hubble time) and efficiently. Galaxy formation occurred in a bottom-up fashion, with successively larger clumps merging together, according to the gravitational instability model for the origin of large scale structure. Elliptical formation was demarcated by major mergers, between objects of comparable mass. This occurred with significant probability in dense regions of the universe, where rich clusters are developing. Consider the universe at the epoch of cluster formation, $z \sim \Omega^{-1} - 1$. The most recent galaxy mergers occurred when the cluster underwent its initial collapse, a phase which, according to hierarchical models, is dominated by mergers of galaxy groups.

In low density regions, the typical mergers are likely to have been minor mergers between objects with mass ratios of 10:1 or more. Minor mergers are likely to trigger less efficient star formation. The preexisting disks can survive, as the shallower potential well of the merging cloud is relatively inefficient at retaining the debris of massive star formation. This provides a supply of gas that can dissipate over a much longer time scale to form a disk.

Mergers or strong tidal interactions drive gas clouds into the inner kiloparsc of the evolving galaxy potential well. Non-circular motions are generated and gas clouds collide and radiate away orbital energy. Prior to the merger, gas clouds formed stars, more or less as inefficiently as they do in the local interstellar medium. The merger drives agglomeration of giant gas clouds that will be unstable to star formation and capable of forming stars more efficiently. The high pressure environment where the central gas concentration develops provides global efficiency by inhibiting disruption of molecular clouds.

Let us then consider a typical major merger, the event that led to the formation of an elliptical. This must have occurred at modest redshift, when galaxy clusters were undergoing their initial collapse. The overdensity in a cluster and the frequency of galaxies with relatively low collision velocities mean that mergers were inevitable (Mamon 1992). Additional mergers occur for galaxies within subclusters just prior to virialization. A merger model provides a simple means of accounting for the dependence of elliptical galaxy fraction on local density and radial distance from the cluster center.

S0's would have formed by minor mergers. They also are concentrated in clusters, and a minor merger origin can account for the broader distribution, as a function of local density and distance from cluster center, of the S0 population relative to the ellipticals. The minor mergers are more frequent and on the average occurred more recently, because of the higher frequencies of minor mergers relative to major mergers in regions where one has not already depleted the dwarf galaxy population. Such depletion occurs near a centrally dominant galaxy, where dynamical friction operates. The most recent mergers in a cluster are expected to be S0 precursors. This enables one to understand the so-called $E + A$ phenomenon (Dressler & Gunn 1992).

Galaxies with a spectral distribution like that of an ordinary elliptical but containing A-type spectral features underwent a starburst some $1 - 2 Gyr$ ago. The present day starburst galaxies are rare in present day rich clusters, but frequent at $z \gtrsim 0.5$. The blue galaxy fraction is also much larger in these distant clusters than in present-day clusters. The present-day counterparts of the blue galaxies as well as the $E + A's$ are plausibly the S0's. HST measurements suggest that the blue galaxies have disk-like morphologies, and similar claims have been made for nearby $E + A's$. The population of galaxies in the outer parts of nearby clusters contains an enhanced fraction of blue and $E + A$ galaxies, relative to cluster cores, consistent with a merger history, since the outer part of a nearby cluster has been assembled relatively recently. Isolated $E's$ similarly show existence of major mergers, in the form of shells and dust lanes that occurred several Gyr ago (Schweizer and Seitzer 1992).

The E-forming merger cannot have been too recent, since the stellar populations are $\gtrsim 10 Gyr$ old, although there may in some cases be an intermediate age admixture. In this latter case, an age of $\sim 6 Gyr$ is possible, for what may have been a significant ($\gtrsim 10$ *percent*) addition of stars in the most recent merger event (Charlot & Silk 1994). On the other hand, major mergers cannot have occurred at high redshift, when such events would have been exceedingly rare according to our best estimates of the density fluctuation power spectrum that determines the hierarchy of structure formation.

Early type galaxies show little or no luminosity or spectral evolution in deep redshift surveys carried out to $z \sim 1$. It is likely that the typical ellipticals formed at $z \approx 2-3$: this estimate must be increased if $\Omega < 1$ by a factor $\sim 1/\Omega - 1$. Coincidentally this is also the epoch where the quasar population peaks. It has often been suggested that the quasar phenomenon may be identified with events that demarcate the formation of the (active) nucleus of an elliptical galaxy, such as formation of a massive black hole, and accretion of gas or disruption of stars by the central hole.

The clustering of galaxy clusters is much stronger than that of elliptical galaxies, suggesting that for a Gaussian density field, clusters formed from $\sim 3\sigma$ fluctuations. One can have a simple fit to the power spectrum, with variance approximately $\sigma(M) \propto M^{-\frac{1}{2}-\frac{n}{6}}$ and $n \approx -2$ on galaxy scales, to estimate that if galaxies represent rms 2σ density peaks, then

$$1 + z_{cluster} = [3\sigma(M_{cluster})/2\sigma(M_{galaxy})](1 + z_{galaxy}) \approx \frac{1}{2}(1 + z_{galaxy}).$$

Since the merging hierarchy on galaxy scales is going to spread over at least a factor of 2 in background expansion factor, it seems inevitable that mergers accompany cluster formation.

The injection of energy into the interstellar medium during a starburst is observed to drive galactic winds. Nearby examples such as NGC 253 and M82 require a considerable number of supernovae to drive a wind. It has been argued that in the case of M82, the IMF must be top-heavy, weighted towards massive stars relative to the solar neighborhood, to account for the observed infrared luminosity and wind (Doane & Mathews 1993). However, these are sub-L_* galaxies. For a giant protoelliptical undergoing a starburst during its final gas-rich merger to develop a wind and leave it gas-poor certainly requires a top-heavy IMF during the burst. Wind-driven gas loss is the most likely explanation of the low gas content of early-type galaxies. Since ellipticals inside and outside clusters have similarly old stellar populations with little recent star formation, one cannot appeal to ram pressure stripping by the intergalactic medium to exhaust the gas supply. A merger will concentrate much of the gas until the starburst develops.

There is more than a hint of fossil evidence that giant ellipticals once underwent a top-heavy starburst and were stripped of gas by a wind. Consider first the nucleosynthetic evidence. Giant ellipticals have an excess by about a factor of 2 in Mg/Fe relative to the solar value (Worthey, Faber & González 1992). This is precisely what one sees in old halo stars at $[Fe/H] < -1$. One understands this as the SN II contribution, from exploding massive stars, to stellar abundances, and this dominates over the SN I contribution for the first few Gyr of galactic chemical evolution. The apparent signature of a massive star-dominated IMF in giant ellipticals can be explained as a top-heavy burst, following which much of the remaining gas was ejected. This could help suppress any extended duration of the period of low mass star formation that could contribute to the SN I rate, as well as any SN I's that a starburst with a normal IMF would eventually have generated. Most of the mass loss in the form of SN I ejecta cannot be retained in the halo of the elliptical without exceeding the halo Fe/H abundance inferred from X-ray observations (Forman *et al.* 1993). A top-heavy burst generates a wind that is capable of cleaning out the interstellar gas. Provided that the bursts occur after *most* of the stars in the present-day elliptical have formed, later SN I's, exploding in an almost interstellar medium-free elliptical, will still pollute the gaseous halo, but contribute only a small fraction of the Fe generated either in the burst, or prior to the burst.

Moreover there is fossil evidence that winds occurred in early-type galaxies. Metallicity correlates with galaxy luminosity. However, the correlations of metallicity with central velocity dispersion over a wide luminosity range (Bender 1992) and local escape velocity for giant E's (Franx and Illingworth 1990) have significantly less dispersion. Such correlations are most simply understood if spherical systems, from dwarfs to giants, all underwent mass loss by galactic winds. The correlation of metallicity with galaxy mass does not have a unique explanation. It may require that galaxies of increasing mass eject systematically less gas in a wind than do less massive galaxies. However this would presumably lead to the more massive ellipticals being younger, in terms of mean age of the stellar population, and bluer, than lower mass counterparts. This prescription fails to account for the color–magnitude correlation of early type galaxies. I will describe below a more exotic explanation of the metallicity–mass correlation that incorporates some of the preceding ideas about starbursts and does result in giant old, and red, galaxies. However it is necessary to first describe the properties of the intracluster gas, which provide important constraints and tests of the past evolution of ellipticals.

3. The Intracluster Medium

All stellar ejecta from stars in ellipticals, if not recycled during the starburst, are ejected, as argued above, by a wind, and collect in the intracluster medium. The intracluster gas iron abundance is well measured in clusters to be about 35% of the solar value (Ohashi 1995). Since the cluster gas fraction is about $8h^{-3/2}$ percent of the total cluster mass (White *et al.* 1993), one infers that there is about as much iron in the gas as in the stars. Arnaud *et al.* (1992) found that the iron mass in the gas is proportional to the luminosity for early-type (E + S0) galaxies, with

$$M_{Fe} \approx 2 \times 10^{-2} L_v.$$

The source of the iron must be from galaxies, and specifically from giant ellipticals and S0's. The proportionality between light and iron abundance suggests this, as does the dominance of early-type galaxies in clusters. In fact most of the stellar iron is in the giant galaxies, and most of the iron synthesized in the past and ejected from the galaxies must have originated in these galaxies. The observed intracluster iron abundance per unit mass in stars is about 5 times higher than in the Milky Way. The intracluster iron mass requires a source that is indicative of a top-heavy IMF, boosting heavy element yields by up to an order of magnitude without affecting the present epoch luminosity. The temperature of the intrastructure gas is measured to be higher than the equivalent kinetic temperature that corresponds to the galaxy velocity dispersion, $T_{gas} \approx 2T_{gal}$. This suggests that supernova-driven galactic winds indeed both heated and enriched the intracluster gas.

Galaxy groups are generally less enriched than clusters, by a factor $\sim 2-3$, although there is a large spread in abundance, and abundance determinations are far less certain at ASCA spectral resolution for gas at $T \lesssim 1 keV$. Isolated ellipticals contain gas fractions of a few percent, and the gas iron abundance is considerably less than the stellar iron abundance (near solar in giant ellipticals). There are examples of locally enhanced gaseous iron abundances, around central cDs, e.g.. in Centaurus, M87/Virgo. These deeper potential wells are likely to retain more of the gas ejecta from supernovae than would be trapped in the halos of normal E's.

4. The Galaxy Connection

Elliptical galaxies, and more generally, spheroids, define a plane in the parameter space of velocity dispersion, half-light radius and absolute magnitude. This so-called fundamental plane minimizes the dispersion of the observational parameters. In combination with the virial theorem, the fundamental plane is equivalent to

$$\frac{M}{L} \propto L^{1/5},$$

for the visible light band. The fundamental plane has a dispersion within a given galaxy cluster amounting to $\lesssim 15-20$ percent that presumably reflects the initial conditions that determined the relations between galaxy structural and dynamical parameters.

The fact that the relation reduces to a weak dependence of M/L on L strongly suggests that an explanation must be sought within the context of star formation. One possibility is that more luminous galaxies are more metal-rich and hence would have generated a higher fraction of stellar remnants than less luminous galaxies. However this fails quantitatively to reproduce the observed M/L variation with L. Moreover such a model would predict that more luminous galaxies have recycled more gas and metals for a given gas fraction before termination of the star formation process than less luminous galaxies. This would suggest that the stellar content is systematically younger, in the mean, for the most processed galaxies. More massive, luminous galaxies form stars more recently and are younger, yielding a luminosity-color relation that would be too flat in slope, in the opposite sense to what is observed.

A starburst model accounts for the fundamental plane relation provided that the IMF in the starburst is top-heavy. There is no late-time contribution to the light, but the starburst contributes enrichment and remnants. This approach fails however if the starburst occurs when the galaxy forms, since in this case the galaxy overrichens its own stars.

One can arrange to produce the requisite metallicity required for the ICM and the remnants need to account for galactic M/L, without overproducing stellar metallicities if the starbursts are delayed, only occurring after a certain amount of star formation has already been underway. In fact, delayed starbursts are mandatory if the starburst IMF is top-heavy. Dust formation would occur as the heavy elements are formed, during the minor mergers that precede the final dramatic starburst, which involves a central concentration of interstellar matter generated by the merger and could consequently be very dusty.

The merger history anticipated in a bottom-up sequence of structure formation suggests that the duration of star formation is extended over a series of minor and major mergers. The typical mass of merging remnants increases with time. The last, most significant merger would have been a major merger that resulted in the formation of an elliptical galaxy. Early, and most, star formation occurred with a normal IMF, culminating in a final major merger and associated top-heavy starburst. A major merger is a critical ingredient in the formation history of an elliptical galaxy. Any preexisting disks are mostly destroyed. Efficient star formation is triggered. The angular momentum transfer as a massive substructure orbits into the center of the merged galaxy efficiently transfers angular momentum. The result is formation of a dense stellar core with anisotropic velocity dispersion.

The typical luminosity of a bright elliptical is $L_* \approx 10^{10} h^{-2} L_\odot$. Several empirical results suggests that similar physics governs elliptical formation down to a luminosity of about $0.03 L_*$. The more luminous ellipticals are strongly supported by the anisotropic pressure of stellar velocity dispersion, whereas the figures of low luminosity ellipticals are not anisotropically supported. The fundamental plane is sharply defined for the more luminous ellipticals, but at low luminosities, there is considerable dispersion in galaxy parameters. Finally, the surface brightness–absolute magnitude correlation for ellipticals peaks towards a brightness of about $0.1 L_*$, with surface brightness decreasing towards both lower and higher luminosities. The metallicity-luminosity relation requires sub-L_* ellipticals to have subsolar metallicity, and so they are unlikely to have played a major role in enriching the intracluster medium. I will argue that ellipticals above $\sim 0.03 L*$ have undergone a similar history that is established as a consequence of a major merger.

5. The Starburst Model for Ellipticals

A major merger that resulted in a starburst with a top-heavy IMF provides a means of simultaneously enhancing the M/L ratio and of producing enriched gas that is driven out of the galaxy by the starburst induced galactic winds. Thus one can understand both the fundamental plane correlation and the enrichment of the intracluster gas. One finds that the mass fraction in remnants inferred from interpreting the $M/L \propto L^{1/5}$ relation as arising from a top-heavy starburst gives just enough iron to account for the observed abundance in the ICM. The dispersion in the fundamental plane is effectively attributed to a systematic increase of starburst fraction with galaxy mass (*i.e.* as $M^{1/6}$) that reflects the merging history of cluster galaxies.

The wind velocity, of order the escape velocity from the elliptical, is sufficient to drive the enriched gas out of elliptical galaxy halos, and to some extent, even from galaxy groups. Thus one would expect to find lower Fe/H in elliptical halos and in galaxy groups than in clusters of galaxies. The rich cluster ICM retains all of the ejecta in elliptical winds. The primary source of cluster Fe is from the metal-rich L_* ellipticals: low luminosity, metal-poor, dwarf ellipticals, although common and easily stripped by winds, can have synthesized only a factor f_E^{-1} more Fe than survives in their stellar component, assuming a normal IMF, with iron fraction f_E relative to the solar value. This would make only a small contribution to the ICM iron mass.

The starburst-driven enrichment hypothesis predicts that the intracluster gas should be enriched by ejecta that are dominated by remnants of Type II supernova. Outside rich clusters, there are enough ellipticals to significantly enrich the intergalactic medium. If 30 percent of ellipticals are outside rich clusters, then one might expect the typical IGM enrichment to attain a level of about 10 percent of the ICM enrichment. This assumes that two-thirds of galaxies are spheroids that do not contribute to the enrichment. The enrichment will occur late, perhaps by $z \sim 1$.

There are also consequences for ellipticals themselves. The ellipticals are wind-stripped and gas-poor, but the stars should retain the metallicity abundance anomalies characteristic of Type II supernovae. This would result in luminous ellipticals having enhanced O and Mg relative to Fe by a factor of ~ 2, as seen in old halo stars where Type II SN yields are inferred. The supernova rate per unit luminosity implied by our model of ICM enrichment is a factor of ~ 5 larger than in conventional Galactic models, the ratio of M_{Fe}/M_* in clusters to its value in the Milky Way, where there were approximately 6 SNII's for every SNIa (Tsujimoto et al. 1995). Hence the rate of SNII in protoellipticals is enhanced by a factor of about 6 relative to the SNII rate requirements of the standard chemical evolution model for our Galaxy. Also, as previously remarked, cD galaxies underwent more frequent mergers than normal ellipticals, and one might expect their local environment to retain evidence of some starburst-induced enrichment that is diluted by later SNI ejecta. Such an enhanced Fe abundance is indeed observed in the vicinity of nearby cD's, and the model advocated here predicts that this gas should reveal abundance ratios intermediate between those of Type I and Type II supernova yields. Isolated ellipticals are expected to be rare, forming via late major mergers in the field. These objects should have an intermediate age population characterizing the last merger event. It is this population, rather than the underlying old stellar population, that is expected to have the characteristic SNII yield signature.

Finally, I discuss an especially intriguing prediction of the model. The synthesis of the intracluster metals requires a stellar energy source whose contribution to the diffuse background light is difficult to hide. In fact, given that starbursts are dusty, one would expect the resulting radiation to appear as a diffuse far infrared background. One can estimate its intensity as follows. The diffuse background light generated in synthesizing a metallicity Z in massive stars at redshift z_f is

$$\nu i_\nu = (4 \times 10^{-3} c^2 erg\, g^{-1})[Z]\rho_o \frac{c}{4\pi}(1+z_f)^{-1}.$$

This results in the following prediction for the diffuse radiation background:

$$\nu i_\nu = 20h^{-1}\left(\frac{\Omega_{ICM}/\Omega}{0.1h^{-3/2}}\right)\left(\frac{M_{Fe}/L_v}{0.02h^{-3/2}}\right)\left(\frac{10h}{M_*/L_v}\right)\left(\frac{M_Z/M_{Fe}}{10}\right) \times \quad (1)$$

$$\times \left(\frac{\Omega}{0.2}\right)\left(\frac{4}{1+z_f}\right) \quad nwatts/m^2 sr, \qquad (2)$$

perhaps half of which may plausibly be in the far infrared spectral region. For comparison, Puget et al (1995) report the detection of the cosmic far infrared diffuse background near 300 microns at a level of $\nu i_\nu \approx (3-10) nwatts/m^2 sr$. It is tempting to tentatively associate this flux with the predicted diffuse far infrared background, previous calculations of which, based on backwards evolution of IRAS source counts, lie in a similar range (Franceschini *et al.* 1994) to that estimated here. Further consequences (Zepf and Silk 1995) include the likely clustering of such a background, if associated with elliptical formation, and the implications for submillimeter searches for individual sources, expected to be at the mJy level, and of such a high early universe supernova rate for cosmic ray interactions and the diffuse gamma ray flux.

Acknowledgments. I thank Steve Zepf for collaborating on much of the work described here. I also acknowledge useful discussions with J.-L. Puget.

References

Arnaud, M., Rothenflug, R., Boulade, O., Vigroux, L. and Vangioni-Flam, E. 1992, A&A, 254,49.
Bender, R. 1992, in *The Stellar Populations of Galaxies*, ed. B. Barbuy (Dordrecht: Kluwer), p. 267.
Charlot, S. & Silk, J. 1994, ApJ, 432, 453.
Doane, J.S., & Mathews, W.G. 1993, ApJ, 419, 573
Dressler, A. & Gunn, J. E. 1992, ApJS, 78,1.
Forman, W., Jones, C., David, L., Franx, M., Makishima, K. & Ohashi, T. 1993, ApJ, 418, L55.
Franceschini, A., Mazzei, P., Dezotti, G. and Danese, L. 1994, ApJ, 427, 140.
Franx, M. & Illingworth, G. 1990, ApJ, 359, L41.
Mamon, G. A. 1992, 1990, ApJ, 401, L3.
Ohashi, T. 1995, in Dark Matter, ed. S.S. Holt & C.L. Bennett (New York: A.I.P.), 255
Puget, J.-L. *et al.* 1995, A&A, submitted.
Schweizer, F. and Seitzer, P. 1992, AJ, 104, 1039.
Tsujimoto, T., Nomoto, K., Yoshii, Y., Hashimoto, M., Yanagida, S., & Thielemann, F.-K. 1995, MNRAS, in press.
White, S. D. M., Navarro, J. M., Evrard, A. E. and Frenk, C. S. 1993, Nature, 366,429.
Worthey, G., Faber, S.M., & González, J.J. 1992, ApJ, 398, 69
Zepf, S. and Silk, J. 1995, ApJ, submitted.

Cosmic Abundances
ASP Conference Series, Vol. 99, 1996
Stephen S. Holt and George Sonneborn (eds.)

Metal Enhancements in the X-Ray Gas Around Central Cluster Galaxies

Andreas Reisenegger

Institute for Advanced Study, Princeton, NJ 08540, USA

E-mail: andreas@ias.edu

Abstract. The X-ray emission by hot gas around the central galaxies of galaxy clusters is commonly modeled assuming the existence of steady-state, multiphase cooling flows. The inflowing gas will be chemically enriched by type Ia supernovae and stellar mass loss occurring in the outer parts of the central galaxy. This may give rise to a substantial metallicity enhancement towards the center, whose amplitude is proportional to the ratio of the central galaxy luminosity to the mass inflow rate. The metallicity of the hotter phases is expected to be higher than that of the colder, denser phases. The metallicity profile expected for the Centaurus cluster is in good agreement with the iron abundance gradient recently inferred from ASCA measurements (Fukazawa et al. 1994). However, current data do not rule out alternative models where cooling is balanced by some heat source. In either case, the enhancement expected from injection by type Ia supernovae is roughly as observed. Most of this work is described in more detail in Reisenegger, Miralda-Escudé, & Waxman (1996).

1. Introduction

Clusters of galaxies are observed to contain hot gas which is detected due to its X-ray emission. In the central ~ 100 kpc of most clusters, the gas is so dense that (in the absence of any heating mechanism) it will cool to low temperatures in less than a Hubble time (Fabian 1994). This motivated the idea of a "cooling flow," i.e., that the gas flows into the center of the cluster as its decreasing entropy decreases the pressure support (Cowie & Binney 1977). However, a homogeneous, steady-state cooling flow predicts an X-ray emissivity profile that is more centrally peaked than observed. This can be avoided if one assumes that at each radius there is a distribution of gas phases at different temperatures (but in pressure equilibrium), which are thermally insulated from each other, but at the same time forced to comove, by a magnetic field (Nulsen 1986). Since the cooling time is shorter for the cooler phases, these will cool to very low temperatures at a finite distance to the center, dropping out of the flow.

Outside this "cooling region," the metallicity of the hot gas is fairly uniform at $\sim 0.3 Z_\odot$, with relative abundances roughly consistent with injection by type II supernovae (Loewenstein & Mushotzky 1995; K. Arnaud, this volume). However, the iron abundance in the center of some clusters rises up to $\sim 1 Z_\odot$

(Fukazawa et al. 1994; K. Arnaud, this volume). Here, I discuss work reported in more detail in Reisenegger, Miralda-Escudé, & Waxman (1996), which explores whether the central enhancement can be understood in terms of metal injection by stars of the central galaxy into the inflowing gas.

Reisenegger et al. (1996) use a steady-state, inhomogeneous, self-similar cooling flow model (Nulsen 1986; Waxman & Miralda-Escudé 1995), where the gas in the central region of the cluster is close to hydrostatic equilibrium, but flows radially inwards as it cools and its pressure support decreases. The distribution of gas in phases of different densities is inferred from the X-ray brightness profile. The gas has a uniform initial metallicity, and additional metals are injected into the inflowing gas at a rate proportional to the local (V-)luminosity density of the stars in the central galaxy.

2. General results

Since metals are injected into the flow, the metallicity must increase with decreasing radius. The central metallicity enhancement is approximately proportional to the ratio of the optical luminosity, L_{opt} of the central galaxy to the mass inflow rate, $\dot{M}$. Since $\dot{M}$ varies much more strongly than L_{opt} among different clusters, the former should determine the observability of the metallicity enhancement, as long as it has not been erased by a merger. This would be supported, at least qualitatively, if it is confirmed that observable metallicity enhancements tend to occur in clusters with weak or absent cooling flows and no obvious sign of a merger (see, e.g., Fujita & Kodama 1995).

It also follows that the cooler, denser phases are less metal-enriched than the hotter phases, which occupy a larger volume, and therefore capture more metals, for a given mass of gas. However, this prediction does not appear to be testable with the present data.

3. The Centaurus cluster as an example

This model is tentatively applied to the Centaurus cluster, in whose cooling region ($r < r_{cool} \approx 6' \approx 60h^{-1}kpc$, assuming a pure Hubble flow with Hubble parameter $H_0 = 100 h km s^{-1} Mpc^{-1}$) a substantial enhancement of the iron abundance was suggested by ROSAT (Allen & Fabian 1994) and confirmed by ASCA (Fukazawa et al. 1994). As shown in Fig. 1 of Reisenegger et al. (1996), the optical luminosity in and around this region is strongly dominated by the central galaxy, NGC 4696, suggesting that this galaxy is the source of the excess iron.

Since the gravitational potential around NGC 4696 (dominated by dark matter) is not well determined, we considered two steady-state cooling-flow models which approximately span the range of allowed possibilities: 1) A total enclosed mass profile $M(r) \propto r^{1.625}$, which gives a homogeneous cooling flow with the metallicity profile shown as the dashed curve in Fig. 1, and 2) an isothermal mass profile $M(r) \propto r$, which results in a multiphase flow, with metallicities in different phases lying between the two thin, solid lines. For comparison, we showed 3) the result of injecting metals for a Hubble time into a static gas (in

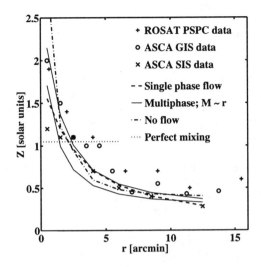

Figure 1. Iron abundance as a function of radius in the region inside and surrounding NGC 4696. The points correspond to observations with three different instruments, the ROSAT PSPC (Allen & Fabian 1994), and the ASCA GIS and SIS (Fukazawa et al. 1994). The lines correspond to different models; see text for details. [After Reisenegger et al. 1996; ©1995 by The Astrophysical Journal.]

which the flow is stopped by some hypothetical energy source; dot-dashed line) and 4) the result of perfectly mixing all the metals of model 3 in the cooling region. It can be seen that the cooling flow models produce a metallicity profile of similar general shape as observed, while the static model might be too peaked at the center. However, even if this turned out to be a significant discrepancy, it could easily be fixed with some turbulent mixing.

Fitting the metallicity curves to the ASCA SIS data (except the innermost point which might be strongly affected by the finite spatial resolution), one obtains iron injection rates $\eta = \eta_{-13} \times 10^{-13} M_\odot L_\odot^{-1} yr^{-1}$, with $\eta_{-13} = 7.4$ and 3.2 for models 1 and 2, and $H_0 \int \eta_{-13} dt = 1.3$ for case 3. For injection dominated by type Ia supernovae, each injecting $\sim 0.6 M_\odot$ of iron, and with a rate $\theta_{SN} R_{vT}$, where $R_{vT} = 0.87 h^2/(100yr)/(10^{10} L_{B,\odot})$ is the fiducial rate for elliptical galaxies from van den Bergh & Tammann (1991), our results imply $\theta_{SN} = 1.4 h^{-2}$ and $0.6 h^{-2}$ for the two cooling-flow models, and $H_0 \int \theta_{SN} dt = 0.25 h^{-2}$ for the static model. For comparison, Turatto, Cappellaro, & Benetti (1994) found an observed rate $\theta_{SN} = 0.24 \pm 0.12$, and estimates from the observed iron abundance in the hot gas halo of some ellipticals not in cluster centers give numbers as low as $\theta_{SN} < 0.05$ (Serlemitsos et al. 1993; M. Loewenstein, this volume).

4. Discussion and Conclusions

It was shown that models in which metals are steadily injected into the hot intracluster gas by the stars of the central cluster galaxy can produce a radial iron abundance gradient that resembles that observed in the Centaurus cluster. At present, it does not seem possible to discriminate between models with and without cooling flows. As long as no substantial mixing has occurred (e.g., through a merger), the amplitude of the central enhancement is proportional to the ratio of the central galaxy luminosity to the mass inflow rate. For Centaurus, it is roughly consistent with expectations from observed rates of type Ia supernovae in ellipticals, but inconsistent with the low rates inferred from the iron abundance in the hot gas of ellipticals which are not at cluster centers. Type Ia supernovae would be ruled out if a similar gradient is confirmed in other metals, such as S or Si. Enrichment by stars is a viable alternative only if the stellar metallicity is substantially above solar. J. Silk (this volume) proposes that the enhancement might come from ejecta of early type II supernovae blown out by a galactic wind. If a cooling flow is present in such a scenario, it will tend to smooth the initial metallicity gradient. Thus, since gradients are still observed now, the boundary between enriched and unenriched material at early times must have been fairly sharp and roughly coincident with the current cooling radius.

Acknowledgments. I am very grateful to J. Miralda-Escudé and E. Waxman for introducing me to the subject of cooling flows and for their collaboration in the work discussed here. I also thank K. Arnaud, N. Bahcall, M. Carollo, M. Currie, U. Hwang, M. Loewenstein, R. Mushotzky, and M. Strauss for interesting and useful discussions. This work was supported by NSF grant PHY 92-45317 and by a grant from the Ambrose Monell Foundation.

References

Allen, S. W., & Fabian, A. C. 1994, MNRAS, 269, 409
Cowie, L. L., & Binney, J. 1977, ApJ, 215, 723
Fabian, A. C. 1994, ARA&A, 32, 277
Fujita, Y., & Kodama, H. 1995, ApJ, 452, 177
Fukazawa, Y., Ohashi, T., Fabian, A. C., Canizares, C. R., Ikebe, Y., Makishima, K., Mushotzky, R. F., & Yamashita, K. 1994, PASJ, 46, L55
Loewenstein, M. & Mushotzky, R. 1995, ApJ, in press
Nulsen, P. E. J. 1986, MNRAS, 221, 377
Reisenegger, A., Miralda-Escudé, J., & Waxman, E. 1996, ApJ (Letters), in press
Serlemitsos, P. J., Loewenstein, M., Mushotzky, R. F., Marshall, F. E., & Petre, R. 1993, ApJ, 413, 518
Turatto, M., Cappellaro, E., & Benetti, S. 1994, AJ, 108, 202
van den Bergh, S., & Tammann, G. 1991, ARA&A, 29, 332
Waxman, E., & Miralda-Escudé, J. 1995, ApJ, 451, 451

Cosmic Abundances
ASP Conference Series, Vol. 99, 1996
Stephen S. Holt and George Sonneborn (eds.)

Abundances in the Intra-Cluster Medium

K. A. Arnaud[1]

Code 662, Goddard Space Flight Center, Greenbelt, MD 20771

Abstract. We are at the beginning of a new era for the measurement of abundances in the intra-cluster medium (ICM). The first examples of the new results that this will bring are just becoming available. Clusters show varied behaviour in their spatial distribution of heavy elements. Some have central peaks, others do not. The iron abundance in the ICM has been measured out to $z \sim 0.5$ with no evidence of variation with redshift. Reliable abundances of α-burning elements have been measured in selected clusters. The results demonstrate that SNII were an important source of the heavy elements now seen in the ICM.

1. Introduction

This article is concerned with the measurement of elemental abundances in the intra-cluster medium. The importance of such measurements has been summarized by Renzini *et.al.* (1993) in the following words :

"...an uncommon variety of astrophysical issues converges on the quest for iron in clusters, from the formation and evolution of elliptical galaxies and clusters of galaxies to supernova physics and astrophysics and from nucleosynthesis and chemical evolution in the halo and disk of our Galaxy to the evolution of gas flows and X-ray properties of elliptical galaxies, with topics such as star formation, dark matter, AGN activity, galaxy interactions with the environment, and many others coming into play."

As I will show we can now expand the reference to iron to include other elements. This will provide new insights into many of the issues listed by Renzini *et.al.* . In this article I will only be able to touch on one or two of these issues. I will mainly concentrate on the observations and their interpretation, with particularly emphasis on possible systematic errors in the derivation of elemental abundances.

2. The Intra-Cluster Medium

Clusters of galaxies comprise three components, the individual galaxies, a hot plasma occupying the space between the galaxies, and something else providing most of the mass. The hot plasma has temperatures of $10^6 - 10^8$ K and densities

---

[1] Astronomy Department, University of Maryland, College Park, MD 20742

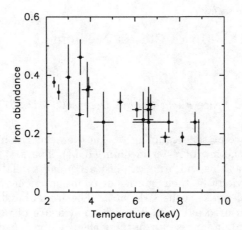

Figure 1.   Iron abundances measured using the Ginga satellite

of $10^{-2} - 10^{-5}$ electrons/cm^{-3}. The plasma is collisional with X-ray emission from optically-thin thermal bremsstrahlung and emission lines. X-ray luminosities range from 10^{43} to a few times 10^{45} ergs/sec. In rich clusters the mass in plasma exceeds that in stars by a factor of five. Most of this plasma must be primordial, having been part of the cluster when it first collapsed or having been accreted subsequently. The plasma is static except in the cores of clusters where cooling flows are seen and in clusters that are undergoing mergers or accreting sub-clusters.

3. Pre-ASCA history

To see the changes over the last few years in our knowledge of elemental abundances in the ICM it is instructive to review what was known at the time of the last major conference on cosmic abundances, that in Minneapolis in 1988. The review by Mushotzky (1989) quoted two basic results. Iron abundances for about two dozen clusters were available, measured to an accuracy of, at best, 20%. These iron abundances were measured using detectors with a large beam so these abundances were emission-weighted averages for the entire cluster. In addition, an handful of clusters had measurements of the abundances of elements other than iron but these derivations had large uncertainties and possible systematic errors. Since Mushotzky's (1989) article the Ginga satellite has improved the accuracy of emission-weighted average abundances for about twenty-five clusters (Yamashita 1992). The best determined abundances[1] are plotted in figure 1, which shows a trend of increasing average iron abundance with decreasing average ICM temperature.

---

[1] This article quotes iron abundances relative to $n(Fe)/n(H) = 4.68 \times 10^{-5}$. The numbers quoted are always linear, not logarithmic, fractions

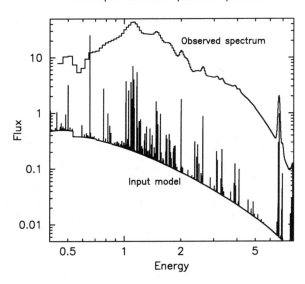

Figure 2. An illustration of the line blending seen using current technology spectrometers

4. Issues in the measurement of abundances

4.1. What can we measure ?

Between the energies of 0.5 and 10 keV and for a plasma at the temperatures characteristic of clusters of galaxies the main emission lines observed are from transitions to the $n = 1$ state (K-shell) for He-like and H-like ions. The exception is Fe which has a large number of emission lines to the $n = 2$ state (L-shell) for ion stages from Fe XVI to Fe XXV. These cover the energy range $0.7 - 1.5$ keV.

The elements with strong enough lines to be observed with present technology are O, Ne, Mg, Si, S, Ar, Ca, Fe, and Ni. The Fe K-shell lines have large equivalent widths and are in an isolated part of the spectrum so Fe is observationally easy and reliable to measure. Si, S, Ar, and Ca are all in isolated parts of the spectrum but have smaller equivalent widths so these are reliable but harder to measure. Mg, and Ne are in the energy range dominated by the Fe L-shell lines. With current X-ray spectrometer resolutions these elements cannot be measured independently of the iron lines. Thus, abundance measurements for these elements are more susceptible to systematic problems. O is isolated from the Fe L-shell problems but CCD detectors have a deep instrumental neutral O edge and this reduces the efficiency at O line energies and requires accurate calibration to avoid systematic uncertainties. Ni measurements should be reliable but the Ni K lines have smaller equivalent widths than Fe K and current telescopes and detectors have low efficiencies in this energy range.

4.2. What physics should we worry about ?

To accurately measure abundances we need to be confident that the plasma is not over- or under-ionized for its continuum temperature. For instance, in supernova remnants the plasma is usually under-ionized making accurate determinations of abundances difficult (Petre, these proceedings). However, for the temperatures and densities seen in the ICM both ionization and recombination timescales are $\leq 10^9$ yrs so it is safe to assume ionization equilibrium.

To convert measured equivalent widths to abundances requires accurate knowledge of individual line transition strengths. Most of the important lines are K-shell transitions of He-like or H-like ions. These are well understood theoretically and extensively observed in the Solar corona. The Fe L-shell lines between 0.7 and 1.5 keV are less secure and may have systematic problems (Fabian et.al. 1994; Liedahl, Osterheld & Goldstein 1995).

Finally, radiation transfer effects should be considered. The continuum emission is optically-thin everywhere in the cluster. There may be significant effects due to resonant absorption and electron scattering in the very central parts of clusters (Wise & Sarazin 1993). So, provided these central regions are excluded no radiation transfer effects need be included in the interpretation of spectra.

4.3. What are the observational issues ?

The measurement of abundances requires that the ICM be isothermal or that the temperature structure be well determined. So, it is a good idea to avoid the centers of clusters, where there are cooling flows causing gas to be at a range of temperatures. It is also a good idea to avoid those clusters showing signs of recent merger or accretion events since these may have complicated thermal structures and be out of ionization equilibrium.

Current spectral resolutions mean that many lines are blended together. In particular, Mg and Ne are in amongst the Fe L-shell lines. This is illustrated in figure 2, which shows an input plasma emission model and what we would actually observe using the ASCA CCD detector. In the 1-2 keV region accurate plasma emission models are required to allow deductions of line strengths.

Detectors have a number of spectral features associated with elements in the X-ray path or the reflection surface of the optics. The ASCA telescopes produce Au-M edge features while the CCDs show both O and Si edges from the gate structure.

5. New Results from ASCA

5.1. Spatial Distribution of Iron in the ICM

Prior to the launch of ASCA very little was known about the spatial distribution of elements in the ICM. Spatially-resolved iron abundance measurements were available only for a few clusters, mostly using indirect methods with controversial results (Ponman et.al. 1990; Koyama, Takano & Tawara 1991; Arnaud et.al. 1992; Hughes et.al. 1993; White et.al. 1994). The ability of ASCA to obtain X-ray spectra with a spatial resolution of a few arcminutes has put these measurements on a firmer footing. The first results available show a range of behaviours.

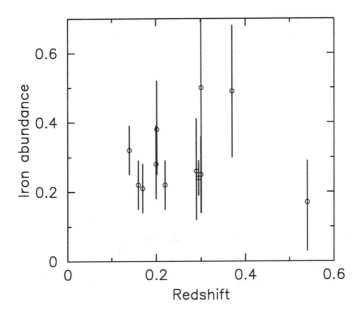

Figure 3. ICM iron abundance vs. Redshift

In the Perseus cluster the iron abundance declines slowly with distance from the cluster center with no evidence for a strong central peak (Arnaud et.al. 1994). By contrast, the iron abundance distribution in the Centaurus cluster is sharply peaked in the center of the cluster (Fukazawa et.al. 1994). Fukazawa et.al. suggest that the iron in the core of this cluster comes from the central galaxy, NGC 4696. This scenario has been worked out in more detail by Reisenegger, Miralda-Escudé & Waxman (1996; see also Reisenegger, these proceedings).

Ohashi (1995) has summarized the iron abundance distribution statistics for clusters observed using ASCA. Six clusters have been observed to have central iron concentrations while approximately twenty others do not. Those clusters with observed central iron concentrations are nearby, low luminosity systems and similar central concentrations would not be observed even if present in the higher redshift systems. However, the nearby system Abell 1060, which in other properties is similar to the Centaurus cluster, has a uniform iron distribution (Tamura et.al. 1995) so there are real differences between clusters.

5.2. The Redshift Distribution of Iron in the ICM

A consequence of ASCA's high sensitivity and low background is that it is now possible to measure the average iron abundance of the ICM in clusters at redshifts > 0.1. Mushotzky (1995) has analyzed all the available ASCA data for high-redshift clusters and his results are shown in figure 3. There is no evidence yet for any evolutionary effect in the iron abundance implying that the enrichment of the ICM occured at redshifts greater than one half. This conclusion

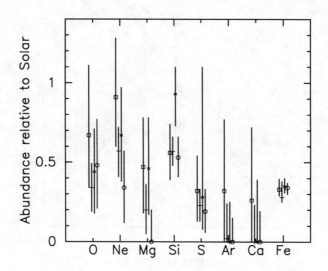

Figure 4. Measured abundance with 90% confidence errors. The boxes are A496, crosses are A1060, stars are A2199, and circles are AWM7.

could be wrong if the iron becomes more centrally concentrated in clusters at higher redshift. The reason for this is that we observe an emission-weighted average abundance. If the iron is concentrated where the emission peaks then this abundance will overestimate the true average abundance. So, either the iron abundance stays constant with increasing redshift or the iron abundance decreases and becomes more centrally concentrated.

5.3. Relative Elemental Abundances

ASCA observations provide the first opportunity to measure reliable relative elemental abundances in the ICM. Mushotzky *et.al.* (1995) have measured abundances in four clusters, selected for their lack of evidence of recent mergers and for their isothermality outside the cluster core. Figure 4 shows the measured abundances with 90% confidence ranges for the ICM in these clusters. These abundances were measured from spectra extracted by excluding the cluster cores, where complicated thermal structures are likely.

The four clusters show similar abundances, the only outlier being Abell 2199 which appears to have a higher Si abundance than the other three. Similar results have been obtained for the Perseus cluster (Arnaud, unpublished), Abell 1795 (Mushotzky, priv. comm.) and MKW3s (Hatsukade, Kawarabata & Takenaka 1995).

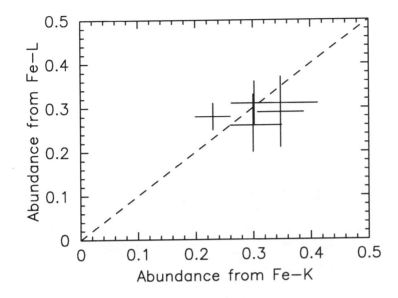

Figure 5. Iron abundance measured from the Fe-K and Fe-L lines

These ASCA observations can also be used to test for some of the systematic effects worried about above. Figure 5 shows the iron abundance measured independently using the Fe-K lines and the Fe-L lines. There is no evidence for a systematic offset implying that, at least at this gross level, the uncertainties in the Fe-L physics do not effect the measurement of the abundance.

Since ASCA has two different types of detector it is also possible to check for any systematics introduced by instrumental effects. Figure 6 shows the Si abundance as measured by the SIS and the GIS on ASCA. There does appear to be a slight systematic shift towards higher abundances in the GIS although individual abundances are consistent within the statistical uncertainties.

Figure 7 shows the average relative abundances for the four clusters from Mushotzky et.al. (1995). The relative abundances are shifted to give one for Fe. The uncertainty ranges plotted are the union of the individual cluster uncertainty ranges. As such they are a conservative measure of the uncertainty if all four clusters are drawn from the same population of relative abundances.

Arnaud et.al. (1992) showed that there is a correlation between the iron mass in the ICM and the light from elliptical galaxies in the cluster. From this they deduced that the heavy elements in the ICM were produced by supernovae in the cluster ellipticals. However, with only iron abundances available they could not tell whether SNI or SNII were responsible. Figure 7 shows the expected relative abundances for both types of SN (type I abundances from Nomoto, Thielemann & Yokoi 1984; type II abundances from Hashimoto et.al. 1994 assuming a Salpeter initial mass function from 10 – 70 Solar masses). These

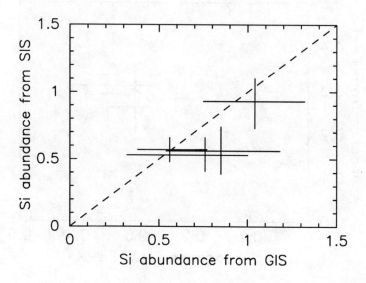

Figure 6. Silicon abundance measured using the SIS and GIS

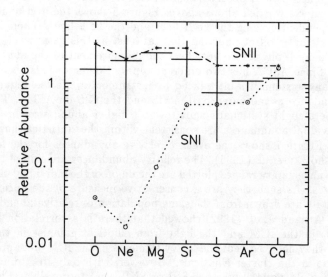

Figure 7. Average relative abundances for the ICM

relative abundances are normalized to match the data at Fe. The observed relative abundances are consistent with production by SNII although the S, Ar, and Ca abundances are all low.

Assuming that the ICM is homogeneous requires 10^{12} SNII per cluster to make the observed mass in metals (Renzini et.al. 1993). This requires the IMF to be biassed towards high mass stars, which is most easily accomplished by supposing a "starburst" phase which has not been observed. If present day ellipticals were responsible for the metals observed then they must have lost approximately half of their baryonic mass. A more detailed look at the theoretical implications of these results are given in Silk's article in these proceedings and in Loewenstein & Mushotzky (1995).

6. Prospects

Approximately 150 clusters have been observed by ASCA so far. All of these will yield at least an iron abundance. Many will enable measurements of other elements. All this data will end up in the public archive, accessible electronically. In the longer term Astro-E (launch in 2000) will provide 10 eV resolution spectroscopy. Among other advantages this will mitigate the problems associated with the Fe L lines. Abundance measurements of other, less common, elements may become possible with resolutions of 2 eV (perhaps with NGXO - White et.al. 1995). Isotopic abundances are likely to be impossible because of thermal broadening of lines.

Acknowledgments. This article benefited from many conversations with Mike Loewenstein and Richard Mushotzky.

References

Arnaud, K.A. et.al. 1992, in Ginga Memorial Symposium, F. Makino & F. Nagase, ISAS, 1992, 114

Arnaud, K.A. et.al. 1994, ApJ, 436, L67

Arnaud, M., Rothenflug, R., Boulade, O., Vigroux, L., Vangioni-Flam, E. 1992, A&A, 254, 49

Fabian, A.C., Arnaud, K.A., Bautz, M.W & Tawara, Y. 1994, ApJ, 436, L63

Fukazawa, Y., Ohashi, T., Fabian, A.C., Canizares, C.R., Ikebe, Y., Makishima, K., Mushotzky, R.F. & Yamashita, K. 1994, PASJ, 46, L55

Hashimoto, M., Nomoto, K., Tsujimoto, T. & Thielemann, F. 1994, preprint

Hatsukade, I., Kawarabata, K. & Takenaka, K. 1995, in The Eleventh International Colloquium on UV and X-ray Spectroscopy of Astrophysical and Laboratory Plasmas, in press

Hughes, J.P., Butcher, J.A., Stewart, G.C. & Tanaka, Y. 1993, ApJ, 404, 611

Kowalski, M.P., Cruddace, R.G., Snyder, W.A., Fritz, G.G., Ulmer, M.P., Fenimore, E.E. 1993, ApJ, 412, 489

Koyama, K., Takano, S. & Tawara, Y. 1991, Nature, 350, 135

Liedahl, D.A., Osterheld, A.L. & Goldstein, W.H. 1995, ApJ, 438, L115

Loewenstein, M. & Mushotzky, R.F. 1995, ApJ, in press
Mushotzky, R.F., 1989, in Cosmic Abundances of Matter, C.J. Waddington, AIP, New York, 1989, 325
Mushotzky, R.F. 1995, in Röntgenstrahlung from the Universe, in press
Mushotzky, R.F., Loewenstein, M., Arnaud, K.A., Tamura, T., Fukazawa, Y., Matsushita, K., Kikuchi, K & Hatsukade, I. 1995, ApJ, in press
Nomoto, K., Thielemann F. & Yokoi, K. 1984, ApJ, 286, 644
Ohashi, T. 1995, in Röntgenstrahlung from the Universe, in press
Ponman, T.J., Bertram, D., Church, M.J., Eyles, C.J. & Watt, M.P. 1990, Nature, 347, 450
Renzini, A., Ciotti, L., D'Ercole, A., Pellegrini, S. 1993, ApJ, 419, 52
Tamura, T., Day, C.S.R., Isamu, Y., Ikebe, Y., Makishima, K., Mushotzky, R.F., Ohashi, T., Takanake, K. & Yamashita, K. 1995, PASJ, in press
White, N.E. et.al. 1995, preprint
White, R.E., Day, C.S.R., Hatsukade, I. & Hughes, J.P. 1994, ApJ, 433, 583
Wise, M.W. & Sarazin, C.L. 1993, in The Evolution of Galaxies and Their Environment, NASA Ames Research Center, 1993, 271
Yamashita, K. 1992, in Frontiers of X-ray Astronomy, Y. Tanaka & K. Koyama, Universal Academy Press, Tokyo, 1992, 475

From Solar Flares to the Big Bang: Rapporteur Paper on the Cosmic Abundances Conference

Reuven Ramaty

Laboratory for High Energy Astrophysics, Goddard Space Flight Center, Greenbelt, MD 20771

1. Introduction

The study of elemental and isotopic abundances is one of the most fundamental ingredients of all of astrophysics. The information derived from abundances relate to a great variety of astrophysical problems, including the big bang, galaxy formation, galactic evolution, nucleosynthesis, stellar evolution, the interstellar medium, particle acceleration, the origin of the solar system, and the dynamics of the solar atmosphere. It is of course not possible to do justice to all of these topics in a short rapporteur paper. The reader can certainly consult the written versions of the papers that were presented at the Conference, as well as some of the key reviews that I attempted to include in my writeup. Much additional information is also available in Virginia Trimble's introductory paper to the Conference.

The issue of the primordial ^{2}H abundance, discussed in section 2, was perhaps the 'hottest' topic discussed at the Conference. Indeed, the question of the value of ^{2}H/^{1}H in the quasar absorbers observed with Keck remains very much at the forefront of current astrophysical research as it addresses the fundamental questions of the mean baryon density in the universe and the existence of non-baryonic dark matter. I next devoted an entire section to the question of the origin of the light elements, a topic of great current interest to me. Section 4 deals with solar system abundances. These abundances, which can be studied using several independent observational techniques (including nuclear spectroscopy), exhibit great variability from one site to another in the solar atmosphere. In particular, they reveal the as-yet poorly understood FIP (first ionization potential) bias which seems to differentiate the coronal from the photospheric abundances. Meteorites are another important site for solar system abundances. The analysis of interstellar grains in primitive meteorites and the discovery of a variety of extinct radioactivities can delineate the sequence of events leading to the formation of the solar system.

Several other important topics were discussed at the Conference (sections 5, 6 and 7) including theories of nucleosynthesis, the evidence for nucleosynthesis, the interstellar medium, X-ray observations of stars, galaxies and clusters, and the implications of abundances on galaxies in clusters. The recent X-ray observations, e.g. with ASCA, certainly open a new window on the study of extragalactic abundances. In the section on nucleosynthesis I also included a discussion on gamma ray line observations, a very relevant topic, which, however, was not well treated at the Conference. Observations of gamma ray line emission from SN87A, with SMM, balloon borne Ge detectors, and OSSE, and

CasA and Vela with COMPTEL, provide direct evidence for recent nucleosynthesis in supernovae.

2. ^{2}H, ^{3}He, ^{4}He

The abundances of ^{2}H, ^{3}He, and ^{4}He are the diagnostics of Big Bang nucleosynthesis (BBNS). ^{2}H is a unique relic of the Big Bang. Essentially all the observed ^{2}H was produced in the big band, with the subsequent galactic evolution only decreasing the ^{2}H abundance due to destruction is stars. The observation of ^{2}H in QSO absorbers (Songaila et al. 1994 and others) has now provided a measure of the primordial ^{2}H abundance which is about an order of magnitude larger than the galactic ^{2}H abundance. This higher primordial ^{2}H abundance is in much better agreement with the ^{4}He and ^{7}Li abundances predicted by BBNS. On the down side, however, there remains the question of the validity of the identification of the absorption features as due to ^{2}H, and the problem of the Galactic evolution of ^{2}H which could lead to an overabundance of ^{3}He.

These issues were discussed at the Symposium by David Schramm, C. Hogan, Gary Steigman, M. Rugers and E. Jenkins and continue to be the topic of intense further discussions. There are also recent relevant publications (Dar 1995; Hata et al. 1995; Copi, Schramm & Turner 1995). At the time of the present writing (December 1995) the validity of the ^{2}H detection in QSO absorbers is still not fully established and the galactic evolution of ^{2}H and ^{3}He still remains to be fully understood. Information on the baryon density independent of BBNS could be obtained from observations of the cosmic microwave background with sub-degree angular resolution. This topic was discussed at the Conference by M. Kamionkowski. The direct ^{2}H abundance observations that will be carried out with FUSE were discussed by S. Friedman.

3. Li, Be, B

^{6}Li, Be and B are the products of nuclear interactions of accelerated particles with ambient matter. ^{7}Li is produced by BBNS as well as nucleosynthesis in stars (see Reeves 1994 for details). Observations of Li in the interstellar medium were reviewed at the Conference by M. Lemoine.

The production of ^{6}Li, Be and B by galactic cosmic rays (GCR) interacting with interstellar matter was originally proposed by Reeves, Fowler & Hoyle (1970) and worked out in detail by Meneguzzi, Audouze & Reeves (1971) and Mitler (1972). This GCR origin, however, appears to have two flaws: it cannot account for the solar system ^{11}B/^{10}B and it predicts a more rapid increase of [Be/H] and [B/H] with increasing [Fe/H] than observed. (The square-bracket enclosed ratios are the logarithms of the actual abundances). These issues were reviewed at the Symposium by M. Cassé (see also Cassé, Lehoucq, Vangioni-Flam 1995).

D. Duncan presented new HST data which show a clear linear dependence of [Be/H] and [B/H] on [Fe/H]. As emphasized by both Duncan and Cassé at the Conference, this linear dependence is much more consistent with light element production by an accelerated particle population which is depleted in protons and α-particles than by the GCR. This is because for accelerated particles which are depleted in protons and α particles the Be and B production is mostly due to the breakup of accelerated C and O on ambient H and He; in this case both the target and projectile abundances could remain constant, leading to a linear growth of the Be and B abundances. On the other hand, the GCR would produce much of the isotopes from the breakup of C and O in the ambient medium whose abundances increase with time, leading to a quadratic growth. Evidence for the existence of a low energy cosmic ray population which is distinct from the GCR was provided by gamma ray line observations from Orion with COMPTEL on CGRO (Bloemen et al. 1994). The connection between these gamma ray observations and the problem of the origin of the light elements was treated by Cassé et al. (1995), Ramaty, Kozlovsky & Lingenfelter (1995a; 1996a) and Ramaty (1996).

4. Solar System Abundances

Abundances in the solar system provide the benchmark against which all other abundance studies are compared. N. Grevesse reviewed the standard solar abundances, emphasizing new results, particularly that for Fe. For details see Anders & Grevesse (1989) and Grevesse & Noels (1992). Abundance studies in other stars should definitely complement the solar observations. X-ray observations of stellar abundances were reported at the Conference by S. Drake.

The very complex solar atmosphere was discussed by J.-P. Meyer who reviewed some of the methods of abundance determinations: atomic spectroscopy (optical, EUV, X-ray observations) and direct observations, both in the solar wind and in solar energetic particles. Reviews of these subjects are available in the literature (Meyer 1992; Reames, Meyer, & von Rosenvinge 1994; Reames 1995; Widing & Feldman 1995). The new technique of abundance determinations using gamma ray spectroscopy was discussed by G. Share and R. Ramaty (see also Share & Murphy 1995; Share, Murphy, & Skibo 1996; Ramaty et al. 1995b; 1996b) Perhaps the most important result of these solar abundance determinations concerns the abundance variations that are correlated with the first ionization potential (FIP) of the various elements. The low-FIP to high-FIP abundance ratios are higher in the corona than in the photosphere. The origin of this FIP bias is not well understood, although obviously it must be established in a relatively low temperature region ($<10^4$K). As the gamma ray data also show the FIP bias, and since the gamma rays are probably produced quite deep in the atmosphere, the bias must be established relatively close to the photosphere.

Meteorites provide one of the most important sources of cosmic abundances. This topic was reviewed at the Conference by Ernst Zinner (see also Zinner 1995). Zinner emphasized the abundances obtained from interstellar grains in primitive meteorites which provide unique information on processes of nucleosynthesis. Abundance studies in meteorites also revealed the existence of now-extinct radioactive isotopes that were live in the protosolar nebula or the early solar system: ^{26}Al (Lee, Papanastassiou and Wasserburg 1977); ^{41}Ca (Srinivasan, Ulyanov & Goswami 1994); ^{53}Mn (Birck & Allegre 1985; Lugmair, MacIsaac & Shukolyukov 1992; 1994); and ^{60}Fe (Lugmair, Shukolyukov & MacIsaac 1995). Various origins for these radioisotopes have been proposed, including accelerated particle interactions and nucleosynthesis in a supernovae or an AGB star in the proximity of the interstellar cloud from which the protosolar nebula formed (see Clayton 1994; Wasserburg et al. 1995; Cameron et al. 1995; Ramaty et al.1996a).

5. Nucleosynthesis

Theories of nucleosynthesis and chemical evolution were dealt with at the Conference by B. Meyer, S. Starrfield, S. Woosley, F. Timmes and R. McCray. The r and s processes were reviewed by Meyer (see also Meyer 1994) who emphasized the role of nucleosynthesis in high entropy per baryon environments (in supernova explosions in the vicinity of neutron stars). Starrfield discussed nucleosynthesis in nova outbursts on O-Ne-Mg white dwarfs showing that these explosions can produce interesting amounts of ^{22}N and ^{26}Al. The latter isotope plays a very important role in gamma ray astronomy (see below). Woosley and Timmes presented detailed calculations of galactic chemical evolution (see also Timmes, Woosley, & Weaver 1995). B. Pagel discussed globular cluster ages derived from abundance observations and S. Balachandran presented metal abundances in stars, including abundances in stars with very low metallicity.

Observational evidence for the products of nucleosynthesis in supernovae was discussed by R. Kirshner and R. Petre. The ASCA X-ray observations of supernova remnants reported by Petre have been used to determine the progenitor type of the supernova. There are also important gamma ray data that provide evidence for ongoing nucleosynthesis. These include observations of live ^{56}Co and ^{57}Co in supernova 1987A (Matz et al. 1988; Kurfess et al. 1992), of ^{44}Ti in CasA (Iyudin et al. 1994) and of ^{26}Al at many galactic sites including the Vela supernova remnant (Diehl et al. 1995). See also the recent reviews by Ramaty & Lingenfelter (1995) and Prantzos (1996).

6. Interstellar Medium

A variety of observations of the interstellar medium were discussed. J. Mathis considered the role of dust and showed evidence for important differences between solar and local interstellar abundances. H. Dinerstein discussed galactic abundance gradients based on observations of gaseous nebulae, A. Rudolph showed IR data which also reveal such gradients, and B. Savage discussed abundances in the halo.

A talk on the 'anomalous component' of the cosmic rays was presented by R. Mewaldt. These cosmic rays are a direct sample of the local interstellar medium. The accepted scenario for their origin (Fisk, Kozlovsky, & Ramaty 1974) starts with neutral interstellar atoms (He, N, O, Ne and others) penetrating into the solar cavity where they are ionized and picked up by the solar wind to be transported to the termination shock where they are accelerated to energies of a few tens of MeV/nucleon. These particles can then be observed throughout the solar system. They are anomalous by virtue of their energy spectrum, their composition (which reflects neutral to ionized element abundance ratios in the local interstellar medium), and their charge states (they are singly charged). The fact that they are singly charged affords them easy penetration into the magnetosphere which thus constitutes a storage region of local interstellar matter (Adams et al. 1991). Using the trapped magnetospheric anomalous cosmic rays, Mewaldt reported new measurements of the Ne isotopic ratio in the local interstellar medium.

7. Extragalactic Abundances

ASCA X-ray observations of abundances were reported by K. Arnaud for clusters and by M. Loewenstein for galaxies (mostly ellipticals). These X-ray observations are complementary to optical observations. The X-rays of course sample the hot gas and allow relatively model-independent derivations of abundances (see Loewenstein & Mushotzky 1996). The intracluster Fe abundance has been measured out to a redshift of about 0.5, demonstrating that SNII were an important source of heavy elements now seen the galaxy clusters. The implications of the abundance determinations, both of the galaxies in clusters and the intracluster gas, were discussed by J. Silk. Silk emphasized the role of major mergers of successively larger clumps in triggering starbursts that may be necessary to explain the intracluster Fe. Abundance observations related to AGN and Quasars were discussed by J. Krolik.

References

Adams, J. H. et al. 1991, ApJ, 375, L45
Anders, E. & Grevesse, N. 1989, Geochim. et Cosmochim. Acta, 53, 197
Birck, J-L. & Allegre, C. J. 1985, Geophys. Res. Lett., 12, 745
Bloemen, H. et al. 1994, A&A, 281, L5
Cameron, A. G. W., Höflich, P., Myers, P. C., & Clayton, D. D. 1995, ApJ, 447, L53
Cassé, M., Lehoucq, R., & Vangioni-Flam, E., 1995, Nature, 373, 318
Clayton, D. D. 1994, Nature, 368, 222
Copi, C. J., Schramm, D. N., & Turner, M. S. 1995, ApJ, 75, 3981
Dar, A. 1995, ApJ, 449, 550
Diehl, R. et al. 1995, A&A, 298, L25
Fisk, L. A., Kozlovsky, B., & Ramaty, R. 1974, ApJ, 190, L35
Grevesse, N. & Noels, A. 1992, in Origin and Evolution of the Elements, eds. N. Prantzos et al. (Cambridge: Cambridge Univ. Press), 14
Hata, N. et al. 1995, Phys.Rev.Lett, 75, 3977
Iyudin, A. F. et al. 1994, A&A, 284, L1
Kurfess, J. D. et al. 1992, ApJ, L137
Lee, T., Papanastassiou, D. A., & Wasserburg, G. J., 1977, ApJ, 211, L107
Loewenstein, M. & Mushotzky, R. F. 1996, ApJ, in press
Lugmair, G. W., MacIsaac, Ch., & Shukolyukov, A. 1992, Lunar and Plan. Sci., 23, 823

Lugmair, G. W., MacIsaac, Ch., & Shukolyukov, A., 1992, Lunar and Plan. Sci., 25, 813
Lugmair, G. W., Shukolyukov, A., & MacIsaac, Ch. 1995, in AIP Conf. Proc. 327, Nuclei in the Cosmos III, ed. M. Busso, R. Gallino, & C. A. Raiteri (New York: AIP), 591
Matz, S. M. 1988, Nature, 331, 416
Meneguzzi, M., Audouze, J., and Reeves, H. 1971, A&A, 15, 337
Meyer, B. S. 1994, ARA&A, 32, 153
Meyer, J-P. 1992, in Origin and Evolution of the Elements, eds. N. Prantzos et al. (Cambridge: Cambridge Univ. Press), 26
Mitler, H. E. 1972, Ap&SS, 17, 186
Prantzos, N. 1996, A&A, in press
Ramaty, R. 1996, A&A, in press
Ramaty, R., Kozlovsky, B., & Lingenfelter, R. E. 1995a, Annals New York Acad. Sci., 759, 392
Ramaty, R., Kozlovsky, B., & Lingenfelter, R. E. 1996a, ApJ, in press
Ramaty, R. & Lingenfelter, R. E. 1995, in The Analysis of Emission Lines, eds. R. E. Williams and M. Livio, (Cambridge: Cambridge Univ. Press), 180
Ramaty, R., Mandzhavidze, N., Kozlovsky, B., & Murphy, R. J. 1995b, ApJ, 455, L193
Ramaty, R., Mandzhavidze, N., & Kozlovsky, B., 1996b, in High Energy Solar Physics, R. Ramaty, N. Mandzhavidze, X.-M. Hua, eds. (New-York: AIP), in press
Reames, D. V. 1995, Adv. Space Res., 15, (7) 41
Reames, D. V., Meyer, J.-P., & von Rosenvinge, T. T. 1994, ApJS, 90, 649
Reeves, H. 1994, Revs. Modern Physics, 66, 193
Reeves, H., Fowler, W. A., & Hoyle, F. 1970, Nature, Phys. Sci, 226, 727
Share, G. H. & Murphy, R. J. 1995, ApJ, 452, 933
Share, G. H., Murphy, R. J., & Skibo, J. G. 1996, in High Energy Solar Physics, R. Ramaty, N. Mandzhavidze, X.-M. Hua, eds. (New-York: AIP), in press
Songaila, A., Cowie, L. L., Hogan, C. J., & Rugers, M. 1994, Nature, 368, 599
Srinivasan, G., Ulyanov, A. A., & Goswami, J. N. 1995, ApJ, 431, L67
Timmes, F. X., Woosley, S. E., & Weaver, T. A. 1995, ApJS, 98, 617
Wasserburg, G. J. et al. 1995, ApJ, 440, L101
Widing, K. G. & Feldman U. 1995, ApJ, 442, 446
Zinner, E. K. 1995, in Nuclei in the Cosmos II, eds. M. Busso, R. Gallino, and C. M. Raiteri, (AIP: New York), 567

CONFERENCE PROGRAMME

Monday, October 9, 1995

0915	Session #1 -- Welcome and Introduction	Chair: S. Holt
0930	V. Trimble - Historical perspective and outlook	
1030	Coffee	
1100	Session #2 -- Big Bang Nucleosynthesis I	Chair: J. Mather
1100	D. Schramm - Primordial nucleosynthesis	
1130	C. Hogan - Primordial D/H	
1200	S. Friedman - Anticipated D/H measurements with FUSE	
1210	M. Kamionkowski - Future CMB constraints to the baryon density	
1220	Discussion	
1230	Lunch	
1400	Session #3 -- Big Bang Nucleosynthesis II	Chair: D. Spergel
1400	G. Steigman - Statistical tests of primordial nucleosynthesis	
1430	E. Jenkins - QSO absorption line systems and primordial nucleosynthesis	
1500	M. Lemoine (for R. Ferlet) - Light elements in the local ISM	
1530	M. Rugers - Primordial D/H from new Keck measurements	
1540	Discussion	
1600	Tea	
1630	Session #4 -- Solar System	Chair: A. Poland
1630	N. Grevesse - Standard abundances	
1700	J-P Meyer - Solar coronal abundance anomalies	
1730	E. Zinner - Isotopic abundances in stars as inferred from the study of presolar grains in meteorites	
1800	G. Share - Gamma-ray measurements of flare-to-flare variations in ambient solar abundances	
1810	Discussion	
1830	Wine and cheese until 10 PM for poster perusal	

Tuesday, October 10, 1995

0830	Session #5 -- Stars I	Chair: D. Duncan
0830	J. Thorburn - Lithium abundances in stars	
0900	S. Balachandran - Metal abundances in stars	
0930	S. Drake - X-ray measurements of coronal abundances	
1000	Discussion	
1030	Coffee	
1100	Session #6 -- Stars II	Chair: N. Gehrels
1100	B. Meyer - r-, s-, and p-processes	
1130	S. Starrfield - Nucleosynthesis in novae	
1200	S. Woosley - SN nucleosynthesis	
1230	R. Ramaty - Abundance determinations from gamma-ray measurements	
1240	Discussion	
1300	Lunch	

Tuesday, October 10, 1995 (continued)

1400	Session #7 -- Stars III	Chair: G. Sonneborn

1400 R. Kirshner - Observational evidence for nucleosynthetic products
1430 R. Petre - New X-ray measurements of supernovae
1500 R. McCray - Abundance inferences from measurements
1530   Discussion
1600   Tea
1630   Session #8 -- Galaxy I   Chair: J. Ormes
1630 F. Timmes - Galactic nucleosynthesis
1700 B. Pagel - Abundances and globular cluster ages
1730 M. Cassé (for E. Vangioni-Flam) - Energetic Particles and Li/Be/B
1800 R. Mewaldt - Anomalous cosmic rays: a sample of the local interstellar medium
1810   Discussion
1830   Banquet
      E. Salpeter

Wednesday, October 11, 1995

0830   Session #9 -- Galaxy II   Chair: V. Rubin
0830 J. Mathis - Elemental abundances in interstellar dust
0900 H. Dinerstein - Abundances in gaseous nebulae
0930 A. Rudolph - Abundance measurements in the outer galaxy
0940   Discussion
1000   Coffee
1030   Session #10 -- Galaxy III   Chair: C. Bennett
1030 B. Savage - Abundances of the galactic halo gas
1100 J. Silk - Galaxies in clusters
1130 K. Arnaud - Intergalactic gas in clusters
1200 M. Loewenstein - ASCA observations of hot gas in ellipticals
1210 J. Krolik - Can elemental abundances be derived from AGN emission lines?
1220   Discussion
1230   Lunch
1330   Session #11 -- Conclusion   Chair: S. Holt
1335 R. Ramaty - Rapporteur

CONFERENCE PARTICIPANTS

Arge, Nick	Bartol Research Institute	arge@bartol.udel.edu
Arnaud, Keith	NASA/GSFC	kaa@genji.gsfc.nasa.gov
Audley, Damian	NASA/GSFC	audley@lheavx.gsfc.nasa.gov
Balachandran, S.	National Air & Space Museum	suchitra@wright.nasm.edu
Barbier, Louis	NASA/GSFC	barbier@lheavx.gsfc.nasa.gov
Barrett, Paul	NASA/GSFC	barrett@piglet.gsfc.nasa.gov
Bennett, Chuck	NASA/GSFC	bennett@stars.gsfc.nasa.gov
Binns, W. Robert	Washington Univ., St. Louis	wrb@howdy.wustl.edu
Bludman, Sidney	Univ. of Pennsylvania	bludman@bludman.hep.upenn.edu
Boldt, Elihu	NASA/GSFC	boldt@lheavx.gsfc.nasa.gov
Borkowski, Kazimierz	North Carolina State U.	kazik@rosserv.gsfc.nasa.gov
Brandt, John C.	LASP/University of Colorado	
Casse, Michel	Service d' Astrophysique, CEA-Saclay	casse@iap.fr
Cavallo, Robert	University of Maryland	rob@astro.umd.edu
Cheung, Cynthia	NASA/GSFC	ccheung@nssdc.gsfc.nasa.gov
Chi, L. K.	U.S. Naval Academy	chi@scs.usna.navy.mil
Choudhury, Latif	Elizabeth City State U.	choudhal@alpha.ecsu.edu
Cline, T. L.	NASA/GSFC	cline@lheavx.gsfc.nasa.gov
Connell, Jim	Univ. of Chicago	connell@odysseus.uchicago.edu
Copi, Craig J.	University of Chicago	copi@oddjob.uchicago.edu
Cowan, Ron	Science News	
Crannell, Carol Jo	NASA/GSFC	crannell@stars.gsfc.nasa.gov
Dinerstein, Harriet	Univ. of Texas-Austin	harriet@astro.as.utexas.edu
Drake, Stephen	NASA/GSFC	drake@lheavx.nasa.gsfc.gov

DuVernois, Mike	Univ. of Chicago	duvernoi@odysseus.uchicago.edu
Duncan, Douglas	Univ. of Chicago	duncan@oddjob.uchicago.edu
Dutil, Yvan	Laval University	dutil@phy.ulaval.ca
Fahey, Richard P.	NASA/GSFC	fahey@stars.gsfc.nasa.gov
Felten, James E.	NASA/GSFC	felten@stars.gsfc.nasa.gov
Fichtel, Carl	NASA/GSFC	sccef@scfvm.bitnet
Friedman, Scott	Johns Hopkins University	scott@pha.jhu.edu
Frost, Kenneth	NASA/GSFC	frost@nssdca.gsfc.nasa.gov
Gehrels, Neil	NASA/GSFC	gehrels@lheavx.gsfc.nasa.gov
Gendreau, Keith	NASA/GSFC	
Glanz, James	AAAS/Science	
Grevesse, Nicolas	Institut d'Astrophysique	
Gull, Ted	NASA/GSFC	gull@stars.gsfc.nasa.gov
Hall, Charles	NASA/GSFC	
Harding, Alice	NASA/GSFC	harding@lheavx.gsfc.nasa.gov
Hartman, Bob	NASA/GSFC	rch@sage0.gsfc.nasa.gov
Hasan, Hashima	NASA Headquarters/JPL	hhasan@hq.nasa.gov
Hill, Vanessa	DASGAL/Observatoire, Paris-Meudon	hill@memaga.obspm.fr
Hogan, Craig	University of Washington	hogan@astro.washington.edu
Holt, Steve	NASA/GSFC	holt@lheavx.gsfc.nasa.gov
Hua, Xin-Min	NASA/GSFC/NRC	hua@rosserv.gsfc.nasa.gov
Hunter, Stan	NASA/GSFC	sdh@sage0.gsfc.nasa.gov
Jahoda, Keith	NASA/GSFC	jahoda@lheavx.gsfc.nasa.gov
Jenkins, Edward	Princeton University Obs.	ebj@astro.princeton.edu
Kallman, Tim	NASA/GSFC	tim@xstar.gsfc.nasa.gov
Kamionkowski, Marc	Columbia University	kamion@phys.columbia.edu

Kazanas, Demosthenes	NASA/GSFC	kazanas@lheavx.gsfc.nasa.gov
Keohane, Jonathan	NASA/GSFC & U. of Minn	jonathan@cassiopeia.gsfc.nasa.gov
Kimble, Randy	NASA/GSFC	kimble@stars.gsfc.nasa.gov
Kingsburgh, Robin	Univ. Nacional Autonoma de Mexico	robin@bufadora.astrosen.unam.mx
Kirshner, Robert	Harvard University	kirshner@cfa.harvard.edu
Kobulnicky, Chip	University of Minnesota	chip@astro.spa.umn.edu
Kondo, Yoji	NASA/GSFC	kondo@iue.gsfc.nasa.gov
Kozlovsky, Ben Zion	NASA/GSFC	kozlovsky@lheavx.gsfc.nasa.gov
Krolik, Julian	The Johns Hopkins University	jhk@gauss.pha.jhu.edu
Landsman, Wayne	Hughes STX	landsman@stars.gsfc.nasa.gov
Lemoine, Martin	Institut d'Astrophysique, Paris	lemoine@iap.fr
Loewenstein, Michael	NASA/GSFC	loewenstein@lheavx.gsfc.nasa.gov
Lu, Limin	Caltech	ll@troyte.caltech.edu
Lubowich, Don	Hofstra Univ. & AIP	dal@aip.org
Ma, Chung-Pei	Caltech	cpma@tapir.caltech.edu
Macomb, Daryl	COSSC/USRA	macomb@grossc.gsfc.nasa.gov
Maran, Stephen P.	NASA/GSFC	hrsmaran@stars.gsfc.nasa.gov
Marshall, Frank	NASA/GSFC	marshall@rosserv.gsfc.nasa.gov
Massa, Derck		
Mathis, John	University of Wisconsin	mathis@madraf.astro.wisc.edu
Mattox, John	University of Maryland	mattox@astro.umd.edu
McCray, Richard	University of Colorado	dick@jila.colorado.edu
Mewaldt, Richard	Caltech	dick@citsrl.caltech.edu
Meyer, Bradley	Clemson University	brad@cosmo.phys.clemson.edu
Meyer, Jean-Paul	Service d'Astrophys, Saclay	meyer@sapvxg.saclay.cea.fr

Michalitsianos, Andy	NASA/GSFC	michalits@torte.gsfc.nasa.gov
Mitalas, R.	University of Western Ontario	mitalas@uwovax.uwo.ca
Mitchell, John W.	NASA/GSFC	mitchell@lheavx.gsfc.nasa.gov
Moos, Warren	Johns Hopkins University	hwm@pha.jhu.edu
Mushotzky, Richard	NASA/GSFC	richard@xtelab.gsfc.nasa.gov
Neff, Susan	NASA/GSFC	neff@stars.gsfc.nasa.gov
Norman, Dara	University of Washington	norman@astro.washington.edu
Norris, Jay P.	NASA/GSFC	norris@grossc.gsfc.nasa.gov
Ormes, Jonathan	NASA/GSFC	ormes@lheavx.gsfc.nasa.gov
Pagel, Bernard	NORDITA	pagel@nordita.dk
Peimbert, Manuel	University of Mexico	peimbert@astroscu.unam.mx
Petre, Robert	NASA/GSFC	petre@lheavx.nasa.gsfc.gov
Pisarski, Ryszard	NASA/GSFC	pisarski@rsdps.gsfc.nasa.gov
Poland, Arthur	NASA/GSFC	apoland@solar.standord.edu
Polidan, Ronald	NASA/GSFC	polidan@aesop.gsfc.nasa.gov
Primas, Francesca	University of Chicago	primas@oddjob.uchicago.edu
Ptak, Andy	NASA/GSFC	ptak@lheavx.gsfc.nasa.gov
Quintana, H.	Universidad Catolica-Chile	hquintana@astro.puc.cl
Raimann, Gerhard	The Ohio State University	raimann@mps.ohio-state.edu
Ramaty, Reuven	NASA/GSFC	ramaty@lheavx.gsfc.nasa.gov
Rebull, Luisa	University of Chicago	rebull@oddjob.uchicago.edu
Reisenegger, Andreas	Institute for Advanced Study	andreas@sns.ias.edu
Rho, Jeong Hee	NASA/GSFC	rho@xray-1.gsfc.nasa.gov
Roman, Nancy Grace	Hughes/STX	nancy.g.roman@gsfc.nasa.gov
Rose, William K.	University of Maryland	
Rubin, Vera	Carnegie Institute of Washington	rubin@gal.ciw.edu

Rudolph, Alex	Harvey Mudd College	rudolph@hmc.edu
Rugers, Martin	University of Washington	rugers@astro.washington.edu
Saba, Julia	Lockheed @ GSFC	jlrs@sdac.nascom.nasa.gov
Salpeter, Edwin	Cornell University	hann@astrosun.tn.cornell.edu
Savage, Blair	University of Wisconsin	savage@madraf.astro.wisc.edu
Schramm, David N.	University of Chicago	dns@oddjob.uchicago.edu
Schweizer, Francois	Carnegie - DTM	schweizer@bmrt.ciw.edu
Seckel, David	Bartol Research Institute	seckel@bartol.udel.edu
Serlemitsos, Peter	NASA/GSFC	pjs@astron.gsfc.nasa.gov
Shafer, Richard	NASA/GSFC	shafer@stars.gsfc.nasa.gov
Share, Gerald	Naval Research Laboratory	share@osse.nrl.navy.mil
Shrader, Chris	COSSC/NASA/GSFC	shrader@grossc.gsfc.nasa.gov
Silberberg, Rein		
Silk, Joseph	University of California	silk@ucbast.berkeley.edu
Silverberg, Robert	NASA/GSFC	silverberg@stars.gsfc.nasa.gov
Singh, K. P.	NASA/GSFC -- TIFR	kps@rosserv.gsfc.nasa.gov
Smith, Eric	NASA/GSFC	esmith@hubble.gsfc.nasa.gov
Snowden, Steve	NASA/GSFC	snowden@lheavx.gsfc.nasa.gov
Sofia, U. J.	NASA/GSFC	sofia@stars.gsfc.nasa.gov
Sonneborn, George	NASA/GSFC	sonneborn@stars.gsfc.nasa.gov
Spergel, David	Princeton University Obs.	dns@astro.princeton.edu
Spergel, Martin	York College/CUNY	cosmic@neptune.york.cuny.edu
Stahle, Caroline	NASA/GSFC	stahle@lheavx.gsfc.nasa.gov
Starrfield, Sumner	Arizona State University	starrfie@hydro.la.asu.edu
Stecher, Ted	NASA/GSFC	stecher@uit.gsfc.nasa.gov
Stecker, Floyd	NASA/GSFC	stecker@lheavx.gsfc.nasa.gov

Steigman, Gary	Ohio State University	steigman@ohstpy.bitnet
Stiller, Bertram		bstiller@capaccess.org
Streitmatter, Robert	NASA/GSFC	streitmatter@lheavx.gsfc.nasa.gov
Summers, Frank	Princeton University	summers@astro.princeton.edu
Swank, Jean	NASA/GSFC	swank@lheavx.gsfc.nasa.gov
Sweigart, Allen	NASA/GSFC	sweigart@lheavx.gsfc.nasa.gov
Szymkowiak, Andrew	NASA/GSFC	andrew.szymkowiak@gsfc.nasa.gov
Teplitz, Doris Rosenbaum	U. of Maryland	teplitz@phyvms.physics.smu.edu
Thompson, Dave	NASA/GSFC	djt@sage0.gsfc.nasa.gov
Thorburn, Julie	University of Chicago	thorburn@yerkes.uchicago.edu
Timmes, F. X.	University of Chicago	fxt@nova.uchicago.edu
Trimble, Virginia	University of Maryland	vtrimble@astro.umd.edu
Tsuru, Takeshi	Kyoto University	tsuru@cr.scphyskyoto-u.ac.jp
Tyler, Pat	NASA/GSFC/HEASARC	tyler@lheavx.gsfc.nasa.gov
Ulmer, Andrew	Princeton University Obs.	andrew@astro.princeton.edu
Vandegriff, Jon	The Ohio State University	jonv@mps.ohio-state.edu
Varosi, Frank	Hughes STX	varosi@idlastro.gsfc.nasa.gov
von Rosenvinge, Tycho	NASA/GSFC	vonrosen@nssdca.gsfc.nasa.gov
Waddington, Jake	University of Minnesota	waddington@physics.spa.umn.edu
Waller, William H.	Hughes STX & NASA/GSFC	waller@stars.gsfc.nasa.gov
White, Nick	NASA/GSFC	white@heagip.gsfc.nasa.gov
Woodgate, Bruce	NASA/GSFC	woodgate@uit.gsfc.nasa.gov
Woosley, Stan	U. of California-Santa Cruz	woosley@ucolick.org
Zhang, William	NASA/GSFC	zhang@lheavx.gsfc.nasa.gov
Zinner, Ernst	Washington University	
Zucker, Daniel	University of Washington	zucker@astro.washington.edu

ASTROPHYSICAL CONSTANTS

CONSTANT	SYMBOL	MKS		CGS		OTHER
astronomical unit	AU	$1.50 \cdot 10^{11}$	m	$1.50 \cdot 10^{13}$	cm	
	AU/year					4.74 km/s
parsec	pc	$3.09 \cdot 10^{16}$	m	$3.09 \cdot 10^{18}$	cm	3.26 LY
solar mass	$M_\odot$	$1.99 \cdot 10^{24}$	kg	$1.99 \cdot 10^{33}$	gm	
solar luminosity	$L_\odot$	$3.90 \cdot 10^{26}$	J/s	$3.90 \cdot 10^{33}$	erg/s	
solar effective temperature	$T_{\text{eff}\odot}$	5780	K	5780	K	
solar radius	$R_\odot$	$6.96 \cdot 10^{8}$	m	$6.96 \cdot 10^{10}$	cm	
Earth radius	$R_\oplus$	$6.38 \cdot 10^{6}$	m	$6.38 \cdot 10^{8}$	cm	
Earth mass	$M_\oplus$	$5.98 \cdot 10^{24}$	kg	$5.98 \cdot 10^{27}$	gm	
Earth density	$\rho_\oplus$	5520	kg/m^3	5.52	gm/cm^3	
Jansky	Jy	$1.0 \cdot 10^{-26}$	W/m$^2 \cdot$Hz	$1 \cdot 10^{-23}$	erg/s$\cdot$cm$^2 \cdot$Hz	
Hubble constant	H_0	$3.24h \cdot 10^{-18}$	s^{-1}	$3.24h \cdot 10^{-18}$	s^{-1}	$100h$ km/s$\cdot$Mpc
critical density (=$3H_0^2/8\pi G$)	ρ_0	$1.88h^2 \cdot 10^{-26}$	kg/cm^3	$1.88h^2 \cdot 10^{-29}$	gm/cm^3	
plasma frequency						$8.98\sqrt{n_e(\text{cm}^{-3})}$ kHz/gauss
radian						$57.29578° = 206,265"$
CMB photon density	n_γ	$4.15 \cdot 10^{5}$	m^{-3}	415	cm^{-3}	

TABLE OF PHYSICAL CONSTANTS

CONSTANT	SYMBOL	MKS		CGS		OTHER
speed of light	c	$3.00 \cdot 10^8$	m/s	$3.00 \cdot 10^{10}$	cm/s	(2.997925)
electron charge	e	$1.60 \cdot 10^{-19}$	coul	$4.80 \cdot 10^{-10}$	esu	
Planck constant	h	$6.63 \cdot 10^{-34}$	J·s	$6.63 \cdot 10^{-27}$	erg·s	
	$\hbar$	$1.05 \cdot 10^{-34}$	J·s	$1.05 \cdot 10^{-27}$	erg·s	
	hc	$1.99 \cdot 10^{-25}$	J·m	$1.99 \cdot 10^{-16}$	erg·cm	
	$\hbar c$	$3.15 \cdot 10^{-26}$	J·m	$3.15 \cdot 10^{-17}$	erg·cm	200 MeV·fm
Boltzmann constant	k	$1.38 \cdot 10^{-23}$	J/K	$1.38 \cdot 10^{-16}$	erg/K	$8.6 \cdot 10^{-5}$ eV/K
	k/h	$2.08 \cdot 10^{10}$	s^{-1}/K	$2.08 \cdot 10^{10}$	s^{-1}/K	
	k/hc	69.5	m^{-1}/K	0.695	cm^{-1}/K	
Gravitational constant	G	$6.67 \cdot 10^{-11}$	$N \cdot m^2/kg^2$	$6.67 \cdot 10^{-8}$	$dy \cdot cm^2/gm^2$	
Gas constant	R	8.314	J/K·mole	$8.31 \cdot 10^7$	erg/K·mole	
Avogadro's number (= R/k)	N	$6.02 \cdot 10^{26}$	amu/kg	$6.02 \cdot 10^{23}$	amu/kg	$6 \cdot 10^{23}$ molecules/mole
electron mass	m_e	$9.11 \cdot 10^{-31}$	kg	$9.11 \cdot 10^{-28}$	gm	0.51 MeV
proton mass	M_p	$1.67 \cdot 10^{-27}$	kg	$1.67 \cdot 10^{-24}$	gm	938 MeV
neutron mass	M_n	$1.67 \cdot 10^{-27}$	kg	$1.67 \cdot 10^{-24}$	gm	939 MeV
pion mass (=270·m_e)	m_π	$2.46 \cdot 10^{-28}$	kg	$2.46 \cdot 10^{-25}$	gm	140 MeV
muon mass (=207·m_e)	m_μ	$1.89 \cdot 10^{-28}$	kg	$1.89 \cdot 10^{-25}$	gm	106 MeV
classical elect radius (=e^2/mc^2)	r_c	$2.82 \cdot 10^{-15}$	m	$2.82 \cdot 10^{-13}$	cm	
Compton wavelength (=h/mc)	λ_c	$2.43 \cdot 10^{-12}$	m	$2.43 \cdot 10^{-10}$	cm	0.02 Å

Quantity	Symbol	Value	Units	Value	Units
Thomson cross-section	σ_T	$6.65 \cdot 10^{-29}$	m^2	$6.65 \cdot 10^{-25}$	cm^2
Planck length ($=\sqrt{\hbar G/c^3}$)	l_{Pl}	$1.61 \cdot 10^{-35}$	m	$1.61 \cdot 10^{-33}$	cm
Planck time ($=\sqrt{\hbar G/c^5}$)	t_{Pl}	$5.39 \cdot 10^{-44}$	s	$5.39 \cdot 10^{-44}$	s
Planck density ($=c^5/\hbar G^2$)	ρ_{Pl}	$5.16 \cdot 10^{96}$	kg/m^3	$5.16 \cdot 10^{93}$	gm/cm^3
Bohr radius ($=\hbar^2/me^2$)	r_B	$0.53 \cdot 10^{-10}$	m	$0.53 \cdot 10^{-8}$	cm	0.5 Å
Fine structure constant ($=e^2/\hbar c$)	α	$7.30 \cdot 10^{-3}$		$7.30 \cdot 10^{-3}$		1/137
Bohr magneton ($=e\hbar/2m_e c$)	μ_B	$9.27 \cdot 10^{-24}$	J/T	$9.27 \cdot 10^{-21}$	erg/gauss
Nuclear magneton ($=e\hbar/2M_p c$)	μ_N	$5.05 \cdot 10^{-27}$	J/T	$5.05 \cdot 10^{-24}$	erg/gauss
Permitivity of vacuum	ε_o	$8.85 \cdot 10^{-12}$	fd/m			$1/4\pi\varepsilon_o = 9.0 \cdot 10^9$
Permeability in vacuum	μ_o	$4\pi \cdot 10^{-7}$	Hen/m		
Stefan-Boltzmann constant	σ	$5.67 \cdot 10^{-8}$	$W/m^2 \cdot K^4$	$5.67 \cdot 10^{-5}$	$erg/s \cdot cm^2 \cdot K^4$
Rydberg ($=m_e e^4/2\hbar^2$)	R_∞	$2.18 \cdot 10^{-18}$	J	$2.18 \cdot 10^{-11}$	erg	13.6 eV
1 amu		$1.66 \cdot 10^{-27}$	kg	$1.66 \cdot 10^{-24}$	gm	931.5 MeV
1 calorie		4.19	J	$4.19 \cdot 10^7$	erg
1 year		$3.16 \cdot 10^7$	s	$3.16 \cdot 10^7$	s
1 atmosphere		$1.01 \cdot 10^5$	N/m^2	$1.01 \cdot 10^6$	$dyne/cm^2$	14.2 lbs/in^2
		$1.01 \cdot 10^5$	Pascal			760 Torr
1 eV		$1.6 \cdot 10^{-19}$	J	$1.6 \cdot 10^{-12}$	erg	11,605 K
		$1.24 \cdot 10^{-6}$	m	$1.24 \cdot 10^{-4}$	cm	
		1 Tesla		10^4	gauss	

AUTHOR INDEX

A
Arnaud, K. A. — 409

B
Balachandran, S. C. — 188
Barlow, T. A. — 105
Bell, R. A. — 207
Blondin, J. M. — 294
Boesgaard, A. M. — 179, 184
Borkowski, K. J. — 294
Brown, J. A. — 203
Brown, J. S. — 231

C
Cassé, M. — 366
Cavallo, R. M. — 207
Coble, K. A. — 179
Copi, C. J. — 59, 63
Crotts, A. P. S. — 199
Cummings, J. R. — 381

D
Deliyannis, C. P. — 179, 184
Dinerstein, H. L. — 337
Drake, S. A. — 215, 227
Duncan, D. K. — 175, 179, 184
Dutil, Y. — 389
DuVernois, M. A. — 385

E
Erickson, E. F. — 358

F
Ferlet, R. — 78
Fich, M. — 358

Friedman, S. — 109

G
Grevesse, N. — 117

H
Haas, M. R. — 358
Hill, V. — 211
Hobbs, L. M. — 179, 184
Holt, S. S. — 1
Hogan, C. J. — 67, 100
Hubeny, I. — 199

J
Jenkins, E. B. — 90
Jungman, G. — 74

K
Kamionkowski, M. — 74
Keohane, J. W. — 362
King, J. R. — 179, 184
Kingsburgh, R. L. — 350
Kirshner, R. P. — 263
Kosowsky, A. — 74

L
Landsman, W. B. — 199
Lanz, T. — 199
Lemoine, M. — 78
Leske, R. A — 381
Loewenstein, M. — 393
Lopez, J. A. — 350
Lu, L. — 105
Lubowich, D. — 114
Luo, N. — 231

M
Malumuth, E. M. — 354

Mathis, J. S.	327
McCray, R.	273
Mewaldt, R. A.	381
Meyer, B. S.	231
Meyer, J.-P.	127
Moos, W.	109

N

Noels, A.	117

O

O'Connell, R. W.	199
Oegerle, W.	109

P

Pagel, B. E. J.	307
Parker, J. W.	354
Peimbert, M.	350
Peterson, R. C.	175
Petre, R.	284
Primas, F.	175, 179

R

Ramaty, R.	377, 419
Rebull, L. M.	179, 184
Reisenegger, A.	405
Rose, W. K.	196
Rudolph, A. L.	358
Rugers, M.	100
Ryan, S.	179, 184

S

Sarazin, C. L.	294
Sargent, W. L. W.	105
Sauval, A. J.	117
Savage, B. D.	315
Schramm, D. N.	36, 59, 63

Sembach, K. R.	315
Silk, J.	397
Simpson, J. P.	358
Singh, K. P.	227
Sparks, W. M.	242
Spergel, D. N.	74
Spergel, M. S.	162
Starrfield, S.	242
Stecher, T. P.	199
Steigman, G.	48
Sweigart, A. W.	207
Szymkowiak, A. E.	294

T

Timmes, F. X.	298
Thorburn, J. A.	165, 175
Trimble, V.	3
Truran, J. W.	242
Turner, M. S.	59, 63

V

Vangioni-Flam, E.	366

W

Waller, W. H.	354
Wallerstein, G.	203
White, N. E.	227
Whitney, J.	199
Wiescher, M.	242
Woosley, S. E.	253

Y

York, D.	109

Z

Zinner, E.	147
Zucker, D.	203

SUBJECT INDEX

Page numbers refer to the *first* page of the contribution in which the subject appears

A

Abundances
Abundance ratios
- D/H 3,36,48,67,78, 90,100,109,114
- Li/H 36,48,59, 78,114,165,366
- B/H 3,114,179,184,366,419
- Be/H 78,165,175,366,419
- C/H 90,327
- N/H 211,298,315, 327,337,358
- O/H 3,90,114,165,196,211, 327,337,354,358,362,389
- Mg/H 315
- Si/H 298,315
- S/H 315
- Cr/H 315
- S/H 315
- Mn/H 315
- Fe/H 3,105,165, 179,184,188,211, 298,307,315,397,419
- Ni/H 315
- Zn/H 298,315
- B/Be 179,253,366
- C/O 188,196,337,366,377
- N/O 78,105,337,358
- Ne/O 127,337,377
- Mg/O 127,377
- Si/O 337,377
- S/O 337,377
- Ar/O 337
- O/Fe 3,188,211,307
- C/Fe 188,211
- Si/Fe 105,393
- Mg/Fe 397
- Cr/Fe 105
- Mn/Fe 105
- Zn/Fe 105,298
- α/Fe 211,307
- B star 327
- Coronal 117,127,215, 227,377,419
- Solar 3,117,227,253,284, 298,315,327,366,419

Tables of abundances
- Galactic halo 315
- Globular cluster (ω Cen) 203
- H II regions 358
- LMC,SMC (CNO) 211
- Orion 337
- Planetary nebula 350
- Novae 242
- Solar 117
- Supernovae (II) 253

Absorption lines 3,36,67,90,105, 109,117,298,315,327,362

Accretion 242,393,397
Asymptotic giant branch (AGB) 78, 147,162,165,188,196, 199,203,231,307,366,419
Active Galactic Nuclei (AGN) 109, 114,315,409
Advanced Satellite for Cosmology and Astrophysics (ASCA) 36, 127,215,227,284,294, 362,393,397,405,409,419

B
Baryon density 36,63,67, 74,100,419
Beryllium 3,78,117,165,366
Big Bang nucleosynthesis 3,36,59, 63,67,74,165
Boron 3,78,114,117,165, 179,184,253,366,377
Binary stars 215,242

C
Carbon 3,117,127,188, 211,253,327,366
Carbon stars 147,196
Chromosphere emission 78
Clusters of galaxies 36,67, 74,397,405
  Intracluster gas 393,409,419
  Virgo 337
Circumstellar material 284,294
Compton Gamma Ray Observatory (CGRO) 242,419
Copernicus satellite 78,90,109,358

Cosmic microwave background 74
Cosmic rays (*see also* Spallation)
  Anomalous 381,419
  Galactic 78,127,162,165, 284,377,385,419

D
Dark matter 3,36,74,100, 393,397,405,409,419
Depletion 36,78,105, 117,127,147,165,199, 215,298,315,327,337,377
Deuterium 3,36,48,67,78,90, 100,109,114,165,366
Diffusion 117,165,175,199,307
Dust grains
  Circumstellar 147
  Interstellar 327

E
EUV spectra 127,215,227
Extreme Ultraviolet Explorer
  (EUVE) 215,127,377,419

F
Flares
  Solar 127,377,419
  Stellar 165,215
Far Ultraviolet Spectroscopic
  Explorer (FUSE) 78,109,419

G
Galaxies (*see also* Milky Way)
  Abundance gradient 263, 389,393,405
  Ellipticals 3,393,397,405,409
  Spirals 3,67,337,389

Gamma Ray Observatory
(GRO) 242,419
Globular clusters 3,165,199,
203,207,307
Goddard High Resolution
Spectrograph (GHRS) 78,109,
175,179,184,315,327
Gunn-Peterson effect 36,90

H

H II regions 48,109,188,211,
273,337,354,358
Hubble Space Telescope (HST) 36,
78,90,175,179,184,242,
315,337,354,366,397,419

I

Infrared emission 196,273,327,337
Intergalactic material (IGM) 36,90,
393,397
International Ultraviolet Explorer
(IUE) 78,109,199,242,315
Interstellar medium (ISM)
 Local ISM 36,78,117,147
 Milky Way 63,67,78,109,
284,315,327,362,419
 Other galaxies 389,393,397
Interstellar shocks 284,
315,337,366
Iron
 Extragalactic 393,397,405,409
 Interstellar 315
 Nebular 263,284

Solar 117
Stellar 3,188,203,211,
242,253,263,298,307
Isotopes
Al 3,162,203,207,242,253,419
Ar 3,242
B 3,78,179,253,366,377,419
Be 3
C 3,78,114,196,203,207,
231,242,253,298,362,366,377
Co 253,263
Cr 253
D (*see Deuterium*)
Fe 3,36,231,253,377,419
H (*see Deuterium*)
He 36,48,59,63,
78,109,165,231,419
N 3,242,253,298,381
Ni 3,231,253,263
O 3,63,78,196,
242,253,377,381
Mg 3,162,196,203,
207,242,253,377
Si 3,242,377
Zn 253

L

Lithium 3,36,48,67,78,
100,114,117,227,298
Lyman alpha 78,90,298,327
Lyman limit 67,90
Lyman series 67,90,100,109

M

Magellanic clouds 109,211,284, 298,315,337,354
Microlensing 3,36
Milky Way
   Abundance gradient 3,78, 315,337,358
   Disk 3,105,165, 179,188,298,315
   Halo 3,78,109,165,179, 298,307,315,327,366
   Population I 3,165,327
   Population II 3,165,215
Molecules
   H_2 78,196,315,327,362
   O_2 114,188,327
   CO 188,196,327
   HD 78
   HCN 78
   DCN 78,114
   H_2CO 362
   NH_3 362
   SiO_2 327
   Fe_2O_3 327

N

Nebular diagnostics 273, 337,350,389
Neutrinos 3,36,48,59,74,114, 105,165,231,298,366
Nitrogen
   Extragalactic 3
   Interstellar 327
   Nebular 337
   Solar 117
   Stellar 188,199, 207,211,242,253
Novae 3,147,165,196,242,419
Nuclear reactions 3,36,48,196, 207,242,253,298,377
Nucleosynthesis
   Primordial 3,36,48,59,63,67, 74,78,109,114,165,366,419
   Stellar 3,147,165,188,207, 196,231,242,253,284,298,419
   Cosmic ray 162,165,377,419
   p-Process 3,231,298
   r-Process 3,196,231, 263,298,366
   s-Process 3,147,196,231,298

O

Open clusters 165,354
Oxygen
   Extragalactic 63,389
   Interstellar 147,315,327
   Nebular 263,273,284,337
   Solar 117
   Stellar 188,196,211,242, 253,263,298,307,366

P

Planetary nebulae 67,78,327, 337,350,405

Q

Quasi-stellar objects (QSO)
- Absorption lines 3,67, 90,109,298
- GC0636+68 67,100
- HS1700+6263 90
- Q0014+813 67,90,100,109
- Q0127-003 90
- Q0127-0019 90
- Q0420-388 67
- Q0956+122 67
- Q1937-1009 67,90,109
- 3C 273 3,315

R

Röntgen Satellite (ROSAT) 36, 215,227,242,284,362,393,405

S

Solar Maximum Mission (SMM) 377
Spallation 3,78,114,165,179, 184,253,366,381,385
Space Telescope Imaging Spectrograph (STIS) 90,337
Stars
- BD-13 3442 184
- BD+26 3578 179
- HD 18100 315
- HD 22586 (ξ Per) 315
- HD 29647 327
- HD 38666 (μ Col) 315
- HD 76932 179
- HD 93521 315
- HD 103095 188
- HD 116852 315
- HD 120086 (ζ Oph) 315
- HD 117583 184
- HD 149881 315
- HD 167756 315
- HD 197890 227

Supernovae (SN)
- Ejecta 147
- Shocks 127,231, 253,273,284,294
- SN 1987A 263,273,298,366
- SN 1993J 263,273,284
- Type Ia 3,36,105,188,231,253, 263,284,298,393,397,405
- Type II 3,105,63,147,165,211, 231,253,263,397,405,409

Supernova Remnants (SNR) 263, 273,284,294,362,419
Synchrotron radiation 284

W

White dwarfs 3,78,109, 199,242,273,284,419

X

X-ray processes 74,273,284, 362,397,409,419
X-ray spectra 127,215,227,284, 294,377,393,405,409,419